International Conference on Complex Equipment of Quality Control Laboratories 2019

Journal of Physics: Conference Series Volume 1384

St. Petersburg, Russia
14 - 17 May 2019

Editors:

Vasilyeva Maria Alexandrovna
Nagornov Dmitriy Olegovich
Kremcheev Eldar Abdollovich

ISBN: 978-1-7138-1139-8
ISSN: 1742-6588

Printed from e-media with permission by:

Curran Associates, Inc.
57 Morehouse Lane
Red Hook, NY 12571

Some format issues inherent in the e-media version may also appear in this print version.

Institute of Physics
Dirac House, Temple Back
Bristol BS1 6BE UK

Phone: 44 1 17 929 7481
Fax: 44 1 17 920 0979

techtracking@iop.org

Additional copies of this publication are available from:

Curran Associates, Inc.
57 Morehouse Lane
Red Hook, NY 12571 USA
Phone: 845-758-0400
Fax: 845-758-2633
Email: curran@proceedings.com
Web: www.proceedings.com

TABLE OF CONTENTS

Journal of Physics: Conference Series **1384** (2019) 011001 doi:10.1088/1742-6596/1384/1/011001

The special issue "Integrated equipping of quality control laboratories" includes research articles, scientific reviews and notes submitted by the participants of the international scientific and practical conference held at Saint-Petersburg Mining University.

In 2019, scientists, representatives of educational organizations, instrument-makers, specialists in the operation of measuring and testing equipment, quality control specialists from different regions of the Russian Federation and from other countries came together at the oldest Russian technical university to share their experience and knowledge. The participants of the conference, round tables and seminars discussed trends in the modern instrument-making industry, the potential of new measuring equipment, the latest changes in the regulatory framework for quality control and ensuring the uniformity of measurements.

The materials presented in this special issue cover a wide sphere of technical knowledge on industrial electronics, metrological support of measurements, and instrument making. Among the authors of the papers are well-known scientists from the leading universities of the Russian Federation, young scientists and specialists from the manufacturing sector. A significant part of the issue was prepared in collaboration with members of international professional organizations, societies and research institutions.

In the preparation of the papers and the compilation of the issue, the MineCite intellectual analytical system was used.

MINECITE

Editors:

Vasilyeva Maria Alexandrovna, Candidate of Technical Sciences, Associate professor of the Mining transport machines department, Senior Deputy of the Chairman of young scientists of Saint-Petersburg Mining University

Nagornov Dmitriy Olegovich, Candidate of Technical Sciences, Associate professor of the Geo-ecology department, Saint-Petersburg Mining University

Kremcheev Eldar Abdollovich, Doctor of Technical Sciences, Head of the Metrology and quality control department, Saint-Petersburg Mining University

Peer Review Statement

All papers published in this volume of *Journal of Physics: Conference Series* have been peer reviewed through processes administered by the proceedings Editors. Reviews were conducted by expert referees to the professional and scientific standards expected of a proceedings journal published by IOP Publishing.

Journal of Physics: Conference Series **1384** (2019) 012001 doi:10.1088/1742-6596/1384/1/012001

Diagnostic methods and tools for truck parts and components

E K Abdulaev, P N Makharatkin and A I Shihov
Saint-Petersburg Mining University, 21st line 2, Saint Petersburg, Russia, 199106

E-mail: ehldarabdulaev@mail.ru

Abstract. This article discusses a variety of diagnostic parameters and different control methods applied to trucks which are operated in open-pit mines. It contains the classification of truck parts and components based on their functions and the description of the main diagnostic parameters for each classification group. It analyzes equipment health monitoring practices and methods of predicting mean time between failures (MTBF) for the reduction drives of wheel motors produced by different companies. It describes one of the operational factors determining the choice of a nondestructive testing (NDT) method for the selected equipment. It analyzes the cost-effectiveness of using NDT methods and describes effective ways to reduce costs when using OTR tires taking into account the dependence of tire life on tire pressure and temperature.

1. Introduction

During the operation of mining vehicles in a harsh environment, where operations are associated with high dynamic stresses, it is necessary to adopt a set of equipment health monitoring measures aimed at the creation of an adaptive maintenance and repair system for truck parts and components.

The failures occurring in trucks are accompanied by significant costs which may eventually exceed the cost of diagnostic equipment that makes it possible to calculate wear rate for parts in contact with each other and to find internal defects and stresses in metal frames with sufficient accuracy and efficiency in accordance with methods approved by State Technical Inspection [1].

When analyzing methods and means of health monitoring equipment, it is necessary to take into account many factors determining the effectiveness of selecting testing equipment and approaches to testing [2].

In the case of vehicles with rubber tires, the vibration analysis method and the spectral analysis method applied to oils are considered to be the most effective. They make it possible to assess the wear rate of a working surface, gap sizes in joints, and processes occurring in diesel engines and hydraulic drives.

It is necessary to deal with vibration in machines and mechanisms in order to increase their reliability and durability. It is impossible to develop effective means of reducing vibration without knowledge of why, where from, and how vibration propagates in mechanisms and structures, which requires that vibration analysis tools and equipment should be developed. Vibration analysis is also one of the main elements in the equipment of life cycle extension, which is a set of measures aimed at extending overhaul periods and reducing the likelihood of sudden equipment failure during operation. The emphasis is put on identifying causes of defects, assessing dangers associated with defects, and finding ways to reduce vibrations in the object. One of the main sources of increase in vibration in mechanisms being operated is defects in rolling-element bearings.

In the mining industry, oil analysis has been used in the health monitoring of reduction drives in wheel motors. As there are different operation modes for dump trucks, reduction drives will have different wear rates, which means that MTBF values can fluctuate over a wide range (depending on the manufacturer). BelAZ-7521 failure mileage varies from 80 to 200 thousand kilometers, while that of M200 (Unit Rig) ranges from 40 thousand to 100 thousand kilometers and that of HD1200 (Komatsu) varies from 60 thousand to 150 thousand kilometers. It can be seen from the data presented above that the behavior of BelAZ dump trucks is the most difficult to predict because there is a wide range of failure mileage values.

2. Methods

At Yakutugol enterprise, it was found that early failures happen due to technical reasons. The studies performed by the vehicle diagnostic service in the company showed that there were metal particles in reduction gear oil whose sizes reach tens and even hundreds of micrometers. The greater their concentration, the greater the wear rate of the reduction drive.

Today, there are several trends in the development of nondestructive methods of testing physical and mechanical characteristics of materials and structures:

1) finding correlations between physical and mechanical parameters of materials;

2) assessing empirical correlations and statistical data on the strength and a given physical parameter of an object under study;

3) developing multiparameter control methods;

4) creating nondestructive methods of testing the integral strength of objects and structures [3].

When choosing NDT methods for testing specific parts and components, the following factors should be taken into account: the character of discontinuity and its location, sensitivity of the testing method, conditions in which a mechanism is operated, technical specifications, type of material, surface roughness, part size and shape, and ease of access for testing [4].

NDT methods include:

• visual testing. This method relies on the visual inspection of an object with the use of a magnifying tool.

• liquid penetrant testing. This method is aimed at finding defects by applying liquids to the surface of an object which penetrate into cracks, pores, and welding defects.

• magnetic particle testing. This group of methods is based on detecting anomalies in flux patterns above defects or determining the magnetic properties of an object being tested.

• electromagnetic testing (eddy current testing). This method is based on detecting changes in the eddy current density induced by the coil in an electrically conductive object being tested.

• radiographic testing. This group of methods relies on the interaction between electromagnetic radiation and a solid object.

• ultrasonic testing. This method is based on analyzing ultrasonic wave propagation through an object being tested.

3. Results

The main test parameters depend on the types of parts and components which can be divided into several groups. Table 1 and Figure 1 show test parameters and part groups for a mining dump truck [5-8].

Table 1. Test parameters for a mining dump truck.

Equipment	Main parts and components	Main test parameters
Engine and its components	Air supply system	Vacuum parameters; manifold pressure; exhaust gas temperature; compressor run-down time after the engine stops; oil pressure in the centrifugal compressor.
	Fuel supply system	Fuel injection advance angle before TDC is reached; fuel consumption; exhaust gas temperature; manifold pressure; nozzle vibration values.
	Cooling system	The temperature of the coolant in the engine cooling system and the temperature of the air; pressure in the engine cooling system and boost air pressure.
	Exhaust system	Turbine exhaust pressure.
	Engine	Oil parameters; fuel consumption; maximum engine power; differences in oil flow between cylinders.
Electric machines	Support bearings	Vibration parameters; noise parameters; heating temperature.
	Armature winding	Armature current; current and voltage in the power circuit.
	Electric commutators	Assessment of electrical commutation characteristics
Electric circuits and devices	Power supply system	Voltage maintained by the voltage regulator during generator operation.
	Engine starting system	Voltage maintained by the batteries when the engine is started; starter motor current.
	Power transmission	Current and voltage of the electric motor depending on the engine speed; voltage in the synchronous generator and the exciter.
Mechanical transmission	Wheel motor reducer; power takeoff; fan drive reducer.	Oil parameters; drum brake free play; vibration values.
Wheel	Tire; wheel hub; rim.	Standard pressure; tire temperature; rim temperature; wheel hub temperature.

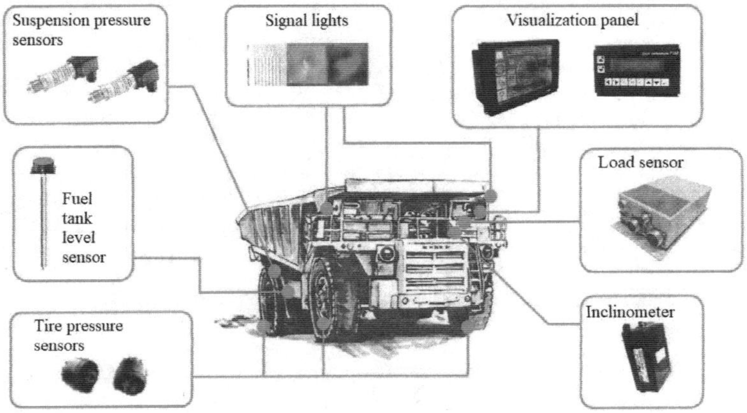

Figure 1. Nondestructive testing devices

The fact that there is a wide variety of parameters to be tested requires that various testing methods including NDT ones should be used. For example, one of the most important operational factors concerning vehicles with rubber tires is the violation of tire pressure norms which can cause such consequences as uneven tread wear, a decrease in tire tread puncture resistance, structural (physical) damage to the tire carcass, and an increase in heat generation. It is not by chance that costs for purchasing and maintaining OTR tires are among the major cost items for open-pit operators. Taking into account the impact of OTR tires on the cost structure of the company and the production process itself, leading producers and consumers of OTR tires are interested in finding ways to manage tire maintenance more effectively [10].

Pressure values for different types of tires lie within limits indicated in specifications and depend on external loads and operating conditions. When there is excessive pressure, the main load is distributed to the central part of the tire contact patch, while low pressure means that the main load is distributed to the peripheral parts, resulting in the so-called bridging effect, which in both cases leads to an increase in tread wear in these zones [11].

The effective way of reducing costs when using OTR tires is constant performance monitoring during operation. For this purpose, tire-pressure monitoring systems (TPMS) are used. They promptly detect deviations from the recommended values of tire pressure, which helps to prevent emergencies, reduce wear, and use tires to their full potential. The decrease in tire wear will cause a reduction in a negative impact on the environment as a result of a decrease in the number of harmful particles emitted in the air. Tire-pressure monitoring systems are quite easy to operate and usually contain sensors transmitting signals to a monitor located in the driver's cabin. The system can also be integrated into the dispatch system (if it is used by the company), which will make it possible for an operations controller to monitor parameters remotely.

According to data provided by Michelin, if a tire pressure value is higher or lower by 1% than the recommended values, it will result in a 1.5% decrease in tire life. If there is a bigger deviation (more than 15%), the tire carcass starts to disintegrate. This is a critical situation in which the tire is most susceptible to wear, punctures, and sidewall cuts, and, as a result, to an explosion, which will mean downtime due to tire loss [12].

Each tire manufacturer guarantees that their products will operate properly if recommended pressure values are maintained. If there is a slight deviation from these values, it will cause the increase in tread wear.

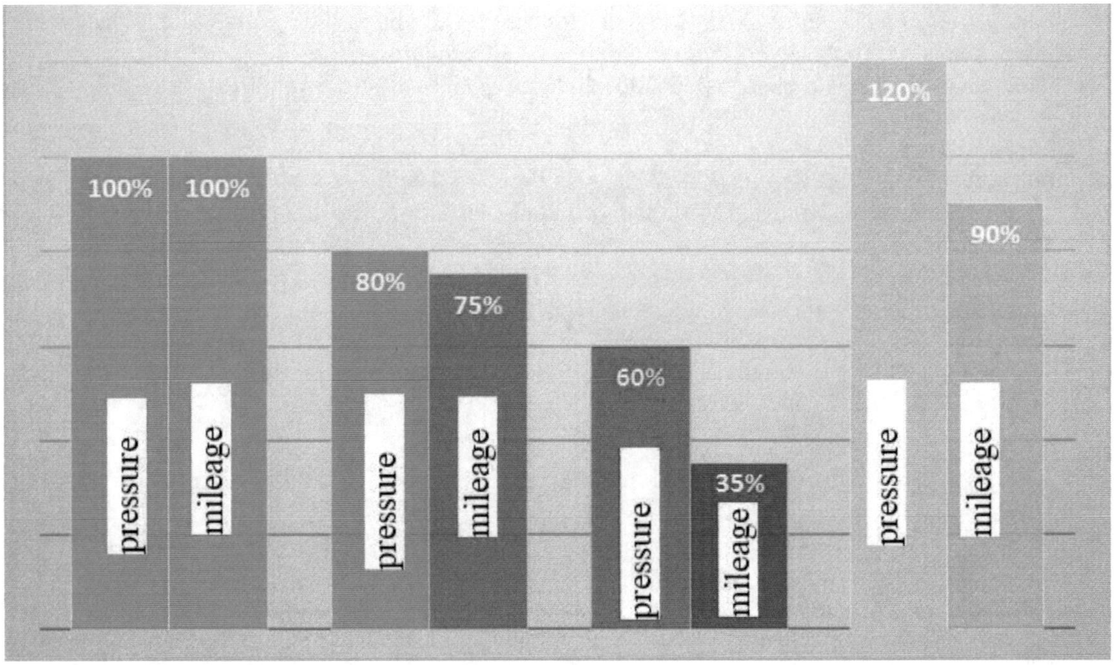

Figure 2. Dependence of tire life on internal pressure.

Attention should also be paid to tire operating temperature, which is one of the main factors affecting tread wear. At the moment, there are no recommendations describing optimal temperature ranges for OTR tires. This issue requires further study. However, a temperature of 120 °C is considered to be critical for modern tubeless tires. At higher temperatures, the tire will start to disintegrate. This is why temperature monitoring is also one of the ways to reduce costs when using OTR tires.

The main method of dump trucks maintenance used today is planned preventive maintenance (PPM). However, OTR tire maintenance usually involves only scheduled tire replacement and surface repair. Condition-based maintenance (CBM) seems to be a better strategy for this type of objects [12].

4. Conclusion

Condition-based maintenance systems which involve the application of several NDT methods depending on equipment part types are already used by companies operating mining dump trucks. Information systems being implemented can not only decrease the probability of failure but also increase truck fleet availability by monitoring the parameters of truck parts and components.

References

[1] Koptev V Y 2017 *IOP Conf. Ser.: Mater. Sci. Eng.* **177** 012005
[2] Aleksandrov V I and Avksentiev S Y 2016 *Indian Jour. of Sc. and Tech.* **9(5)** 87590
[3] Potapov A I 2018 *Zapiski Gornogo instituta* **230** 176–84
[4] Loshkarev I Y and Chernyshov A S 2013 *IV Int. Conf. "Topical Issues in the Energy Sector of the Agricultural Industry"* 184–86
[5] Tarasov P I et al 2006 *Diagnostics of Electromechanical Transmission in Open Pit Diesel-Electric Dump Trucks* (Yekaterinburg: Russian Academy of Sciences)
[6] Kudrevatykh AV 2017 *IOP Vestnik of Kuzbass State Technical University* **2(120)** 88–93
[7] Medvedev V A and Yanchelenko V A 2013 *Dvigatelestroyeniye* **1(151)** 28–30
[8] Doronin S V and Dontsova T V 2012 Journal of Siberian Federal University **5** 703–14

[9] *Loading and Fuel Testing* (2019, May 15). Retrieved from https://www.vistgroup.com.ua/18/
[10] Zenockin M Yu 2009 *Mining Informational and Analytical Bulletin* **11** 56–64
[11] Dolgushin A A and Cherkasov A B 2007 *Bulletin of Novosibirsk State Agrarian University* **1(6)** 48–49
[12] *Michelin Technical Databook* 2012
[13] Fokin A S 2012 *Proc. of the 10th Int. Sc. and Practical Conf. "Mineral Resources of the North: Problems and Solutions"* (Vorkuta: Vorkuta Mining Institute) 407–10

Journal of Physics: Conference Series **1384** (2019) 012002 doi:10.1088/1742-6596/1384/1/012002

Resonant-frequency properties of low-dimensional junction of semiconductor-metal-semiconductor and calculation methodology

D P Valiukhov[1], I S Baklanov[1], E V Shtab[1], A V Shtab[1], R V Pigulev[1], A S Iliasov[1]

[1]North Caucasus Federal University, 1 Pushkina St, Stavropol, 355017, Russia

E-mail: i_s_baklanov@mail.ru

Abstract. The article describes frequency characteristics of semiconductor–metal–semiconductor Si–Ag–Si structures (ohmic contacts) having metal nanolayers as well as features of resonance phenomenon and sandwich structure energy interaction. Mathematical simulation of mentioned above processes is provided. The limiting factor for resonance process is determined. It is found that the resonance frequency is a subject of semiconductor-metal junction parameters while metal layer thickness impact is minor.

1. Introduction

Small size of such sandwich structures provides a capability to use them as either a resonator or CL-element microelectronics or nano-electronics.

Meanwhile the structures of this type are not appropriate for the production of reactive elements, resonators and resonator-based circuits, passive frequency filters in microelectronics and/or nanoelectronics. Active filters are designed as integrated electronic components based on direct and back coupling operating amplifiers.

The issue is non-availability of microelements and nanoelements providing features of such components as reactive elements (inductance), resonators, frequency filters.

The results of a semiconductor-metal-semiconductor structure with an ultrathin metal layer investigation on amplitude-frequency properties are described hereby.

It is explained that the structure can be used in high-frequency and modulated signals processing as a frequency active element setting a fixed frequency (similar to a quartz resonator) as well as an element having LC characteristics.

2. Materials and methods

The following equipment was used for frequency characteristics analysis:

- Pulsed-oscillator Г5-56
- Oscilloscope C1-93

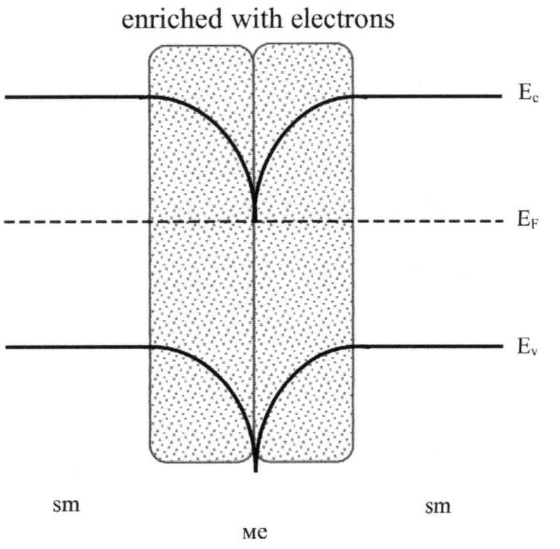

Figure 1. Energy band diagram of semiconductor-metal-semiconductor structure.

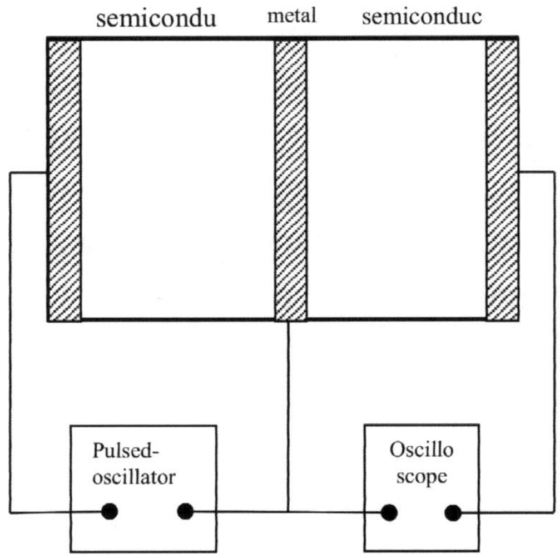

Figure 2. Connection diagram for the study of the frequency characteristics of the structure.

Connection diagrams are shown in the Figures 2 and 4. All the measurements are performed under normal conditions.

3. Results

The main feature of the structure is ultrathin metal layer (silver). Metal layers thicknesses in study samples are 0.7nm and 1.2nm. Semiconductor layers thicknesses (i-type silicon) are 70-80 nm. The energy band diagram is shown in Figure 1.

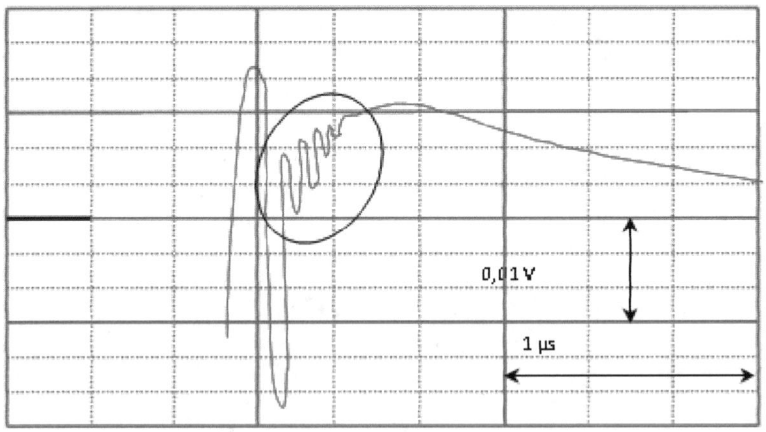

Figure 3. Transients phenomena in the structure at a rectangular pulse.

Energy zones bending takes place at semiconductor areas while at metal areas it is not observed due to screening effect. The only role of ultrathin metal layers in the structure is metal-semiconductor contact formation resulting in the energy well appearance.

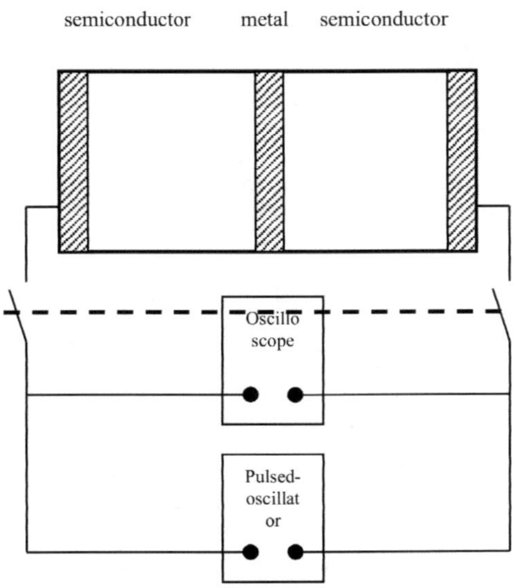

Figure 4. Connection diagram for frequency characteristics analysis in case of free metal layer.

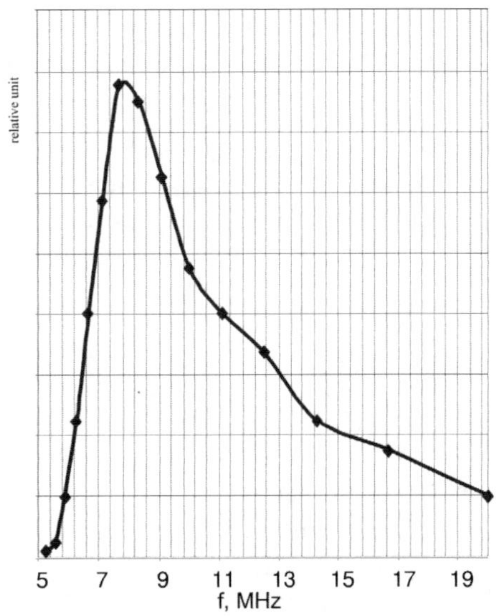

Figure 5. Amplitude-response curve for Si-Ag-Si structure (80 nm-0.7nm-80 nm).

Upon an ohmic contact formation a number of electrons passes from metal areas into semiconductor areas creating enhanced regions. The enhanced regions are considered as an electron gas containing chaotically moving electrons able to pass through ultrathin metal layers since thicknesses of the metal layers are less than mean free path of electrons. For the electrons passing through a potential well with energy up to 5eV (in case of silver ultrathin metal layers) the mean free path is thought to be from 2.5 to 5 nm. [2]. In case of equilibrium, the number of electrons on one side of the contact is equal to the number of electrons on the other (in case of semiconductor layers of the same material). Once voltage is applied to one of the semiconductor regions, then the equilibrium is upset. Thus some electrons will pass through a thin metal layer to another semiconductor layer. Once the voltage is removed the electron gas returns to the state of equilibrium. This results in electron gas oscillations relative to equilibrium state (Langmuir oscillations of electrons) [3] (Figure 3).

The frequency of these oscillations depends on the total energy of the system and the type of potential well. In other words it depends on features of the ohmic contact

.

Figure 5 shows the amplitude-frequency characteristic of the Si-Ag-Si structure (80 nm-0.7nm-80 nm; i-type silicon (undoped)). Resonance frequency for this structure is 8 MHz.

One-electron approach is used for frequency characteristics analysis due to low electron density.

In case of electron oscillation in a potential well U (x) with oscillation limits x_1 and x_2, then the oscillation period is determined as follows [4]:

$$T(E) = \sqrt{2m}\int_{x_1}^{x_2} \frac{dx}{\sqrt{E-U(x)}} \quad (1)$$

where:
m – electron mass;
E – total electron energy;

$U(x)$ – potential electron energy;
Oscillation limits x_1 and x_2 are the roots for the equation $U(x) = E$

According to the research work [5] , where a is the characteristic length (the screening length – the limits of electron density change from n_0 (semiconductor intrinsic electron density) to n_k (electron density at semiconductor-metal contact). In the general case, the width of the ohmic contact:

$$a = \left(\varepsilon k\mathrm{T} \frac{1}{2\pi n_k e^2} \right)^{1/2} \tag{2}$$

where:
ε – dielectric capacity;
k – Boltzmann constant;
T – Absolute temperature;
e – Elementary charge;
n_k - electron density at semiconductor-metal contact, which is determined by the following equation [3]

$$n_k \exp\left(\frac{eu_k}{k\mathrm{T}} \right) = n_0 \tag{3}$$

where:
u_k – contact potential.
According to the contact potential value u_k is measured distantly from semiconductor-metal contact. It is used for calculations only and not shown in the figure. The contact potential is associated with the electron work function and in this case leads to energy structure formation [5].
Since metal layer thickness is much less than potential well width the metal layer shall be considered as entirely transparent to electrons. Thus we omit influence of the metal layer on electron gas oscillations.
Taking into account that $U(x)$ is symmetric about zero, the oscillation period is determined as follows:

$$T(E) = 2\sqrt{2m} \int_0^{x_2} \frac{dx}{\sqrt{E - 2k\mathrm{T} \ln\left(1 + \frac{x}{a}\right)}} \tag{4}$$

For small-amplitude oscillations $x << a$, so:

$$\ln\left(1 + \frac{x}{a}\right) \approx \frac{x}{a}$$

Then [1]:

$$T(E) = 2\sqrt{2m} \int_0^{x_2} \frac{dx}{\sqrt{E - 2k\mathrm{T} \frac{x}{a}}} \tag{5}$$

Taking into account that $U(x) = E$ we have the following equation for x_2:
$x_2=Ea/2k\mathrm{T}$.
This results in the following oscillation period to electron energy function:

$$T(E) = 2\sqrt{2m}\int_0^{\frac{Ea}{2kT}} \frac{dx}{\sqrt{E - 2k\mathrm{T}\frac{x}{a}}} = \frac{4}{3}\frac{\sqrt{2\pi m}}{\frac{2k\mathrm{T}}{a}}\sqrt{E} \tag{6}$$

In order to estimate applied voltage value influence on oscillation period and frequency it is necessary to take into account that the electron energy shall be defined as $E = e(\phi_1 - \phi_2)$ where: φ_1 and φ_2 are potential values at points x_1 and x_2, respectively.

X_1 and x_2 are turning points of electronic gas oscillation. The distance between them is much bigger than metal layer thickness. The screening length in the area of the ohmic contact is considered [4].

Upon electrons energy decrease (relaxation process is shown in selected region of Figure 3) frequency increases until electronic gas becomes stable.

The calculated maximum frequency value for Si-Ag-Si structure is 10 MHz in case undoped silicon is used (n_k = 5.8 * 1015 m-3) as well as the thickness of the metal layer is 1nm. The calculated value is close to the experimentally obtained value (Figure 5).

The limiting oscillation frequency (period) shall be calculated with the condition that the electron oscillation energy is equal to thermal oscillation energy. It means (since one-dimensional case is considered):

$$T(E) = \frac{4}{6}\frac{\sqrt{2\varepsilon m}}{e\sqrt{n_k}} \tag{7}$$

Thus we have a mechanical resonance of aggregated electronic gas of enriched regions.

As resulted from formula (7) the maximum frequency of the structure is proportional to the electron density in the contact area. In case dopant concentration is increased up to 10^{15}-10^{16} m^{-3}, the maximum frequency value will increase by 10^5 and reach THz.

This reactive element has a certain resonant oscillation frequency. This frequency should be as high as possible, since the working area is below the resonant frequency.

4. Conclusion

Taking into account mentioned above data it can be concluded that the resonance frequency in the semiconductor-metal-semiconductor structure depends on ohmic contact characteristics (i.e. electron density in the enriched semiconductor-metal junction regions) and does not depend on metal layer dimensional characteristics (in this case metal layer thickness must be less than electron mean free path in current metal) [6].

Basing on the obtained results we can summarize as follows:

- to solve a number of problems in telecommunications it is proposed to use HTS elements as high-frequency filters (high-temperature superconductors). Their use is limited by maximum application temperature (70 K) and crock of structure (planar film structure with ring resonators using small surface areas) [7, 8].

Use of SMS-structure resonators provides the following capabilities:

- filtering of any spectrum signals;
- structure micro-size;

- operating at normal conditions (room temperature);
- at the same time, SMS structure considered as a microelectronics and/or nanoelectronics element can be used for manufacturing of nonlinear filters [9], ADSL filters [10], correlation filters [11], matched filters (spectral filter) [12], band-pass filters [13, 14], analogue of the quartz resonator.

Nowadays such filters are used in:
- passive elements (R, L, C):
- operational amplifiers with feedback;
- integrated electronic components (multipliers, integrators, signal generators, etc.).

The implementation of mentioned above devices allows:
- to use them as reactive elements in micro- and nanoelectronics;
- to simplify considerably designing of active and mixed filters;
- to reduce the cost of the final product [15, 16, 17].

References

[1] Blank T V, Goldberg I A 2007 Mechanisms of current flow in ohmic metal–semiconductor contacts, *Physics and technology of semiconductors* **41**(11) 402-407

[2] Feldman L, Mayer D 1989 *Fundamentals of surface analysis and thin films* (Moscow: Mir) p 344

[3] Vladimirov V V, Volkov A F, Meilikhov E. 1979 *Plasma of semiconductors*

[4] Landau L D, Livshits E M 1965 *Mechanics"* (Moscow: Nauka)

[5] Bonch-Bruevich V L, Kalashnikov S G 1990 *Physics of semiconductors* (Moscow: Nauka)

[6] Kargi N I , Pagnuev Y I, Svistunov I V, Sinelnikov B M, Shtab A V , Shtab E V *Thin-film semiconductor electronic resonator* Patent RU 2293397

[7] Emelyanov V 2001 Microelectronic microwave components based on high-temperature superconductors *Components and technologies* **6** 1-6

[8] Emelyanov V 2001 Microelectronic microwave components based on high-temperature superconductors *Components and technologies* **7** 1-2

[9] Pogorelov A M 2003 Analysis of FPGAs implementing non-linear digital filters, *Journal of Radioelectronics* **2** URL: http://jre.cplire.ru/koi/feb03/1/text.html

[10] Kruk B, Popantonopulo V, Shuvalov V 2003 *Telecommunication systems and networks.* (Moscow.: Hotline-Telecom) p 647

[11] Evtikhiev N N, Zakhartsev A V, Ivanov P A et all 2005 Synthesis and study of invariant filters with a linear phase coefficient for the problems of optical-electronic correlation discrimination of images *Science-intensive technologies* **6(5)**

[12] Baskakov S I *Radio circuits and signals* (Moscow: Higher school, 2nd edition) p 446

[13] Johnson D, Johnson J 2003 *Active Filter Reference* (Moccow: Energoatomizdat) p 128

[14] Besedin V 2007 Bandpass filter for RF. *Radio World* **7** 24- 25

[15] Baklanov I S , Baklanova O A, Erokhin A M , Ponarina N N, Akopyan G.A 2018 Myth as a means of ordering and organizing social reality, *TarihKültürveSanatAraştırmaları* **7**(2) 41-47

[16] Ivashova V, Goncharov V, Erokhin A., Kolosova O, Berkovsky V 2018 Life strategies of graduates of regional universities: the reality of Russian society, *International Journal of Civil Engineering and Technology* **13** 398-404

[17] Kulaev E V, Gritcay, D I , Danilov M V, Serikov S S, Ivashova V A 2018 Updating of CDIO standards in agricultural engineering education *International Journal of Civil Engineering and Technology* **9(13)** 1055-1062

Development of automated measuring device with magnetic fluid

B A Baranov[1a] *, A Yu Sokolov[1b], I V Malakhov[1c]

[1] Plekhanov Russian University of Economics, 36 Stremyanny Pereulok, Moscow, 117997, Russia

E-mail: [a]bba@bk.ru, [b]alrs@inbox.ru, [c]malahov.igor2016@yandex.ru

Abstract. The article presents theoretical principles of the methods and devices for water activity (a_w) measurement. The purpose of the research is the increase of the validity of water activity indicator measurement on the basis of a direct method of material test; the optimization of the structure of an automated measuring device to provide the possibilities for a simultaneous study of several samples (for example, food materials) with various water activity levels. A developed measuring device allows a simultaneous determination of indicators in several samples (8) located circumferentially and includes thermal sensors. The sensors in this device are represented by induction coils. They are installed on both leg pipes (made of non-magnet materials) of a liquid level manometer. The device contains a special silicon organic oil with the pressure of partial steams within 10-6 Pa (at normal conditions) used as a manometric liquid. While measuring the device is vacuumized and thermostated. The method and device approbation resulted in making calibration charts at various temperatures, approximated with the help of software. The device provides the validity of measurements in water activity range from 0.45 to 1.0.

1. Introduction

The relevance of improving the methods and devices of quality measurement for various materials, including food systems, is without any doubt.

The chemical analysis and technologies include the parameter "solution activity" while the food industry has the parameter "water activity". The indicator a_w correlates with the development of microorganisms in food products [1, 2].

At the device elaboration the authors used the unit designed in our country and further improved as a prototype. The prototype drawback is the presence of a mechanical sensor, significantly reducing the reliability and accuracy of the measuring device which, in its turn, makes the device structure more complicated. In addition, this device allows measuring water activity in reservoirs only in succession.

The measuring device, developed in the research, can be used for the control and measurement in industrial areas with the account of the tool base improvement concept and the improvement of methods&devices properties. The results of water activity measurement provide forecasting the safety of raw materials and products in scientific research in the area of chemical technologies, food industry, retail, etc. [2, 3].

2. Methods of water activity measurement

The purpose of the developed device is improving measurement reliability and accuracy due to a simplified measurement unit design, providing for a possibility of a simultaneous research of a large sample number (up to 8) with various water activity as well as increasing the accuracy of measurements and software data processing.

The principles and methods of water activity measurement a_w are versatile. To determine water activity, the food industry uses the direct or indirect determining of water steam equilibrium pressure in closed systems; certain methods are recognized as official and/or international ones [1, 8]. The authors developed the following methods to determine $a_{w\ 1}$) direct pressure manometric measurement and its varieties [2, 4, 5]; 2) dew point measurement; 3) determining the change of the condenser capacity; 4) determining the change of the electrolyte electrical conductivity; 5) measuring the gyroscopic line length; 6) defining sorbing agent weight increase; 7) measuring temperature changes at establishing the equilibrium in closed systems; 8) defining the change of the freezing point in the experimental system, etc.

At the moment water activity is the key quality (safety) indicator of food products. To control the product quality, the tool base of methods for the analysis of the indicator a_w is improved, for example, the biospeckle laser [7] is reported to be used for testing food materials (starches, pectins, sugars) at various temperatures (from 0 up to 25 °C as well as at negative temperatures). For the production of certain types of products it is suggested using a special microbioreactor to trace the changes of a_w at maturing of biotechnological media in cheese manufacturing [6].

The present research purpose and objectives are resolved due to the reservoirs for the researched products installed at the base of the device intended for a direct (i.e., manometric) determination of water activity. These reservoirs are equipped with thermal sensors connected to a corresponding number of the measuring tubes of a liquid manometer. These reservoirs are located circumferentially. The liquid manometer measuring tubes are made of non-magnetic materials and the sensors (induction coils or electrodes) are installed on them; these sensors are connected to the external registering device. As a manometric liquid magnetic fluids on the basis of silicon organic oils are used. The device is equipped with a vacuum system.

The magnetic fluid application provides for a possibility to increase the measurement accuracy and reliability due to the conversion of direct working fluid movements into an electrical signal.To measure the change of induction, the tubes should be made of non-magnetic materials (glass or plastics). As the basis for magnetic fluids one can use silicon organic oils - polyphenyl methylsiloxanes with partial steam pressures at 20°C within 10-6 Pa.

3. Development of the automated measuring device and the operation principles

Figure 1 shows the diagram of the device containing the reservoirs for the studied product (1), installed in the thermostat, with thermal sensors inside (2). Each reservoir is connected to one of the liquid manometer leg pipes (3). The second free manometer leg pipes are interconnected and also connected to the reservoir for distilled water (4) and the vacuum system (5). Both leg pipes of the liquid manometer (made of non-magnetic materials) have the sensors - induction coils (6, 7) - installed on them. The sensors are situated on the manometer leg pipes and have the length (100-300 mm) providing for the registering of pressure drops within the water activity range 0.5-1.0 at the temperature 0 - 80°C. The liquid manometer is filled with a magnetic fluid on the basis of silicon organic oils (8).

The pipelines are blocked with magnetic vacuum valves (9, 10, 11, 12). The whole unit is located on a flat round foundation (13) and is placed in the thermostat (14).

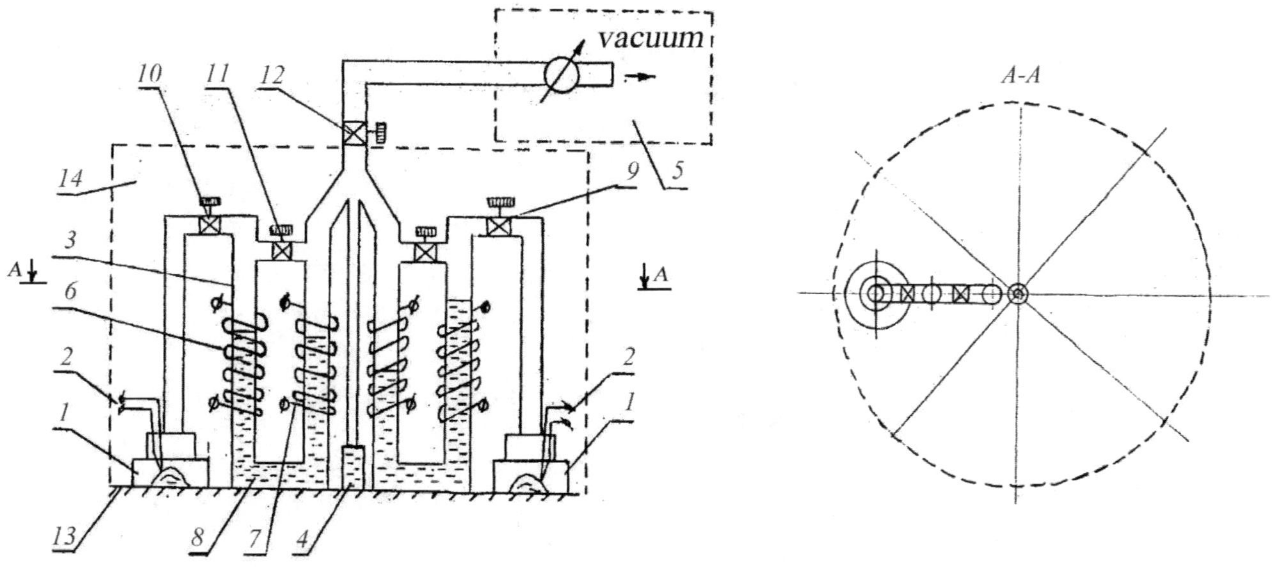

Figure. 1. Diagram of the automated device with a magnetic manometric liquid to define a_w: 1 - reservoirs for the researched product; 2 - thermal sensors; 3 - liquid manometer; 4 - distilled water reservoir; 5 - vacuum system; 6, 7 - induction coils; 8 - magnetic fluid; 9, 10, 11, 12 - electromagnetic vacuum valves; 13 - foundation; 14 - thermostat

At the device operation the difference between water steam pressure over the product and distilled water causes the change in the liquid level inside the manometer; this change is registered by the induction coil. The electromotive force change causes the electrical signal change. The electrical signal is registered by an external device with the indication in water activity units. The device is calibrated on the solutions of saturated salt solutions with the known water activity values. The device design allows simultaneous studying of several (8) samples as each measuring reservoir has its own manometer.

The reservoir (1) contains the researched products, the reservoirs are sealed. Further, vacuum valves are opened (9, 10, 11, 12). The vacuum system is turned on and pumps air until residual pressure reaches 10-1 Pa. The device is thermostated. The temperature is registered by sensors (2). As a result of the difference in pressures over water and products, the magnetic fluid (8) moves inside the manometer (3), and this motion is registered by the sensors – induction coils (6) and (7) and is transferred to the external device. After the measurement the valves (10) open first and, after pressure equalizing, the valves (9, 11, 12) open simultaneously while the vacuum is released.

The valves (9) serve for disconnecting the device gas medium from the atmosphere in case not all reservoirs are in use. This device advantage is a higher measurement accuracy and reliability.

Magnetic fluid application allows decreasing the device overall dimensions as the density of silicon organic oils is equal to 0.9-1.1 g/cm3, while the oil-based magnetic fluid density– 4.0-6.5 g/cm3. The most reliable way is the use of magnetic fluids for which the criteria P_M and P_H (P_M characterizes a solid substance volumetric fraction with magnetic properties, P_H is the ratio of hydrodynamic concentration and solid phase concentration) shall be close to the values 0.67 and 2.8, correspondingly. Otherwise, one can expect a complicated nature of the dependence of magnetic fluid physical properties on the magnet field, the motion state and the time of liquid storing which is not permissible for measurement devices. The problem of the magnetic fluid free surface sustainability in the magnetic bias field of the excitation coil in case the intensity of a longitudinal homogeneous magnet field is equal to H = 3.5-7.0 kA/m while the magnetic fluid magnetization Mo =45-60 kA/m.

The second option of the device design have the electrodes of the wire with a higher resistance in place of induction sensors inside the manometer leg pipes, in this case pressure drops cause the resistance change registered with the help of a common bridge circuit.

Fig. 2 shows a calibration curve to define the water activity on the liquid manometer data (ΔH, mm) at various temperatures within the interval 15 - 25°C. As the magnetic fluid density is 5 times higher than the one of a silicon organic fluid, the obtained values are correspondingly lower.

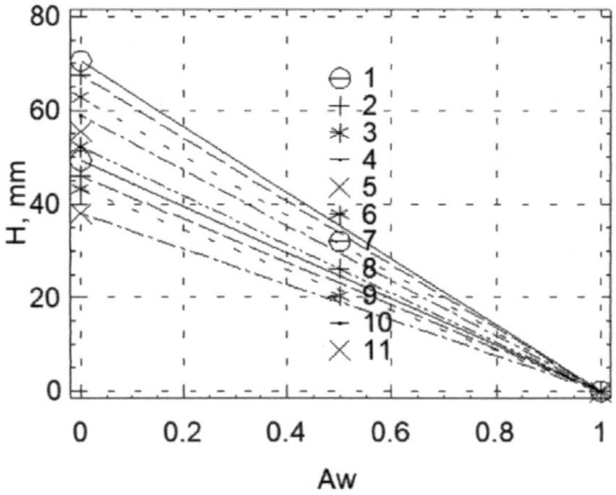

Figure 2. Calibration of oil manometer indications (ΔH) on the temperature (°C): 1 – 25; 2 – 24; 3 – 23; 4 – 22; 5 – 21; 6 – 20; 7 – 19; 8 – 18; 9 – 17; 10 – 16; 11 – 15

The device advantage is the possibility of a remote and continuous change as the registering device is outside the thermostat in which the device is installed. The device is reliably operated in the water activity range from 0.45 to 1.0.

For the unit calibration and obtaining measurement errors the authors used saturated salt solutions with the known water activity values at the temperature 25°C. The thermostat temperature was maintained at the level 25±0.5°C. The water activity values, obtained by the experiment, were compared with the reference ones [5].

The results of computer statistical processing of the experimental data are presented in Figure 3. The figure data show that the accuracy of measurements reduces with the growth in the residual pressure inside the unit and the decrease in the water activity level in the studied material.

The dependence of measurement error on the residual pressure level and the water activity level (s) is expressed by the following formula (1):

$$s = 0.6233 - 1.300\ a_w + 0.0129 \cdot P_o, \qquad (1)$$

where P_o – residual pressure, Pa.

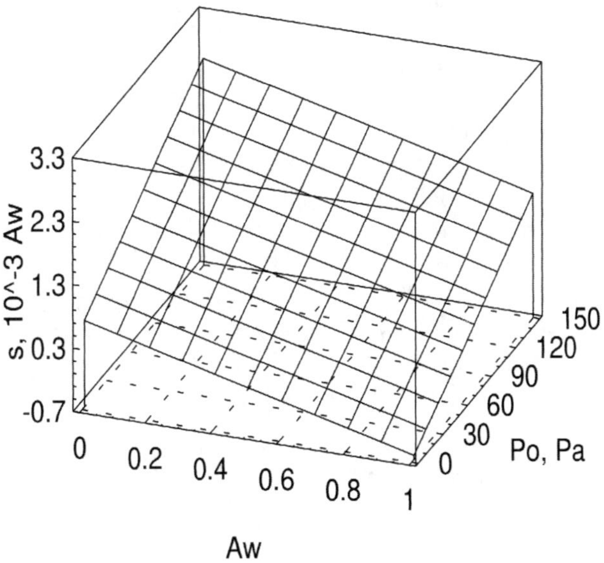

Figure 3. Dependence of the device data error (s) on the residual pressure (Po) and water activity (a_w)

It is necessary to note that the units of this design do not provide the opportunity to find out the water activity in the products where microbiological processes and enzymatic reactions can occur with gas emission. It neither makes it possible to research the products containing easily volatile components (for example, alcohols, ammonia, etc.) The measurement accuracy is influenced by the water activity level. At the lower values of water activity (lower than 0.45) the accuracy of measurement reduces because of water losses at air pumping.

It is known that a high residual pressure of the gas medium in the unit (more than 79, 300 Pa) leads to a slowdown of the water steam diffusion process while too low pressure (lower than 13.3 Pa) results in the increase of the length of molecule free path, which also influences the rate of sorption - desorption processes and results in the increase of the measurement duration, that is why residual pressure in the unit should be equal to 40-50 Pa.

4. Conclusion

This paper presents principal and technical aspects of the automated unit design for testing materials in terms of water activity. It also provides scientific&practical results demonstrating the prospects for the method upgrade and material (including food products) test instrumentation. The authors identified the advantages of the developed measuring device, in particular, improved technical and metrological characteristics comparing with similar devices. The paper points out some limits of the device application in the research and quality control of various materials.

References

[1] State Standard ISO 21807-2015. Microbiology of food products and feed items. Water activity definition (URL: www.gost.ru Date of entry 01.07.2017)

[2] Baranov B A, 2000 Theoretical and Applied Aspects of Indicator "Water Activity" in Food Product Technology, the author's abstract of Doctor of Technical Science, Saint-Petersburg.

[3] Parshina S A, Sokolov A Yu, Savchenko E O 2018 Technological processes obtain of monolithic polydisperse protein-base systems *Solid State Phenomena* **284** 133-138

[4] Rodel W, 2001 Water Activity and Its Measurement for Food Products. Tools and Sensors for

Food Industry, 2nd edition. Kress-Rogers E. and Brimelow C.J.B. (eds). Woodhead. Cambridge

[5] Zanoni B, Peri C, Giovanelli G, Pagliarini E, 1998 Design and setting up of a water vapour pressure capacitance manometer for measurement of water activity *Journal of Food Engineering*, **38 (4)** 407-423.

[6] Le Page J F, Dachraoui O, Daudin J D, Mirade P S 2012, Time-course estimation of evaporated water fluxes together with mean surface water activity in uncooked pressed cheeses during ripening using a purpose-built micro-bioreactor, *International Dairy Journal*, **22 (1)** 66-72.

[7] Silva S H, Lago A M T, Rivera F P, Prado M E T, Braga R A, de Resende J V 2018 Measurement of water activities of foods at different temperatures using biospeckle laser, *Journal of Food Measurement and Characterization* **12 (3)** 2230-2239.

[8] Information on https://www.aoac.org/

Evaluation of the traction-chain properties of a wheel tractor in dry peat deposit

G A Basalay[1], E A Kremcheev[2], D O Nagornov[2]

[1] Belarusian National Technical University, Independence Avenue 65, Minsk, 220013, Belarus
[2] St. Petersburg Mining University, 2, 21-line VO, St. Petersburg, 199106, Russia

E-mail: nagornov_d_spmi@mail.ru

Abstract. The paper highlights the results of studies on the creation of a measuring complex suitable for experimental studies of the towing and coupling and dynamic properties of wheeled tractors operated in difficult geological conditions of peat deposits of various degrees of drainage. The conceptual solutions for the layout of the measuring complex, which allows carrying out field measurements of the kinematic and dynamic parameters of machine-tractor units, to measure the hook force and torque on the power take-off shaft, are presented. Solutions are proposed to support the testing process with auxiliary devices that create hook loads on tractors. The requirements for the organization of a technological site for conducting traction tests of wheeled tractors are given. As a result of the research work, a test procedure was developed; the issues of metrological support of research and processing of experimental data were resolved.

1. Introduction

The productivity of machine-tractor units (MTU) in the main technological operations for the extraction of peat depends both on the design parameters of the machines and on the properties of the deposits of the developed field [1, 2, 3, 4,11]. The physicomechanical properties of the peat deposit are decisive in the formation of the tangential traction force by the tractor wheel propulsion. These properties are mainly determined by the type of peat, the degree of decomposition, and humidity along the depth of the peat massif [5, 6,12].

Due to the significant variability of factors that have a direct impact on the cross-country ability of vehicles with wheeled engines in the conditions of a peat deposit, when changing the operating modes of the units, adjusting the technological scheme, changing the type of tires or switching to the use of new machine-tractor units, issues of assessing the patency of wheeled vehicles come to the fore in all operating conditions in the real conditions of the field being exploited. The creation of mobile measuring systems, which allow carrying out field measurements of the kinematic and dynamic parameters of machine-tractor units, to measure the hook force and torque on the power take-off shaft, to develop unified requirements for the organization of pilot sites and the test process as a whole, will make it possible to carry out comparative tests to obtain objective information allowing reasonably make technical and technological decisions, with a guarantee of their applicability in the conditions of a particular enterprise.

2. Materials and methods

In ordfer to conduct research using the available exploration data, five technological sites were selected for milling peat extraction. Three of them are sites on the lowland deposits with a surface layer humidity in the range from 65% to 94%, and two sites with a transitional and upper deposits with an average operational humidity in the range of 78-81%.

Sampling for humidity control was carried out according to the sample scheme shown in (Figure 1). Humidity was determined in laboratory conditions by a standardized method (GOST 278-54). To determine the bearing capacity, a densitometer was used [7,13].

At each of the five sites, five test plots were marked. The lengths of the control sections, as well as the sections for accelerating and stopping the tractor between the experiments were selected according to the available technical requirements for traction tests.

Currently, a wide range of tractor tires with various treads is available. Despite the conditional separation of tires by application depending on design features, the effectiveness of a particular tire in different enterprises in terms of patency is not guaranteed by the manufacturer, and additional field tests are required on a particular wheeled tractor mover.

In the territory of the CIS countries, a major producer of tractor tires is the «Belshina» production association. The tires of this manufacturer are equipped with movers of wheeled tractors of traction classes 1.4 and 2. Tires are very diverse both in external geometric dimensions and tread, and in operational parameters. To carry out comparative traction and technological tests of the MTZ-852 tractor at the peat field under development, three sets of tires were selected that are used to complete the front and rear drive axle propulsors in various modifications: 1) serial: FDA - (8.3 / 8-20) V-105, RDA - (15.5-38) F-2A; 2) experienced: FDA - (16.20-20) F-64, RDA - (18.4-34) F-11; 3) rice-growing: FDA - (16.0-20) F-76, RDA - (18.4-34) F-44. The features of the 2nd and 3rd tire sets compared to the serial are the increased width and outer diameter of the tires, and the difference, including among themselves, is determined by the difference in the height of the lugs and the density of the tread pattern.

Based on the selected three sets of tires, five layout schemes of a tractor propulsion unit with a 4x4 wheel arrangement were created under conditions of peat mining (Table 1).

Table 1. Ways to install wheels on a tractor

Drive axle	Tire Layout Number				
	1	2	3	4	5
Front	(8.3/8-20) V-105	(16.20-20) F-64	(16.0-20) F -76	Dual (1+1)	(16.0-20) F -76
Rear	(15.5-38) F -2A	(18.4-34) F -11	(18.4-34) F -44	Dual (1+1)	Dual (3+1)

The front wheels were doubled according to the 4th scheme by installing the disk of the added wheel directly on the bolts of the main wheel disk in the order: left - to the right, right - to the left. The rear wheels in the 4th and 5th layout schemes were doubled through transitional standard hubs.

The pressure in the tires was set as low as possible according to the technical characteristics of the tires: front in the range from 0.14 to 0.16 MPa; rear tires in the range from 1.0 to 1.2 MPa.

One of the additional tasks in the process of traction testing was the development of a method and device for creating a hook load of a tractor. At present, several methods for creating a tractor hook load during traction tests are known and widely used. In particular, the hook load can be realized using the appropriate selection of technological machines that would provide a stepped increase in traction to the maximum realized by the tractor. On solid soils and roads of various categories, transport trucks with variable load are used. In the study of transient and transient modes of operation, as well as in special tests, the hook load is created by braking the mover of the towed tractor. The main task when choosing a method for creating a hook load is the possibility of its stepwise change over a wide range up to the maximum possible in accordance with the traction class of the tractor.

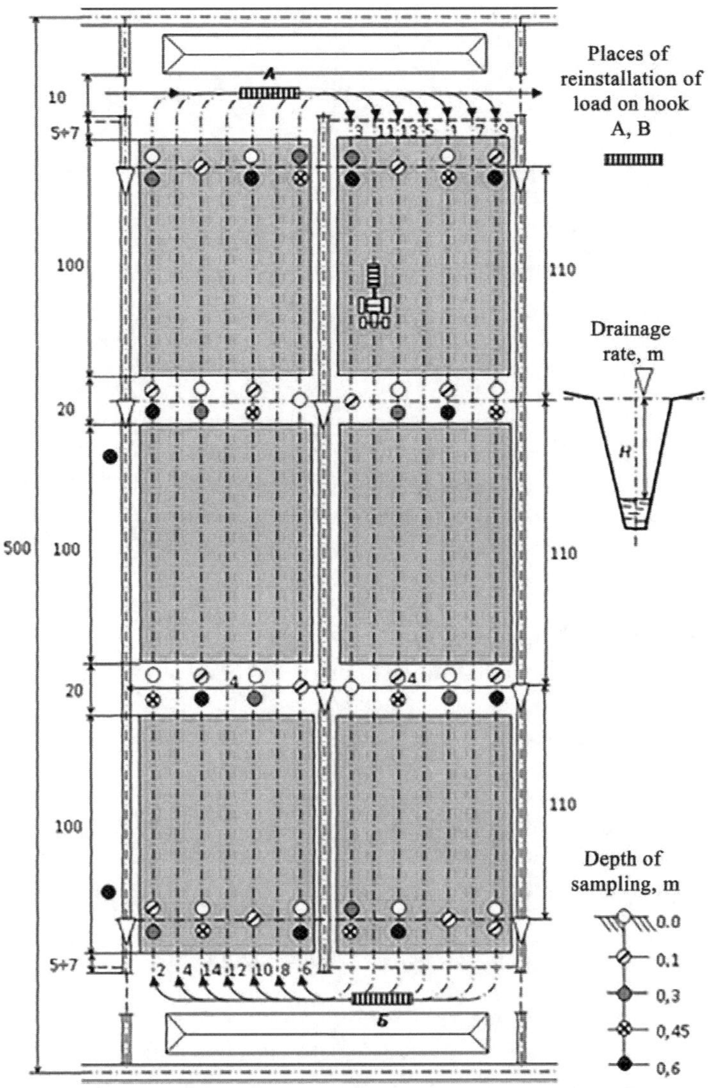

Figure 1. Scheme of the technological site for traction testing of wheeled tractors

During the traction tests of the MTZ-852 tractor at the production sites for milling peat milling of the «Zelenoborskoye» enterprise (Republic of Belarus, Minsk region), the hook load was created by a set of spent track caterpillars of the MTF-43 milling peat harvesting machine. Caterpillar tracks were grouped into seven sets of twenty links, which allowed ensuring the entire range of the hook load realized by the tractor. Link sets 1 (Figure 2) were connected in three rows to the tow hitch 2 and, using steel cables 3 (length 2.5-3.0 m), to the link 4, mounted on the cross-section of the tractor rear linkage.

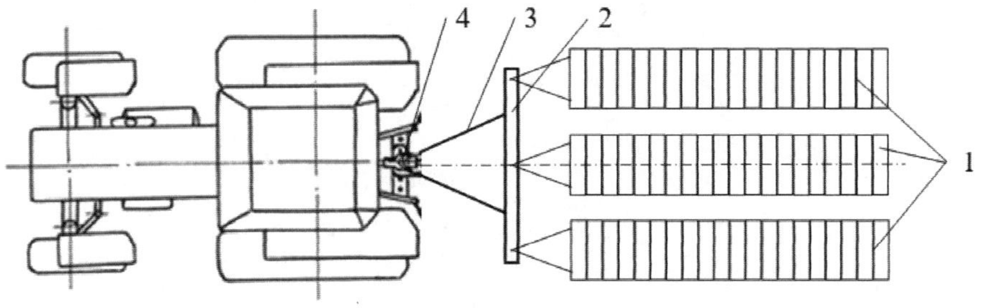

Figure 2. Scheme of formation of the hook load of the tractor using sets of links of the caterpillar chain

The track links of the MTF-43 machine do not have lugs, which made it possible to carry out traction tests for a long time without significant disturbance to the relief and surface structure of technological platforms, on which milling peat extraction machines operated simultaneously.

3. Results and Discussions

During the traction tests of the MTZ-852 tractor on a peat base, tractor operating modes were created that made it possible to more fully study the traction and coupling characteristics of the propulsion device, as well as the tractor's energy capabilities when aggregating it with the main technological machines. These modes were provided by tests of a tractor equipped with alternating propulsors in accordance with the accepted layout schemes (table 1) at five technological sites under conditions of variation of parameters in terms of humidity and bearing capacity of the peat base. The change in operating modes was achieved by connecting the front drive axle automatically or forcefully, as well as automatically locking the differential of the wheels of the rear drive axle, in the range of tractor speeds provided by gears from the 3rd to the 7th gear change box with the reduction gear turned off.

Traction tests were carried out in series with the equipment of the tractor mover according to one of the layout schemes (table 1).

During the testing the following processes were carried out:

1. Registration of controlled parameters of the machine during forward and reverse tractor movement in five control sections and in five gears from 3rd to 7th with automatic inclusion of the front drive axle without hook load. After this, the front drive axle was forcedly turned on and a series of experiments was repeated.
2. Repeat a series of experiments with successive loading of the tractor with hook force due to the connection of one, two, three, etc. sets of loads up to the limit values for the traction class of the tested tractor. These experiments were carried out before loading the tractor with the maximum possible hook force in this mode.
3. Registration of tractor idle motion parameters when towing it by another tractor.
4. Conducting experiments on points 2 and 3 on the remaining four sites.
5. Sequential equipment of the tractor by the movers of the remaining layout schemes (table 1) and the repetition of the experiments in paragraphs 2, 3, 4.
6. Conducting experiments according to paragraphs 2, 3, 4 with the layout schemes of the propulsion device 3, 4, 5, which ensured the patency of the tractor at the same five technological sites with a surface layer moisture depth of 100-250 mm up to 94%, and in some cases more.

The layout schemes of the mover were completed in the sequence 1, 2, 3, 5, 4, which is due to the smaller amount of work on the installation, dismantling and re-installation of elements of the movers.

Processing of experimental data was carried out in two stages. At the first stage, the values calculated by the procedure were calculated from the measurement results. The necessary statistical analysis of the data was carried out in the second stage.

The kinematic characteristics of tractor movement in both traction and technological tests were established in the following sequence:

- the actual speed of movement $v = L/t$ was determined, where $L = 2\pi Rn$ is the distance traveled, measured using the "fifth" wheel; R - is the kinematic rolling radius of the "fifth" wheel; n - is the number of revolutions of the "fifth" wheel during the experiment t;

- determined the theoretical speed of the tractor in a given mode. The speed value was taken equal to the actual speed when the machine moves without load (idle tractor passage):

$$v_t = v_r = \frac{2\pi R n_0}{t_0} \quad (1)$$

where n_0, t_0 - the number of revolutions of the "fifth" wheel and the experiment time at idle tractor, respectively;

- the conditional coefficient of tractor slipping was determined by the formula:

$$\delta = \frac{v_t - v}{v_t} \cdot 100\% \quad (2)$$

- slip coefficients for the drive wheels, or slip for the driven wheels were calculated by the formulas:

$$\delta_k = \frac{L_t - L}{L_t} \cdot 100\%, \qquad \varepsilon_k = \frac{L - L_t}{L} \cdot 100\%, \quad (3)$$

where $L_t = 2\pi R_t n_r$ - is the length of the path that the wheel would go, making n_r revolutions, with a rolling radius equal to the rolling radius R_t at idle; n_r is the number of revolutions of the wheel during the working passage;

- traction power and traction efficiency of the tractor were determined by the formulas:

$$N = Pv, \eta = \frac{N}{N_e} \quad (4)$$

where P - is the mathematical expectation of the hook effort in the experiment; N_e - is the installed power of the tractor engine.

Statistical processing of experimental data was carried out in order to calculate estimates of the mathematical expectation and standard deviation of the values of the determined values, as well as to test hypotheses about the laws of their distribution. This part of the data processing was carried out according to standard programs from the computer software.

In order to ensure field testing of tractors, a measuring complex was developed that is installed directly on the test machine with the ability to control and perform control functions at the workplace of the machine operator.

For testing the MTZ-852 tractor in conditions of peat mining, the measuring complex included a power source, a control panel, a personal computer, distribution blocks, current collectors, strain gauge sets, strain gages and traction strain gages, angular velocity sensors, pulse and potentiometric sensors, devices for installing devices on the tractor, as well as connecting cables. For the purpose of unification, measuring channels were included in the structure of the measuring complex for recording the stress state of the parts, their rotation frequency, and also the relative angular displacement of the mating parts [8, 9]. Such a layout of the complex made it possible to carry out remote synchronized measurements of the recorded parameters using sensors located at different points of the MTU.

The measuring channels of the complex are formed as follows.

The channel for determining the loads according to the stress state of the parts includes a strain gauge bridge with additional and adjusting resistances. It is used to measure the torques on the wheels and the power take-off shaft. In the strain gauge connection schemes, balanced four-arm bridges are used, the loads are determined by the unbalance value of which. To ensure operation on an amplifier-free circuit, the shoulders of the bridge are active, several strain gages are connected to each of them.

The measurement of the rotational speed of the tractor wheels and the "fifth" wheels is made by pulse sensors, which together with the resistance to adjust the second type of measuring channel. The third type of measuring channel serves to measure the speed of the power take-off shaft. It uses a tachogenerator.

The registration of the hook load during the tests was carried out using a special traction link, in which the sensor is a potentiometer. It makes up the measuring channel of the fourth type.

Figure 3 shows the installation diagram of devices and measuring and recording equipment on a wheeled tractor.

As a power source used batteries 1 large capacity, providing a stable voltage on-board network. To configure and adjust the measuring channels, turn the equipment on, the control panel 2 is used. The information received from the measuring channels is displayed and stored on a personal computer 3.

The torques on the wheels of the front and rear driving axles are measured by tensor 4 and 5, and the wheel speeds are measured by impulse sensors 6 and 7. The load on the hook is measured by traction link 9, the torque on the power take-off shaft and its speed by sensors 8. The tractor moves cornering is fixed by a sensor 10 mounted on the elements of the steering trapezoid of the front wheels.

Measuring instruments are mounted on damping devices in the form of spiral springs and rubber gaskets.

To fix the kinematic parameters of tractor operation, the system includes two piano-type wheels15 (fifth wheels), which are mounted on the tractor frame using articulated parallelograms 12 and extendable transverse pipes 11. The forks of the "fifth" wheels are equipped with impulse sensors 14, which are wheel revolution counters and are used to measure the distance traveled. On the tractor frame there are pulse sensors 6 and 7 for reading the number of revolutions of the wheels of the front and rear axles. According to the readings of these sensors and a time counter, the actual speed of movement and slipping of the propulsion device are determined. In the vertical hinges of the "fifth" wheels, potentiometric sensors 13 are fixed, which register the angular deviations of the wheels, which makes it possible by conversion to find the kinematic characteristics of the movement of the MTU in the corner. The deformation of the carrier base of the mover is determined by the readings of potentiometric sensors 16 located in the horizontal joints of the movable parallelograms 12.

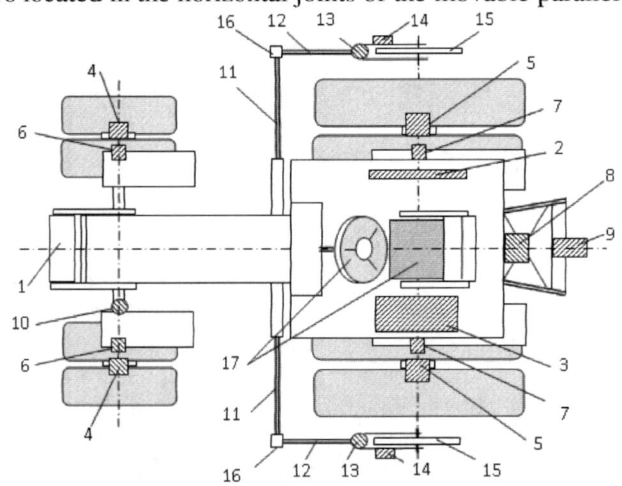

Figure 3. Layout of measuring and recording equipment on a wheeled tractor

The tractor is controlled and the measuring system is operated from the driver's workplace 17, which is played by an experimental engineer with a license to operate the tractor.

Processing of experimental data is performed on a computer using PowerGraph software [10, 14]. PowerGraph software supports a wide range of digital measurement equipment. A high-quality and productive system of information visualization, mathematical processing of analysis, import and export of data allows organizing an effective data recording system with automatic control.

The hook force in the process of traction and technological testing was measured by a specially made traction link.

The main elements of the traction link (Figure 4) are a coil spring 1, a body and a rod, and a potentiometer. The housing consists of a main fork 2 with four studs 3, on which the thrust wall is fixed 4. The rod 5, with the front wall 6 and the hooked fork 7 fixed thereon, makes up the movable part of the device. A potentiometer 8 is fixed on the abutment wall 4, on the input shaft of which is mounted a gear 9, which engages with a rack 10 mounted on the front wall. The potentiometer is connected to the measuring circuit with a connector 11 fixed to the fork 2. The traction link from the side of the main fork with two fingers 12 is mounted on the diameter of the tractor hitch. The towing fork with the help of the pivot 13 is the coupling of the tractor with the trailed machine.

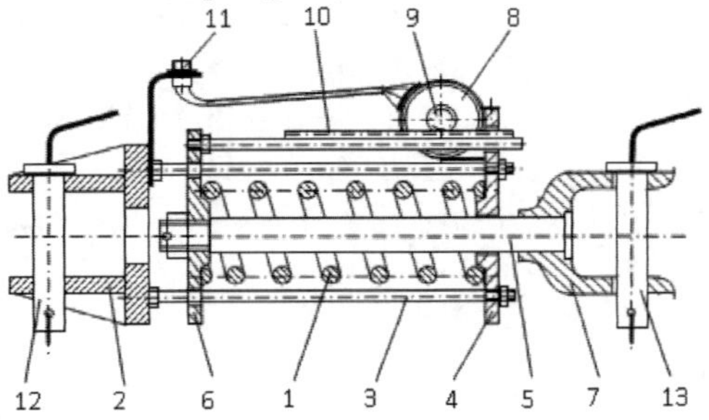

Figure 4. Traction link

The traction link developed for the measuring complex works as follows. Under the action of the hook load, the rod acts on the spring 1 and compresses it. In this case, the distance between the front and persistent walls changes by the amount of spring deformation. Together with the front wall, the rail 10 mounted on it moves, rotating the gear of the potentiometer mounted on the thrust wall [15, 16].

To measure the torque on the power take-off shaft and its rotation frequency, device 8 was used (Figure 3), consisting of an intermediate shaft placed between the PTO shaft and the driveshaft, with load cells glued on it, connected by a bridge circuit, and also connected in parallel with flexible transmission of the tachogenerator. To feed the circuit and remove the electric signal from the rotating shaft, a continuous mercury-amalgamated TRAP type current collector was used.

Torque measurements were carried out when aggregating a tractor with a MTF-13 milling drum, as one of the most energy-intensive trailed machines.

4. Conclusion

To ensure field testing of tractor units in difficult mining and geological conditions of peat deposits of various degrees of drainage, a concept has been developed and a mobile measuring system has been created that allows field measurements of the kinematic and dynamic parameters of the tractor, hook force and torque on the power take-off shaft. Methodological support was developed for testing tractors at standardized experimental sites. As a result of the research work, the issues of metrological support of research and processing of experimental data were resolved. The research results were tested in the conditions of the existing «Zelenoborskoye» peat enterprise (Republic of Belarus, Minsk region).

References

[1] Bazin E.T., Kopenkin V.D., Kosov V.I. 1992 Technical analysis of peat (Moscow: Nedra).

[2] Gamajunov S.N., Misnikov O.S., Puhova O.V. 1999 *Gornyj zhurnal* **10** 41-44.

[3] Alexandrov G., Iablonev A. 2018 Rental relations as a factor of mining investment prospects *E3S Web of Conferences* ***41***

[4] Mikhailov A V 2017 Open-pit mining of lignin waste storage *Journal of Mining Institute* **223** 66-69.
[5] Misnikov O S 2006 *Theoretical Foundations of Chemical Technology* **40** 4 423-430.
[6] Misnikov O S 2016 *Mires and Peat* **18** 1-15.
[7] Misnikov O S Dmitriev O V, Popov V I 2016 The use of peat-based water repellants to modify fire-extinguishing powders *Polym. Sci. Ser. D.* **9** 133.
[8] Alexandrov G., Yablonev A. 2019 Rental and investment regulation of mining companies in subsoil developing industries *E3S Web of Conferences* **105**
[9] Nagornov D. O., Kremcheev E. A. 2018 Framework to evaluate efficiency of peat processing using roller-disc grinding machines *Journal of Physics: Conference Series* **1118**(1).
[10] Shishlyannikov D I, Romanov V A, Zvonarev I E 2019 Determination of the operating time and residual life of self-propelled mine cars of potassium mines on the basis of integrated monitoring data *Journal of Mining Institute* **237** 336-343.
[11] Aghajani A, Osanloo M 2007 Application of AHP-TOPSIS method for loading-haulage equipment selection in open pit mines *Proc. of XXVII International Mining Convention* (Mexico).
[12] Gorlov I V, Rakhutin M G 2017 *Gornyy informatsionno-analiticheskiy byulleten* **12** 139–145.
[13] Kazachenko G V, Basalai G A 2014 *Gornyy informatsionno-analiticheskiy byulleten'* **3** 89–95.
[14] Wang J J, Jing Y Y, Zhang C F, Shi G H, Zhang X T 2008 A fuzzy multi-criteria decisionmaking model for trigeneration system *Energy Policy* **36** (10) 3823-3832 DOI: 10.1016/j.enpol.2008.07.002
[15] Burger J A et al 1985 Impact of tracked and rubber-tired tractors on a forest soil *Trans ASAE* **28** 369-373
[16] Sinitsyn V F, Kopenkina L V 2017 *Gornyy informatsionno-analiticheskiy byulleten'* **2** 181–194.

Method of measuring vibratory condition of a drill string

K A Bashmur[1], E A Petrovsky[1], V V Bogachev[1], E V Tsygankova[1], V S Tynchenko[1,2] and V V Bukhtoyarov[1,2]

[1] Siberian Federal University, 79, Svobodny Av., Krasnoyarsk, 660041, Russia
[2] Reshetnev Siberian State University of Science and Technology, 31 Krasnoyarsky Rabochy Av., Krasnoyarsk, 660037, Russia

E-mail: bashmur@bk.ru

Abstract. The article discusses issues related to the improvement of the operational reliability of downhole measurement systems. The circumstances of insufficient readings accuracy due to severe operating conditions are shown. The inherent negative factors are given, which jointly affect the reliability of the systems and measuring instruments in use. In particular, special attention is paid to vibration loads and the need for their measurement. The urgency of development and research of new reliable systems is demonstrated. A developed method for measuring the vibratory condition of a drill stem and a developed device for the implementation of this method are presented, and operation of the latter is analyzed.

1. Introduction
The improvement and development of technical and economic indicators for drilling oil and gas wells is directly related to the need for measuring and monitoring the drilling operations profile. Along with the increase in the number of monitored parameters, more and more requirements are imposed on the reliability of measurement results. This is due to the increase in drilling speed, depth of directional wells, high cost of the equipment and magnitude of the responsibility for management decisions in the technological process [1]. Therefore, development of reliable and efficient devices and methods for measuring and monitoring well parameters is one of the most important tasks for the oil and gas industry.

2. Urgency of development and investigation of reliable methods for measuring performance parameters
Modern borehole measurement tools and systems, such as MWD (Measurement While Drilling) and LWD (Logging While Drilling), are usually implemented using electronic systems [2]. The well deepening process is inevitably related to the occurrence of various kinds of dynamic loads. These are, first of all, vibration loads and temperature stresses increasing with the depth of the well, which have an adverse effect on measuring systems as a whole and their electronic components in particular [3, 4]. Abrasive and physico-chemical wear also affects the drill string (DS) equipment and, therefore, the measuring systems contained in them [5]. In particular, the geothermal gradient imposes certain restrictions on the use of some measuring devices in a reliable range, for example, devices with permanent magnets at great depths of wells [6].

The above mentioned factors can result in damage to reliability of the measured parameter values, abnormal operation and failure of drilling parameter measuring and control systems [3].

The dynamic loads, which occur when the downhole tool (bit) interacts with the rock, are probably the most detrimental loads that occur during drilling. The influence of vibrations on DS equipment includes loss of stability of the drilling modes, deterioration of the well stem drilling and lower resource of the well equipment [1].

Various control systems are used to control DS vibratory condition, which incorporate sensors measuring amplitude and frequency of the acting oscillations [7, 8]. As it was mentioned before, the sensors contain electronic components that have low operational reliability under high dynamic loads, and often require power sources, which is not always reasonable in a borehole due to increasing general risks for the device performance.

In this regard, development and investigation of reliable means of measuring vibrations, which would depend little on external factors, seems to be an urgent task. In particular, such principles can be implemented on the basis of mechanical devices containing electronic components.

3. New method of measuring vibratory condition of the DS equipment

The authors developed a method for measuring vibrations in downhole conditions, which can be used for a continuous assessment of the vibratory condition without relying on electronic components as the primary analyzing system [9].

The essence of the method is as follows. When longitudinal and/or torsional vibrations occur in the drill string, the bit breaks off from the rock causing a shift of the rigidly bound drill string bottom (DSB) assembly. A hydromechanical vibration sensor, which perceives the vibration load, is installed above drill collars (DC) in the area subject to DS tensile loads (Figure 1).

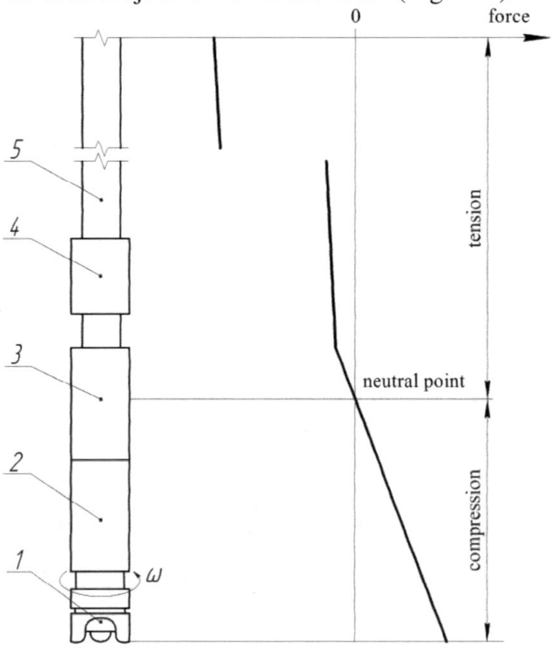

Figure 1. Method for measuring DS vibrations in a well: 1 – a bit; 2 – a mud motor; 3 – MWD; 4 – a hydromechanical vibration sensor; 5 – a drill string.

The sensor redistributes part of the mud flow into the sensor cavity according to the amplitude of the acting vibrations. The resulting change in the flow rate through the DS main stem is either recorded by a downhole flow meter or affects the mud motor speed, which, in turn, is recorded by a tachometer.

4. Development of device for measuring vibratory condition of downhole equipment

For the implementation of the proposed method of measuring vibrations, the authors developed a hydromechanical vibration sensor (Figure 2) [9].

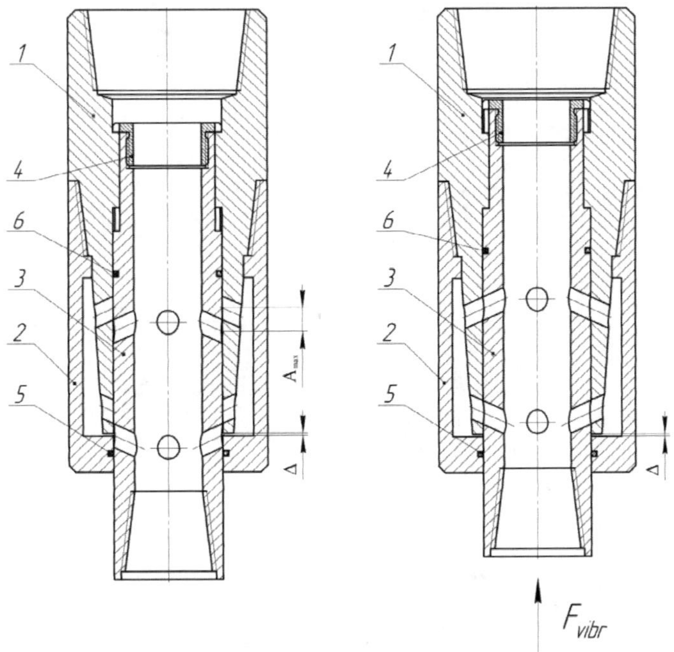

Figure 2. Hydromechanical vibration sensor: 1 – a case; 2 – a cup; 3 – a stem; 4 – an adjusting sleeve; 5, 6 – packing rings.

The hydromechanical vibration sensor is a mechanical device installed in the DS assembly, which perceives the vibration load from the downhole tool and generates a signal on occurrence of vibrations.

The sensor works as follows. When the bit is detached from the bottom, the vibration force F_{vibr} affects the stem 1 of the sensor and moves it along the guide slots in the case 2 to an intermediate or extreme position at maximum amplitude of vibration $A_{\max}$. There is an intersection or coincidence of holes in the sensor, through which part of the mud flow moving within the DS cavity passes into the cavity formed by the case 1 and the cup 2. As a result, the mud flow rate in the DS main stem changes, which is registered by bottom sensors. Thus, the proposed method is implemented.

5. Analysis of the hydromechanical sensor performance

In order to assess performance of the developed sensor, it is necessary to obtain a mud flow rate $Q(A)$ through the hole dependence on the amplitude A of the acting vibrations. In general, the flow through the hole is determined as follows:

$$Q(A) = \mu \cdot S_{\mathrm{h}}(A) \cdot \sqrt{2g \cdot H} \quad (1)$$

where μ – is flow coefficient; S_{h} – is opening area of the hole; H – is height of the mud hydrostatic column above the hole.

The $S_{\mathrm{h}}(A)$ dependence is determined by solving a geometric task of finding an area formed by the intersection of two circles (Figure 3).

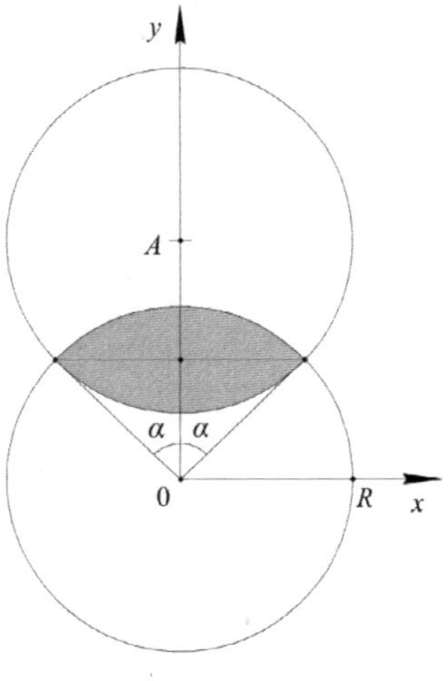

Figure 3. Calculation scheme for determining the area of two intersecting circles.

A system of equations of two circles is put down as follows:

$$\begin{cases} x^2 + y^2 = R^2, \\ x^2 + (y - A)^2 = R^2. \end{cases} \tag{2}$$

Transformation of the system of equations (2) provides an obvious solution $y = \frac{A}{2}$.

The area formed by intersection of two circles will be equal to twice the area of a circle segment:

$$S_{\text{op}} = R^2 \cdot (2\alpha - \sin 2\alpha) \tag{3}$$

Using obvious geometric transformations, the expression for S_{op} is brought to the following form:

$$S_{\text{op}}(A) = 2R^2 \cdot \arccos\left(\frac{A}{2R}\right) - \frac{A}{2} \cdot \sqrt{4R^2 - A^2} \tag{4}$$

Then the S_{h} value can be put down as follows:

$$S_{\text{h}}(A) = \pi R^2 - 2R^2 \cdot \arccos\left(\frac{A}{2R}\right) + \frac{A}{2} \cdot \sqrt{4R^2 - A^2} \tag{5}$$

Substitution of expression (5) into formula (1) gives the $Q(A)$ dependence:

$$Q(A) = \mu \cdot \left(\pi R^2 - 2R^2 \cdot \arccos\left(\frac{A}{2R}\right) + \frac{A}{2} \cdot \sqrt{4R^2 - A^2}\right) \cdot \sqrt{2g \cdot H} \tag{6}$$

The flow coefficient μ is determined from an empirical expression depending on the Reynolds number. When Re >10000 [10]:

$$\mu = 0.592 + \frac{5.5}{\sqrt{\text{Re}}} \tag{7}$$

According to the formula (6) and data close to the actual ones from table 1, a graph can be built up to present the dependence of mud supply through the hydromechanical sensor hole on the amplitude of acting vibration (Figure 4).

Table 1. Parameters for plotting a graph of Q through the hydromechanical sensor hole dependence on the amplitude of the acting vibrations A

Parameter	R, mm	Re	μ	H, m
Value	8	24519	0.627	60

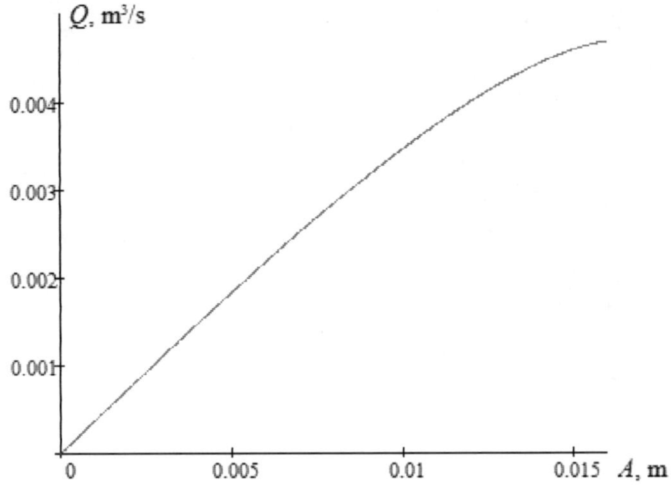

Figure 4. The dependence of the mud flow rate through the hydromechanical sensor hole on the vibration amplitude.

Thus, one of the advantages of the created method is that it is can be quite easily analyses to estimate the vibration parameters and does not require additional downhole analyzing devices.

The simulation was performed in the SolidWorks Flow Simulation software module to determine the distribution of velocities and pressure flows at the hydromechanical sensor working position. The simulation results are shown in Figure 5.

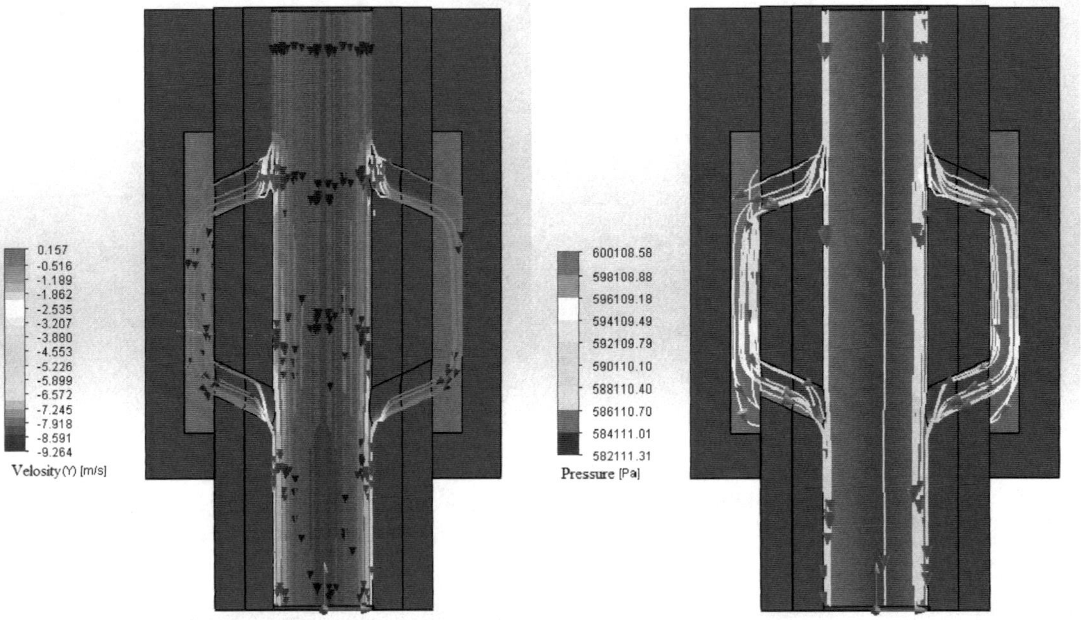

Figure 5. Model of mud flow velocity (left) and pressure (right) distribution in the hydromechanical sensor.

It can be seen from the obtained flow distributions that the return flows are practically absent, and the created hydraulic resistances have little effect on the drilling mud flow. Thus, the proposed method and device for measuring the DS vibratory condition are operable and can be used to assess the vibration parameters.

6. Conclusion

The urgency of development and investigation of reliable methods for measuring DS vibrations in a well has been revealed. A method for measuring vibrations based on the use of a hydromechanical sensor that perceives the vibration load from the bit, which results in a change in drilling mud supply through the DS main stem, is proposed. A design of a hydromechanical sensor is proposed that implements this method of vibration measurement. The analysis of the method performance was carried out, and the flow rate dependence on the acting vibrations amplitude was obtained to prove that the developed method can be easily applied in analysis to estimate the vibration parameters and does not require additional downhole analyzing devices. The obtained models of flow distributions through a hydromechanical sensor show that there are practically no reverse flows therein, and the created hydraulic resistance has little effect on the drilling mud flow.

References

[1] Petrovskii E A, Bashmur K A, Nashivanov I S 2019 *Chem Petrol Eng* **54(9-10)** 711–716
[2] Prensky S 2006 *World Oil* **227(3)** 69–75
[3] Zhan S, Ahmad I, Heuermann-Kuehn L and Baumann J 2010 *SPE Annual Technical Conference and Exhibition (ATCE 2010)* **1** 241–250
[4] McCluskey P, Grybowski R R, Condra L, Das D, Fink J, Jordan J, Torri T 1998 *Proceedings of the High-Temperature Electronic Materials, Devices and Sensors Conference* 199–206
[5] Iyasara A C, Ovri J E O 2015 *Int. Journal of Scientific Research Engineering & Technology (IJSRET)* **4(9)** 917–923
[6] Kutasov I M, Caruthers R M, Targhi A K, Chaaban H M 1988 *Geothermics* **17(4)** 607–618
[7] Dong G, Chen P 2016 *Shock Vib.* **2016(1)** 1–34

[8] Márquez M B S, Boussaada I, Mounier H, Niculescu S-I 2015 Analysis and Control of Oilwell Drilling Vibrations Springer International Publishing 294
[9] Patent application RU 2019116885, 31.05.2019
[10] Altshul A D 1982 Hydraulic resistance [In Russian – Gidravlicheskie soprotivleniya] Nedra 224

Journal of Physics: Conference Series **1384** (2019) 012006 doi:10.1088/1742-6596/1384/1/012006

Nature-like equipment for backfilling the mine chambers to reduce the risks of technogenic seismic impact

M A Vasilyeva[2], V S Bochkov[1]

[1] Ural State Mining University, Department of Mining Machines and Complexes, Khokhryakova st., 85A, Yekaterinburg, Sverdlovsk Region, 620014, Russia
[2] St. Petersburg Mining University, 2, 21-line VO, St. Petersburg, 199106, Russia

E-mail: saturn.sun@mail.ru

Abstract. The priority task in the mineral industry is to increase the efficiency of the exploited and infrastructure, equipment, including the creation of transport systems that have high performance while reducing the cost of transporting mineral raw materials and products of its processing. One-third of the world's mining companies use their mining systems at their enterprises to create a space. Backfilling activities are carried out in order to improve the safety of mining operations, namely, to control mining pressure, reduce possible mineral losses during excavation, prevent the occurrence of underground fires, as well as sudden outbursts of coal and gas, protect against destruction of objects in residential areas. Particular attention should be paid to the use of hydro-folding as a measure to prevent technogenic-tectonic earthquakes during long-term and continuous development of deposits. Taking into account the peculiarities of the materials used as a base when preparing filling mixtures, the specificity of the process of moving them to the bookmark site, necessary for obtaining a backfill array of required characteristics, is becoming increasingly important to use fundamentally new transport technologies for filling mixtures.

1. Introduction

The laying of the spent space allows controlling the mining pressure and improve the safety of mining operations, conducting mining at the same time by underground and open methods, increasing the percentage of mineral extraction, and also reducing the negative impact on the environment.

The factors that have the main influence on the increase in risks from the man-made seismic regime of mine workings are their age and the number of depleted layers, as well as the fact of the backfilling and the type of filling material used [1, 2].

2. Materials and methods

The laying of the worked-out space is used to reduce the deformations of the water barrier to protect against the flooding of workings and to reduce the subsidence of the earth's surface under residential areas. In addition, it is possible to increase the volume of recoverable mineral resources, improve the conditions for ventilation of existing mine workings, reduce the area of land allocated for dumps and thus reduce the amount of land intended for remediation [3]. Waste enrichment with impurities of insoluble materials, as well as clay-salt wastes are used as filling material [4]. Nowadays, about 85% of the bookmark is held hydraulically and only 15% is mechanical [5]. According to the research work [6], the laying of voids in the workings as a measure to prevent flooding of the developed space is

Journal of Physics: Conference Series **1384** (2019) 012006 doi:10.1088/1742-6596/1384/1/012006

mandatory on 2 layers with 3 development, as well as under the condition of conducting excavation work under the planning structure of the city.

a b c d

Figure 1. Surface collapse caused by failure of mine workings:
1, 2 - Solikams; 3, 4 – Novokuznetsk

After the completion of the laying of waste voids in one of the ways, during a certain time, the array shrinks in the vertical and horizontal directions. In addition, the formed massif under the pressure of its own weight is additionally compacted and deposited in vertical projection. To illustrate this fact, the authors used the data of B.A. Borzakovsky, who writes that the shrinkage of the array during the dry laying of the VKMKS ends mostly after 8-12 months. After shrinkage, an increase in the strength of the array occurs. This process is much slower. Hydraulic laying gives a more rapid decrease in the average density of the seismic energy released, as opposed to dry (about 10 years faster) [7].

The study of the long-term impact of work on the laying on the seismic potential of the workings shows that these works contribute to the stabilization of the disturbed rock mass not by a one-time decrease in the intensity of its destruction, but as a decrease in the period of seismic activity after the laying. The study of workings with the use of backfill materials of various compositions allowed to establish that the hydraulic folding leads to earlier (about 10 years earlier) attenuation of seismicity (Fig. 2) [8].

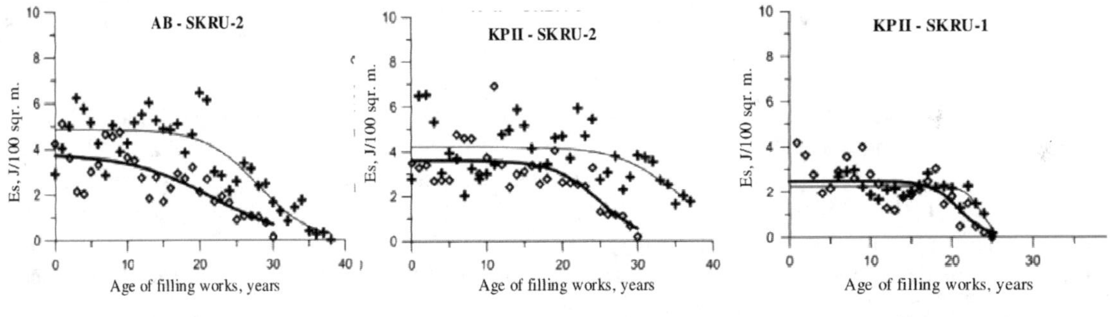

Figure. 2 Dependence of seismic energy release density on the age of the filling works:

◇ - hydraulic laying, ✚ - dry laying

When transporting the filling mixture into the produced chamber, it is impossible to feed it under the roof of the mine, which leads to the appearance of unfilled cavities in (Fig. 3). The consequence of incomplete bookmarks is the subsidence of the roof of the mine, disruption of the continuity of the backfill array being formed, as well as the formation of channels for water breakthrough [9,18].

Figure. 3 Filled output with an air cavity at the stage of feeding the filling mixture

The bearing capacity and stability of backfill arrays is not sufficiently realized due to the lack of full-fledged contact with the overlying mass over the entire surface of the mine. The incompleteness of the laying of the developed rock volumes is one of the main factors contributing to the displacement of the masses being worked on [10].

The use of measures for the laying of voids formed during the extraction of minerals requires the construction of special filling complexes, which is associated with significant financial investments. In addition, the use of the complex on the laying of workings significantly complicates the technological scheme of the process of mining. The stowage complex is a complex system introduced into the composition of a mining enterprise with its own technological processes, the violation of one of which can lead to a malfunction of the entire enterprise.

The lack of efficiency of the complex on the tab inhibits the growth of enterprise productivity and access to new capacity. For example, at the Komsomolsky mine (Russia, Norilsk), an increase in production volumes led to the need to increase the production of backfill mixes from 0.95 million tons to 1.25 million tons, which naturally required the modernization of the backfill complex.

The construction of a complex for laying a developed space is not able to solve all the problems on its own, as the question arises of transporting the filling mixture to the developed space [11]. The development of mining operations leads to the increase in the distance between the place of laying the mixture and the place of its preparation, which leads to the increase in the length of its transportation. For example, at the Buribayevsky GOK (Russia, Bashkortostan), the development of mining operations led to the removal of the clearing site from the backfill complex to a distance of about 3 km. Such removal made it extremely problematic to work with the use of a system with a bookmark at this mine, and it was ultimately decided to abandon it.

3. Discussion

The transportation of backfill material into the open space is the most important technological link in the chain of processes associated with the extraction of mineral resources. When pumping the filling mixture, a number of difficulties arise. For example, at the Gaysky GOK and the mines of the MMC Norilsk Nickel, transport to deep levels caused damage to the pipelines due to the occurrence of air strikes. In addition, when feeding slurry over large vertical distances, the question arises of sharp pressure fluctuations associated with starting and stopping pulp feed modes.

When moving the filling material to large horizontal distances, it becomes necessary to use various auxiliary equipment, since the kinetic energy of the filling mixture flow is insufficient due to high pressure losses. Moving the filling mixture through pipelines leads to their rapid deterioration due to the high abrasivity of the solid slurry fraction. At the same time, when the backfill material is moved with an insufficient flow rate, its separation and silting of the pipelines can occur, which leads to an uneven feed of the backfill material and, accordingly, deterioration of the quality of the monolithic backfill.

For the transportation of hardening mixtures the pipes made of steel with a diameter of 76-220 mm and a wall thickness of 4-14 mm are usually used. The use of polyplastic pipes, which are characterized by high wear resistance and lower friction coefficient, is promising. At Uralkali, the use of polyplastic pipes made it possible to increase the service life of pipelines and increase the length of the horizontal section of the gravity flow pipeline, which led to a decrease in the cost of filling works [12].

In some designs of backfill complexes, the vibration method is used to prevent the occurrence and elimination of pipeline blockages caused by the separation of the transported material [13]. In an emergency case (with an increase in pressure in the line), the mechanisms that create a longitudinal vibration are activated at a particular section of the pipeline. As a result of vibration exposure, near-wall friction is sharply reduced, which makes it possible to eliminate blockages and siltation [14].

For the timely prevention of the occurrence of air strikes, which arise as a result of the increase in pressure when the backfill material vertically moves under its own weight to great depths, various devices are used. One example is a bleed dumping device used at the Thompson Mine (Canada) mine. Another example of a mechanism to prevent the occurrence of pneumatic shocks when transporting to great depths and sharp pressure fluctuations associated with starting and stopping modes during pulp feed is the use of a device to release excess air. To prevent emissions of filling mixture from the mouth of the pipeline at the mine "Tara Mine" (Ireland), a device is used that allows you to remove air from the pipeline.

It is also not uncommon in the production of a filling mixture of a situation where an excessive amount of water is added to it in order to increase its transportability, this leads to the separation of the mixture during transportation and, consequently, to a decrease in the strength of the filling mass. Thus, at the mines of the company "Outokumpu" (Finland) with the transition of cleaning works to deeper horizons, the side walls of the artificial pillars collapsed. In order to eliminate the negative effects of excess water the consumption of the binder increased in the mixtures, which led to an increase in the cost of production.

To improve the rheological properties, various chemical additives are introduced into the monolithic tab - plasticizers, catalysts (retarders and activators) of hardening. They allow increasing plasticity with less water consumption and a binder component, preventing the mixture from splitting, blocking the transport pipeline, adjusting too fast or slow hardening, helping to increase the mechanical strength of the array [15].

Most of the existing hydrotransport systems operate at low concentrations of solid material in the flow of the filling mixture, which leads to a decrease in the technical and economic efficiency of the hydraulic transport system, as well as to a significant increase in water inflow in the workings. In addition, in hydrotransport technologies, the dosage of material supply to the pipeline is insufficiently used, as well as adjustment and control of parameters, as a result, as a rule, the hydrotransport systems of mining enterprises operate in low-cost and unstable modes.

Efficient transportation of pasty filling mixtures by ground pumps causes difficulties due to the existing non-linear relationship between efficiency and solid phase concentration, as well as the kinematic viscosity coefficient.

Taking into account the peculiarities of the materials used as filling mixtures, an effective transportation process required to obtain a backfilling array of specified characteristics should be carried out under the following conditions:

• providing an intense physical impact on the mixed components to organize their movement along mutually intersecting trajectories;

• compliance with the optimal residence time of components in the transport line, to maintain the homogeneity of the filling mixture.

To implement such tasks, a fundamentally new technology of transportation of hardening filling mixtures with simultaneous impact on the material is necessary.

4. Conclusion

1. The well-known advantages of underground geo-technology of the development of mineral deposits with the laying of the worked-out space in some cases are not realized and require the improvement of the existing technologies used during the laying. Recently, the development of technology and related technologies, whose work is based on the principle of nature-similarity, has been actively developed.
2. The design of pumping equipment, in which the speed of movement of the filling material and the pressure developed by the pump, does not depend on the viscosity of the substance, but is directly dependent on the electrical parameters of the drive, will allow for efficient filling work in remote areas.

5. Acknowledgments
Research is supported by the Russian Science Foundation grant 19-79-10151.

References

[1] Limanskiy A V, Vasilyeva M A 2016 Using of low-grade heat mine water as a renewable source of energy in coal-mining regions. *Ecological Engineering* **91** 41-43
[2] Liss N Y, Kremcheev E A, Afanasiadi K I 2017 Reclamation of salinized land at tailings ponds at ALROSA *Gornyi Zhurnal* **10** 65-69.
[3] Verkholantseva T V, Dyagilev R A 2016 Study of the influence of backfilling on the seismic regime of potash mines *GIAB* **12** 222-224.
[4] Starkov L I, Zemskov A N, Kondrashev P I 2007 *The development of mechanized development of potash ores* (Perm: PSTU publishing house).
[5] Russian Ministry of Industry 2008 Guidelines for the protection of mines from flooding and the protection of underworked facilities in the conditions of the Verkhnekamskoye potassium salt deposit (process regulations) (St Petersburg: VNIIG).
[6] Borzakovskiy B A, Papulov L M 1994 Laying works at Verkhnekamsk potash mines (Moscow: Nedra).
[7] Shulakov D Y, Verkholantseva T V 2013 Relation between microseismic activity and parameters of mining in the Verkhnekamskoye potash deposit In *Proceeding of the 8th International Symposium* pp. 505–510 (Obninsk: Perm).
[8] Yermolovich E A 2012 Dosakladka waste cameras *GIAB* **3** 13-15.
[9] Weisberg L A 1981 General principles of the design of vibration machines. Machines for the vibration separation of bulk mixtures. Vibrating screens. Vibrations in technology: Reference 6 t.t. T.4. Vibration processes and machines (Moscow: Mechanical Engineering)
[10] Blekhman I I, Blekhman L I, Weisberg L A 2015 Vibration overcomes gravity - non-linear effects, simple models In *Proc. of Dynamics of vibro-impact (highly non-linear) systems. XVIII International Symposium dedicated to the 100th anniversary of the birth of Dr. Sc., Prof. A.E. Kobrin.* (pp. 64-72).
[11] Vasilyeva M A 2015 *Key Engineering Materials* **685** 291-294
[12] Vasilyeva M A 2016 *Materials Science Forum* **870** 13-19
[13] Andreev M N, Boguslavsky E I 2010 Development of the composition of the filling material and testing its strength properties *Zapiski Gornogo instituta* **189**

Journal of Physics: Conference Series **1384** (2019) 012007 doi:10.1088/1742-6596/1384/1/012007

Determination of main parameters of clay grinder

N N Dubinin, S A Mikhailichenko, S I Goncharov and L S Uralskaya

Belgorod State Technological University named after V.G. Shukhov, 46, Kostyukova St., Belgorod, 308024, Russia

E-mail: nndubinin@mail.ru, mihaylichenko.sa@bstu.ru

Abstract. The article presents the main technological and power parameters of the clay grinder which depend on a number of geometric parameters, and various rheological characteristics of the processed material made it possible to derive pressure dependencies based on the flow of ductile masses which will allow designing such machines with rational geometric and kinematic parameters.

1. Introduction

The clay grinder is a machine with a complex effect of the working body on the clay mass. Plastic clay in the working channel between the blade and the body of the grinder is mixed and dispersed and simultaneously pressed through holes in the gratings of the body.

The performance, quality of processing and energy consumption are affected by a large number of different parameters, including bowl size, diameter and number of holes in the body, the number of revolutions of the impeller, the number of blades and their geometry, and the minimum gap between the blade and the bowl.

The main parameter on which the performance of the clay grinder depends is the pressure arising in the working channel between the blade and the wall of the body. The pressure should depend not only on the geometry of the working channel, but also to a large extent on the rheological properties of the processed plastic clays.

The processing of clay in the clay grinder is poorly understood and there is no method for calculating this machine.

The objective of this article is to determine the law of pressure change in the “working channel” of clay grinder, average pressure on the blade, friction force on the surface of the blade, torque, velocity gradient and some other parameters.

2. The nature of the movement of the plastic mass in the chamber of variable cross section

Due to the large size of the machine body and the small gap between the blade and the body the task can be simplified by considering the body deployed on the plane relative to which the blade moves (Figure 1).

The boundary conditions of the problem:

1. On the "lower" plane of the body

$$y = 0, u_x = 0, u_y = 0. \quad (1)$$

2. On the surface of the blade

$$y = \mathrm{h}, \mathrm{u_x} = \mathrm{V}, \mathrm{u_y} = 0.$$

3. $$P = p_0 \text{ for } x = a.$$

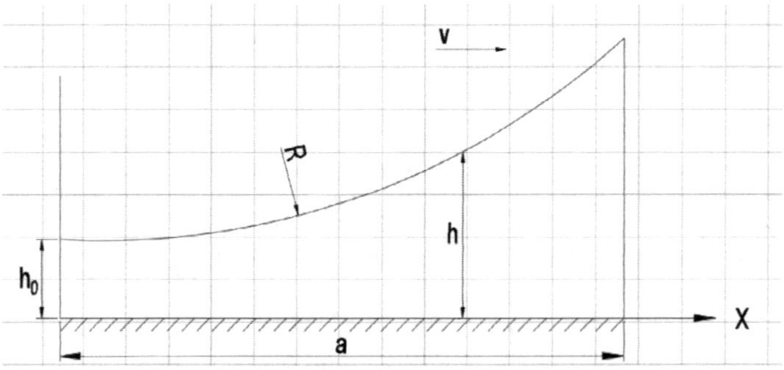

Figure 1. Calculation scheme of clay grinder

The movement of the clay mass in the working channel formed by the wiper blade and the perforated clay grinder body occurs in the direction indicated in Figure 2. As a consequence, in the initial equations of motion of an incompressible viscous fluid [1–4] it is possible to discard terms containing u_z and derivatives with respect to z. Neglecting the action of mass forces with an insignificant height of the channel for the deformation of the clay mass (F = 0) and also considering the process of movement steady, i.e.

$$\frac{\partial u_x}{\partial t} = 0$$

we obtain the Navier Stokes system of equations describing the plane-parallel motion of plastic material

$$\frac{\partial^2 u_x}{\partial x^2} = \frac{1}{\mu} \cdot \frac{dP}{dy}$$

$$\frac{\partial u_x}{\partial x} + \frac{\partial u_{xy}}{\partial y} = 0 \tag{2}$$

Integrating equation (2) twice, we get

$$u_x = \frac{1}{2\mu} \cdot \frac{dP}{dx} y^2 + Ay + B \tag{3}$$

$$V_x = \frac{1}{2\mu} \cdot \frac{dP}{dx} yh^2 + Ah$$

From where

$$A = -\frac{1}{2\mu} \cdot \frac{dP}{dx} h + \frac{V_x}{h} \tag{4}$$

Then

$$u_x = \frac{1}{2\mu} \cdot \frac{dP}{dx} y^2 + \left(\frac{V_x}{h} - \frac{1}{2\mu} \cdot \frac{dP}{dx} h\right) y \tag{5}$$

and finally

$$u_x = \frac{1}{2\mu} \cdot \left(\frac{dP}{dx}\right) \left(y^2 - h_y\right) + \frac{V_{xy}}{h} \tag{6}$$

$\frac{dP}{dx}$ – in equation (6) it can be found using the fact that the "flow" of material Q through any cross-section of the working channel is the same:

$$Q = \int_0^h u_x dy = -\frac{1}{2\mu} \cdot \frac{dP}{dx} \frac{h^3}{6} + \frac{V_x h}{2}$$

and then

$$\frac{dP}{dx} = -\frac{12}{h^3}\mu\left(Q - \frac{V_x h}{2}\right) = \frac{6\mu V_x}{h^3}\left(h - \frac{2Q}{V_x}\right) \tag{7}$$

We introduce:

$$h_1 = \frac{2Q}{V_x} \tag{8}$$

then the final equation for the pressure gradient takes the form

$$\frac{dP}{dx} = \frac{6\mu V_x}{h^3}(h - h_1) \tag{9}$$

Substituting equation (9) into equation (6), we obtain

$$u_x = \frac{3V_x}{h^3}(h - h_1)\left(y^2 - h_y\right) + \frac{V_{xy}}{h} \tag{10}$$

$$\frac{dP}{dx} = \frac{dP}{dh}\cdot\frac{dh}{dx} \tag{11}$$

Because $h = h_0 + \frac{x^2}{2R}$ (Figure 2).

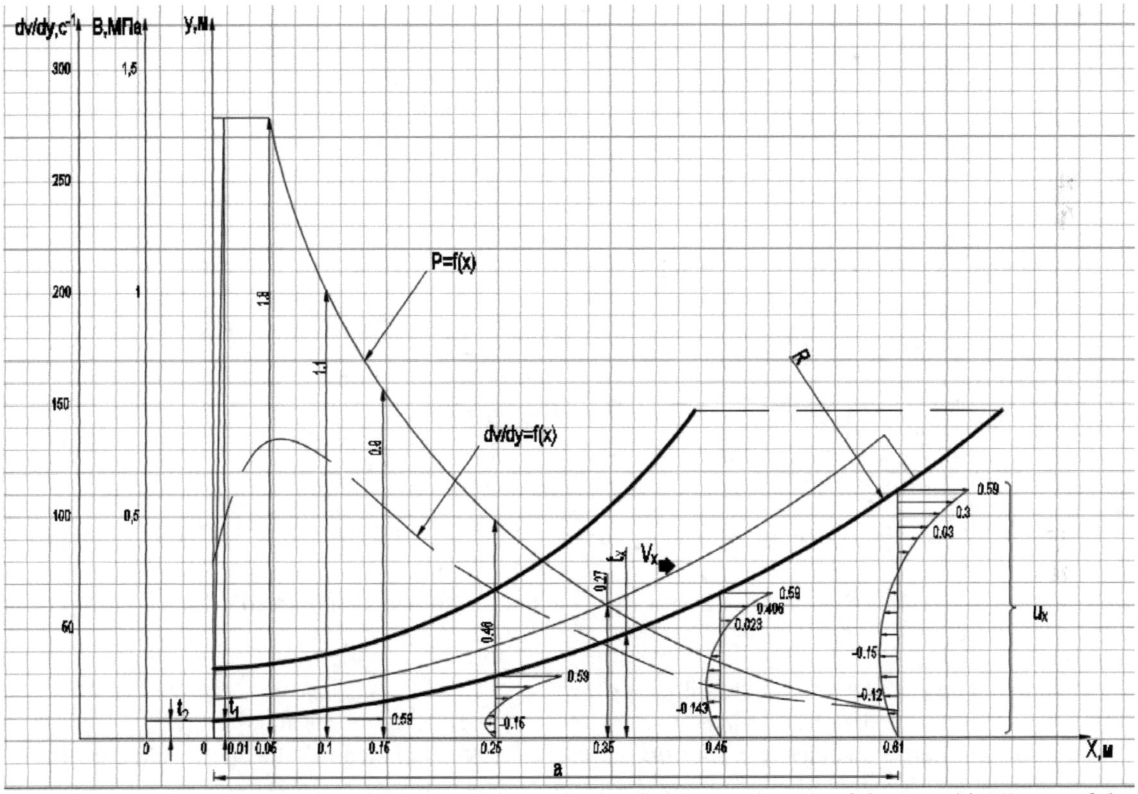

Figure 2. Graphs of pressure, velocity gradient, velocity field ersus length of the “working” part of the blade for V=0.59 m/s; p=0.006 m

From here, after differentiation, we get

$$\frac{dh}{dx} = \frac{x}{2R} \tag{12}$$

and therefore

$$\frac{dP}{dx} = \frac{dP}{dh}\cdot\frac{x}{R} \tag{13}$$

and

$$\frac{dP}{dx} = \frac{6\mu V_x}{h^3}(h - h_1)$$

Then

$$\frac{dP}{dx} = \frac{6\mu V_x}{h^3}(h - h_1)\cdot\frac{R}{x} \tag{14}$$

Using the expression x from (11) you obtain

$$\frac{dP}{dh} = \frac{6\mu V_x(h-h_1)}{h^3\sqrt{2R}\sqrt{h-h_0}} \tag{15}$$

Integrating equation (15):

$$\int\frac{dP}{dh} = \frac{6\mu V_x R}{\sqrt{2R}}\left(\int\frac{dh}{h^3\sqrt{h-h_0}} - \int\frac{h_1 dh}{h^3\sqrt{h-h_0}}\right)$$

As a result of integration, we obtain

$$P = \frac{6\mu VR}{\sqrt{2R}}\left\{\left[\frac{\sqrt{h-h_0}}{h_0 h} + \frac{1}{2h_0}\left(\frac{2}{\sqrt{h_0}}arctd\sqrt{\frac{h-h_0}{h_0}}\right)\right]\times\left(1-\frac{3h_1}{4h_0}\right) - \frac{h_1\sqrt{h-h_0}}{2h_0h^2}\right\} + b \tag{16}$$

In order to find the integration constants b and h_1, we use the boundary conditions (1):

- for $X = 0\ P = P_0, h = h_0$ (17)
- for $X = a\ P = P_0, h = h_0 + \frac{a^2}{2R}$

We find that

$$b = P_0 \tag{18}$$

Taking into account (18) from equation (16), we obtain after transformations

$$h_1 = \frac{\dfrac{a}{\sqrt{2R\left(h_0+\dfrac{a^2}{2R}\right)}} + \dfrac{1}{h_0}arctg\dfrac{a}{\sqrt{2Rh_0}}}{\dfrac{a}{2\sqrt{2R\left(h_0+\dfrac{a^2}{2R}\right)^2}} + \dfrac{3}{4}\left[\dfrac{a}{\sqrt{2R\cdot h_0\left(h_0+\dfrac{a^2}{2R}\right)}} + \dfrac{1}{h_0\sqrt{h_0}}arctg\dfrac{a}{\sqrt{2Rh_0}}\right]}$$

Given the constants, the integration of equation (16) takes the form

$$\mathrm{P} = \frac{6\mu V_x R}{\sqrt{2R}}\left\{\left[\frac{\sqrt{h-h_0}}{h_0 h} + \frac{1}{h_0\sqrt{h_0}}arctg\sqrt{\frac{h-h_0}{h_0}}\right]\left(1-\frac{3h_1}{4h_0}\right) - \frac{h_1\sqrt{h-h_0}}{2h_0h^2}\right\} + P_0 \tag{20}$$

Operation of clay grinders with clearances established by manufacturers is recommended within

$$h_1 \approx 1.33h_0 \tag{21}$$

which allows for engineering calculations to write equation (20) in a simple form:

$$P = \frac{2\mu V_x\sqrt{2R(h-h_0)}}{h^2} \tag{22}$$

where h_1 for exact calculations is determined by the equation (19), and h by the formula (11). The pressure on the surface of the blade in any section allows calculating the formula (22).

To determine the average pressure on the curved surface of the working part of the blade, we integrate equation (22) along the length of the working blade in the range from 0 to a.

$$\int_0^{x=a} p\cdot dx = \int_0^a \frac{3\mu V_x h_1 b\cdot xdx}{2h_0\left(h_0+\frac{x^2}{2R}\right)^2} = \frac{3\mu V_x h_1 b\cdot xdx}{2h_0}\times\int_0^a\frac{4R^2\cdot xdx}{\left[\left(\sqrt{2Rh_0}\right)^2+x^2\right]^2}$$

after transformations we get

$$P_{\text{cp}} = \frac{3\mu V_x h_1 R a^2 b}{2h_0^2(2Rh_0+a^2)} \tag{23}$$

To determine the velocity gradient on the surface of the blade we differentiate by the height of the channel the equation (16)

$$\left(\frac{dU}{dy}\right)_{y=h} = \frac{h}{2\mu}\frac{dP}{dx} + \frac{U}{h} \tag{24}$$

Substituting the value of dP/dx from equation (19) into equation (25) and performing the transformations, we obtain

$$\left(\frac{dU}{dy}\right)_{y=h} = \frac{h}{2\mu}\frac{6\mu V_x}{h^3}(h-h_1) + \frac{V_x}{h} = \frac{4V_x}{h} - \frac{3V_x h_1}{h^2} \quad (25)$$

Substituting the values of h into equation (25), we obtain the equation for determining the velocity gradient at any point on the surface of the blade:

$$\left(\frac{dU}{dy}\right)_{y=h} = \frac{4V_x}{h_0+\frac{x^2}{2R}} - \frac{3V_x h_1}{\left(h_0+\frac{x^2}{2R}\right)^2} \quad (26)$$

Determining the average value of the velocity gradient on the surface of the blade allows you to perform the optimal geometric shape of the blade associated with the design of clay grinder body. To determine the average velocity gradient, we integrate equation (26) along the length of the working zone of the blade [5–7]:

$$\langle \left.\frac{dU}{dy}\right|_{y=h} \rangle_{[0,a]} = \frac{4V}{a}\int_0^a \frac{dx}{h_0+\frac{x^2}{2R}} - \frac{3Vh_1}{a}\int_0^a \frac{dx}{\left(h_0+\frac{x^2}{2R}\right)^2} = \frac{VR}{2h_0 a}\left(\frac{16h_0-6h_1}{\sqrt{2h_0R}}\right) arctg\frac{a}{\sqrt{2h_0R}} - \frac{6h_1 a}{2h_0R+a^2} \quad (27)$$

The design of the drive of the clay grinder and the strength calculations of the units and parts of the machine require the determination of the friction force during the processing of the material. The shear stress acting on the blade in the zone of material capture is determined by the dependence

$$\tau = \mu\left(\frac{dU_x}{dy}\right)_{y=h} \cdot \cos a \quad (28)$$

where $\left(\frac{dU_x}{dy}\right)_{y=h} \cdot \cos a$ – projection of the velocity gradient on the normal of the blade surface.

The total value of the friction force acting on the unit width of the blade is equal to

$$F_{\text{тр}} = \int_0^l \mu\left(\frac{dU}{dy}\right)_{y=h} \cdot \cos a \cdot dl \quad (29)$$

where l – blade length in the area of material capture.

But $l \approx a$ and $dl \cdot \cos a = dx$ in range $0 \le x \le a$; therefore,

$$F_{\text{тр}} = \mu\int_0^a \left(\frac{dU_x}{dy}\right)_{y=h} dx \quad (30)$$

Integration of the expression $\left(\frac{dU}{dy}\right)_{y=h}$ was made earlier in (27), finally the formula for calculating the friction force of the material on the blade can be written as

$$F_{\text{тр}} = \frac{\mu V_x R\cdot b}{2h_0}\left(\frac{16h_0-6h_1}{\sqrt{2h_0R}} arctg\frac{a}{\sqrt{2h_0R}} - \frac{6h_1 a}{2h_0R+a^2}\right) \quad (31)$$

where b – blade width.

3. Conclusion

The calculated dependencies given in the article can be used in the design of clay processing machines such as clay grinders. In addition to finding the strength characteristics of the working bodies of the machines the dependences take into account the rheological properties of plastic clays which sometimes differ by an order of magnitude and allow designung machines with optimal parameters.

4. Acknowledgments

This research is conducted as a part of the Program of flagship university development on the base of the Belgorod State Technological University named after V G Shukhov, using equipment of high Technology Center at BSTU named after V. G. Shoukhov.

References

[1] Rebinder P A 1966 Physico-chemical mechanics of dispersed structures Physico-chemical mechanics of dispersed structures (Moscow: Science) p 16

[2] Dubinin N N 2013 Brushless machines for the formation of plastic masses (Publishing House of BSTU V G Shukhov) 113 p

[3] Vyalov S S 1978 Rheological foundations of soil mechanics (Moscow: High School) 447 p
[4] Ilina T N, Sevostyanov V S, Shkapetkin E A and Sevostyanov M V 2011 Study of the conditions of the process of microgranulation in disperse systems *Bull. of BSTU named after V G Shukhov* **1** 81–6
[5] Dubinin N N, Evtuschenko E I, Nemets I I, Nosov O A and Osokin A V 2014 Rotary Machines for Production of Ceramic Wall Materials *Res. J. of Pharmaceutical, Biological and Chemical Sci.* **5(5)** 1710–8
[6] Dubinin N N and Uralskaya L S 2015 Analytical methods for studying the movement of clay masses in working bodies *Bull. of BSTU named after V G Shukhov* **4** 75–9
[7] Dubinin N N, Mikhailichenko S A and Uralskaya L S 2016 Productivity of rotary machines with a camera of variable section *Bull. of BSTU named after V G Shukhov* **4** 102–5

Analysis of the mixture of rare earth elements by atomic spectroscopy

N V Dzhevaga, O L Lobacheva

Saint-Petersburg Mining University, 2, 21st line, Vasilyevsky Island, Saint-Petersburg, 199106, Russia

E-mail: dzhevaga331@mail.ru

Abstract. The authors of this paper give the characteristic and the principle of operation of the ICPE-9000 spectrometer, proposed for the chamber residue analysis that is one of the products of ion flotation, for individual rare earth elements and their mixture content by atomic spectroscopy. The chamber residue was obtained by conducting the ion flotation from nitrate solutions, containing individual rare earth elements of the cerium group and their mixture. The results of the analysis of the solutions by optical emission spectroscopy are shown in the form of molar concentration. According to the obtained experimental data, a tendency for the change of distribution coefficients, separation factors and pH values of maximum recovery at the different values of equilibrium aqueous phase pH value was identified. The possibility of the most effective separation of rare earth elements of the cerium group at different concentrations of indifferent ions was revealed, as well as the forms of recovered compounds were determined.

1. Introduction

In the research of a more effective removal of various components scientists study rare earth elements (REE), ferrous, non-ferrous, and amphoteric metals; use different models (GANN, MLR, CCD and others); analyze composition and structure of extracted substances.

The articles of Jensen, Bond, Chiarizia described removal of different lanthanides and actinides. In the extraction of trivalent lanthanide (Sm^{3+}, Eu^{3+}) and actinide (Am^{3+}, Cm^{3+}) cations from nitrate systems into n-dodecane solutions of phosphinic acids (HC272, HC302, HC301) it is stated that in all cases the dependence of the distribution ratios on the total extractant concentration is approximately second power. The effect of aggregation is considered for every extractants, and the stoichiometries of the extracted complexes are $M(H(C272)_2)_3$, $M(C302)_3$, and $M(C301)_3$ [1].

The paper of Jensen et al. [2] is related to separating americium from curium and all the lanthanides by solvent extraction. The sterically constrained, macrocyclic, aqueous soluble ligand N,N′-bis (6-carboxy-2-pyridyl)methyl-1,10-diaza-18-crown-6 (H2BP18C6) was studied. Pairing H2BP18C6, which favors complexation of larger f-element cations, with acidic organophosphorus extractants that favor extraction of smaller f-element cations, solvent extraction systems with good Cm/Am selectivity, excellent trans-lanthanide selectivity (K_{ex}, Lu/K_{ex}, La = 10^8), but poor selectivity for Am against the lightest lanthanides was created.

The article of Lutskiy et al. analyzes the information on modern developments in the utilization of phosphogypsum demonstrating the possibility of using phosphogypsum as a raw material for obtaining rare-earth elements [3].

The papers of Götze J, Arsentyev V A [4, 5] describe using of the analytical combination of cathodoluminescence microscopy and spectroscopy, electron paramagnetic resonance spectroscopy for characterizing trace element compositions and defect structures in quartz samples.

The authors of the article [6] were determined the difference between zircons of the 1st, 2nd and 3rd group by contents of trace elements by electron microscopy, mass-spectrometry of secondary ions, Raman spectroscopy.

For ion flotation and extraction Hosenian and others have used the algorithms called GANN and MLR to remove Zn as well as Ni. Research related to REE encompasses various directions: ion flotation, liquid extraction, filtration and others. In their research modelling process of ion flotation of nickel is carried out [7]. The purpose of this study was to model the Zn(II) flotation using a hybrid neural network/genetic algorithm (GANN) and multivariate linear regression (MLR). The results showed that this model could be useful in predicting the Zn(II) removal in response to changes in the input parameters.

In another research [8] a new predictive model based on the hybrid neural genetic algorithm (GANN) was developed to predict the Ni(II) ion removal and water removal during the process from aqueous solutions using ion flotation. The input variables of the model were pH, collector concentration, frother concentration, impeller speed and flotation time. The outcomes revealed that the proposed GANN models could be used to predict the Ni(II) removal and water removal during ion flotation.

According to the articles [9, 10] ion flotation is a green and cost-effective technology where separation of molecules exploits the difference in surface affinity. In the work of Ghosh and others [11] a batch ion foam fractionation system was designed and optimized for the separation of trace Cr(VI) from aqueous solutions. The effect of surfactant head groups (collectors) on the adsorption dynamics was analyzed. Cetyl trimethyl ammonium bromide (CTAB), a cationic surfactant demonstrated high efficiency for the removal of Cr(VI) from aqueous solutions. The experimental investigation of the effect of different operational parameters on the separation characteristics has been presented.

The effects of chemical conditions on the recovery of Zn(II) and water during the ion flotation process were assessed utilizing a central composite design (CCD) of response surface methodology [12]. The optimum effective parameters including pH, collector and frother concentration were determined. The results demonstrated that the pH and collector concentration were effective factors for the efficiency of Zn(II) flotation. The effects of collector and frother concentration on the characterization of sublate and the complexation of sodium dodecyl sulphate with Zn(II) were studied using scanning electron microscopy coupled with energy dispersive X-ray (SEM/EDX) spectroscopy and Fourier transform infrared (FTIR) spectroscopy.

Apart from researching ways of increasing effectiveness of chemical processes, the attention of researchers is focused on developing environmental accountability in different spheres of companies' work, especially in the sphere of risk management [13, 14]. The article [15] demonstrates the effectiveness of employing case study analysis in English in order to teach would-be oil and gas engineers on environmental accountability.

Risks can be prevented by means of environmental integrated monitoring system in the stage of liquidation, reclamation of mining operations [16], the wastewater disposal system modernization by creating a recycled water supply [17]. Another way of monitoring system can be the one based on Unmanned Aerial Vehicles and utilized in air and water pollution [18], also the research of evaluation of the alluvial anthropogenic array - tailings storage of apatite-nepheline ores [19].

2. Materials and methods

All chemical reagents used in experimental studies of this paper are chemically pure and have quality certificates. The nitrate solutions of rare earth elements with concentration 0.01 M were prepared by the dissolution of rare earth elements with 99.9% purity in distilled water. All process solutions were prepared from this solution by dilution to 0.001 M concentration. The initial solution concentration was checked by sample trilonometry titration by EDTA solution with Xylenol orange indicator. Sodium dodecyl sulphate salt for biochemistry (NaDS, $C_{12}H_{25}SO_4Na$) manufactured by Acros organics (USA) with 99% purity was used as surfactant. During the test, a sample of NaDS was added to the working solution. Only fresh nitric acid and sodium hydroxide for adjusting the required pH value to the required value was used.

The recovery of rare earth elements from aqueous nitrate solution to froth pulp was carried out by using the ion flotation technique [20-23].

The diluted solution of individual rare earth elements with concentration of 0.001 M was used as an initial solution; its volume was set to 200 ml. Sodium dodecyl sulphate was added to the solution in accordance with a stoichiometry of the following reaction:

$$Ln^{3+} + 3C_{12}H_{25}SO_4^- = Ln(C_{12}H_{25}SO_4)_3$$

and its concentration was 0.003 M. PH value was adjusted to the required value by a batchwise addition of NaOH 1 N alkaline solution or diluted nitric acid $HNO_{3diluted}$. The aqueous solutions pH value was determined to an accuracy of 0.01 by using the "pH-150 MA" pH-meter. The flotation process was conducted on the mark 137 V-FL laboratory flotation machine with the cell volume of 1 dm^3 and the agitator rotational speed of 1,000 RPM for 5 minutes. The selection of duration and rotational speed was conducted experimentally and determined by obtaining the highest process efficiency indices. The froth pulp was destroyed by 2 N solution of H_2SO_4.

3. The analysis of ion flotation products by AES

In the studies of Pusz, Picayo, Galico, Zapala and others XRD technique, simultaneous thermogravimetry and differential scanning calorimetry (TG–DSC) were utilized for characterization of complexes. New solid lanthanide(III) (Ln) compounds of Tb(III), Ho(III), Er(III) and Yb(III) with sodium salt chrysin-4í-sulfonate (Nachsa-4í) were obtained [24]. Solid products showed $Ln(OH)(Nachsa\text{-}4í)_2\sum nH_2O$ stoichiometry based on elemental analysis and thermogravimetric studies. The complexes were characterized in solution by UV/Vis and infrared as well as by the ^1H NMR spectroscopies.

Understanding and optimizing the kinetics of rare earth separations represent a frequently overlooked approach to increasing the throughput of rare earth separations [25]. The kinetics of solvent extraction can be incredibly complex. In order to stimulate kinetic studies of rare earth separations, we provide an overview of liquid extraction chemistry relevant to rare earth separations, the equipment and methods for collecting kinetics data, and fundamental kinetic models for dealing with extractions controlled by diffusion or slow chemical reactions. Published kinetic studies available for rare earth extractions using the most viable and industrially applied systems are analyzed.

Thermogravimetry (TG), differential thermal analysis (DTA), differential scanning calorimetry (DSC), powder X-ray diffractometry (XRD), elemental analysis, complexometric titration, middle (MIR) and near (NIR) infrared region spectroscopy were used to characterize solid-state [La(Nap)3(H2O)], in which Nap is naproxen; the proposed stoichiometry is confirmed [26]. The experimental results provided information concerning the chemical composition, dehydration, coordination modes of the ligand, crystallinity of the sample, thermal behavior, thermal decomposition and polymorphism. Cyclic DSC and XRD in different temperatures showed that the complex presents five temperature dependent polymorphic forms. MIR spectroscopy suggest that the naproxen coordinates through the carboxylate group as a chelating ligand.

New binuclear complexes of La(III), Ce(III), Pr(III) and Nd(III) ions with N- phenylanthranilic acid were synthesized and characterized by elemental analysis, infrared spectroscopy, XRD technique and thermogravimetric analysis [27, 28]. The obtained results show that in complexes only the carboxyl group is involved in the coordination of the rare earth ions in the bidentate mode. The calculated dehydration enthalpies are low for all studied complexes what indicates that it is crystallization water. Moreover, the thermal stability and the decomposition path of these compounds depend on the metal ion and the used atmosphere. The EGA data allowed the identification and comparison of gaseous products evolved during pyrolysis and oxidative decomposition of the compounds which were considered.

The new ternary complexes of the formula $Ln_2(nfa)_5(phen)(OH)\cdot nH_2O$ (Ln = La(III), Ce(III), Pr(III) and Nd(III)) were synthesized and characterized by elemental analysis, complexometry, infrared spectroscopy XRD technique, simultaneous thermogravimetry and differential scanning calorimetry (TG–DSC). Low values of the dehydration enthalpies indicate that dehydration is connected with the release of crystallization water [29]. Using the coupled TG–FTIR–MS techniques the decomposition pathways of the synthesized compounds were analyzed and the gaseous products released during pyrolysis and thermal oxidation of the complexes were identified.

New binuclear complexes of La(III), Nd(III), Sm(III), Eu(III), Gd(III) and Tb(III) ions with mefenamic acid were obtained in solid phase [30]. For all compounds considered, complexation occurs at pH 6-6.5. The analyzed complexes were characterized on the basis of chemical and elemental analyses, molar conductance, spectroscopies as well as thermoanalytical techniques. According to the research results, the molecular formula of the complexes is $Ln_2(C_{15}H_{14}NO_2)_3(OH)_3\times nH_2O$, with the exception of the La(III) complex which has the formula $La_2(C_{15}H_{14}NO_2)_5(OH)$. The received results give information on the composition, thermal stability, thermal decomposition, as well as the gaseous products that evolved during the thermal decomposition of the considered compounds.

The ion flotation products (chamber residue) was analysed for rare earth elements content by optical emission spectroscopy on ICPE-9000 spectrometer at the analytical research department of Saint Petersburg Mining University (SPMU).

The analysis of atomic emission spectral (AES) is based on the thermal excitation of atoms or ions that are in the vaporous or gaseous state, and the registration of optical spectres (qualitative analysis) or the measurement of the intensity of separate spectral lines of the determined elements (quantitative analysis).

The atomic spectres appear during the emission or absorption of electromagnetic radiation by free or weakly bound atoms (in gases and vapours) and consist of separate spectral lines.

Emission spectres are obtained when the atoms get excited. A lifetime of the excited state is short and is 10^{-7}-10^{-8} seconds. During this time frame the atom losses the excess energy by emitting the electromagnetic radiation quantum.

The measured values of emission intensity are then compared with the values of the intensity of standards with known concentration to obtain the element concentration value in the unknown sample.

4. ICP burner and high-frequency generator

ICP is a plasma discharge that is excited in the argon flow and sustained by impact of high-frequency electromagnetic field on ionized argon.

The plasma temperature may reach 10,000 K. which ensures complete atomization of sample elements and minimizes chemical effects of interference.

The ICP burner consists of well-centred quartz tubes. The induction coil is connected to a high-frequency generator with a frequency of 27 - 56 MHz and an input power of several kW. The plasma-supporting argon gas is fed to the tubes with different velocity: the inner analyte aerosol carrying flow has a velocity of 1 l/min, the intermediate axial flow has a velocity of approx. 1 l/min. and the outer flow – 10 - 20 l/min. The outer flow is fed to the tube tangentially and forms vortexes when moving up. Originally under the impact of high-frequency spark the argon ionization occurs, after which the high-frequency generator is activated automatically and discharge is created at the expense of the electromagnetic field interaction with flowing gas, the electrical conduction of which is provided at the expense of a high degree of ionization. A plasma jet is blown outside. The burning of plasma is sustained at the expense of induction heating of the gas. The aerosol carrying gas flow is supplied to the plasma base and passes through the toroidal high-temperature zone, forming a cooler jet over the bright plasma. For analytical purposes, a jet that is sustained at the specified height over the burner by an intermediate axial flow is used.

5. Optical system and spectrometer

Radiation emitted by the ICP carries the information about the elemental composition of the analysed sample. However, to retrieve the obtained information is possible only after the dispersion of spectrum or the separation of a narrow spectral range that corresponds with the analytical line.

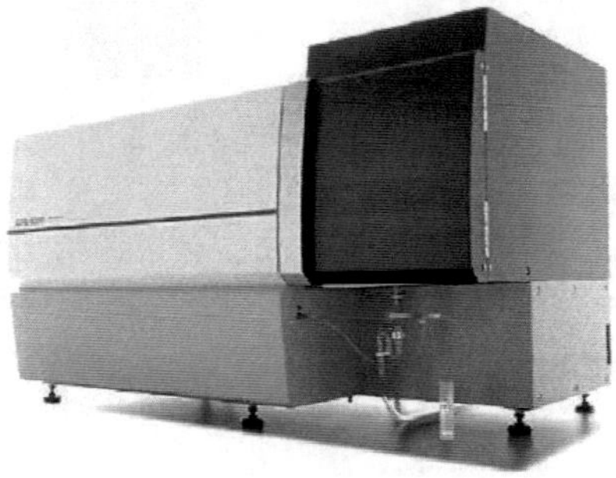

Figure 1. ICPE-9000 spectrometer

The ICPE-9000 uses an optical echelle scheme and a CCD detector on solid-state semiconductor arrays (CCD – charge coupled device). It represents a two-dimensional array, consisting of photosensitive semiconductor p-n junctions (pixels) ≤ 10 μm in sizes that are isolated from each other. Under the impact of electromagnetic radiation the alternation of electrical charge at the p-n junction occurs, that is proportional to luminous flux intensity and exposure time. Information reading from the array is performed by using a computer.

6. Results and discussion. Ion flotation in the joint presence of samarium (III) and europium (III)

The results of the conducted studies of ion flotation in the joint presence of cerium group elements (samarium (III) and europium (III)) are specified in tables 1 and 2.

Table 1. Dependence of the concentration of rare earth elements in chamber residue and froth pulp on pH at their flotation in the joint presence

	$[Ln^{3+}]$ chamber residue, mmol/l		$[Ln^{3+}]$ froth pulp, mmol/l	
pH	Sm	Eu	Sm	Eu
4.0	0.054	0.066	0.851	0.841
5.0	0.053	0.070	0.758	0.744
6.0	0.041	0.054	0.852	0.841
6.7	0.006	0.008	0.795	0.794
7.1	0.006	0.008	0.895	0.893
7.5	0.005	0.007	1.106	1.103
7.8	0.005	0.006	1.106	1.104
8.5	0.022	0.024	1.087	1.084

After analysing the experimental data in table 2 we came to the following conclusions. When floating the mixture of rare earth elements the samarium cations have the maximum distribution coefficient of 221.1 at pH 7.5-7.8 that is significantly higher than in the case of individual flotation recovery, at which the maximum K_{distr} is 81.9 at pH 6.7. In the case of europium, the maximum recovery is observed during the individual flotation recovery and the pH of maximum recovery is shifted towards the smaller values range compared to the flotation of the mixture of rare earth elements. Thus during the flotation of the mixture, the maximum distribution coefficient for europium is 184.1 at pH 7.8. In the case of individual flotation the maximum K_{distr} of europium is 208.9 at pH 7.1.

Table 2. Distribution coefficients of rare earth elements at their flotation in the joint presence

pH	Sm		Eu	
	$K_{distr\Sigma}$	K_{distr}	$K_{distr\Sigma}$	K_{distr}
4.0	15.8	14.1	12.7	5.0
5.0	14.3	11.5	10.6	10.2
6.0	20.8	15.1	15.6	16.6
6.7	132.5	81.9	99.2	204.1
7.1	149.1	81.5	111.6	208.9
7.5	221.1	81.5	157.6	194.0
7.8	221.1	81.6	184.1	182.6
8.5	49.4	92.0	45.2	176.8

The experimental data are shown in fig. 2 - 3.

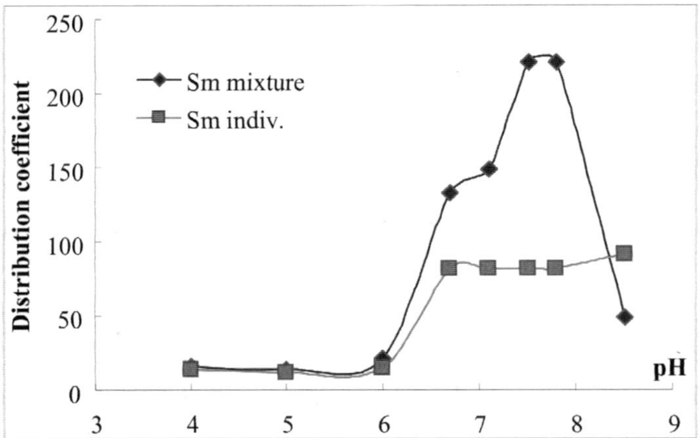

Figure 2. Samarium (III) cations distribution coefficients as a function of pH of the solution for the cases of their individual flotation and in the joint presence

Thereby in the case of europium, the K_{distr} and separation are decreased in the joint presence of cations, and in the case of samarium, the K_{distr} and separation are on the contrary increased.

7. Conclusions

In the process of ion flotation of rare earth elements in the presence of indifferent ions, different trends of distribution coefficient changes are observed, and accordingly the recovery ratio and pH of maximum recovery as a function of the composition of the floating system.

When floating the mixture of rare earth elements by the example of cerium group elements (samarium (III) and europium (III)) it was determined that the distribution coefficient of samarium cations increased significantly in the case of rare earth elements flotation in the joint presence, and the pH of maximum recovery shifted towards the higher values range. In the case of europium cations, the relation is inverted. The maximum distribution coefficients are observed in the case of flotation of individual elements, and the pH of maximum recovery during the flotation of the mixture of rare earth elements also shifts towards the higher values range.

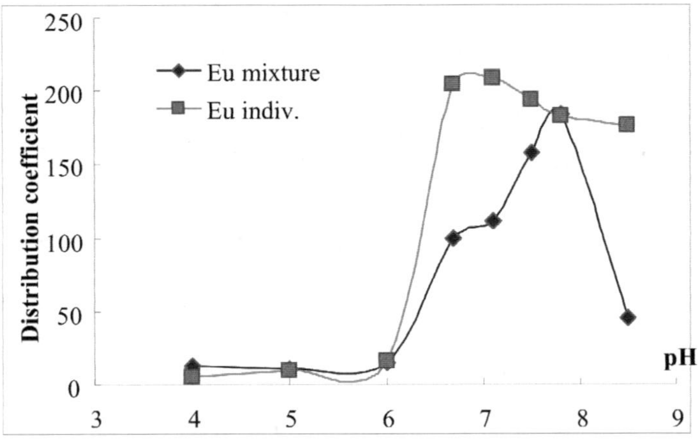

Figure 3. Europium (III) cations distribution coefficients as a function of pH of the solution for the cases of their individual flotation and in the joint presence

The grant of the President of the Russian Federation to young scientists and post-graduate students, project SP-347.2019.1 "An innovative approach to the efficient removal of rare earth elements by ion flotation with the possibility of their individualization in the processes of concentration of diluted solutions of technogenic genesis".

References

[1] Jensen M P, Bond A H 2002 *Radiochim. Acta* Influence of aggregation on the extraction of trivalent lanthanide and actinide cations by purified Cyanex 272, Cyanex 301 and Cyanex 302 **90** 205-209.

[2] Jensen M P, Chiarizia R, Ulicki J S, Spindler B D, Murphy D J, Hossain M M, Roca-Sabio A, Rodríguez-Blas T 2015 *Solvent Extr. Ion Exch.* Solvent extraction separation of trivalent americium from curium and the lanthanides **33** 329-345.

[3] Lutskiy D, Litvinova T, Ignatovich A, Fialkovskiy I 2018 *Journal of Ecological Engineering* Complex processing of phosphogypsum - A way of recycling dumps with reception of commodity production of wide application **19 (2)** 221-225.

[4] Götze J, Pan Y, Müller A, Kotova E L, Cerin D 2017 *Minerals* Trace element compositions and defect structures of high-purity quartz from the southern ural region, Russia **7 (10)** 189.

[5] Arsentyev V A, Gerasimov A M, Kotova E L 2017 *Obogashchenie Rud* Sylvinite ore thermochemical modification by means of super high frequency thermal treatment **6** 3-8.

[6] Lobach-Zhuchenko S B, Skublov S G, Egorova Y S, Prishchepenko D V, Galankina O L 2018 Zapiski Rossiiskogo Mineralogicheskogo Obshchestva The composition and structure of

zircon from the harzburgite inclusion in granulites of the Pobuzhsky complex, Ukrainian shield **147 (6)** 22-40.

[7] Hoseinian F S, Rezai B, Kowsari E, Safari M 2018 *Minerals Engineering* Kinetic study of Ni(II) removal using ion flotation: Effect of chemical interactions **119** 212-221.

[8] Hoseinian F S, Rezai B, Kowsari E 2017 *Journal of Environmental Management* The nickel ion removal prediction model from aqueous solutions using a hybrid neural genetic algorithm **204** 311-317.

[9] Nikolaeva N, Aleksandrova T, Romashev A 2018 *Mineral Processing and Extractive Metallurgy Review* Effect of grinding on the fractional composition of polymineral laminated bituminous shales **39 (4)** 231-234.

[10] Kulikov B P, Baranov V N, Bezrukikh A I, Deev V B, Motkov M M 2018 *Metallurgist* Preparation of Aluminum-Scandium Master Alloys by Aluminothermal Reduction of Scandium Fluoride Extracted from Sc_2O_3 **61 (11-12)** 1115-1121.

[11] Ghosh R, Sahu A, Pushpavanam S 2019 *Journal of Hazardous Materials* Removal of trace hexavalent chromium from aqueous solutions by ion foam fractionation 589-598.

[12] Hoseinian F S, Rezai B, Kowsari E 2018 *J. Environ. Manage.* The main factors effecting the efficiency of Zn (II) flotation: optimum conditions and separation mechanism **207** 169-179.

[13] Goman I V 2017 *17th International Multidisciplinary Scientific Geoconference //Albena: SGEM World Science* Teaching writing skills in the foreign language to future petroleum engineers specialising in oil and gas development and operation **17** 195 – 203.

[14] Goman I V, Oblova I S 2018 *18th International Multidisciplinary Scientific Geoconference //Albena: SGEM World Science* Analysis of company's corporate social responsibility performance as a way to develop environmental ethics for students specializing in oil and gas activity **18** 11 - 19.

[15] Goman I V 2017 *17th International Multidisciplinary Scientific Geoconference //Albena: SGEM World Science* Case study analysis as a way of developing the environmental accountability of future oil and gas engineers **17** 17 – 27.

[16] Danilov A S, Pashkevich M A, Petrova T A 2019 *Innovation-Based Development of the Mineral Resources Sector: Challenges and Prospects - 11th conference of the Russian-German Raw Materials* Environmental integrated monitoring system at reclamation of large open-cast coal mine **2018** 189-194.

[17] Strizhenok A V, Korelskiy D S, Kuznetsov V S 2019 *Journal of Ecological Engineering* The wastewater disposal system modernization during processing of amber deposit as a way to reduce the anthropogenic load on the baltic sea ecosystem **20 (3)** 30-35.

[18] Kremcheev E A, Danilov A S, Smirnov Yu D 2019 *Journal of Mining Institute* Metrological Support of Monitoring Systems Based on Unmanned Aerial Vehicles **235** 96-105.

[19] Strizhenok A V, Korelskiy D S 2015 *Pollution Research* Assessment of the anthropogenic impact in the area of tailings storage of the apatite-nepheline ores **34 (4)** 809-811.

[20] Dzhevaga N V, Lobacheva O L 2018 *International Multidisciplinary Scientific GeoConference Surveying Geology and Mining Ecology Management, SGEM* Separation of the rare-earth elements (REE) by flotation approaches **18 (5.1)** 801-808.

[21] Lobacheva O, Dzhevaga N 2018 *International Multidisciplinary Scientific GeoConference Surveying Geology and Mining Ecology Management, SGEM* Flotation technologies in the removal of the Pr(III) salts from aqueous solutions **18 (5.1)** 415-422.

[22] Lobacheva O, Dzhevaga N 2017 *Journal of Ecological Engineering* Rare-earth elements recovery on the example of Europium (III) from lean technogenic raw materials **18 (6)** 122-126.

[23] Chirkst D, Lobacheva O, Berlinskii I, Dzhevaga N 2011 *Russian Journal of Applied Chemistry* Effect of chlorides on cerium(III) and samarium(III) ionic flotation **84 (2)** 341-344.

[24] Pusz J, Pieniazek E, Woznicka E, Ciszkowicz E, Lecka-Szlachta K, Wolowiec S, Byczy-Ski L 2018 *Acta Poloniae Pharmaceutica - Drug Research* Synthesis and antibacterial activity of

Tb(III), Ho(III), Er(III) and Yb(III) complexes of chrysin-4í-sulfonate **75** **(3)** 613-623.

[25] Picayo G A, Jensen M P 2018 *Handbook on the Physics and Chemistry of Rare Earths* Rare Earth Separations: Kinetics and Mechanistic Theories **54** 145-225.

[26] Gálico D A, Fraga-Silva T F C, Venturini J, Bannach G 2016 *Thermochim. Acta* Thermal, spectroscopic and in vitro biological studies of the lanthanum complex of naproxen **644** 43-49.

[27] Zapała L, Kosińska M, Woźnicka E, Byczyński Ł, Zapała W, Kalembkiewicz J 2017 *J. Anal. Appl. Pyrolysis* Spectroscopic study, thermal investigation and evolved gas analysis (EGA) during pyrolysis and oxidative decomposition of new binuclear complexes of La(III), Ce(III), Pr(III) and Nd(III) with N-phenylanthranilic acid **123** 1-11.

[28] Gostishchev V V, Kim E D, Ri E H 2018 *Solid State Phenomena* Synthesis of aluminum-matrix alloys with Al-Zr-W by metallothermy of oxides **284** **SSP** 465-469.

[29] Zapała L, Kosińska M, Woźnicka E, Byczyński Ł, Zapała W, Kalembkiewicz J 2018 *Thermochimica Acta* Preparation, spectral properties and thermal decomposition of new ternary complexes of La(III), Ce(III), Pr(III) and Nd(III) ions with N-phenylanthranilic acid and 1,10-phenanthroline **659** 242-252.

[30] Zapała L, Kosińska M, Woźnicka E, Byczyński Ł, Zapała W 2016 *J. Therm. Anal. Calorim* Synthesis, spectral and thermal study of La(III), Nd(III), Sm(III), Eu(III), Gd(III) and Tb(III) complexes with mefenamic acid **124** 363-374.

Journal of Physics: Conference Series **1384** (2019) 012009 doi:10.1088/1742-6596/1384/1/012009

Determination of the composition of chalcogenid glasses AS_xSE_{1-x} by the method of X-ray fluorescent analysis

A Yu Egorova, E S Lomakina and A N Popova

Saint-Petersburg Mining University, 2, 21 line, Saint-Petersburg, 199106, Russia

E-mail: lomakina_es@pers.spmi.ru

Abstract. The features of X-ray fluorescence analysis of chalcogenide glassy semiconductors are considered in the research. The standard method was used to determine the concentrations of arsenic and selenium As_xSe_{1-x}alloys. The use of this method allows determining the quantitative composition of glasses with an accuracy of ± 0.0002.

1. Introduction

The peculiarity of chalcogenide glassy semiconductors is that they belong to the group of compounds of variable composition, thus, in the glassy state it is possible to get homogeneous materials in a wide range of compositions. However, the absence of long-range order in glasses excludes the arsenic-selenium types of the non-destructive method of X-ray phase analysis from the control methods of the chemical composition of the systems [3, 4, 6–8].

The research objects were glassy and crystalline alloys of the As_xSe_{1-x} systems ($As_{0.02}Se_{0.98}$, $As_{0.1}Se_{0.9}$, $As_{0.286}Se_{0.714}$, $As_{0.4}Se_{0.6}$, $As_{0.417}Se_{0.583}$, $As_{0.5}Se_{0.5}$, $As_{0.6}Se_{0.4}$). The compositions of these samples are given by the composition of the initial mixture, determined with an error of ± 0.001.

Glassy alloys of arsenic-selenium systems were synthesized at a temperature of 700 °C. During the synthesis of alloys containing selenium, the temperature in the furnace was raised to a maximum at a rate of 4 °C per minute and the melts were kept at this temperature for 6 hours. The synthesis of alloys containing sulfur was carried out in two stages. Initially, the ampoules in the furnace were heated at a rate of 3 °C per minute to a temperature of 450 °C (melting of sulfur and its interaction with arsenic occur with the formation of As_2S_3). At 450 °C, the ampoules were kept for 5 hours. In the second stage, the heating rate of the ampoules to a maximum temperature was 4 °C per minute and the melts were kept at this temperature for 6 hours. The melt was hardened in the air [1, 5, 9–15].

2. Experiments

Glassy specimens weighing ~ 5 grams were monolithic ingots; the top of the ampoule did not contain traces of sublimation. The criteria for the vitreous state were a fibrous fracture, the absence of lines on the patterns, the absence of inclusions and irregularities when viewing polished surfaces in a MIM-7 metal microscope and an MIK-1 infrared microscope. Glassy alloys were not subjected to special machining (grinding, polishing, etc.).

X-ray fluorescence spectra were recorded on an X-ray M-analyzer, which was designed to determine the qualitative and quantitative elemental composition of solid, liquid and powdery substances. The effective excitation of the fluorescence of chemical elements in the analyzed sample (from Mg to U) was carried out by the X-ray flux generated by the X-ray tube. To obtain qualitative, semi-quantitative and quantitative information about the analyzed sample, the spectra were processed using software functions.

The spectrometer guarantees the most accurate, reliable and high-performance automatic microanalysis of samples of any shape and size from 0.1 mm and more. The spectrometer allows detecting and determining impurities with a detection threshold of ~ 10^{-4} %. The size of the analyzed surface (adjustable) is from 1 to 10 mm. It is possible to scan the surface of a sample of any size using an external analytical unit. The time of one measurement is from 15 seconds to 24 hours (depending on the task). Operating temperatures range from –35 to 35 °C. Continuous operation time is at least 30 days.

X-ray fluorescence spectra were measured at anode voltage U values from 12 to 50 kV.

The presence of arsenic in the most important from a practical point of view alloys leads to significant difficulties in determining their composition of classical methods of chemical analysis (presence of a special chemical laboratory, analysis time and large errors in determining the content of chemical elements). All this stimulates the development of new methods.

That is why in order to determine the quantitative composition of the glasses used in the study we applied the method of X-ray fluorescence analysis (XRF). The dependence of the fine structure of the spectra on the installation parameters (anode voltage and cathode current) was studied so that the ratio of the lines of arsenic atoms and chalcogens in the X-ray spectra corresponded to the ratio of atoms of arsenic and chalcogens in glasses. Since the intensity ratio of the K_α lines of arsenic and selenium for the studied samples depended on many factors (fluorescence yields, absorption cross sections for both primary and fluorescent radiation by all atoms), it was impossible for quantitative analysis to directly use theoretical ratios for line intensities in X-ray fluorescence spectra.

In order to determine the concentrations of arsenic and selenium As_xSe_{1-x} alloys, we used the standard method.

The standard method is a traditional X-ray phase analysis method, where the ratios of the intensities of the lines of two atoms for a known compound are first calibrated, and then these calibration ratios are used to determine the composition of the unknown alloy.

If the anode voltages, cathode current, the distance between the detector and the irradiated surface, as well as the detector loading, are fixed, the relative intensity of the arsenic K_α-line increases with the x index in As_xSe_{1-x} alloys, and this confirms the possibility of using the X-ray fluorescence method to determine the composition of these alloys.

For this purpose, X-ray fluorescence spectra of standards were measured - $As_{0.4}Se_{0.6}$ glass (at U = 40 kV, cathode current 5 μA, distance between the detector and the irradiated surface 5 mm, detector loading 3800 counts / s). The choice of As_2Se_3 alloys as standards is explained by the fact that they are reliably established compounds in As_xSe_{1-x} systems, which can be obtained in stoichiometric composition by the method of fusion of the initial components.

For a given alloy composition, the spectrum was measured at an arbitrarily selected point on the surface five times (to calculate the average value and determination error).

The characteristic X-ray spectra of glassy and crystalline alloys As_xSe_{1-x} films contained all lines of the K-series of arsenic, sulfur and selenium, the most intense lines being the superposition of their $K_{\alpha1}$ and $K_{\alpha2}$ components.

For the spectra of standards, the areas under the $K_{\alpha1,2}$ lines of arsenic and selenium S_{Se} were determined and according to the ratios

$$x_{RFA}=S_{As}/(S_{As}+S_S),\ x_{RFA}=S_{As}/(S_{As}+S_{Se}), \qquad (1)$$

The atomic fractions of arsenic were determined (here, the "x_{RFA}" index means that the x_{PFA} values are determined from X-ray fluorescence analysis data with an error of ± 0.0002). Then, the calibration dependences of $x_{RFA} = f(x)$ were built on three points: for $x = 0$ (for it $x_{PFA} = 0$ was taken), for $x = 1$ (for $x_{RFA} = 1$ was taken) and for x standard.

The indicated calibration dependences for the As_xSe_{1-x} systems are presented in Fig. 1. These dependences can be approximated by polynomials of the second degree. These polynomials and their graphs can serve as calibration ratios for determining the composition of targets from the x_{RFA} value obtained from the spectra.

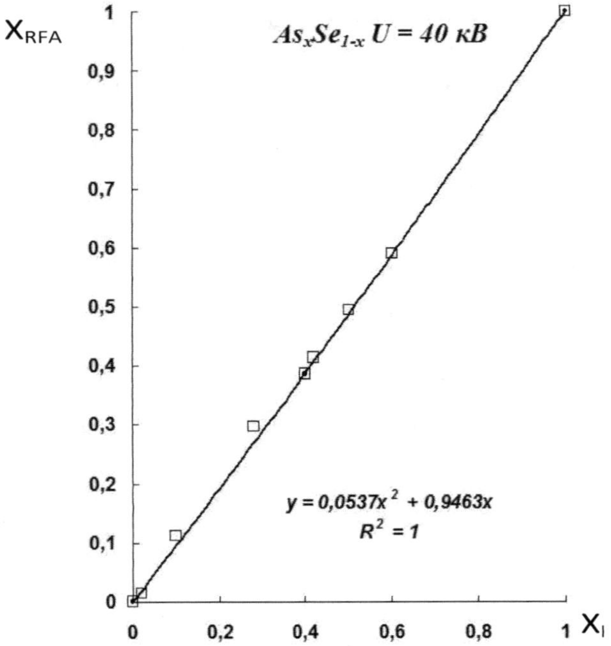

Figure 1. Dependence x = f (x_{RFA}) for As_xSe_{1-x} glasses. The dots show the value of x and x_{RFA} used to construct the calibration dependence (it is drawn with a solid line). Square symbols show x and x_{RFA} values for As_xSe_{1-x}binary compositions. The spectra were measured at U = 40 kV

In order to demonstrate this feature, in Figure 1 there are additional experimental X_{RFA} values for As_xSe_{1-x} glasses ($As_{0.02}Se_{0.98}$, $As_{0.1}Se_{0.9}$, $As_{0.286}Se_{0.714}$, $As_{0.4}Se_{0.6}$, $As_{0.417}Se_{0.583}$, $As_{0.5}Se_{0.5}$, $As_{0.6}Se_{0.4}$), In Figure 2 it is possible to see typical X-ray fluorescence spectra of the corresponding alloys. It can be seen that the experimental data fit well the calibration relations between the values of x and x_{RFA}. Some scatter of data is explained by the greater error in determining x from the composition of the initial charge (not better than ± 0.0005) as compared to the error in determining x_{RFA} (not worse than ±0.0002).

Similar measurements for the As_xSe_{1-x} system were carried out at an anode voltage of 40 kV).

3. Results and discussion

Let us calculate the errors in determining the composition of $As_{1-x}Se_x$ alloys using the calibration ratios. We denote the x_{RFA} letter y. This parameter is measured experimentally and it is necessary to determine the parameter x for it.

The approximation will be represented as a parabolic function

$$y = x + \alpha x (1 - x)$$

This parabola has only one parameter α, its ends are fixed at the points (0,0) and (1,1):

$$\alpha = (y_0 - x_0)/x_0(1 - x_0)$$

Calibration for the case of U = 40 kV has the values x_0 = 0.60 and y_0 = 0.61, the value of the parameter α = 0.041.

Z

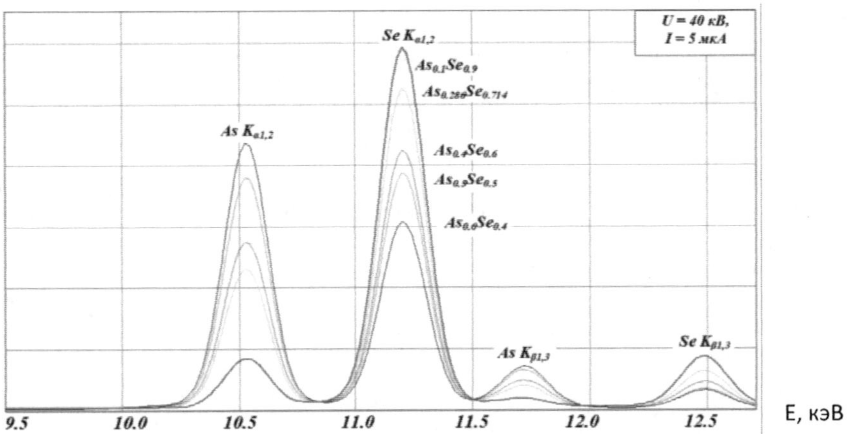

Figure 2. X-ray fluorescence spectra of As_xSe_{1-x} glasses measured at an anodic voltage of 40 kV

The value of x_0 is considered to have no error and, therefore,

$$\Delta\alpha = \Delta y / x_0(1 - x_0) = 4.1 \cdot 10^{-4}.$$

The relative error $\delta\alpha = \Delta\alpha / \alpha = 0.01$.

The composition parameter x is extracted from a quadratic equation:

$$\alpha x^2 - (1 + \alpha)\, x + y = 0,$$

thus,

$$x = (1/2\alpha)\,(1 + \alpha - (1 + \alpha)^2 - 4\alpha y)^{1/2}$$

We decompose this expression in a row:

$$x \approx y/(1 + \alpha) + \alpha y^2/(1 + \alpha)^3$$

Calculations using this formula give:

y	*x*
0	0
1	0.997
0.61	0.599

In this formula, the approximate values are y and α, and according to the formulas for the errors:

$$\Delta x = \mathrm{Sqrt}((y/(1+\alpha))^2((\Delta y/y)^2+(\Delta\alpha/(1+\alpha))^2)+(\alpha^2 y^4/(1+\alpha)^6)((\Delta\alpha/\alpha)^2+4(\Delta y/y)^2+9(\Delta\alpha/(1+\alpha))^2)$$

The calculated errors do not exceed 5.10^{-4}(see table)

y	Δx
0.1	$1.03 \cdot 10^{-4}$
0.5	$1.9 \cdot 10^{-4}$
0.9	$4.5 \cdot 10^{-4}$

The increase in the relative intensity of the line of the light component (arsenic) (compared to the content of this component in the alloy) in the region of small values of the anode voltage U and to its decrease with increasing U is demonstrated. This is explained by the shift of the X-ray tube bremsstrahlung spectrum to high energies. It was concluded that it is most appropriate to measure X-ray fluorescence spectra of As_xSe_{1-x}alloys at an anode voltage of ~ 30–50 kV.

4. Conclusion

The relative intensities of the lines related to the two components of the alloy also depend on the cathode current of the x-ray tube (which controls the cathode temperature and, consequently, the intensity of the bremsstrahlung of the x-ray tube) – the increase in the cathode current (at a fixed anode voltage) increases the relative intensity in the spectrum light components for As_xSe_{1-x}alloys and an increase in the relative intensity of the heavy components for As_xSe_{1-x}alloys. This is explained by the peculiarities of the shift of the spectrum of bremsstrahlung radiation of the x-ray tube to the high energy region [2].

The above-mentioned data scatter for As_xSe_{1-x} glasses is explained not only by the greater error in determining x in the composition of the initial mixture compared to the error in determining x_{RFA}, but also by the uneven distribution of components over the volume of glass. It turned out that the spread of x_{RFA} for three arbitrarily selected points is ± 0.001 for $As_{0.5}Se_{0.5}$ glasses, ± 0.0015 for $As_{0.4}Se_{0.6}$glasses and ± 0.004 for $As_{0.4}Se_{0.6}$, which significantly exceeds the value of the mean-square measurement errors ± 0.0002. In other words, all As_xSe_{1-x} glasses turn out to be heterogeneous in composition, and the heterogeneity noticeably increases with increasing selenium content in the glass.

Thus, we have demonstrated the high efficiency of the x_{RFA} method for determining the quantitative composition of glassy arsenic selenides.

References

[1] Marchenko A, Dashina A, Dzemidko I and Kozhokar M 2011 Determination of Composition of Chalcogenide Glasses and Films by X-ray Fluorescence Spectroscopy. *News of the Russian State Pedagogical University* **138** 45–54

[2] Paima K, Rabchanova T, Rasnyuk A and Seregin P 2015 Dependence of Fine Structure of Fluorescence Spectrum Glasses As $_x$Se $_{1-x}$, Excited By X-Ray Bremsstrahlung on the Parameters of the X-Ray Tube *News of the Russian State Pedagogical University* **173** 45–50

[3] Bolobov V I, Gabov V V, Gembitskaya I M, Ponurova I K and Aleksandrov M A 2008 To the Application of the Method of X-ray Fluorescence Spectroscopy to Determine the Steel Grade of a Rock Cutting Tool *Notes of the Mining Institute* **178** 173–4

[4] Kondrasheva N K, Rudko V A and Povarov V G 2017 Determination of Sulfur and Trace Elements in Petroleum Coke by X-ray Fluorescent Spectrometry *Coke and Chemistry* **60(6)** 247–53

[5] Bordovskiy G A, Marchenko A V, Kozhokar M YU, Nasredinov F S and Seregin P P 2013 Determination of the Composition of Multicomponent Chalcogenide Glassy Semiconductors by X-ray Fluorescence Analysis *Physics and Chemistry of Glass.* **39(4)** 545–52

[6] Kozyukhin S A, Kupriyanova T A and Vargunin A I 2007 Molecular Structure of As_xSe_{100-x} Glasses Studied by X-ray Spectroscopy *Inorganic Materials* **43(8)** 897–900

[7] Nalivaiko V I 2009 Nanostructured Layers of Chalcogenide Semicomductors – Structure, Optical Properties, Application *Geo-Siberia* **5(1)** 210–4

[8] Lezal D, Zavadil J and Prochazka M 2005 Sulfide, Selenide and Telluride Glassy Systems for Optoelectronic Applications *J. of Optoelectronics and Advanced Materials* **7(5)** 2281–2291

[9] Terukov E I, Seregin P P, Marchenko A V, Zhilina D V and Bobokhuzhaev K U 2015 X-ray Fluorescence Analysis of $Ge_{1-x}Se_x$, $As_{1-x}Se_x$, and $Ge_{1-x-y}As_ySe_x$ Glasses Using Electronic Excitation (Article) *Semiconductors* **49(10)** 1352–6

[10] Bordovsky G A, Marchenko A V, Seregin P P and Bobokhuzhaev K U 2015 X-Ray Fluorescence Analysis of Ge-As-Se Glasses Using X-Ray and Electron-beam Excitation *Inorganic Materials* **51(9)** 939–43

[11] Marchenko A V, Bobokhuzhaev K U, Nikolaeva A V, Nasredinov F S and Seregin P P 2015

X-ray Fluorescence Analysis of Chalcogenide Glass with Electron Beam Fluorescence Excitation *Glass Physics and Chemistry* **41(5)** 474–7

[12] Bordovskii G A, Marchenko A V, Nikolaeva A V, Seregin P P and Terukov E I 2014 Determination of the Composition of Multicomponent Chalcogenide Semiconductors by X-ray Fluorescence Analysis *Semiconductors* **48(2)** 257–62

[13] Bordovskii G A, Marchenko A V, Kozhokar M Y, Nasredinov F S and Seregin P P 2013 Composition Determination of Multicomponent Chalcogenide Glassy sSemiconductors with X-ray Fluorescence Analysis *Glass Physics and Chemistry* **39(4)** 377–81

[14] Bordovsky G A, Marchenko A V, Seregin P P, Smirnova N N and Terukov E I 2010 Determination of the Composition of Binary Chalcogenide Glasses by X-ray Fluorescence Analysis *Semiconductors* **44(1)** 24–7

[15] Bordovskii G A, Gladkikh P V, Eremin I V, Marchenko A V, Seregin P P, Smirnova N N and Terukov E I 2011 X-ray Fluorescence Analysis of the Composition of As-Ge-Se Glasses and Films *Technical Physics Letters* **37(3)** 250–2

Journal of Physics: Conference Series **1384** (2019) 012010 doi:10.1088/1742-6596/1384/1/012010

System of process control for associated concentrate nanostructures production

V A Ershov[1], V O Gorovoy[1], S A Nebogin[1], A S Govorkov[1], A S Suhanov[1]

[1] Irkutsk National Research Technical University, 83 Lermontov Street, Irkutsk, 664074, Russia

E-mail: ershov@istu.edu

Abstract. The article presents developed automated process control system of the associated concentrate nanostructure production and the technical and system software, which includes a local process control system and software and hardware that are shared by the technological purpose. The main technological parameters controlled and managed by the APCS are reflected. The article presents the composition of lower-level control system of the plant for production line nanomaterial products, software and hardware complex heat-exchanger and coagulator. The requirements to the functioning modes of the APCS ACNP are presented. It is shown that the developed automated control system of technological process of associated concentrates nanostructures provides the opportunity for the system structure development, including the increase in the number of measurement points using a margin of 10% on signals of input / output. While software and information support of automated control systems CNAA involves changing the configuration of the system and possibility of its expansion through modernization and the introduction of new tasks (complexes of tasks and algorithms).

1. Introduction

The existing methods of production of nanostructures in the form of spherical silicon dioxide and fullerene-like carbon require special power, increased consumption of raw materials and energy [1-6]. Microsilicon is an integral by-product of silicon and ferrosilicon production [7-12]. It is obtained by the purification of gaseous emissions of the metallurgical industry, which improves the environmental situation at the plants, but creates a significant problem of utilization [13-21]. In Norway, after the official adoption of technological decisions in the form of standards and relevant specifications, microsilica was used as a modifying agent in concrete since 1980 [18]. At present, nanosilica presents the greatest interest to researchers as a modifier of the concretes strength and building mixtures [19-20]. For the mass use of nanomaterials, it is important to develop technological solutions for the creation of high-tech production. At the same time, the technology should be based on the selection of nanostructures in the form of fullerene-like carbon particles and spherical silicon dioxide, not microparticles.

The analysis shows that it is appropriate to use the associated technology of obtaining concentrates of nanostructures in the form of fullerenelike carbon by trapping nanosized dust from electric arc ore-thermal furnaces in the production of silicon as well as enrichment and conditioning of concentrates from stale stamicarbon gas cleaning systems sludge [22-23].

Journal of Physics: Conference Series **1384** (2019) 012010 doi:10.1088/1742-6596/1384/1/012010

In this paper, an automated process control system of the associated concentrate nanostructure production (APCS ACNP) is considered.

2. Purposes of the development of APCS ACNP

In the development of APCS ACNP as the main objectives were taken:

- operational control and decision-making management based on analysis of the incoming information about the state of the process;
- prevention of erroneous actions of personnel;
- automatic prevention of emergency situations;
- selection of technological modes based on rapid adjustment modes according to laboratory tests and on the indications of process analyzers that are installed on the flows;
- adaptability to possible changes in the process and control algorithms, reducing the time spent on the orientation of personnel in the regime and operational environment, timely detection of problems and deviations;
- analysis of technological parameters and forecasting of optimal equipment operation modes.

In addition APCS ACNP must be operated in the following modes:

a) The automatic mode is the main operation mode at serviceable technical means and a normal condition of technological object of management. This mode includes all the functions of the upper and lower levels of the system, control functions, automatic regulation, blocking. In this case, the possibility of operative operator’s intervention in the operation of the system is provided.
b) Remote mode - auxiliary control mode of the system. This mode allows the operator-technologist remotely (from the operator's workplace) controlling individual elements of electrical equipment in accordance with the priority and level of access. In this mode all functions of control, management and blocking of the devices which are not brought to the mode of remote control are automatically carried out.
c) Local "manual" mode – auxiliary mode of the system. It is included by technological personnel at the time of preventive or repair work on the object of. This mode provides a level of control parameters (due to indicating instruments, set in place, the operator panel with control parameters fixing on the server), alarm and local control of basic drives and mechanisms.

The operator's workstation shall provide the following modes of operation:

a) automatic;
b) automated.

In automatic mode, the operator's workstation must perform the following tasks:

a) collection, processing and presentation of information for process and equipment control and management;
b) identification and signaling of equipment malfunctions;
c) processing and archiving of operational information on the state of technological equipment and technological process controlled parameters values;
d) diagnosis of the work of APCS TP;
e) synchronization of a single astronomical time with the APCS TP for obtaining silicon in electric arc furnaces.

Automated workstation operation should provide the operator with the following tasks:

a) process equipment management;
b) selection of regulators operating modes (manual, automatic);
c) selection of required information;
d) work with archival data and background information;
e) changes of settings of regulators.

3. Automation object

The automation objects presented in Figure 1 to be considered in general case:

a) installation of process gases from ore-thermal furnaces thermal energy extraction;
b) nanosized dust pre-coagulation installation;
c) nanodispersed dust capture module;
d) nano-containing products production line shop.

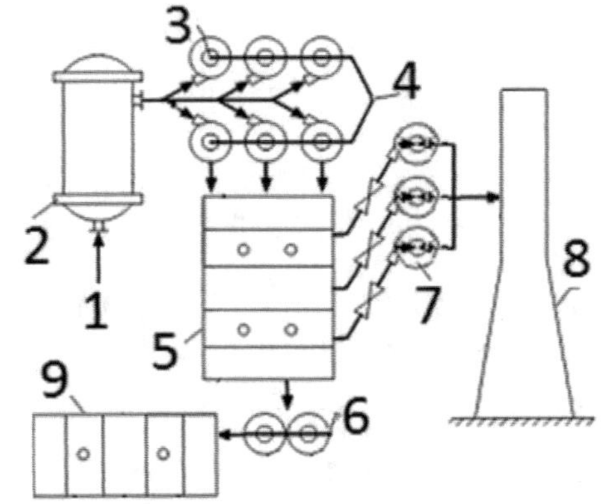

Figure 1. Technological scheme of dust capture from ore-thermal furnaces gases:

1 - gases from the OTF; 2 - heat exchanger; 3-coalescer; 4 - unloading in special vehicles; 5 - bag filter; 6 - transportation system; 7 - exhaust fans; 8 - release of purged gases into the atmosphere; 9 - the nanomaterial products shop.

Table 1. Shop equipment for the nanocarbon line products manufacturing.

№	Equipment	Purpose
1	Receiving hopper	Initial dust auger feed in repulper
2	Water container	Storage and supply of water in repulper
3	Regulator	Mixing the initial dust from the receiving hopper with water
4	Flotation machine	Fullerenelike flotation of carbon particles and deposition of spherical silica
5	Capacity for flotation agents	Storage and supply of flotation agents (kerosene, pine oil) to the flotation machine
6	Drum vacuum filter	The flotation machine chamber product primary drying
7	Calcination rotating oven	Calcination of the machine product chamber
8	Reactor	The hydrofluoric acid washing of the foam product from the flotation machine
9	Acid capacity	Storage and supply of hydrofluoric acid to the reactor
10	Water capacity	Storage and supply of water to the washing foam product tank
11	Revolver vacuum filter	Acid and foam product separation after reactor
12	Foam product washing tank	Water washing of the dried foam product from the flotation machine to remove the reaction residues with hydrofluoric acid
13	Coagulator	Condensation of solid particles under the influence of coagulant
14	Coagulant tank	The coagulant flow into the coagulator
15	Revolver vacuum filter	Primary drying of the carbon-containing product from the coagulator
16	Oven	Carbon-containing product calcination
17	Packing line	Products packaging

Journal of Physics: Conference Series **1384** (2019) 012010 doi:10.1088/1742-6596/1384/1/012010

Table 2. Shop equipment for the nanocarbon line products manufacturing.

Object of control	Controlled parameter	Value unit	Parameter value
Vibrations			
Main exhausts bearings	The running gear bearings vibration	mm/sek	0-7,1
Temperature			
Flue gas at the heat exchanger inlet	Gas temperature	°C	150-200
Flue gas at the outlet of the heat exchanger	Gas temperature	°C	90-140
Piping at the inlet of the heat exchanger	The temperature of the accumulative fluid	°C	20-30
Pipeline at the outlet to the heat exchanger	The temperature of the accumulative fluid	°C	80-100
Flue gas at the entrance to the coagulator	Gas temperature	°C	0-350
Flue gas at bag filter inlet	Gas temperature	°C	0-300
The flue gas duct before the exhaust	Gas temperature	°C	0-300
The ducting after the exhaust	Gas temperature	°C	0-250
Main exhaust bearings	bearing lubricant temperature	°C	0-90
Chamber product temperature	Temperature in rotary kiln	°C	600-650
The carbonaceous product temperature	The temperature in the furnace	°C	350-400
Pressure			
Flue gas at the heat exchanger inlet	Gas pressure	kPa	1,5-2,0
Flue gas at the outlet of the heat exchanger	Gas pressure	kPa	1,5-2,0
Coagulator	Differential pressure	kPa	3,4-4,3
The flue gas duct before the exhaust	Gas pressure (vacuum)	kPa	4,9-6,3
The ducting after the exhaust	Gas pressure (vacuum)	kPa	2,6-1,2
Flotation machine	Air supply line pressure	MPa	0,2-0,4
Expense			
Piping at the inlet of the heat exchanger	Consumption accumulation of fluid	m^3/h	140-150
Flotation machine	Flow rate of air supplied to the flotation machine	m^3/h	1,4-1,5
Water container	The consumption of water from network	m^3/h	0,2-1,3
Level			
Bag filter	The level of collected dust	min-max	10-90
Coagulator	The level of collected dust	min-max	10-90
Receiving hopper	Raw material level	%	20-80
Water container	Water level	%	70-85
Regulator	The pulp level	%	60-85
Capacity for flotation agents	The level of flotation reagents	%	25-95
Concentration			
The clean flue gas after bag filter	Dust concentration in purified gases	mg/ m^3	0-50
Shop for the production of nano-containing products	Concentration of SiO2 dust in the working area air	mg/ m^3	0,4-0,5
Density			
Repulper	The pulp density of repulper	kg/ m^3	1060-1120
Humidity			
Packing line	Humidity SiO_2	%	2-3

More detailed description of the process gases thermal energy recovery unit is presented in [24], a description of the pre-coagulation unit and the nanodispersed dust capture module is presented in [15].

Shop for the nano-containing products line production includes a number of equipment presented in table 1. Table 1 also presents the purpose of the shop equipment for the nanocarbon line products manufacturing.

Table 2 presents the main technological parameters characterizing the technological process of associated production of nanostructures concentrates.

Due to the fact that automation objects are separate processing facilities, in this regard, APCS ACNP includes a local process control system and program-technical complexes (PTC), partial for technological purposes:

a) heat exchanger PTC;
b) PTC coagulator nanosized dust;
c) an automated trap control system module of nanodispersed dust;
d) automated production control system of nano-containing products.

Omitting the details of obtaining nanostructures concentrates physical and chemical processes, we consider the technical and software aspects and controlled automation objects parameters.

4. Technical aspect of the system

The lower (field) level of the software and hardware complex of the heat exchanger consists of the following purchased devices and automation equipment:

- • accumulating liquid temperature sensor at the inlet of the heat exchanger;
- • accumulating liquid temperature sensor at the outlet of the heat exchanger;
- • gas temperature sensor at heat exchanger inlet;
- • gas temperature sensor at the outlet of the heat exchanger;
- • gas pressure sensor at heat exchanger inlet;
- • gas pressure sensor at the outlet of the heat exchanger;
- • pressure sensor (discharge) of electrolysis gases in the flue at the inlet to the FCM;
- • accumulating fluid flow sensor in the heat exchanger;
- • control accessories in complete with electric actuators.

Field-level program-technical complex of coalescent nanodispersed dust consists of following purchase devices and automation sources:

- gas temperature sensors at the inlet of the coagulator;
- differential pressure sensors in the coagulator;
- collected dust level sensors in the hopper coagulator;
- control accessories complete with electric actuators.

Collection and processing of signals at the lower level is carried out with the help of Siemens Simatic ET-200 purchased distributed input/output station. The control station consists of the following purchased elements:

• Siemens SM 331, Siemens SM 332 modules analog signal processing input/output 4-20 mA (+2 - +10 In) DC signal processing transducers;

• Siemens SM 321, SM 322 Siemens - modules of discrete signal processing input/output 24 V DC;

• Siemens Simatic IM 153-1-interface modules for communication between Siemens Simatic ET-200 distributed input/output station and Simatic S7-1200 base controller unit via PROFIBUS DP network.

The control at the lower level should be carried out by Siemens Simatic S7-1200 PLC. The PLC is comprised of purchased items:

- central processor module CPU 1200C various performance modifications to supply a constant or varying current;
- communication module CM1242-5 to connect the controller to the PROFIBUS DP;
- 4-channel switch CSM 1277 for building Ethernet/Profinet network structures;
- PS 1207 stabilized power supply with input voltage ~ 120/230 V and output voltage =24V;
- operator panel TP 900 Comfort series Simatic HMI Comfort Panel, designed to visualize the modes of operation of the heat exchanger, documenting and storing information about the operation of the equipment, emergency situations and actions of operational personnel;
- additional components in the form of SIMATIC Memory cards, input signal simulators for debugging controller programs and buffer battery module.

The structure of the lower (field) level of the APCS of the nano-containing product line shop for the manufacturing includes the following purchased devices and automation:

a) thermoelectric temperature transducers for temperature control in furnaces;
b) electromagnetic flowmeters designed to measure and control the flow of water, acid, coagulant, pulp;
c) hydrostatic level gauges designed to measure the level of liquid reagents and water in containers;
d) radar level transmitters intended for level measurement and control in technological devices;
e) level switch designed for level monitoring in filters;
f) pH meters designed to measure the pH of hydrofluoric acid;
g) density meters designed to control the density of foam and pulp;
h) gas analyzers designed to monitor and alarm the FCM of hydrofluoric acid and SiO2 dust in the air of the working area;
i) hygrometers designed to measure and control the humidity of SiO2;
j) shut-off and control accessories complete with electric actuators designed to regulate the parameters of the process;
k) programmable logic controller (PLC) SIMATIC S7-300, designed to control the lower level;
l) stations of the distributed input-output ET-200M intended for information interaction with object of management and regulation of parameters of technological process;
m) the control station (CS) intended for PLC placement;
n) the control and management station (CMS), designed to accommodate the distributed input/ output station ET-200M;
o) stations of pumps control intended for control of pumps in the "manual" mode and the alarm of its functioning modes;
p) variable frequency equipment drive control station (VFEDCS) Danfoss designed to accommodate, ensure normal operation and visually monitor the operation of the PSU;
q) the compressor control station designed to control the compressor in "manual" mode and to signal its operating modes.
r) u) control stations for automatic lines including dispensers and packaging equipment.

The upper (Supervisory) level CNATP APCS consists of the following purchased items:

a) the workstation (WS) the operator-technologist intended for providing access to the measured and calculated parameters of technological process, control actions input implementation and control of their performance;
b) the technological server intended for storage of operational information and formation of archives;
c) arm shift supervisor intended to implement the functions of APCS ACNP control and management in general, accounting and documentation formation;
d) engineering station, designed to perform work, to configure and maintain operation of the APCS ACNP in general;
e) network equipment.

The top-level equipment management and control lower-level control system station of the plant for nanomaterial line products production are located in the operator's room.

5. Conclusion

The developed automated process control system for the nanostructure production o associated concentrates provides the possibility of developing the system structure, including an increase in the number of measurement of points, using a margin of 10% for input-output signals. At the same time, the software and information support of APCS involves changing the configuration of the system and the possibility of its expansion through modernization and the introduction of new tasks (complexes of tasks and algorithms).

References

[1] Greenberg I ., Ragozin A V, Tenygin A Y 2006 SPb.: MANEB pp 144-145
[2] Kulikov B P, Istomin S P 2004 St Per\tersburg: MANEB pp 477-478
[3] Suslov K V, Stepanov V S, Solonina N N 2013 *IEEE International Symposium on Electromagnetic Compatibility* pp 841-845
[4] Stotzer M, Styczynski Z, Gronstedt P, Buchholz B M, Suslov K V, Glaunsinger W 2012 *IEEE Power and Energy Society General Meeting* 6345090
[5] Suslov K, Solonina N, Stepanov V 2015 Proceedings - 2015 International Symposium on Smart Electric Distribution Systems and Technologies, EDST **7315218** 260-264
[6] Nguyen AV 2003 *Int. J. Miner. Process* **68** 167–182
[7] Sivtsov A V, Sheshukov O Y, Tsymbalist M M, Nekrasov I V, Makhnutin A V, Egiazar'yan D K, Orlov P P 2018 *Russian Metallurgy (Metally)* **12** 1108-1113
[8] Nekrasov I V, Sheshukov O Yu, Sivtsov A V, Tsymbalist M M, Metelkin A A 2016 Steel in Translation 46 **6** 435-442
[9] Sivtsov A V, Sheshukov O Y, Tsymbalist M M, Nekrasov I V, Egiazar'yan D K 2015 *Metallurgist* 59 **5-6** 380-385
[10] Nekrasov I V, Sheshukov O Y, Sivtsov A V, Tsymbalist M M, Egiazar'yan D K, Metelkin A A 2015 *Metallurgist* 59 **3-4** 300-304
[11] Sheshukov O Y, Nekrasov I V, Sivtsov A V, Tsymbalist M M, Egiazar'yan D K, Metelkin A A 2014 *Steel in Translation* **44** 1 43-46
[12] Sivtsov A V, Tsimbalist M M, Fadeev V I, Ostrovskii Y I, Veselovskii I A, Afanas'ev V I 2012 *Russian Metallurgy (Metally)* **12** 1036-1040
[13] Ivanchik N N, Balanovsky A E, Shtayger M G, Sysoev I A, Karlina A I 2018 *IOP Conference Series: Materials Science and Engineering* vol 411 012035
[14] Yolkin K S, Yolkin D K, Shtayger M G, Kolosov A D, Ivanov N A 2018 *IOP Conference Series: Materials Science and Engineering* **411** 012028
[15] Kondratiev V V, Nebogin S A, Sysoev I A, Gorovoy V O, Karlina A I 2017 *International Journal of Applied Engineering Research* 12 **22** 12809-12813
[16] Nemarov A, Lebedev N, Kondrat'ev V, Kornyakov M, Karlina A 2016 *International Journal of Applied Engineering Research* 11 **20** 10222-10226
[17] Karlina A I, Kondratev V V, Kolosov A D, Balanovskiy A E, Ivanov N A 2019 IOP Conference Series: Materials Science and Engineering 012183
[18] Kondratiev V V, Karlina A I, Guseva E A, Konstantinova M V, Kleshnin A A 2018 IOP Conference Series: Materials Science and Engineering 042068
[19] Rassokhin A S, Ponomarev A N, Figovsky O L 2018 *Engineering and Construction Journal* **3 (79)** pp 132-139
[20] Ponomarev A N, Rassokhin A S 2016 *Engineering and Construction Journal* **8 (68)** pp 45-57
[21] Yolkin K S, Yolkin D K, Nemarov A A, Sysoev I A, Karlina A I 2018 *IOP Conference Series: Materials Science and Engineering* 012029
[22] Kondratiev V V, Govorkov A S, Kolosov A D, Gorovoy V O, Karlina A I 2017 *International Journal of Applied Engineering Research* **12 22** 12373-12377
[23] Berezin A I, Rodnov O O, Mazubowski I V, Stont P D, Klykov V A RF patent № 2204629

[24] Kondrat'ev V, Govorkov A, Lavrent'eva M, Sysoev I, Karlina A I 2016 *International Journal of Applied Engineering Research* **11 19** 9979-9983

Journal of Physics: Conference Series **1384** (2019) 012011 doi:10.1088/1742-6596/1384/1/012011

Intensification carboaluminate technology of ultradeep desiliconization

A Yu Feoktistov[1], D A Pavluk[2], E V Ivanova[3]

[1] Cadfem-CIS, Kondratevskii 15, Saint Petersburg, 195197
[2] Engineering company INTRATOOL, sh. Revolucii, 3, Saint Petersburg, 195027, Russia
[3] GC "PSK"Neftehim", pr. Stachek 15, Saint Petersburg, 198097

E-mail: sevong@mail.ru

Abstract. The article presents the results of the study of the effect of additive hydrogarnet sludge on the process of carboalumination desiliconization of aluminate solution. It is proved that type-ahead hydrogarnet sludge significantly intensifies the process of ultra-deep desiliconization due to the catalytic effect of the surface of the slurry in accordance with the theory of heterogeneous catalysis. A mathematical process model for the process of superdeep desiliconization in increased turnover hydrogarnet sludge is presented.

1. Introduction

As a result of numerous experimental works [1-4], universal principle of activating effect on the deep desiliconization of hydrogarnet sludge introduced into the aluminate solution with some advance before direct interaction response of the gysraulic carbo-aluminate calcium with silicas [5-7]. This principle has been verified on various aluminate solutions and carbo-aluminates synthesized on the basis of carbonate raw material of the automated hydrological complex (Mazulskoye, Kiya-Shaltyrskoye limestone fields), under-gas-cap zone (Pikalevskoye field), a number of foreign fields.

The analysis of the above mentioned works suggests that new processing ways of by-products in a complex processing of nepheline raw material are:

The simplified version of this principle was tested in laboratory and in field simultaneously and for multiple times, with simultaneous introduction of the hydrogarnet sludge and calcium carbo-aluminate, and as a result it was proven demonstratively that there is no activation effect of the deep desiliconization response [8].

The advance introduction of the hydrogarnet sludge significantly intensifies the ultradeep desiliconization process. With the HCAC flow rate of 7 g/l by $CaO_{act.}$ and 100% turnover of the hydrogarnet sludge, ~ 30 g/l on hard, for real time of commercial desiliconization (2 hours) at "PGZ-SUAL" branch reaches the value of silicon module at the level of 4,000-5,000 units, and the alumina release, grade G-000 (SiO_2 – 0.02%; Fe_2O_3 – 0.01% reaches 100%.

Our studies show that due to the increased turnover of the hydrogarnet sludge yo 50 g/l on hard the silicon module can be brought to 10,000 units and more (i.e. to the traces) without increasing the HCAC flow rate.

Figure 1 shows the dependence of the depth of desiliconization at the existing commercial flow rate of carbo-aluminate of 7 g/l by $CaO_{act.}$ From the amount of the white sludge seed.

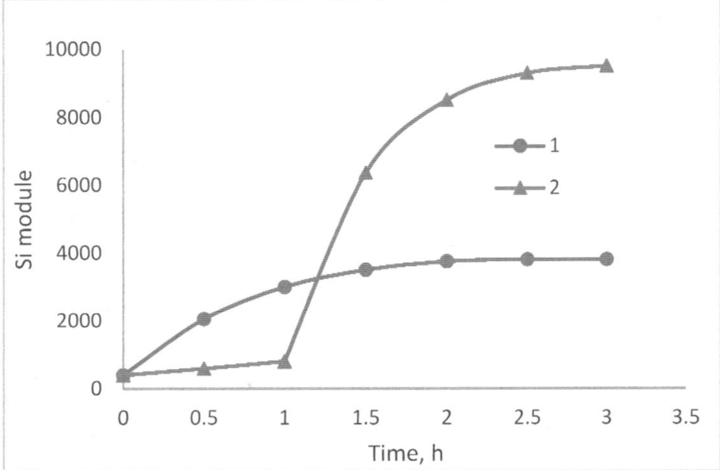

Figure 1. Catalytic properties of hydrogarnet sludge during carboalumination desiliconization
1 – joint input of non-ignition HCAC and circulating hydrogarnet sludge;
2 – advanced circulating hydrogarnet sludge input.
HCAC dosage – 7 g/l by $CaO_{act.}$

It should be noted that bringing the SiO_2 additive in the alumina solution to the traces significantly simplifies the task of producing active sandy alumina (hydroclassification of aluminum hydroxide, efficient use of large and small seeds, respectively, in the soda branch and at the stage of agglomeration in the alkali branch, reduction of α-Al_2O_3 to 6-8%. etc.) The reduction of the SiO_2 addition to the traces increases specific surface of Al_2O_3 from 60 to 80 g/m^3, and the flow ability by 1.5 times (the natural angle of slope reduces from 36 – 37 degree to 33 degree). This is due to the Al (III) hydroxocomplex structure and sharp decrease in the intensity of proton exchange between the Al (III) and Si (IV) ions that facilitates the $Al(OH)_3$ agglomerate growth due to intensified agglomeration of gibbsite radicals by "long" polymerization (Si (IV) hydroxocomplexes break the polymer chains $Al(OH)_6^{3-}$, focusing them in inter-cluster space by dissolving on the alumina solution cluster surface) that corresponds to literature data on structural peculiarities of concentrated alumina solutions.

The intensification of the aluminate solution deep desiliconization by preliminary introduction of return hydrogarnet sludge with further alumina and carboalumina calcium interaction is related, on our opinion, to catalytic interaction of the solid body surface (catalyst), i.e. return sludge. This hypothesis has several features characteristic for a number of elements of the theory of heterogeneous catalysis [9-12]:

1. The apparent activation energy of the deep desiliconization response with the introduction of return hydrogarnet sludge decreases almost twice from 96 kJ/mole to 50 kJ/mole [1].

2. Gibbs energy change for the hydrogarnet formation response [2-3] calculated by structured analogy method [4] is less than 0 (complex heterogeneous catalysis processes can flow in several directions through the interim processes but in any case the condition $\Delta G < 0$ should be observed for the whole process).

3. The main role in accelerating the desiliconization response play active centers of the hydrogarnet sludge; the sludge specific surface is small, ~3.5 m^2/g, it is rather less the specific surface of carbo-alumina phase; even the hydrolysis traces in particular areas of hydrogarnet sludge, when the whole sludge surface is free from passivating aluminum hydroxide films, lead to deactivation of desiliconization response.

4. From the point of view of the theory of heterogeneous catalysis, not only active centers but also their arrangement are important, the so-called principle of geometric conformity requiring the similarity both of distances, and main elements of the symmetry of the formed molecule and catalysts. If there is no such similarity, the process will not accelerate [9].

Actually, during the hydrogarnet sludge change, for example, for active aluminum hydroxide with the high-developed surface or for any other active seed, the intensification effect of the deep desiliconization response disappears.

The role of the geometry principle, entropic factor in heterogeneous catalysis regarding the considered desiliconization process is also confirmed by the accelerating effect of the preliminary electric treatment of aluminate solutions on this process resulting in changing configuration of aluminum (III) and silicon (IV) hydroxocomplexes, thus bringing the system to formation of the relevant quasi-seed domain groups which are similar to hydrogarnet.

When describing the macroprocess of the carboaluminate interaction with alumina in the aluminate solution with the advance introduction of return sludge, it should be clearly represented that the sludge effect on this process does not change its main essence: SiO_2 reaction with hydrated calcium carboaluminate occurs when the rate of isomorphic exchange $4(OH)^- \leftrightarrow [SiO4]^{4-}$ is comparable with the rate of rearrangement of hexagonal metastable carboaluminate lattice in the cubic hydrogarnet lattice.

The return sludge active centres only create the best conditions by forming the interim complex to reduce steric hindrances in the isomorphic exchange mentioned.

2. Mathematical modelling of the aluminate solution of deep desiliconization process with preliminary introduction of hydrogarnet sludge

To prepare the automatic process control system [14-19] of the deep desiliconization process in increased return of hydrogarnet sludge, the corresponding mathematical model was developed [20-23]. In the frameworks of the general desiliconization model, in the presence of HCCA, the desiliconization with preliminary addition of the white sludge can be considered as the interaction of the silicon ion and HCCA occurred in the kinetic area. The hydrogarnet formation kinetics equation corresponds to such process:

$$dM/dt = K \cdot A^m \cdot C^n_{Si}, \quad (1)$$

where M - the amount of the precipitated hydrogarnet, (g/l),

K – chemical reaction rate constant,

A, C_{Si} – concentrations of calcium carboaluminate (for CaO) and silicon (for SiO_2), respectively (g/l),

m, n – reaction orders for CaO and SiO_2.

For the offered mathematical model identification (determination of the reaction rate constant K and reaction orders m, n), the laboratory experimental data were used. The experiments were conducted with aluminate solutions of the chemical composition that correspond to the conditions of stage II desiliconization of "PGZ-SUAL" branch, with the added 30 g/dm^3 of white sludge and 30-60 min. exposed suspension. The HCCA sample weight was varied within the range of 2-5 g/l CaO and the temperature within 70-98°C. Equation integration (1) considering the balance formula:

$$M = C^o_{Si} - C_{Si}, \quad (2)$$

where C^o_{Si} is initial concentration SiO_2,

$$\frac{1}{(n-1)\cdot A^m}\cdot\left(\frac{1}{(C^0_{Si})^{n-1}} - \frac{1}{C^{n-1}_{Si}}\right) = Kt \quad \text{for } n \neq 1 \quad (3)$$

$$\frac{1}{A^m}\cdot \ln\frac{C^0_{Si}}{C_{Si}} = Kt \quad \text{for } n = 1 \quad (4)$$

These expressions were balanced with substitution of variables $C^o_{Si} \rightarrow y$ of the type

$$y = \frac{1}{(n-1)\cdot A^m}\cdot\left(\frac{1}{(C^0_{Si})^{n-1}} - \frac{1}{C^{n-1}_{Si}}\right) \quad \text{for } n \neq 1 \tag{5}$$

$$y = \frac{1}{A^m}\cdot \ln\frac{C^0_{Si}}{C_{Si}} \quad \text{for } n = 1 \tag{6}$$

It appeared that with n=1 in the most experiments, the dependence curves y(t) differ from the straight line and have the distinct up convexity, and with n=2 – down convexity. Thar is why we took the reaction order for SiO_2 as n=3/2. Then, to determine the reaction order for CaO, the nature of families of curves y (t) for various HCCA sample weights at the unchanged temperature of the desiliconization process. m = 3/2 was taken, when the minimum divergence of slope ratios of curves y (t) was observed. Thus, the kinetic equation becomes (see(3))

$$\frac{2}{A^{3/2}}\cdot\left(\frac{1}{(C^0_{Si})^{1/2}} - \frac{1}{C^{1/2}_{Si}}\right) = Kt \tag{7}$$

Or explicit

$$C_{Si}(t) = \frac{C^0_{Si}}{((1+1/2K\cdot A^{3/2}(C^0_{Si})^{1/2}\cdot t)^2} \tag{8}$$

Dependence of the reaction rate constant K and temperature T in the form of the Arrhenius equation.

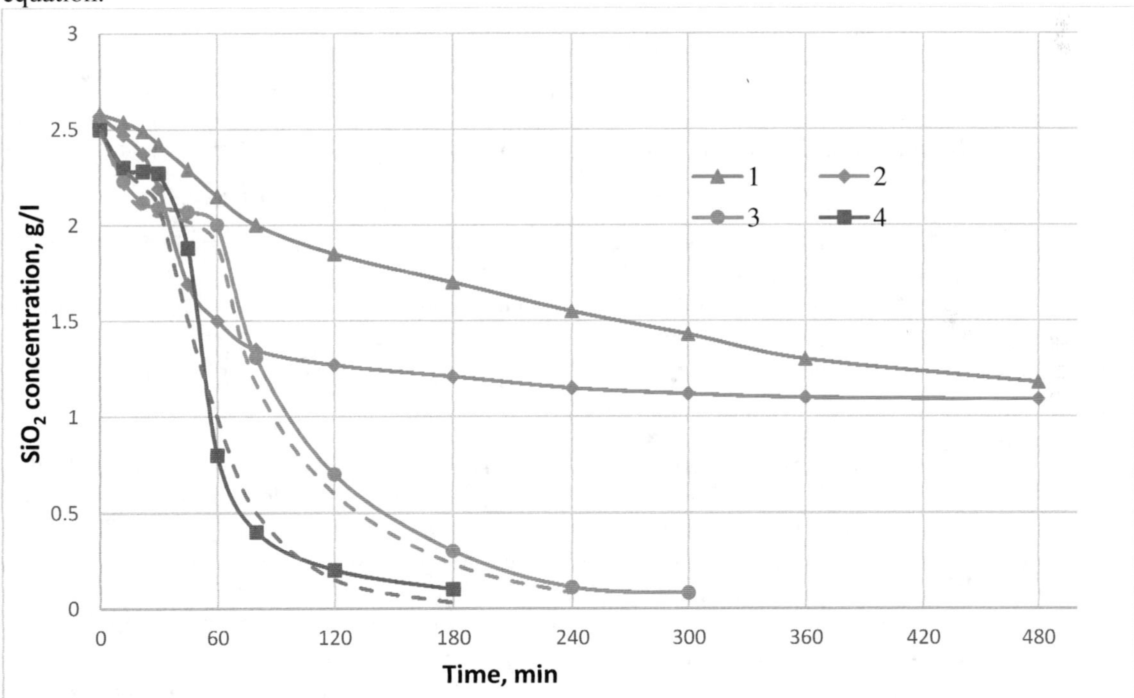

Figure 2. Kinetic regularities of interaction SiO_2 with HCAC, CaO и C_3AH_6 in aluminate solution

Al_2O_3 – 90 g/l, $\alpha_к$ = 1,5; Dosage $\frac{CaO_{act}}{SiO_2} = 6$

1 – C_3AH_6 (75 °C); 2 – CaO (75 °C); 3 – HCAC (75 °C); 4 – HCAC (90 °C);

———————— experimental data

- - - - - - - - - - theoretical data

Processing of the measured values (slope ratios of curves y(t)) for this equation has provided:

activation energy $\Delta E = 50.16$ kJ/mole
ratio $K_0 = (1.32 \pm 0.2)\cdot 10^7$

To compare, it should be noted that the activation energy determined under the same conditions for the hydrogarnet formation response is equal to 55.88 kJ/mol that is rather close to the measured value.

Figure 2 shows the experimental and calculated by equations (8) and (9) kinetic dependences $C_{Si}(t)$ for T = 353K.

The model verification according to the experimental data [16] showed that it is adequate to the describe process according to the Fischer test.

3. Conclusions

As a result of preliminary input additive circulating hydrogarnet sludge may significantly increase the depth and speed of desiliconization of aluminate solutions. The reason for the intensification of the process of ultra-deep desiliconization is a catalytic effect of the surface of the sludge, the active centers which create the best conditions for interaction via formation of an intermediate complex to reduce problems with isomorphic $4(OH)^{-} \leftrightarrow [SiO_4]^{4-}$.

Mathematical model of the process of desiliconization involving hydrogarnet sludge were identified according to laboratory experiments, when it was created was determined magnitude of the rate constant and reaction order.

References

[1] Sizyakov V M 1998 About the formation mechanism of calcium hydrocarboaluminates and its transition to the tricalcium hydroaluminate. *J. of Appl. Chem.*. **71(6)** 390-392

[2] Sizyakov V M, Korneev V I 2000 Scientific basis and technology of new materials with the addition of hydrocarboaluminate calcium *Proc. of the Int.-Conf.* (Moscow: MGU) 515-521

[3] Siziakov V M 2016 Chemical and technological regularities of the processes of sintering of alkaline aluminosilicates and hydrochemical processing specs, *J. of Mining Inst.* **217** 102-112

[4] Sizyakova E V, Sizyakov V M 2017 Synthesis of hydrocarboaluminates of alkali earth metals from natural magnesium-containing feedstock in conditions of alumina production *Source of the Document Russian Journal of Non-Ferrous Metals* **58(1)** 36-43

[5] Siziakova E V, Ivanov P V 2017 Multifunctional coagulants based on hydrocarboaluminates calcium, *J. of Ecolog. Eng.* **18(2)** 16–20

[6] Siziakova E V, Ivanov P V and Boikov A V 2019 Application of calcium hydrocarboaluminate for the production of coarse-graded alumina *Journal of Chemical Technology and Metallurgy* **1(54)** 200 – 203

[7] Siziakova E V, Ivanov P V and Boikov A V 2018 Hydrocarboaliminate calcium application in aluminum processing for production of special cement brands *Journal of Physics: Conference Series* **1118(1)** 012018

[8] Fedorova E R, Firsov A Yu 2018 Red mud flocculation process in alumina production *Journal of Physics: Conference Series* **1015** p 032035

[9] Andreev V V, Korneev V I, Sizyakov V M 1979 Interaction between hydrogarnet sludge of alumina production and aluminum hydroxide. *Cement* **11** 14-15

[10] Cheremisina O, Lutsky D and Fedorov A. 2017 Comparison of extraction methods for extraction of iron, aluminum, manganese and titanium using carboxylic acids and natural vegetable oils from water-salt systems *International Multidisciplinary Scientific GeoConference Surveying Geology and Mining Ecology Management SGEM* (Albena; Bulgaria) **17(11)** 803-810

[11] Lutskiy D Litvinova T, Ignatovich A and Fialkovskiy I 2018 Complex processing of phosphogypsum - A way of recycling dumps with reception of commodity production of wide application *Journal of Ecological Engineering* **19 (2)** 221-225.

[12] Cheremisina O V, Litvinova T E, Lutskiy D S 2019 Separation of samarium, europium and erbium by oleic acisolution at stoichiometric rate of extractant *Innovation-Based Development of the Mineral Resources Sector: Challenges and Prospects - 11th conference of the Russian-German Raw Materials* 413-419

[13] Lutskiy D, Litvinova T, Oleijnik I, Fialkovskiy I Effect of anion composition on the extraction of cerium (Iii) and yttrium (Iii) by oleic acid *ARPN Journal of Engineering and Applied Sciences* **13(9)** 3152-3161

[14] Vasilyeva N V and Ivanov P V 2019 Development of a control subsystem to stabilize burden materials charging into a furnace *Journal of Physics: Conference Series* **1210** 012158

[15] Vasilyeva N V, Koteleva N V and Ivanov P V 2018 Quality analysis of technological process control *International J. for Quality Research* **12(1)** 111 - 128

[16] Vasilyeva N V Fedorova E R, Koteleva N I 2018 Real-time control data wrangling for development of mathematical control models of technological processes *Journal of Physics: Conference Series* **1015** 032067

[17] Vasilyeva N V, Fedorova E R 2018 Statistical methods of evaluating quality of technological process control of trends of main parameters dependence *Journal of Physics: Conference Series* **1118** 012046

[18] Boikov A V, Savelev R V, Payor V A and Vasileva N V 2019 DEM calibration approach: orthogonal experiment *Journal of Physics: Conference Series* **1210** 012025

[19] Boikov A V, Savelev R V, Payor V A 2018 DEM Calibration Approach: Implementing Contact Model *Journal of Physics Conference Series* **1050** 012014

[20] Boikov A V, Savelev R V, Payor V A 2018 DEM Calibration Approach: design of experiment *Journal of Physics Conference Series.* **1015** 032017

[21] Beloglazov I, Petrov P, Martynov S 2018 Application of production processes control algorithm using adaptive control system *International Russian Automation Conference RusAutoCon* 8501801

[22] Beloglazov I I, Petrov P A, Gorlenkov D V 2018 Development of an Algorithm for Control Metallurgical Processes of Fluidized Roasting Using an Adaptive Controller *Journal of Physics: Conference Series* **1059(1)** 012015

[23] Bazhin V Yu, Danilov I V, Petrov P A 2018 Development of automated system based on neural network algorithm for detecting defects on molds installed on casting machines *Journal of Physics: Conference Series* **1015(3)** 032025

Journal of Physics: Conference Series **1384** (2019) 012012 doi:10.1088/1742-6596/1384/1/012012

Mechanical properties measurements with portable hardness testers: advantages, limitations, prospects

K V Gogolinskii[1], V A Syasko[1], A S Umanskii[1], A A Nikazov[2] and T I Bobkova[3]

[1] Saint-Petersburg Mining University, 2, 21st Line, St Petersburg 199106, Russia
[2] Constanta ltd, 21 lit. A, Ogorodniy per., St-Petersburg, 198097, Russia
[3]NRC "Kurchatov Institute" - CRISM "Prometey", 49 Shpalernaya str., St. Petersburg, 191015, Russia

E-mail: nanoscan@yandex.ru

Abstract. The article presents an analysis of the most common hardness measurement methods, implemented by portable hardness testers: rebound or Leeb method, static Portable Rockwell (PR) method and Ultrasonic Contact Impedance (UCI) method. These methods are reviewed in terms of the physical nature of the measurements and the influence of the measured sample properties and the probe parameters on the measurement result. The review contains advantages of each method, possible applications, as well as limitations due to the operation principles of probes. The article also contains the analysis of solving the problem of measurement results validity and their conformity with standardized hardness scales.

1. Introduction

One of the most important mechanical properties, determining material condition, its strength and performance characteristics, is hardness. Hardness measurement is the most common way to control the mechanical properties of materials. Hardness is not a uniquely determined value, but is a complex parameter, associated with the primary mechanical characteristics of materials, depending on test methods [11]. The hardness values measured by static methods (Brinell, Rockwell, Vickers, Knoop) are functionally related to the value of the average contact pressure under the indenter p_d, determined by the ratio of the applied load P to the projected area A of contact of the indenter with test surface:

$$H \sim p_d = P/A\,[N/m^2].$$

The value of the contact pressure under the indenter in case of developed plastic strains for metals is determined by the ratio $p_d = c\sigma_y$, where – σ_y is the yield strength, c– constraint factor [1]. Hence, for static hardness scales $H \sim c\sigma_y$. The value c depends on the material properties and shape of the indenter, and shall be approximately equal to 3 for metals.

The instruments, performing the measurement of the hardness values per static scales, in most cases are stationary laboratory testers. In case of inspection of large parts or equipment elements, it is required to cut special samples, resulting in integrity damage, which is often impossible or unacceptable. A separate task is quality control and inspection of welds and walls of pipelines, oil and gas storage facilities, high-pressure vessels and other metal structures at operating site. The solution of this problem is connected with the development of portable hardness testers. Measuring transducers (probes) of portable hardness testers are characterized by small size and low power consumption, enabling to perform measurements outside the testing laboratories directly at the tested items. When using portable hardness testers, it is necessary to take into account that the applied measurement methods do not correspond to the standardized hardness scales, the values of which are usually specified in the technical documentation (*HB, HV,* or *HRx*), and the hardness values reproduced by them only to some extent correspond to these scales.

2. Leeb dynamic hardness testers

The principle of Leeb method [2] is measurement of the ratio of falling impact body velocity before and after collision with the surface of the test sample. Leeb hardness *HL* is calculated using the equation:

$$HL = (\nu_R/\nu_A) * 1000 ,$$

where ν_R – impact body rebound velocity, ν_A – impact velocity.

The implementation of Leeb method (Figure 1) is based on measurement of velocity of the impact body through the electromotive force (EMF) generated by the magnet installed inside the impact body, passing through the inductor coil mounted on the guide tube of the unit. The induced EMF is proportional to the magnet velocity. Induced EMF signal (Figure 2) is recorded, and the peak values of the induced voltage are used to calculate the Leeb hardness values using the equation:

$$HL = (U_R/U_A) * 1000 ,$$

where U_R and U_A – EMF amplitude, proportional to ν_R and ν_A, respectively.

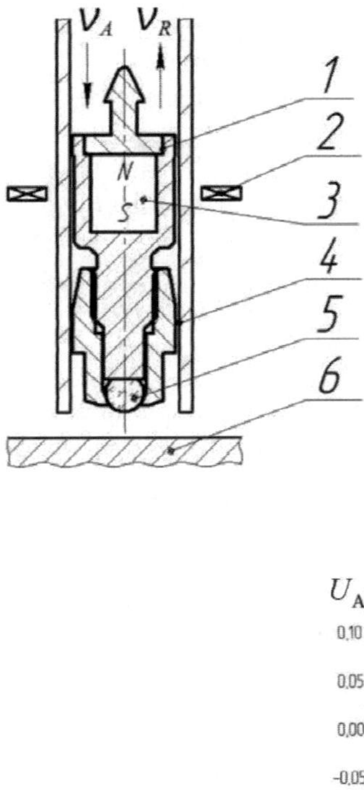

Figure 1. Leeb hardness measurement scheme: 1 - impact body, 2 - inductor coil, 3 - permanent magnet, 4 - guide tube; 5 - indenter ball; 6 - test sample.

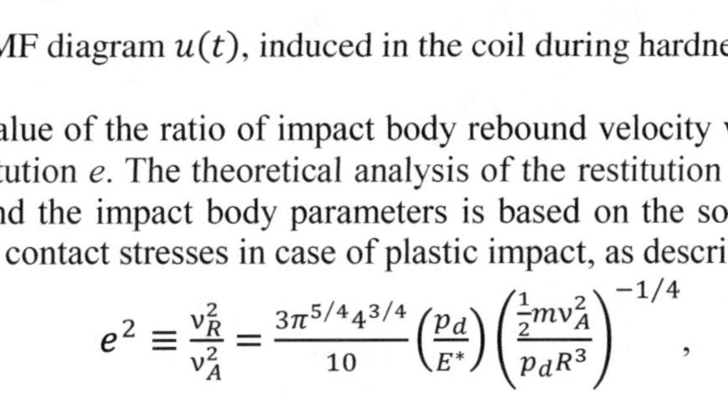

Figure 2. EMF diagram $u(t)$, induced in the coil during hardness measurement.

In the literature, the value of the ratio of impact body rebound velocity ν_R to impact velocity ν_A is called coefficient of restitution e. The theoretical analysis of the restitution coefficient dependence on the material properties and the impact body parameters is based on the solution based on the quasi-static approach to finding contact stresses in case of plastic impact, as described in [3]:

$$e^2 \equiv \frac{\nu_R^2}{\nu_A^2} = \frac{3\pi^{5/4}4^{3/4}}{10}\left(\frac{p_d}{E^*}\right)\left(\frac{\frac{1}{2}m\nu_A^2}{p_d R^3}\right)^{-1/4} ,$$

where m is impact body weight, R is indenter ball radius, p_d - average contact pressure under the indenter, E^* - the reduced plane strain modulus calculated from the ratio:

$$\frac{1}{E^*} = \frac{(1-\mu_m^2)}{E_m} + \frac{(1-\mu_b^2)}{E_b}, \quad (1)$$

where E_m, μ_m and E_b, μ_b are Young's moduli and Poisson ratios of tested material and the indenter ball, respectively. If we take the empirical relation between the yield strength of the tested material and the average contact pressure for metals: $p_d \approx 3.0\sigma_d$ (σ_d - dynamic yield stress), then:

$$e \approx 3.8(\sigma_d/E^*)^{1/2}\left(\frac{1}{2}mv_A^2/\sigma_d R^3\right)^{-1/8}.$$

From this equation it follows that, in general, the restitution coefficient, and, consequently, measured Leeb hardness value, depend on the ratio of the yield strength σ_d and the modulus of elasticity E_m of tested material, as well as on the parameters of measuring probes of hardness testers (so-called "impact device"): impact body weight m, impact body velocity at the moment of impact with the surface v, indenter ball radius R, and its modulus of elasticity E_b. To solve different measuring tasks, different impact device are used, with different impact body weight, impact velocity, indenter ball radius and material, corresponding to different Leeb hardness scales. Impact device parameters for different scales are standardized in [4].

The advantage of Leeb hardness testers is simplicity of measurements. Large area of the indentation reduces the effect of grain, surface layers and surface roughness on measurement dispersion. Availability of international standards [4] and primary hardness standard machines [7],[8] affords to provide metrological traceability per Leeb scales. Different types of impact device is utilized for specific application range. Leeb hardness testers are used to control the hardness of machine parts, process equipment, pipelines, pumps, turbines, etc. The limitation of the method is that the thickness and mass of the sample can affect the measurement result. At the time of contact with the surface of the test specimen the falling impact body transmits part of the energy to it. In case of insufficient thickness or weight of the sample, part of the energy can be spent on the excitation of oscillations in the sample, resulting in loss of energy and reduction of impact body rebound velocity, which in turn affects the measurement result. The minimum weight and thickness values of a sample are specified in all applicable method standards for each type of impact device. In case the test sample does not meet the requirements, heavy support and/or coupling to a solid object is required. Leeb hardness measurement method is standardized on the national and international level: DIN 50156 (1-3), ASTM A956, ISO 16859 (1-3).

3. Ultrasonic contact impedance (UCI) hardness testers

The UCI method is based on the resonance frequency shift of the elastic element (rod) with attached indenter (Vickers diamond pyramid), forced into the surface of the test sample (Figure 3).

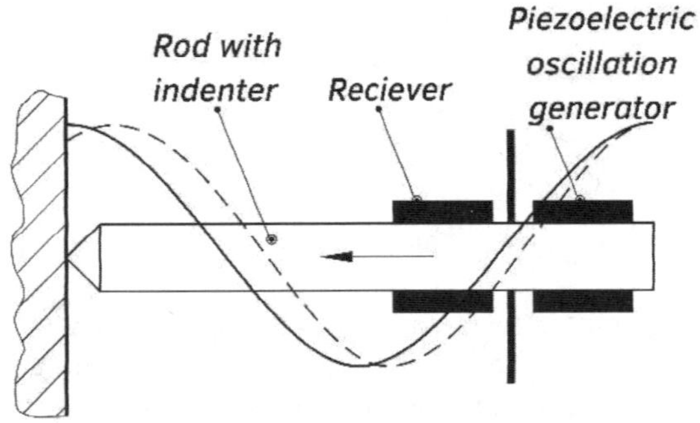

Figure 3. UCI measurement scheme.

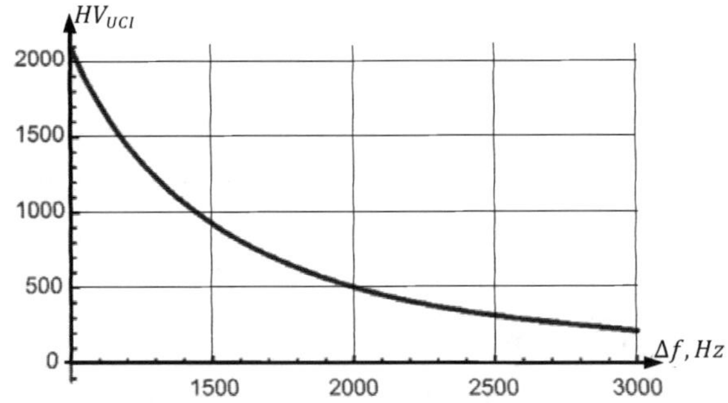

Figure 4. Calibration curve $HV_{UCI}(\Delta f)$.

When the diamond is indented into the test sample, the vibration frequency of the rod increases as the area of contact between the indenter and the surface increases. When the maximum load is reached (1, 5 or 10 kg depending on the modification of the measuring probe), the frequency shift Δf is measured. Hardness values HV_{UCI} are determined based on the calibration curve $HV_{UCI}(\Delta f)$ (Figure 4).

To assess the effect of the test sample material properties on the measurements of UCI tester, we shall analyse the dependence of the frequency of the self-oscillating system based on the rod with the indenter (ultrasonic resonator) as a frequency-setting element on the additional stiffness due to the contact area of the indenter with the sample material. To simplify the solution, we shall reduce the problem to a system with lumped parameters: reduced mass m, spring stiffnessk_0, additional stiffness Δk (Figure 5). Resonance frequency of free oscillations for such system shall be:

$$f_0 = \frac{1}{2\pi}\sqrt{\frac{k_0}{m}}.$$

The effect of additional contact stiffness Δk considering its small value, compared with the stiffness of the whole system ($\Delta k \ll k_0$) is described by the equation:

$$\Delta f = \frac{f_0}{2k_0}\Delta k. \quad (2)$$

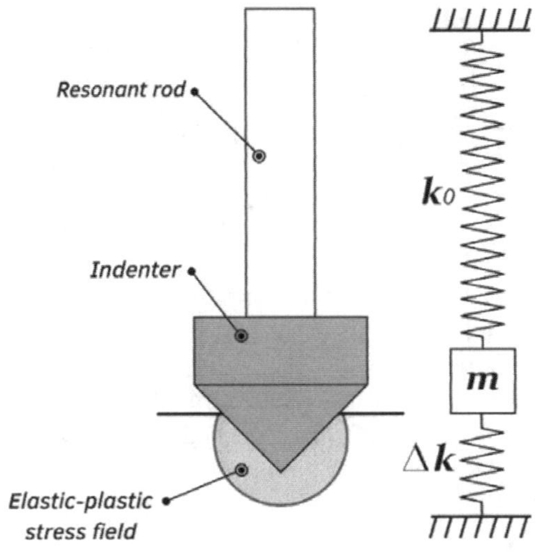

Figure 5. UCI model

The stiffness of the contact area is described by the equation [1]:

$$\Delta k \approx \frac{2}{\sqrt{\pi}}E^*\sqrt{A}\,, \quad (3)$$

where A is the projected area of contact of the indenter with the surface, E^* - the reduced plane strain modulus (1). Vickers hardness is determined by the following ratio $HV = P/A_s$, where A_s is the surface area of the indenter penetrated into the surface. For Vickers indenter (a tetrahedral pyramid with 136° angle between opposite faces) $A = 0{,}927A_s$, hence $A \approx 1{,}079(P/HV)$. Taking into account this ratio, from equations 2 and 3 we obtain the dependence of the measured UCI hardness tester values of Vickers hardness on the material properties and parameters of the ultrasonic resonator: $HV_{UCI} = \frac{1{,}079}{\pi}\left(\frac{f_0}{k_0}\right)^2 P\left(\frac{E^*}{\Delta f}\right)^2$

Since, f_0 и k_0 are instrument characteristics and do not change in the process of measurements, this equation can be represented as:

$$HV_{UCI} \sim P\left(\frac{E^*}{\Delta f}\right)^2$$

The obtained assessment is the same as the solution given in the article [9]. It follows from this equation that the calibration of ultrasonic hardness testers should be carried out on reference test blocks made of material with the same modulus of elasticity as the tested samples, or an appropriate correction should be applied. In general, factory calibration is performed using reference test blocks made of non-alloy or low-alloy steel, therefore, in the case of the control of products made of other materials, it is required to perform calibration of the instrument on samples of tested material.

Hardness testers implementing UCI method are widely used as a portable replacement of stationary hardness testers. In case of application of 1 kg load probes, portable UCI hardness testers produce an indentation of small size and depth, which allows performing measurements of relatively thin coatings, but also raise requirements to surface preparation and material structure (roughness shall not exceed 30% of the indentation depth). Besides, UCI hardness testers are often used to control the heat affected zone (HAZ) of welded joints. Application of ultrasonic hardness testers has a number of limitations. The minimum thickness of the test sample shall exceed 15 mm. In case of failure to comply with that requirement the material can resonate and produce sympathetic oscillations (for instance, thin sheets, pipes, etc.). This problem can be solved by placing the test sample on a heavy metal base, connecting them with viscous paste, lubricant or oil film sufficient to damper elastic oscillations. However, even with the use of such fittings, the thickness of the product less than 2-3 mm is unacceptable. The samples less than 300 grams can go into self-oscillation mode in the process of UCI test, which imposes a limit on the minimum weight of the test sample. UCI method measurements are standardize by DIN 50159 and ASTM A1038.

4. Portable Rockwell hardness testers

Portable Rockwell (PR) method is based on the measurement of the indention depth under static load. Similar to the method of hardness measurement per Rockwell scales, the defined preliminary load (minor load) is applied, followed by main load (major load) (Figure 6).

At the time of preload application, the penetration depth of the indenter h_0 into the test sample is measured. After main load application, the system is held under load for a short time, with subsequent measurement of indenter penetration depth h_a into the material. Then the difference in the indenter penetration depth $\Delta = h_a - h_0$ is determined, which is converted into hardness values. The applied load comprises 1 kg for preload and 5 kg for main load. Diamond truncate cone with 100°± 0.5° cone angle and a flat area diameter of 0.06 mm ± 0.005 mm is used as indenter in PR probes for soft metals, and a sharp cone-shaped indenter with 100°±0.5° angle is used for hard metals. The sharp indenter allows extending the probe operating hardness range. Because the method is static, it does not have limits common to Leeb or UCI methods. Hardness values, measured by PR method, are close, but not fully in line with *HRA* and *HRC* hardness scales. These scales differ in the geometry of the indenter (cone with 120° angle) and applied loads (60 and 150 kg, respectively). At the same time, the measured hardness values will be as close as possible to "true" Rockwell hardness values in case of PR hardness testers calibration using Rockwell test blocks. PR measuring probe are suitable for measuring hardness of small, light, thin, thin-walled or tubular test objects, as well as for measuring the hardness of massive machine parts. Compared with the previously described methods, PR method characterized by higher time consumption of test process, as well as inability to carry out measurements in hard-to-reach places. The thickness of the test specimens shall be 10 times greater than the indentation depth. PR method measurements are standardized by DIN 50157-1:2008 and ASTM B724-00(2006).

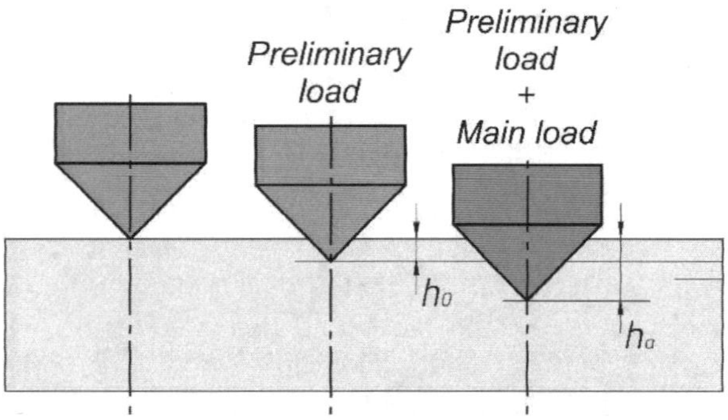

Figure 6. Measurement of Portable Rockwell hardness tester

5. Portable hardness testers application problems and their solutions

The following issues inevitably arise in the course of portable hardness testers application:

1. Is it acceptable to use this type of hardness tester for a specific application?
2. How do portable hardness tester measurements correlated with the standardized hardness scales values?

The answer to the first question can usually be found in the operating instructions. Manufacturers describe in sufficient detail the restrictions related to thickness, weight, surface quality requirements and other properties of tested samples. The answer to the second question is much less obvious, despite the fact that the instrument programs easily allow displaying readings in units of different hardness scales (*HB*, *HV*, *HRx*, etc.). As it is shown above, the result of measurements by Leeb and UCI hardness testers depends on the elastic-plastic properties of tested material and the parameters of the measuring probe itself. It is possible to obtain the measurement result in standardized scale values having calibrated the instrument on reference test blocks of this scale. However, it should be remembered that the instrument readings during measurements, will be correct only in case the reference test blocks used for the instrument calibration are made of material with the same modulus of elasticity. This requirement does not always meet in practice. Leeb or UCI hardness tester, calibrated using reference test blocks made of a particular steel grade will produce incorrect results for duralumin, bronze and even steel of other grades. The correction in compliance with the standard reference data for Young's modulus can also provide unsatisfactory result [10], which can be explained by the discrepancy between reference and real values. The situation becomes even more complicated when the hardness tester is calibrated using reference test blocks of one scale (for instance, *HB*), and the user sets the display output in another scale units (for instance, *HRC*). In this case, the conversion from one scale to another is based on the conversion tables or equations stored in memory. A number of such conversion tables and equations are provided in ASTM E140 and ISO 18265, each defined for a specific material or class of materials. In case the hardness tester calibration is not performed using appropriate reference test blocks, and the conversion table is taken for another material, the measurement result can be arbitrarily far from the real value.

The hardness measurement methods implemented by portable hardness testers do not meet the standards for conventional hardness scales. The results of portable hardness testers measurement are not the true values of these scales, but depend on the calibration of the instrument and the conversion tables used for conversion between the scales. The most effective and comprehensive way to solve this problem, according to the authors, is international standardization of hardness scales of portable hardness testers. The example of such standardization is development of an international traceability chain for Leeb scales, including the adoption of standards for the scales themselves [4], hardness testers calibration procedures [5] and hardness reference test blocks [6], as well as development of national primary references [7],[8]. The availability of such a system allows specifying requirements for materials hardness in Leeb values in the technical documentation, which will automatically eliminate the problems that are described above. In cases when, for one reason or another, it is required to determine the hardness in traditional scales, the calibration of the hardness tester shall be carried out using test blocks made of the same material, certified per required scale. In all other situations, the results of measurements obtained using portable hardness testers should be treated with caution and, if possible, checked for accuracy on test samples in the laboratory.

Complex use of various portable hardness testers can significantly expand the range of materials and objects for hardness measurement. In particular, there are practically no restrictions on the thickness and weight of the test sample for PR measuring probe. By measuring the hardness of a thin-walled pipe, for example, with this probe, it is possible to determine the correction factor for measurement results obtained using Leeb method, for which the wall thickness is a serious limitation. Similarly, the user can refine the correlation dependences between different hardness scales for materials lacking standardized conversion tables.

6. Conclusion

Portable hardness testers allow solving a wide range of problems of mechanical properties control of materials, products and structures in the process of their manufacturing and operation. At the same time, their application requires understanding of the advantages and limitations typical for different types of measuring probes. When selecting a hardness tester, it is required to take into account the influence of the properties of the test objects on the measurement results. Complex use of various probes can compensate for the influence of affecting factors and improve the accuracy and reliability of hardness measurements.

References

[1] Fischer-Cripps A C 2007 *Introduction to Contact Mechanics*, 2nd ed. (US: Springer) p 226
[2] Leeb D 1979 Dynamic hardness testing of metallic materials *NDT International* **12(6)** pp 274-8

[3] Johnson K 1985 Dynamic effects and impact. In *Contact Mechanics* (Cambridge: Cambridge University Press) pp 340-373

[4] ISO 16859-1:2015(E) "Metallic materials - Leeb hardness test - Part 1: Test method".

[5] ISO 16859-2:2015(E) "Metallic materials - Leeb hardness test - Part 2: Verification and calibration of the testing devices"

[6] ISO 16859-3:2015(E) "Metallic materials - Leeb hardness test - Part 3: Calibration of reference test blocks

[7] Herrmann K 2007 Reference measuring instrument for calibration of Leeb-hardness Physikalisch-TechnischeBundesanstalt. Scientific news from division 5, review of the Annual Report http://www.ptb.de/en/org/5/nachrichten5/archiv/2007/nachrichten5_2007.htm

[8] Wei Shi, Wu Zeng and Qingzhong Li 2012 Leeb Hardness Standard with Laser Measuring XX IMEKO World Congress Metrology for Green Growth, Busan, Republic of Korea

[9] Krautkrämer J and Krautkrämer H 1990 Ultrasonic Testing by Determination of Material Properties *Ultrasonic Testing of Materials* (Berlin, Heidelberg: Springer) pp 528-550

[10] Junek M, Janovec J and Ducháček P 2017 Dependence of ultrasonic contact impedance hardness on Young’s modulus of elasticity of creep-resistant steels *Acta IMEKO* **6(1)** article 5

[11] Limanskiy D V, Vasilyeva M A 2018 Application of georadiological subsurface sensing for hydrogeological monitoring, Journal of Physics: Conference Series 1118(1) 012023

Development of the dialogue skills in a foreign language in comparing of oil benchmarks

I V Goman[1]

[1] St. Petersburg Mining University, 2, 21st Line, S. Petersburg, 199106, Russia

E-mail: juliagoman@mail.ru

Abstract. The aim of research is to develop skills of dialogue on professional topics in a foreign language. The objectives of the research are the following: analyze professional goals in programs; identify language skills used in the dialogue; investigate the role of self-study in developing the skills of professional communication; suggest ways of improving the dialogue skills (stimulate asking of questions, giving extended answers). Methods of research include literature analysis on the topics of oil benchmarks, methods of teaching professional communication as well as the interview with an expert in teaching foreign languages. The outcomes of research are planned to be a set of recommended exercises for developing of the dialogue skills.

1. Introduction

The research related to teaching students of oil and gas specialisation ranges from teaching them environmental responsibility [1] to writing skills connected with their specialisation [2] as well as general rhetorical skills [3,4]. The scope of research in St. Petersburg Mining University regarding oil and gas activity is various encompassing international environmental agreements, greenhouse gass emissions, taxation [5, 6, 7], as well as study of metals used in manufacturing of hardwearing oil and gas equipment [8, 9, 10].

Since we plan to teach would-be oil and gas engineers, it is essential to refer to the list of competencies which they are supposed to acquire. The bachelor's student should be able to 'assess risks and suggest measures to provide security of technological processes in oil and gas production; apply principles of rational use of natural resources and environmental protection; take measures of ecological protection when building, repairing, restoring oil and gas wells, collecting and preparing well product, transporting and providing storage of oil and gas' [11].

Having studied the requirements, we see that future engineers have to gain an insight into all stages of oil and gas exploration of resources, development and management. A relevant basis for this could be a study of oil standards as well as their specificities related to a country where it is produced. Other constituents of the requirements' list are also relevant to our purpose. The bachelor's student should be able 'to communicate in written and oral forms both in native and foreign languages…'; to study and analyze scientific and technical information related to professional activity in both native and foreign languages [11]. Thus, future oil and gas engineers should be able to communicate in a foreign language with their partners on the topics of their specialisation. This necessity is taken into account in our research which is supposed to find ways to teach them dialogue skills in a foreign language as a part of professional communication.

2. Language skills in the dialogue

Dialogue skills are a part of speaking skills as well as monologue ones; monologue and dialogue comprise different forms of oral communication. The dialogue is characterized by a constant change of a speaker in the process of exchanging views; the second property is equal participation of both sides in the dialogue. The dialogue skills include the skills to ask questions as well as to answer them. Hence, it is important to train students in performing both functions of dialogue participants: the one who asks the questions as well as the one who answers the questions. The dialogue unit consists of the question and the reply and comprises the unit for developing of the dialogue skills. The question as a component of the dialogue is a multifunctional means. It can be used as a way to monitor the understanding of the material studied; as a way to motivate expression of views by the students; as a way to keep the conversation going. In its first function it is crucial to make up interesting questions so that students feel motivated to respond to them. In its second function they should be thought-provoking in order to cause students’ wish to take part in discussion [12]. In their third function they should be informative so that the partner of communication could use it for asking another question. The reply is another constituent that can consist of a word, a phrase and even an extended utterance. It can be brief, full or extended. Hence, all the types should be trained and used to fulfil a necessary aim.

The skill to ask the questions consists of the following types: the skill to ask various types of questions; the skill to ask the question related to the response of the interlocutor. The skill to answer the questions consists of the skill to give a relevant answer; the skill to give an extended answer that helps to keep the conversation going.

When planning to introduce a dialogue at classes, the preparation before the class is also needed. Hence, it is necessary to turn to the organization of self study in developing of this skill. Self-study is a kind of study activity that is organized and controlled by a teacher by means of study materials, tasks’ formulation without assistance from a teacher. It is inevitable part of study, especially at university as it allows students to choose the method of learning. The exercises provided to students for self-study should be creative in order to stimulate independent decision-making in learning activity. Some study aims are fulfilled in a better way when achieved by means of self-learning; moreover, the combination of self learning and in-class activities can be effective [13].

3. Methods

This paper is based on the theoretical analysis of literature related to oil benchmarks, methods of teaching the dialogue; the interview of an expert teaching technical English to students of oil and gas major. Interviewing as a method of social science research has been regarded by scientists as the one which helps to assess people’s experiences and attitudes in a comprehensive way. The semi-structured type of interview lets the interviewee speak about related experience and touch upon the issues which were not considered by the researcher of the topic. We made a semi-structured interview which was conducted in English. The expert has been teaching foreign languages to the students of oil and gas specialisation at the technical university for four years. The interview lasted about an hour. The questions were divided into three topics: ‘The role of the dialogue skills in teaching professional communication’, ‘The effectiveness of exercises for teaching the skill of asking questions’, ‘The effectiveness of exercises for teaching the skill of answering questions’.

The second method of research was literature analysis of textbooks in academic and professional communication related to the development of the skill of formulation of various types of questions.

4. Results

The aim of the first group of questions was to find out the level of the dialogue skills that students possess; if they realize the importance of this skill in professional communication.

In the first topic, ‘The role of the dialogue skills in teaching professional communication’, we asked two questions. To the question ‘What professional context can we suggest for students to practice the dialogue skills?’ the following response was given: ‘I suggest practicing presentations of research as a way to make a professional context at classes.’ To the question ‘What is the least developed skill in dialogue?' the following response was given: 'The skill to ask the question that relates to the previous answer and is based on the information given in it.'

The aim of the second group of questions was to get an expert's opinion on the most effective exercises for teaching the skill to ask questions. In the second topic, 'The effectiveness of exercises for teaching the skill of asking questions', we asked two questions. The first question, 'Are students good at practicing the skill of asking questions?' was answered in such a way: 'Students do not perform this skill on a high level; they underestimate this skill.' The question 'What are effective ways of developing questions' formulation?' was responded in the following way: 'I consider practicing questions' formulation to be effective. The ones based on the text, e.g. presentation. Both listeners and the speaker should practice making questions: the skill to ask questions to the speaker based on the presentation; the skill to ask the questions to the audience to control understanding of the presentation.'

The aim of the third group of questions was to get an expert's opinion on the most effective exercises for teaching the skill to answer questions. In the third topic, 'The effectiveness of exercises for teaching the skill of answering questions', we asked two questions. For the first question 'Are students good at practicing the skill of asking questions?' such response was given: 'No, answers are not full, sometimes not relevant. There is enough space to improve this skill.' The question 'What are effective ways of developing questions' formulation?' was responded in such a way: 'Both speaker and listeners should participate in practicing this skill. The development of the skill can include varying the amount of sentences in answers.'

The second method used was literature analysis. The authors of textbooks in academic and professional communication suggest making of the following types of questions: multiple, 'good', 'sharp' [14]. The multiple question includes several questions asked at once; related to the topic of presentation. The 'good' question is very much relevant to the content of presentation, continuing its discussion. The aim of the 'sharp' question is to challenge the speaker requiring more arguments for persuasion.

The researchers in the theory of teaching foreign languages suggest using the following exercises to develop the dialogue skills: answer questions using brief, full, extended responses; make the dialogue from the text [12].

5. Discussion

The findings of the expert interview suggest that the interviewee has a very good grasp of issues discussed. The role of the dialogue skills and difficulties that students experienced were clearly stated. The importance of such skills is undoubtedly obvious. Yet, students have not developed those skills on a high level. Hence, there is a room for improvement.

The valuable suggestions of the interviewee related to the exercises for developing dialogue skills were the following: practicing questions and answers by both the speaker and the audience, e.g. the speaker answers the questions of the audience and asks them her questions. The audience asks the questions to the speaker and answers her questions. The second valuable suggestion was to practice answers of different length. It is possible to vary them from one to three sentences depending on the assignment.

Based on the literature analysis and interview we compiled the following list of exercises for developing of the dialogue skills given below with the examples. The dialogue skills presuppose the skills to ask various questions from the speaker to the audience (questions of understanding, multiple questions, 'good' questions, 'sharp' questions); ask questions after presentation: from the audience to the speaker, from the speaker to the audience; ask a relevant question to continue the dialogue; give different answers to questions: brief, full, extended ones.

1. Read the text. Look at the questions of understanding: what density does Brent have? What is sulphur content of Brent? What API gravity does Brent have? Answer them. Practice formulating your own question of understanding.

Brent Crude is a major trading classification of sweet light crude oil that serves as a benchmark price for purchases of oil worldwide. This grade is described as light because of its relatively low density, and sweet because of its low sulphur content. Brent Crude is extracted from the North Sea. Brent is the leading global price benchmark for Atlantic basin crude oils. It is used to price two thirds of the world's internationally traded crude oil supplies. Basket, Dubai Crude, Oman Crude, Shanghai Crude, Urals oil and West Texas Intermediate. Brent blend is a light crude oil. It contains approximately 0.37% of sulphur, classifying it as sweet crude. Brent is suitable for the production of petrol and middle distillates. It is typically refined in Northwest Europe. Brent Crude has an API gravity of approximately 38.06 which is equivalent to a specific gravity of 0.835.

2. Look at the multiple question, the 'good' question, the 'sharp' question based on reading the text. The multiple questions consist of three questions related to each other. The 'good' question helps the speaker to show his research in a detailed way, whereas the 'sharp' question challenges the speaker. Answer them. Practice formulating your own 'good', 'sharp' and multiple questions.

Multiple questions: What other oil benchmarks do you know? Can you give their characteristics? Which one is the most wide-spread? Sharp question: What classifications of oil do you know? Good question: Could you formulate the main idea of presentation in one sentence?

3. Read the dialogue unit. Look at the next question added. Add the question to the dialogue logically related to the theme of discussion.

Dialogue unit: -What was the aim of your presentation?-The aim was to introduce the audience to oil benchmarks, making an example of Brent. The next question: - Why did you decide to start with Brent? (Option: Why is it necessary for students to know about the classifications of oil?)

4. Study a dialogue unit containing an extended answer to the question based on the text. Practice formulating the answer consisting of three sentences when explaining the choice of the topic for your presentation.

-Why did you decide to start with Brent? – I chose this benchmark since it is the leading one in Atlantic basin. Secondly, it is used to price two thirds of oil worldwide. Thirdly, it is relevant to my research interests.

5. Prepare a presentation on one of the oil benchmarks. Formulate questions of understanding based on your presentation.

6. Look at the words written: Brent, Kirkuk, Urals. Explain what they mean. How do oil benchmarks differ?

7. Compare standards of oil using the following expressions: to be (un)like, to be (dis)similar to, likewise, similarity. Make up three sentences.

6. Conclusion

The aim of the research was to suggest ways of the development of the dialogue skills in a foreign language. To reach it, the goals for teaching students in the study programs were analyzed; different resources, both self-study and in-class, were applied. Due to the expert interview the new ways of the development of the dialogue skills were found. As the outcomes of research, exercises for the development of the skills of practicing the dialogue have been suggested.

References

[1] Goman I V 2017 Case study analysis as a way of developing the environmental accountability of future oil and gas engineers *17th International Multidisciplinary Scientific Geoconference Albena: SGEM World Science* **17** 17 – 27.

[2] Goman I V 2017 Teaching writing skills in the foreign language to future petroleum engineers specialising in oil and gas development and operation *17th International Multidisciplinary Scientific Geoconference Albena: SGEM World Science* **17** 195 – 203.

[3] Spyridonova L, Kurbanov A, Goncharko O Y 2017 The dialogue Xenedemos, or voices, by Theodore Prodromo: A critical edition, with English translation *Scrinium* **13 (1)** 227-275.

[4] Shchukina D, Egorenkova N 2017 «Rhetoric Practice» by I. S. Rizhsky (1796) at the Mining

University: History and Contemporaneity *Journal of Mining Institute* **225** 376-384.

[5] Lenkovets O M, Kirsanova N Y, Maksimov S N 2017 Implementation of the paris climate agreement in construction and maintenance of buildings *17th International Multidisciplinary Scientific Geoconference Albena: SGEM World Science* **17** 481-488.

[6] Chvileva T 2017 Russian practice of information disclosure on greenhouse gas emissions in non-financial reporting *International Multidisciplinary Scientific GeoConference Surveying Geology and Mining Ecology Management SGEM* **17 (53)** 775-782.

[7] Smirnova N V, Rudenko G V 2016 Priorities for improving taxation in oil industry in Russia *Indian Journal of Science and Technology* **9 (19)**.

[8] Lobacheva O, Dzhevaga N 2018 Flotation technologies in the removal of the Pr(III) salts from aqueous solutions *International Multidisciplinary Scientific GeoConference Surveying Geology and Mining Ecology Management, SGEM* **18 (5.1)** 415-422.

[9] Lobacheva O L, Berlinskii I V, Dzhevaga N V 2017 *Russ. J. Phys. Chem.* Thermodynamics of complexation in an aqueous solution of Tb(III) nitrate at 298 K **91** 67-68.

[10] Lobacheva O, Dzhevaga N 2017 *Journal of Ecological Engineering* Rare-earth elements recovery on the example of Europium (III) from lean technogenic raw materials **18 (6)** 122-126.

[11] March 12 2015 Order about the acceptance of the Federal educational standard of higher education for the specialty 21.03.01 Oil and Gas Engineering (bachelor level) (Prikazot 12 marta 2015, N 22 obutverzhdenii Federal'nogo gosudarstvennogo obrazovatel'nogo standarta vysshego obrazovaniia po napravleniiu podgotovki 21.03.01 Neftegazovoe delo (uroven bakalavriata).

[12] Galskova N D, Gez N I 2006 *Theory of teaching foreign languages* (Moscow: Academia) 336.

[13] Azimov E G, Shchukin A N 2009 *New dictionary of the terms and concepts (theory and practice of teaching)* (Moscow: IKAR) 268.

[14] Birkin J 2012 *Business Advantage Advanced. Teacher's book* (Cambridge: Cambridge University Press).

Approximation of amplitude-frequency characteristics using equidistants

V E Gozbenko[1, 2], S K Kargapol'tsev[2], Yu I Karlina[3], A I Karlina[3]

[1] Angarsk State Technical University, 60 Chaykovskoy Street, Angarsk, 668535, Russia
[2] Irkutsk State Transport University, 15 Chernyshevsky Street, Irkutsk, 664074, Russia
[3] Irkutsk National Research Technical University, 83 Lermontov Street, Irkutsk, 664074, Russia

E-mail: irgups-journal@yandex.ru

Abstract. Building an equidistant curve to amplitude-frequency characteristic of oscillatory system with one degree of freedom, we introduce the limited damping. The analogous studies were carried out for the systems with the large number of degrees of freedom and analogous conclusions were obtained. It is possible to conclude that the equidistant curve, as a volumetric curve, reflects the results, obtained in the experiment and observed on the screens of different instruments with the study of technical systems.

1. Introduction

Modern technical systems are quite complex objects for theoretical and experimental study and, as a rule, are multicomponent. Therefore, they can be considered from different points of view. One of the main components, of course, is the mechanical part, the basis of the technical system. It is in this aspect that the dynamics and control of technical systems are most often represented, more precisely, at the level of dynamics and control of mechanical systems.

The dynamic properties of these systems are sufficiently described by frequency characteristics. On their basis, it is possible to construct various controls, to estimate the possibilities of introducing structural changes, to evaluate limiting properties, and to ensure dynamic synthesis of the parameters. The characteristics of these systems also have a number of features associated with the existence of discontinuities, asymptotic approximations. At the same time, the determination of properties of systems is associated with the definition of reactions to single and periodic effects, which is related to the calculation of some integral estimates, for example, the areas of the figures.

The study of a fairly complex technical system, as a rule, is accompanied by an experiment, the results of which are important for correcting the initial positions in the formation of a mathematical model, as well as for determining the reliability of information obtained during the analysis. Experimental studies suggest confirmation of both the qualitative and quantitative characteristics of the object and are based on the widespread use of various types of devices. Note that any devices that record changes in the observed parameters are selective; they are characterized by dead zones, desensitization and a certain "erosion" of the results, which is well known in measurement theory. In connection with this, the statements have repeatedly appeared aimed at understanding and taking into account that the parameters of real objects with respect to data from an analytical study are in a certain "erosion" zone, often called a "tube" [1, 2]. Such considerations could not but lead to proposals for the introduction of an equidistant as some kind of dependence, obtained analytically, but containing well-defined information, taking into account the real nature of the object. In turn, the construction of equidistants for the initial dependencies can be considered as one of the methods of approximation, simplification of models, thereby reducing the amount of subsequent experimental studies, often requiring high material costs.

The methodological value of introducing that kind of a specific approximation technique also lies in the fact that the discontinuous characteristics typical of analytical models are smoothed out. This kind of real processes one often has to face in practice.

2. Materials and methods

Experimental methods are widely used: experimental theory, measurement theory, error theory. These methods create methodological and scientific-methodological foundations for the experimental study of properties using modern instrumentation equipment and automated systems. Processing the results of the experiment is associated with the analysis of measurement errors, inaccuracies, taking into account the erosion of the graphs, curves, and phase patterns demonstrated by the devices, which creates prerequisites for understanding the need to introduce tools appropriate to the process. Analytically, the construction of equidistants can be attributed to the integro-differential transformations of the original curves. The equidistant has a certain property of smoothing the original curves, changes their properties, and, in turn, can be seen as the implementation of some approximation procedure over the initial analytical dependence, thus smoothing the transition to the results that can be obtained on an instrument basis in the experiment. The important parameter in these transformations is the height of equidistancing p, which determines the accuracy of the approximation, and further, the degree of closeness of the results to the experimental data.

The concept of equidistant was introduced by N. I. Lobachevsky. There are three main types of movements in his planimetry. For one of them, the notion of equidistants is introduced [3-5].

The lines that are invariant (that is, retain their position) with respect to all the shifts in a single straight line u are not straight lines in Lobachevsky planimetry, as in the case of Euclidean, but are special curves called equidistant curves. N.I. Lobachevsky defined this curve as follows: if we have a sheaf of parallel straight lines and there is a straight line perpendicular to all straight lines of the sheaf, then the curve that lies at a distance from this straight line and intersects each straight line of the sheaf at a right angle is called equidistant. Or: "the equidistant is the locus of points, on one side of the straight line u at one distance from it" [6]. In this case, the straight line u is called the base of the equidistant, and the distance value p is called the height. It is obvious that each straight line can be considered as an equidistant with height $p=0$. Note that the equidistant is a curved line, since the following theorem holds true: "Each straight line has no more than two common points with equidistant." Its proofs are given in the articles [6, 7]. It also holds true that "the height of equidistants is its normal" [3-6]. It is important that the equidistant is symmetric with respect to any of its normals and all its normals are perpendicular to one straight line and, therefore, diverge.

3. Results and discussion

Dynamic characteristics of technical systems, capabilities of management, targeted changes in parameters and the range of possible properties are most often considered using models in the form of oscillatory systems, which implies appropriate attention to research methods and approaches to interpreting results.

Oscillatory systems, by themselves, represent sufficiently informative objects for research, if we bear in mind the parallel study of the properties of oscillatory models, introducing the concept of dynamic constraints, the form and structure of which may differ from classical concepts.

It is known that the amplitude-frequency characteristics of oscillatory systems have features: one - in the case of damped oscillations of a system with one degree of freedom, and two - in the case of undamped oscillations (Figure 1).

It is impossible to calculate the exact area of the figures bounded by these graphs and coordinate axes, since the figures are open and their area is infinite. However in applied problems it is sometimes necessary to know this area. For example, in the tasks of automatic regulation it is used to assess the system regulation quality. The construction of equidistant allows calculating this value with a predetermined accuracy depending on the parameter of equidistant (Figure 2).

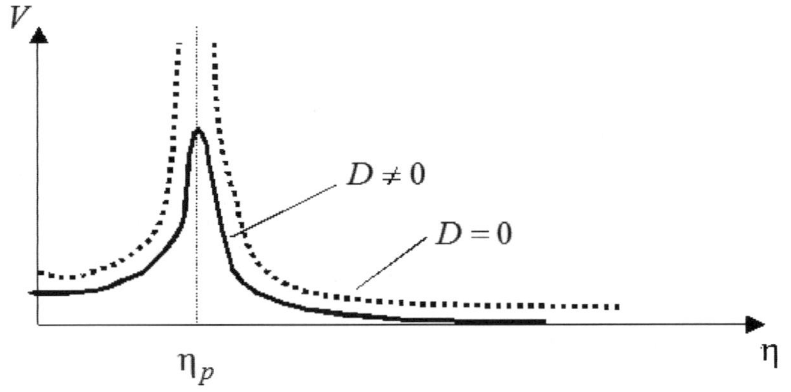

Figure 1. AFRCs of the system with one degree of freedom.

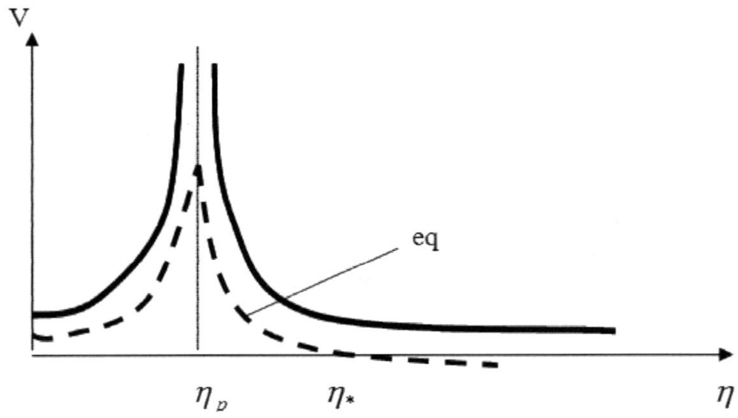

Figure 2. Equidistant to the amplitude-frequency characterictic of a system with one degree of freedom without damping.

However thereby we change the appearance of the amplitude-frequency characteristic. The curve becomes similar to the amplitude-frequency characteristic of a system with friction. That is, we can assume that by introducing an equidistant we thereby introduced some damping.

The direct and inverse dependencies of the parameter of equidistant p and the damping values $D(p)$ are:

$$p = \frac{\sqrt{\left(1-\eta^2\right)^2 + 4D^2\eta^2} - C\left|1-\eta^2\right|}{\left|1-\eta^2\right|\sin\phi\sqrt{\left(1-\eta^2\right)^2 + 4D^2\eta^2}} \tag{1}$$

$$D = \frac{\left|1-\eta^2\right|}{2\eta\left(1-p\left|1-\eta^2\right|\sin\phi\right)}\sqrt{C^2 - \left(1-p\left|1-\eta^2\right|\sin\phi\right)} \tag{2}$$

The studies have shown that the function *p(D)* is bounded above and below. That is, with an unlimited increase of *D*, the value of *p* can not exceed a certain specific value equal to $\frac{1}{|1-\eta^2|\sin\phi}$. Therefore, there is no limit to the function *D(p)* with $p \to \infty$, since *p* can not grow infinitely. Note that with increasing parameter, the equidistant assumes singularities, some of which are described in [8-16]. From all this we can conclude that by building an equidistant to the amplitude-frequency characteristic of a system without friction, we introduce damping, which can not be unlimitedly high.

For the amplitude-frequency characteristic of a system with friction, introducing an equidistant (Fig. 3), we increase the amount of damping of the system. Here the value of *p* is limited by the properties of the equidistant itself, since starting from a certain value of *p*, the equidistant will acquire singularities and it will become ambiguous at point η_p. Moreover, with the growth of *p* these features will increase.

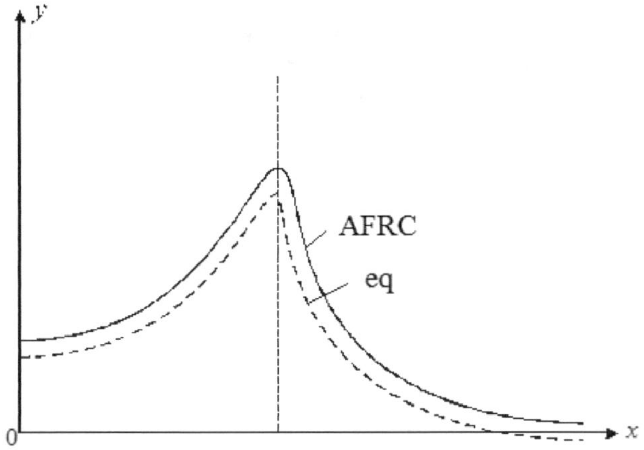

Figure 3. Equidistant curve with amplitude-frequency characteristics of the system with $D \neq 0$.

4. Conclusion

Thus, building an equidistant curve to amplitude-frequency characteristic of oscillatory system with one degree of freedom, we introduce the limited damping.

Analogous studies were carried out and for the systems with the large number of degrees of freedom and analogous conclusions were obtained.

Thus, it is possible to conclude that the equidistant curve, as a volumetric curve, reflects the results, obtained in the experiment and observed on the screens of different instruments with the study of technical systems.

References

[1] Efimov N V, Rozendorn E R 1974 *Linear algebra and multidimensional geometry* (Moscow: Nauka Publ.) p 545
[2] Ilinsky V S 1982 *Protection of radio and electronic devices and precision equipment from dynamic effects* (Moscow: Radiosvyaz' Publ.) p 295
[3] Lobachevsky N I 1945 *Geometric studies on the theory of parallel lines* (Moscow, Leningrad: Publishing House of the Academy of Sciences of the USSR) p 180
[4] Lobachevsky N I 1956 *Selected works on geometry* (Moscow: Publishing House of the Academy of Sciences of the USSR)
[5] Lobachevsky N I 1956 *Three essays on geometry* (Moscow: Gostekhizdat Publ.)
[6] Efimov N V 1971 *Higher Geometry* (Moscow: Nauka Publ.) p 578

[7] Trainin Ya L 1965 *Foundations of geometry* (Moscow: State educ. and paed. publishing house) p 325
[8] Gozbenko V E, Kargapoltsev S K, Kornilov D N, Minaev N V, Karlina A I 2016 *International Journal of Applied Engineering Research* 11(**20**) 10367-10373
[9] Balanovskii A E, Van Huy V 2018 *Journal of Friction and Wear* **39(4)** 311-318
[10] Balanovskii A E 2018 *High Temperature* **56(4)** 486-495
[11] Balanovskii A E 2018 *High Temperature* **56(1)** 3-13
[12] Balanovskii A E 2018 *High Temperature* **56(3)** 329-337
[13] Balanovsky A E, Grechneva M V, Van Huy V, Ponomarev B B 2018 *IOP Conference Series: Materials Science and Engineering* 042010
[14] Balanovskii A E, Grechneva M V, Van Huy V, Zhuravlev D A 2017 *IOP Conference Series: Earth and Environmental Science* 092003
[15] Balanovskii A E, Grechneva M V, Huy V V, Zhuravlev D A 2017 IOP Conference Series: Earth and Environmental Science 092004
[16] Gozbenko V E, Khomenko A P, Kargapoltsev S K, Minaev N V, Karlina A I *2017 International Journal of Applied Engineering Research*. 12 **22** 12369-12372

Journal of Physics: Conference Series **1384** (2019) 012015 doi:10.1088/1742-6596/1384/1/012015

Evaluation of the applicability of measurement techniques to determine the quality of drinking and industrial water used in galvanic production

M I Grichukha and E A Kremcheev

St. Petersburg mining University, 2, Line 21 V.I., Saint-Petersburg, 199106, Russia

E-mail: ilona.g.l@yandex.ru

Abstract. The scientific project «Evaluation of the applicability of measurement techniques to determine the quality of drinking and industrial water used in galvanic production» is devoted to the problems of improving the structure of the methodological base in the production laboratories of enterprises, ensuring its uniformity and efficiency. The paper presents a multi-criteria analysis of measurement methods.

1. Introduction

Nowadays, the regulatory framework for the measurement of water quality indicators is extensive. Sanitary rules and regulations (SRR) 2.1.4.1074-01 [2] and GOST 9.314-90 [6] normalize the content of more than 32 indicators of water quality, which include:

- for drinking water:

1) summary indicators;
2) inorganic substances;
3) organic substances;

- for technical waters:

1) summary indicators;
2) inorganic substances.

In addition GOST 27384-2002 [7] regulates the measurement errors of the indicators of the composition and properties of water.

In order to determine the amount of concentration of normalized components in the samples used standardized and certified measurement techniques. The number of such measurement methods is more than 30 units. Moreover, the methods regulate the procedure for determining the concentrations of both a single component in the sample and a set of components. At the same time, there is no regulated list of methods for measuring the quality of drinking water and technical water used in galvanic production. The difficulty of the list creation is that the use of measurement techniques is voluntary.

2. Water quality

The requirements for the quality of drinking water and industrial water used for galvanic production are different [8]. The determination of the quality of drinking water is measuring indicators such as: hydrogen index, total mineralization, total hardness, oxidation of permanganate, petroleum products, surfactants, phenolic index, aluminum, barium, beryllium, boron, iron, cadmium, manganese, copper, molybdenum, arsenic, nickel, nitrates, mercury, lead, selenium, strontium, sulphates, fluorides, chlorides, chromium, cyanides, zinc and pesticides. In addition, the quality of drinking water is determined by microbiological, parasitological and organoleptic characteristics.

During the determination of the quality of technical water used in galvanic production, the following parameters are measured: hydrogen index, total mineralization, total hardness, sulfates, chlorides, nitrates, phosphates, ammonia, petroleum products, chemically consumed oxygen, residual chlorine, surfactant, iron, copper, nickel, zinc, chromium trivalent, electrical conductivity.

The requirements for the quality of technical water are generally higher than for the quality of drinking water. In addition, the requirements for the quality of technical water are tightened with increasing its category. Thus, for example, the maximum permissible concentration (MPC) of pollutants for the first category of technical water may coincide with the MPC for drinking water, while the MPC for the third category of water is several times tougher.

According to the Federal state budgetary institution "North-West Department for Hydrometeorology and environmental monitoring" [3] the metals and organic substances MPC overage found in samples taken to determine the quality of drinking water in January 2019 at St. Petersburg. These data were obtained in the laboratories of Roshydromet using measurement techniques (MT), which are selected by this organization. The data on the percentage of samples with exceeding the MPC of metals displayed in the source is not correct. Figure 1 shows the result of the analysis of 50 samples taken in January 2019. On the basis of the grades the MPC of polluting substances, which are established to provide guidance to [1] and criteria extremely high pollution and high pollution, which is established by the Order of Russian meteorological service [5], Russian meteorological service established cases of high pollution of surface water, as the concentration of manganese in several samples exceeded the MPC at 30–50 times [4]. It was also noted as general, the excess MPC of heavy metals, phosphorus, easily oxidized organic substances on BPC_5, organic substances on chemically consumed oxygen.

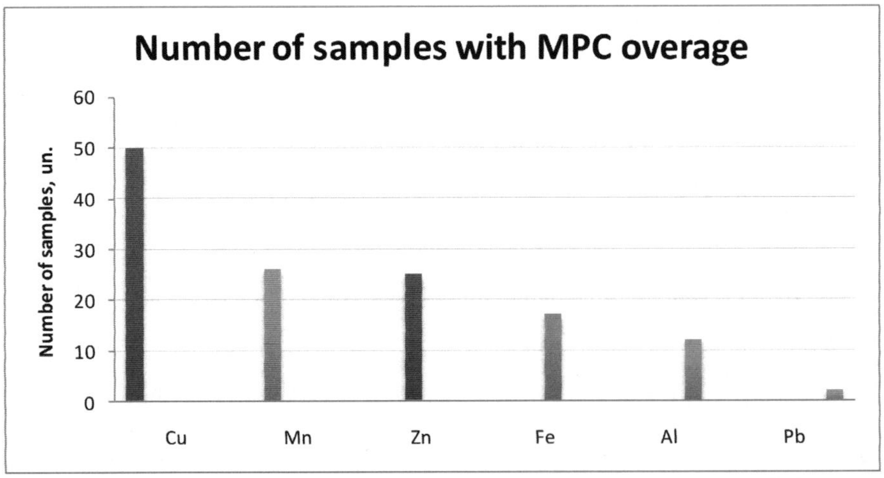

Figure 1. Graphical interpretation of the number of samples exceeding the MPC

According to these data, it can be concluded that the priority for St. Petersburg is the measurement and control of such water indicators as:

- heavy metal;
- phosphorus;
- easily oxidized organic substances according to BPC_5;
- organic matter on chemically consumed oxygen.

3. Water use

Water users include municipal services and various organizations, enterprises and factories. Water supplier at St. Petersburg is SUE Vodokanal St. Petersburg (Vodokanal), which intakes water from different points of the city. There are 5 water intake stations at St. Petersburg:

- Main water supply station (MWSS);
- Northern water supply station (NWSS);
- Southern water supply station (SWSS);
- Volkovskaya water supply station (VWSS);
- Water treatment plants (WTP) g. Kolpino.

Purified water from these supply stations supply utilities and enterprises [9], including galvanic production enterprises [10]. After receiving from Water and Sanitation Authority, industrial consumers purify the water to a technical state and use it in production processes. At the exit from the production process MPC of pollutants in water are exceeded several times, and before discharge into the sewer, it requires pre-treatment. All stages of water transfer and use are accompanied by monitoring of water quality indicators, and, consequently, are needed in the methodological basis for measuring these indicators. Water quality indicators are monitored at the following points:

- at the entrance of the water station;
- at the output of distribution stations;
- at the entrance of the enterprise (where water comes from Water and Sanitation Authority);
- at the outlet of the treatment plant at the enterprise (if the water comes with indicators worse than required in the production);
- at the exit of the production process;
- at the outlet of the treatment plant before the effluent of used water into the sewer.

Effective monitoring of water quality indicators requires a methodological framework that can be used at all stages of monitoring.

The purpose of this study is to select the optimal measurement methods for the determination of metal ions in water used for the needs of housing and communal services and galvanic production enterprises.

According to the guidance document (GD) on "The organization and carrying of regime observations of the state and pollution of surface waters of the land" [1] monitoring of surface waters should be carried out at regular intervals and within a specified time. These parameters are defined by the normative document on Sanitary-epidemiological rules and standards of SRR [2], in addition, the number of samples taken within 1 year is established. In this regard, the cost of monitoring should be economically justified, that is, the assessment of indicators of the quality of drinking water and technical water should be carried out with minimal time and financial resources. Despite the fact that GD [1] and SRR [2] indicate the frequency and timing of surface water monitoring, there is no regulated list of water quality indicators MT.

According to the above mentioned aspects, the measurement methods must meet the following requirements: to ensure ease of experimental data processing, the specified accuracy of the results, technical and economic efficiency.

4. Research methodology

In order to achieve the purpose of the study, we turn to the system analysis. The following tasks are performed during the research:

- selected relevant MT that are appropriate for the accuracy and measuring range;
- the methods of measuring the efficiency are analyzed, that is, the efficiency indicators and efficiency criteria are selected;
- the results are processed;
- the best option is chosen.

5. Research results

As a result of the selection of measurement methods to determine the content of copper and manganese the documents are selected to the analysis of the effectiveness. The list of these documents is presented in tables 1 and 2.

Table 1. Methods for measuring mass concentrations of manganese in water

	Name of the method
1	ERFD* 14.1:2:4.139-98 QCA of waters. Methods of measurement of mass concentrations of cobalt, Nickel, copper, zinc, chromium, manganese, iron, silver, cadmium and lead in samples of drinking water, natural water and wastewater by atomic absorption spectrometry
2	ERFD 14.1:2:4.143-98 QCA of waters. Method of measurement of mass concentrations of aluminum, barium, boron, iron, potassium, cobalt, magnesium, manganese, copper, sodium, nickel, strontium, titanium, chromium and zinc in drinking water, natural water and wastewater by the method of ISP-spectrometry

*Environmental regulatory federal documents.

Table 2. Methods for measuring mass concentrations of copper in water

	Name of the method
1	ERFD 14.1:2:4.139-98 QCA of waters. Methods of measurement of mass concentrations of cobalt, Nickel, copper, zinc, chromium, manganese, iron, silver, cadmium and lead in samples of drinking water, natural water and wastewater by atomic absorption spectrometry
2	ERFD 14.1:2:4.140-98 QCA of waters. Methods of measurement of mass concentrations of beryllium, vanadi, bismuth, cadmium, cobalt, copper, molybdenum, arsenic, nickel, tin, lead, selenium, silver, antimony and chromium in samples of drinking, natural and wastewater by atomic absorption spectrometry with electrothermal atomization
3	ERFD 14.1:2:4.143-98 QCA of waters. Method of measurement of mass concentrations of aluminum, barium, boron, iron, potassium, cobalt, magnesium, manganese, copper, sodium, nickel, strontium, titanium, chromium and zinc in drinking water, natural water and wastewater by the method of ISP-spectrometry

Figure 2 explains the possibility of using the selected techniques at different stages of galvanic processing. In this case, the analysis of content of pollutions in water can be divided into 4 stages:

- stage 1 – at the entrance of the enterprise (where the water comes from the Vodokanal);
- stage 2 – at the outlet of the treatment plant at the enterprise (if the water comes with indicators worse than required in the production);
- stage 3 – at the exit of the production process;
- stage 4 – at the outlet of the treatment plant before the effluent of used water into the sewer.

The color indicates the stages of galvanic production at which it is possible to determine the concentrations of pollutants using appropriate techniques. The figure shows that all of these methods are suitable for use at each stage of galvanic production.

	Stages	1 stage	2 stage	3 stage	4 stage
Elements	Cuprum	1,00	0,02 0,30 1,00	30,00	3,00
	Manganese	0,10	0,10	0,10	0,30
MT	ERFD 14.1:2:4.139-98				
	ERFD 14.1:2:4.140-98				
	ERFD 14.1:2:4.143-98				

Figure 2. Analysis of the applicability of MT at various stages of galvanic production

For clarity and ease of evaluation of measurement methods according to the economic efficiency,

Table 3 presents data on water quality standards, error rates, the total cost of equipment and materials, as well as the time spent on holding a water sample for each MT.

As a result of the MT analysis the following conclusions are made:

- measuring the copper in water samples is more economical in financial and time costs to use ERFD 14.1:2:4.140-98;
- measuring the manganese in water samples is less in financial costs to use ERFD 14.1:2:4.139-98.

Despite the fact that in ERFD 14.1:2:4.139-98 sample holding time exceeds 4 times than in ERFD 14.1:2:4.143-98, the use of ERFD 14.1:2:4.139-98 justified comparative indicator of money savings of almost 1 million rubles.

Table 3. Evaluation of measurement methods by criteria

The criterion	Standard for COPPER	Measurements of the COPPER concentrations			Standard for MANGANESE	Measurements of the MANGANESE concentrations	
		ERFD 14.1:2:4.139 98	ERFD 14.1:2:4.140 98	ERFD 14.1:2:4.143 98		ERFD 14.1:2:4.139 98	ERFD 14.1:2:4.143 98
Measuring range,mg/dm^3	1	01–1	0.01–1	0.04–1	0.1	0.05–0.5	0.05–0.5
Error, %	25	±25	±25	±25	25	±25	±23
Financial costs, ₽/element	–	1 828 588.00	1 561 086.00	3 237 481.00	–	1 828 588.00	3 237 481.00
The time required to extract amples, min	–	40–60	10–15	10–15	–	40–60	10–15

References

[1] GD 52.24.309-2016 The organization and carrying of regime observations of the state and pollution of surface waters of land 2017 (Moscow: Standardinform) 84 p

[2] SRR 2.1.4.1074-01 Drinking water Hygienic requirements to water quality of centralized drinking water supply systems. Quality control. Hygienic requirements for safety of hot water supply systems 2001 (Moscow: Standartinform) 80 p

[3] Water quality of surface water bodies on the territory of St. Petersburg and Leningrad region in January 2019 Retrieved from: http://www.meteo.nw.ru, free

[4] Extremely high pollution help in January-February 2019 Retrieved from: http://www.meteo.nw.ru, free

[5] The order of preparation and presentation of general purpose information about the pollution of the environment the order of the Roshydromet from 31.10.2000 no 156: approved by the order of roshydromet 31.10.2000 2016 (Moscow: Standartinform) 21 p

[6] GOST 9.314-90 Unified system of protection against corrosion and aging (USPACA) Water for galvanic production and washing schemes General requirements 2008 (Moscow: Standartinform) 15 p

[7] GOST 27384-2002 Water The norms of the measurement error for the indicators of the composition and properties 2010 (Moscow: Standartinform) 8 p

[8] Barkan M., Kornev A. 2018 Development of new technological solutions for recovery of heavy nonferrous metals from technogenic waste of electroplating plants and sludge of water treatment systems *Eastern-European Journal of Enterprise Technologies*, Volume 2, Issue 10-92, 17-24

[9] Kuznetsov, V.S.,Suprun, I.K. 2018 The assessment of alumina production waste impact on natural water. *Journal of Ecological Engineering,* Volume 19, Issue 2, 154-158.

[10] Matveeva, V., Danilov, A., Pashkevich, M. 2018 Treatment of multi-tonnage manganese-containing waste water using vermiculite. Journal of Ecological Engineering, Volume 19, Issue 1, 156-162.

Review of applicability of using indicator gas coefficients for determining the temperature of the place of spontaneous combustion of coal

D S Gromyka[1], **E A Kremcheev**[1], **D O Nagornov**[1]

[1] St. Petersburg Mining University, 2, 21-line VO, St. Petersburg, 199106, Russia

E-mail: kremcheev@mail.ru, poloronta@yandex.ru

Abstract. Spontaneous combustion of coal beds is considered to be one of the most hazardous effects of coal mining. The most dangerous factor of this process is its abruptness due to several types of coal possible emergence of reaction at low-temperature conditions and subtle reaction behavior at initial stage of the process, which may turn into environmental disaster. Therefore, on many mines and strip mines, in terms of occupation health and safety program and fire-preventing measures realization, fire monitoring system are introduced, as well as spontaneous combustion source monitoring system to apply fire-preventing measures. The most effective type of monitoring systems is gas monitoring, at which cause in many countries, including Russia, spontaneous combustion source indication methods based on fire gases concentration measurement are developed. Those methods are based on sharp increase of some specific gases, such as carbon monoxide and methane, at some point of spontaneous combustion source temperature, which differs depending on coal type and coal deposit. However, the application of those methods does not always allow the advance detection of coal spontaneous combustion areas, due to its low accuracy and dependence of fire-indication gases from external exposure. For that reason, many authors are advising ratios of fire gases or indicator gases ratios as a means of higher-precise spontaneous combustion sources indication method. In this paper, the analysis of those ratios applicability is conducted as well as existing underground fires indication methods.

1. Introduction

Coal is one of those types of minerals that have the property of spontaneous combustion in contact with oxygen in the air. Extinguishing or isolating the source in an open flame state is commonly used to control spontaneous combustion, which infrequently leads to the desired result. One of the effective measures to prevent endogenous fires is preventive methods, for example, monitoring gases emitted during the combustion of coal. One of the effective measures to prevent endogenous fires is preventive methods, for example, monitoring gases emitted during the combustion of coal. According to some sources [3, 13, 15, 16], coals of any type emit the following gases into the atmosphere at all stages of combustion: CO, CO2, H2S, CH4, H2, SO2, NO2. In addition to these gases, there were also cases of the release of radon [4], lower hydrocarbons C1 – C4 [16, 14], argon [5], and cases of the release of chloride and fluoride compounds from burnt dumps [6].

Gas monitoring of centers of spontaneous combustion is applied both in Russia and abroad. Similar gases are used as the main indicator gases in different ones: CO - In Russia, the USA, Germany, France and Poland, joint analysis of CO and C2H4 - in Japan, England and China [16]. In addition to the main gases, in many countries, the analysis of the emission of specific gas emissions, which, when analyzing the kinetic parameters of coal and determining the temperature of the emission of these gases, can be used to develop preventive measures, has been used to predict changes in the source of spontaneous combustion.

However, it is often not possible to fix the release of specific indicator gases. In such cases, it is advisable to use indicator coefficients or the ratios of the concentrations of various basic gases to identify the early stage of coal spontaneous combustion.

2. Materials and methods

The main indicator coefficient used by many authors to identify the early stages of spontaneous combustion is the Graham coefficient, calculated by the formula [7]:

$$GR = \frac{100(CO_f - CO_i)}{O_f - O_i}, \% \quad (1)$$

where CO_f - is the final concentration of carbon monoxide,%; CO_i - initial carbon monoxide concentration,%; O_f - finished oxygen concentration,%; O_i - initial oxygen concentration,%.

According to the research [10, 11, 12], various values of the coefficient indicate the following condition of the focus:
- less than 0.4% - Normal state;
- from 0.4 to 1% - Unstable state;
- from 1% - Heating;
- from 2% - Open flame.

In addition to equation (1), other coefficient variations are often used, depending on local parameters. For example, when monitoring mines in real time, the following equation is more often used [11]:

$$GR = \frac{100CO_f}{0{,}265N_f - O_f}, \% \quad (2)$$

where 0.265 - is the ratio of oxygen to nitrogen concentration in atmospheric air, N_f - is the final nitrogen concentration,%.

For the initial stage of the oxidation process, the authors of [2] proposed an analytical dependence:

$$T = 456\left(\frac{CO}{\Delta O_2}\right)^{0{,}0076}, K \quad (3)$$

The authors of the work tested this approach at three enterprises: the «Pioneer» mine of Dobropolyeugol LLC, the mine n.a. Artyom GE "Luganskugol" and the mine "Komsomolets Donbass" PJSC DTEK. According to the results of experimental studies, the temperature of the combustion zones slightly differed from that obtained using the ratio of the concentrations of indicator gases characteristic of these mines (unsaturated hydrocarbons).

According to the research [11, 15], the Graham coefficient shows the most representative result with oxygen loss values of more than 0.3%. In addition, the value of this ratio depends on the concentration of carbon monoxide, the increase or decrease in concentration of which, as the authors note [12], weakly correlated with a change in temperature and the source of fire behavior under certain conditions.

Similar to the Graham coefficient is the Young coefficient, where the indicator gas is carbon dioxide:

$$YR - \frac{CO_2}{\Delta O_2}, \quad (4)$$

where CO_2 - is the final concentration of carbon dioxide,%.

The values obtained using the formula (4), which are less than 25 indicate the presence of heating, more than 50 indicate an open flame [12]. The Young coefficient has similar disadvantages: the most representative result is obtained with oxygen loss values of more than 0.3%, the value of carbon dioxide concentration may differ from the true value due to adsorption in water. The coefficient is most often used as an additional one to confirm the presence of spontaneous combustion processes, rather than as an independent indicator [13].

To analyze the status of underground fires, the Jones-Triquet coefficient is often used [5]:

$$JT = \frac{CO_2 + 0{,}75CO - 0{,}25H_2}{\Delta O_2} \quad (5)$$

where H_2 - is the final concentration of hydrogen,%.

If there is a source of endogenous combustion of coal, the value of this coefficient will be in the range from 0.5 to 1.0. With a sharp decrease in the temperature of the focus or its attenuation, the value of the coefficient will not exceed 0.5 [12].

All of the above coefficients use the amount of oxygen loss as a measure of comparison, which is associated with the occurrence of such inaccuracies as the negative value of the coefficient, as described in the works [7, 11], and a non-indicative result for small values of oxygen loss.

To avoid inaccuracies associated with inconsistent levels of oxygen loss, the ratio of carbon monoxide to carbon dioxide is often used. This coefficient allows assessing the degree of combustion or the oxidation process, and by assessing the change in the coefficient with increasing temperature, you can determine the critical temperature of spontaneous combustion (CTSC). For example, in the study [8], the CTSC and crack temperature were estimated at four mines in China; as a result of the studies, graphs of the change in the ratio of gases with increasing temperature were obtained (Fig. 1).

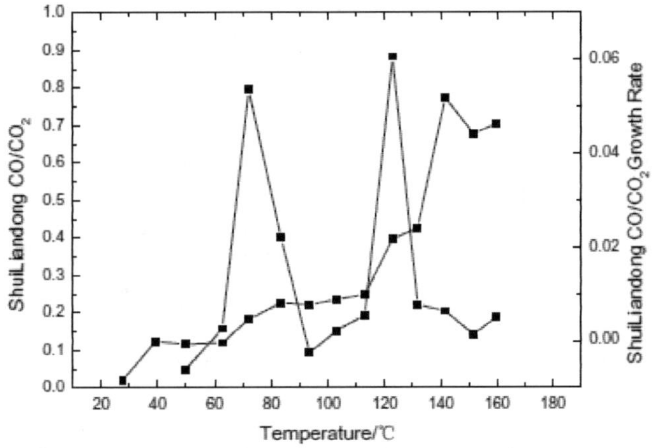

Figure 1. Changes in CO/CO_2 under rising temperatures

An analysis of Figure 1 allows concluding that when the CTSC level and crack temperature are reached, the CO/CO_2 ratio increases sharply. Thus, even at the stage of preliminary studies of changes in the values of a given quantity for a particular field, it is possible to predict with high accuracy the state of the place of spontaneous combustion. However, as noted by the authors of the study [5], the determination by this method becomes possible at later stages of the development of processes preceding the combustion of coal, when the temperature of the center exceeds 200 °C. The authors of the study [12] give a more general value of this ratio for different conditions of the outbreak: for values less than 2 - there is burning in the adjacent territory, and for values of more than 13 - there is active burning.

A similar indicator characteristic is the ratio C2 / C1, where C2 is the concentration of ethane, C1 is methane. Peak values of the coefficient are manifested when certain processes occur in the coal, such as the attainment of CTSC, crack temperature and decomposition temperature (Fig. 2).

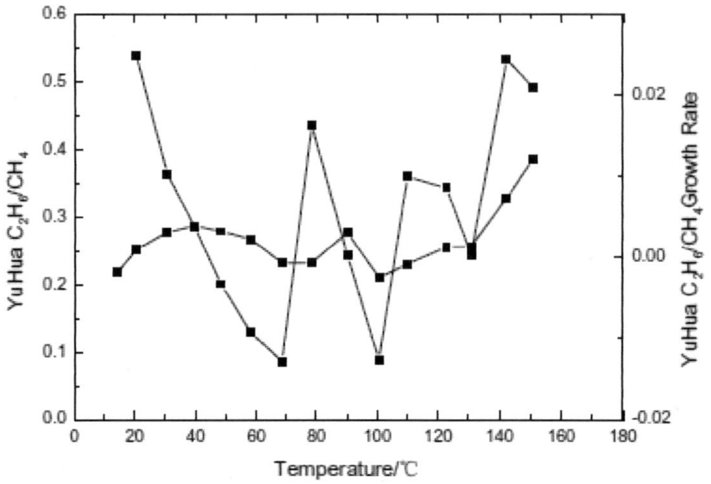

Figure. 2 Changes in C_2/C_1 under rising temperatures

Using the coefficient data allow estimating the temperature of the area, which is in the range of from 50 to 300 ° C [5].

There are also several rarely used gas ratios that are used in mines with certain conditions. For example, the authors of the study [9] proposed using the hydrocarbon coefficient to detect underground fires in depleted mines:

$$R_I = \frac{(1{,}01THC - CH_4) * 1000}{THC - 0{,}01} \quad (6)$$

where THC - is the total concentration of hydrocarbons, ppm; CH_4 - methane concentration, ppm.

When the coefficient values are from 0 to 50, the temperature in the outbreak is below the CCC, from 50 to 100 - a possible temperature increase in the studied area, and more than 100 indicates active heating of coal in the studied area.

This coefficient is not applicable to anthracites due to low hydrocarbon emissions, and also allows using only at a methane concentration of more than 20 ppm [9].

3. Discussion

In the course of the survey study, the practice of using indicator coefficients to predict the behavior of the center of spontaneous combustion and subsequent determination of the temperature of the center was analyzed. Thus, the Graham, Young, and Jones-Tricket coefficients are used in routine monitoring of the state of a self-ignition source, and it is advisable to analyze the joint change in the values of the coefficients to identify changes in the behavior of the source. The CO/CO_2 ratio is recommended to be used in the later stages of spontaneous combustion, when the temperature reaches 200°C or more, and the C_2/C_1 ratio in the initial stages when the temperature reaches 50°C. Some mines use special coefficients, like R1, which prove their effectiveness exclusively under strictly defined conditions of a limited number of fields.

The use of gas indicator coefficients has several advantages over the use of common indicators for predicting the state of a self-ignition source. For example, as it is shown in studies [8, 14, 16], a sharp change in the concentration of common indicator gases begins with a temperature mark of 75-80 ° C, which does not allow timely application of the necessary safety measures in deposits of coal dangerous for spontaneous combustion, which this range.

The advantage of the above coefficients is also their low correlation to factors such as gas flow rate and a sudden change in flow direction, which makes it possible to significantly reduce the measurement uncertainty. However, the studies show that the development of a unified approach to assessing the concentrations of unique indicator gases and the values of indicator coefficients for coals of specific deposits is relevant.

4. Conclusion

Main outcomes of the study are as following:

1. The use of indicator gas coefficients satisfies the conditions of the process of identifying spontaneous combustion and predicting the state of the source at each stage of the process, provided that the adopted model is adequate.
2. The assessment of changes in the center of spontaneous combustion by indicator coefficients is less sensitive to many external factors affecting the effectiveness of field research, compared with an estimate based on an analysis of the concentrations of common indicators, such as CO_2, CO, and others.
3. The joint use of the coefficients will improve the accuracy of the analysis of spontaneous combustion processes.
4. The analysis of the process for the release of special indicator gases to specific deposits (radon, hydrogen, chloride compounds) is characterized by a higher accuracy of the result than the analysis of indicator gas coefficients. However, this method is applicable only if there is information about the isolation of such compounds and preliminary kinetic studies to determine the temperature range over which it is possible to measure the concentration of such substances.

5. Acknowledgments

References

[1] Belavencev L P 2006 Zakonomernosti vydelenija indikatornyh pozharnyh gazov pri nagrevanii uglja *Vestnik nauchnogo centra po bezopasnosti rabot v ugol'noj promyshlennosti* **2** 21-35;

[2] Grekov S P et al 2015 Opredelenie temperatury samonagrevanija uglja po sootnosheniju oksida ugleroda i ubyli kisloroda na avarijnom uchastke *Bezpieczenstwo I Technika Pozarnicza* **3** 119-127;

[3] Igishev V G 2015 Vydelenie indikatornyh pozharnyh gazov pri okislenii uglja na stadijah samonagrevanija i besplamennogo gorenija *Vestnik nauchnogo centra po bezopasnosti rabot v ugol'noj promyshlennosti* **4** 55-59.

[4] Korshunov G I 2012 Kontrol' ochagov samovozgoranija na ugol'nyh shahtah s pomoshh'ju izmerenija plotnosti potoka radona *Gornyj informacionno-analiticheskij bjulleten'* **6** 194-196.

[5] Koshovskij B I 2011 Gazoanaliticheskie metody obnaruzhenija ochagov jendogennyh pozharov i kontrolja za hodom ih tushenija *Nauchnyj vestnik NIIGD "Respirator* **48** 74-83.

[6] Chesnokov B V 2000 Hlor i ftor v gorelyh otvalah Cheljabinskogo ugol'nogo bassejna *UGZh* **3** (15) 153-163.

[7] Brady D 2008 Problems with determining oxygen deficiencies in ratios used for assessing spontaneous combustion activity. *Proceedings from '08 Coal's operators conference* (14-15 February) pp. 209-216.

[8] Deng Jun, Zhao Jingyu, Zhang Yanni, Geng Ruilin 2014 Study on coal spontaneous combustion characteristic temperature of growth rate analysis. *Proceedings from "2014 ISSST", 2014 International Symposium on Safety Science and Technology* **84** 796-805.

[9] Justin T R, Kim A G 1988 Mine fire diagnostics to locate and monitors abandoned mine fires. *Proceedings from Int. Conf. Mine Drainage and Surface Mine Reclamation.*

[10] Morla R, Balusu R, Tanguturi K, Khanal M 2013 Prediction and control of spontaneous combustion in thick coal seams. *13th Coal Operators' Conference, University of Wollongong, The Australasian Institute of Mining and Metallurgy & Mine Managers Association of Australia*, pp. 232-239.

[11] Muller D, Ryan L, Hollyer J and Bajici S 2017 Review of oxygen deficiency requirements for graham's ratio, in Naj Aziz and Bob Kininmonth (eds.). Proceedings of the 17th Coal

Operators' Conference, Mining Engineering, pp. 382-390 (University of Wollongong, 8-10 February)

[12] Ray S K, Singh R P, Sahay N, Varma N K 2004 Assessing the status sealed fire in underground coal mines *Journal of Science & Industrial Research* **63** 579-591

[13] Sloss L L 2015 Assessing and managing spontaneous combustion of coal (UK:IEA Clean Coal Centre).

[14] Wu Yu-guo, Wu Jian-ming 2011 Experimental study on significant gases of coal spontaneous combustion by temperature programmed (TP). *Proceedings of International Symposium on Mine Safety Science and Engineering* **26** 120-125;

[15] Yuan L, Smith A C 2011 CO and CO_2 emissions from spontaneous heating of coal under different ventilation rates *International Journal of Coal Geology* **88**(1) 24-30;

[16] Liang Zhuang, Wang Ji-ren 2011 The Technology of Forecasting and Predicting the Hidden Danger of Underground Coal Spontaneous Combustion. Proceedings from 1st International Symposium on Mine Safety Science and Engineering **26** 2301-2305.

[17] Limanskiy D V, Vasilyeva M A 2018 Application of georadiological subsurface sensing for hydrogeological monitoring, *Journal of Physics: Conference Series* **1118**(1) 012023

[18] Limanskiy AV, Vasilyeva M A 2016 Using of low-grade heat mine water as a renewable source of energy in coal-mining regions *Ecological Engineering* **91** 41-43

Automated defect detection in oil-lubricated parts and units of D30KP/KP-2 aircraft gas turbine engines by results of microwave plasma method

A Y Hodunaev[1], **V G Drokov**[1], **V V Drokov**[1], **V V Murishenko**[1]

[1] Irkutsk National Research Technical University, 83 Lermontov Street, Irkutsk, 664074, Russia

E-mail: dead7simple@gmail.com

Abstract. Classifier between states of "normal/high maintenance/defective" for oil-lubricated parts and units of D30KP/KP-2 aircraft gas turbine engines is developed. The classifier is based on "random forest" machine learning algorithm. It is trained on results of microwave plasma measurements of metallic admixture in oil filter wash samples of engines. Technical state for train set was determined earlier by expert method and was confirmed by factory disassembly study. Classifier result for states "normal/high maintenance/defective" matches expert method in 73 %, 52 %, 66 % respectively.

1. Introduction
Microwave plasma method was developed at Applied Physics Institute of Irkutsk State University and is used for technical state evaluation of oiled parts and units of gas turbine aircraft engines. Diagnostic decision is taken according to measurements of parameters of metallic impurity in oil and oil filter wash samples. Defect can be localized up to assembly unit while using this method.

Large quantity of measured parameters necessitates high proficiency of the evaluator. Hence, automation of the process is preferable.

Diagnostic decision-making algorithm that could be implemented in software was not formalized. Decisions were carried out expertly by comparison of measured values with reference values of statistical model of normally operating engine.

The purpose of this research is the development of an algorithm and machine learning system that determine technical state by results of microwave plasma measurements.

2. The Data
Initial data contains oil filter wash samples from D30KP/KP-2 aircraft gas turbine engines analyzed through 2000-2019. The engines in the data have operating time from 0 (fresh from the acceptance tests) up to 5000 hours. The data contains both engines with regular 50-hour sampling beginning from engine operation and samples, taken after external defect manifestation

Sample distribution by technical state shown in Table 1.

To summarise, the data contains information about defects of back and frontal drive box (FDB, BDB); inner and outer rings and seperator of high pressure turbine (HPT) roller bearing (r/b); intershaft r/b; high pressure compressor (HPC) ball (b/b) and roller bearings. "High maintenance" state means that expert was not able to localize the defective unit, but there were abnormal values in measurements and undetermined defect was possible.

Table 1. Sample distribution by technical state

№	Engine state	Sample quantity
1	Normal	532
2	High maintenance	292
3	Roller bearing HPT	30
4	FDB units defect	33
5	BDB units defect	3
6	ISB defect	2
7	HPC r/b defect	1
8	HPC b/b defect	1

3. Diagnostic features

Microwave plasma method gives the following notable parameters of metallic impurities in samples:

1. Content of element in form of submicron admixture or discrete particles of size up to 2 μm, Cs (ppm);
2. Content of element in discrete particles more than 2 μm in size, Cp (ppm);
3. Discrete particles quantity of specific elemental composition (Al-Cu, Fe-Ni etc.), from 8 possible elements (Al, Cr, Ni, Cu, Fe, Ag, V, Mg), N (cm-1). Particles from one element are called **"simple";** two and more – **"complex"**;
4. Average size of particle for each measured element, D (μm);
5. "**Rating**" of particles of specific elemental composition, R (‰). The rating is ratio of specific particles to total measured particle count. Usually are used:
   1. Ratings of simple particles, containing specific element, $R_{simple.}$(‰), 8 parameters total;
   2. Total rating of particles, containing specific element, $R_{all.}$(‰), 8 parameters total;
   3. Ratings of complex particles, R (‰), 2^8 – 8 – 1 = 247 parameters total.

The work [2] shows, that most information about wear processes is contained in particles, deposited in main oil filter. To measure them, oil wash sample is prepared.

Ratings were developed to account for filter operation time, which correlated with particle counts. Thus, ratings were used almost exclusively in current work. Out of 247 possible complex particle ratings, 221 appeared in the data. It is clear that accounting for such large number of features in a small dataset requires dimensionality reduction, which is done through feature engineering.

Rating of complex particles, containing a particular element, was proposed $R_{complex}$. Next, two groups of features were formed. The **first** contained ratings of simple particles R_{simple} and ratings of complex particles containg particular element $R_{complex}$. The **second** contained ratings of complex particles less than of 5 elements R.

All features were normalized with mean 0 and standard deviation 1 using Yeo-Johnson method (scikit-learn package implementation was used [3]). The distribution parameters were calculated on a subsample of engines with over 500 operating time.

2, 2.5, and 3 standard deviation outlier counts in each group became new features.

Most features have significant dispersion, caused by engine operating conditions [4]. For demonstration purposes, sample dot distribution to rating of complex particles which contain copper is shown on figure 1.

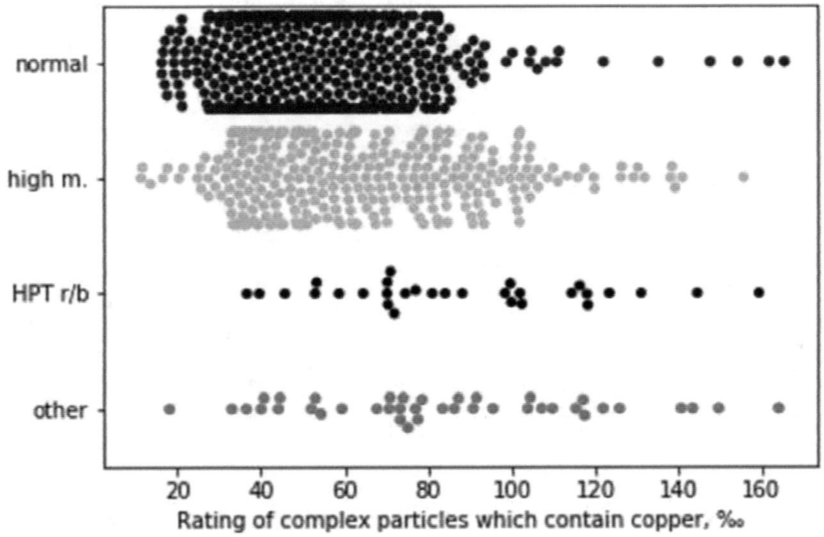

Figure 1. Sample dot distribution relative to rating of complex particles which contain copper.

The dots represent quantity of samples within particular parameter value by technical states; from top to bottom – "normal", "high maintenance", "HPT r/b defect", and "other defects". It is clear that there is significant mixing between the states which makes the classification problematic.

4. Results

The data containing new features was used to train random forest-based classifier (again, scikit-learn implementation was used). Algorithm was chosen due to faster training while offering the same results in comparison to xgboost / neural nets of simple architecture (consisting of several dense layers).

Conditions "normal", "high maintenance", "defective" were chosen as classes. The samples of "high maintenance" class without first or second group 2-standard deviation outliers (20 samples) were transferred to "normal" class.

Train and test set contained half of each class respectively. The samples of rare defects (BDB, ISB, HPC) were transferred to test set from the start.

Confusion matrix averaged on hundred models is shown in table 2.

Table 2. Averaged confusion matrix for model based on outlier counts.

		Classified			
		Normal	High maintenance	Defective	Total samples
	Normal	73 [10] %	25 [10] %	2 [2] %	288
	High maintenance	36 [12] %	52 [10] %	12 [6] %	124
According to expert method	Defective	8 [8] %	27 [14] %	66 [14] %	39

Train-test set splitting was done separately for each model of the hundred. Square brackets contain confidence interval (α=0.95).

According to the table, for classes "normal", "high maintenance", "defective" state can be accurately classified in 73 %, 52 % and 66 % cases respectively. In 8 % cases defect can not be detected by algorithm at wear level, when defective unit can be localized by the expert. In 2 % cases "normal" engine is classified as "defective".

For comparison, Table 3 shows confusion matrix ratings of two-element complex particles.

Table 3. Averaged confusion matrix for model based on ratings of two-element complex particles.

		Classified			
		Normal	High maintenance	Defective	Total samples
According to expert method	Normal	75 [8] %	23 [8] %	2 [2] %	288
	High maintenance	42 [8] %	51 [8] %	7 [6] %	124
	Defective	18 [10] %	39 [14] %	43 [14] %	39

This model detects defects accurately 23 % less often on average, and "misses" it altogether 10 % more often. Using extra non-engineered features for this model does not result in higher classifier accuracy, even with model tuning. This suggests that most of the features have high measurement error, reduction of which is crucial for classification result.

5. Conclusion

The results of the developed classifier match expert method in 73 %, 52 % and 66 % cases for classes "normal", "high maintenance", "defective" respectively. The algorithm uses 2, 2.5, 3 standard deviation outliers of two groups of ratings only.

In 8 % cases defect can not be detected by algorithm at wear level such that defective unit can be localized by expert. In 2 % cases "normal" engine is classified as "defective". Incorrect classification in these cases can be explained by adjustment of statistical model as time passed resulting in variance in margins used by experts. Previously obtained results, evidently, were never revised. This is also corroborated by appearance of "high maintenance" and "defective" classes without statistical outliers in dataset. The same could be said for low (52 %) accuracy of "high maintenance" classification.

Since incorrect classification for said cases is relatively rare (2 % and 8 %), the classifier can be used as rough automated technical state evaluator for oil-lubricated parts and units of D30KP/KU-2 aircraft engines. The decrease in error of measurement of complex particles quantity using microwave plasma method and revision of expert evaluation for the dataset would be primal means to increase the classification result of the algorithm.

6. Acknowledgments

This article was prepared under the framework of the federal target program "Research and development on priority directions of scientific-technological complex of Russia for 2014 – 2020 years." Topic: "Development and creation of software and hardware for microwave plasma system for monitoring, control and safe operation of the oil system of engines for ground and air purposes" The agreement with the Ministry of Education for a grant № 14.577.21.0289, 075-15-2019-1369. Unique identifier is PNIER RFMEFI57718X0289.

References

[1] Technical Note. Engines D-30KP/KU/KU-154, which had malfunctions of oil-lubricated components and assemblies under control of scintillation method 2008 (JSC Saturn: Rybinsk) p 18

[2] Inozemzev A A, Drokov V G, Drokov V V Kazmirov A D and Kaloshin A E 2012 *Control. Diagnostics* **9** 20-28

[3] Pedregosa et al 2011 *JMLR* **12** 2825-2830
[4] Drokov V G 2009 Increase of gas-turbine engines technical state evaluation using scintillation method to decrease emergency rates for aircraft operation. Degree aspiration thesis, Abstract of the dissertation of the doctor of technical sciences, (State Research Institute of Civil Aviation) p 42

The importance of sample preparation in the hydrometallurgical processing of refractory gold concentrates

S A Ivanik[1], D A Ilyukhin[1], A V Shalygin[1]

[1] Saint-Petersburg Mining University, 2, 21st Line, St Petersburg 199106, Russia

E-mail: Ivanik_SA@pers.spmi.ru

Abstract. The article indicates the importance of thorough preparation of concentrate samples before hydrometallurgical operations; the experiments on studying of the factor of the size of the material are described. The comparative analysis of particle size after conventional and fine grinding is carried out. As obligatory stage of preparation of the crushed concentrate before autoclave leaching, the decarbonization process is studied. On the basis of the conducted experiments, the optimal regime of the autoclave leaching process is revealed, which provides the oxidation degree of sulfide sulfur of the order of 99%.

1. Introduction

One of the main trends in the development of the raw material base of the gold mining industry is the decline in the quality of the initial mineral raw materials. This is due to the depletion of rich gold-bearing ores. Therefore, the sphere of production requires a more complex material composition resistant raw materials [1]. The most common reason for the persistence of gold ores is the fine impregnation of gold into rock-forming minerals. Refractory ores account for more than 30% of the world's total reserves [2, 8].

The method of hydrometallurgical autoclave oxidation of the latter in an acidic environment is a effective method of gold extraction from sulfide concentrates. Most often, fine gold is associated with sulfide minerals. Therefore, autoclave technologies are used for oxidation of sulfide minerals in the processing of pyrite and arsenopyrite refractory gold ores. The content of gold in pyrite-arsenopyrite ores is about 3.5-4.0 g/t, sulfides (mainly pyrite) – 2-3%. Sulfide content in raw materials should provide the necessary dissipation to maintain autogenetic process of pressure oxidation.

Technological scheme of processing of refractory gold ore is based on the use of flotation enrichment to obtain concentrates. After that, the concentrates are subjected to autoclave opening and subsequent sorption cyanidation of oxidation products.

The actual direction of development of autoclave technology is application of processes of leaching of sulfide ores and concentrates at rather low parameters. The attractiveness of the technology based on low-temperature leaching is due to the low cost of autoclave equipment. This allows reducing capital costs in the realization of the industrial process [9].

The successful realization of the autoclave leaching process largely depends on the preparatory stages. These operations include grinding and decarbonization. The most complete processing of gold-containing minerals requires fine and ultra-fine grinding, where 80% of the particles have a fineness of less than 10-15 microns [3, 4, 7]. The process of ultra-fine grinding of the material leads to a high degree of deformation set into the crystal lattice of minerals. This increases the activity of minerals and promotes their leaching [10]. Therefore, the question of thorough preparation of the concentrate sample for processing is very important.

2. Characteristics of raw materials

Granulometric characteristics of fine and fine-disperse gold of the initial ore sample (size range from 3 to 25 microns) showed that the proportion of gold of this size in the total content of the precious metal in the sample is 90-92%, the average diameter of the gold is 7 microns. The scheme of preparation of such ore includes one-stage crushing and two-stage grinding. Grinding is carried out in mills operating in semi-autogenous mode at the first stage and ball grinding – at the second stage. According to the test results, the mass fraction of the class - 0.074 mm in the ore sample in the flotation concentrate grinding unit was 95%.

According to mineralogical data, pyrite concentrate is represented by rock components such as quartz (25.1%) and feldspar (16%). The main ore mineral is pyrite, the content of which is 44.5%, there are arsenopyrite (1.5%) and pyrrhotite (1%). Arsenopyrite concentrate is the components of rocks (quartz – 9.5%; feldspar – 2%. carbonate to 2.4%; mica-hydrologist components – 26.3%). The main ore minerals are pyrite (38%) and arsenopyrite (17.8%).

The general scheme of pyrite-arsenopyrite concentrates processing includes the following main processes:

- grinding of flotation gold-containing concentrate in a ball mill;
- acid processing (prior to the removal of the carbonate compounds - the process of decarbonization);
- autoclave oxidative leaching;
- thickening and filtration of oxidized pulp;
- neutralization of the solution with limestone and lime;
- sorption cyanidation of cake.

3. Research methods

Grinding. Fine grinding is necessary to prepare the concentrate for the low-temperature process [5]. In laboratory studies, the required size of the grinding is achieved by using a planetary mill "Pulverisette 6" (Fritsch, Germany). The size of the source and ground material is measured on the laboratory analyzer "Analysette 22" (Fritsch, Germany).

In the study of the size factor, experiments were carried out to grind the sample in two modes:

- conventional (P_{80} = 34-35 microns);
- fine grinding (P_{80} = 8-9 microns).

Grinding was carried out in a planetary ball mill. Loading material, balls and rotation speed were identical for both modes. Time of stay in the mill varied [6].

The size of concentrates at all stages of the experiment is presented in table 1.

Table 1. Size of concentrates at all stages of experiments

Stydy stage	Fineness	P_{80}, microns
The initial concentrate	not ground	47-48
Grinding	conventional	34-35
	finely divided	8-9
Autoclave oxidative leaching	conventional	30-32
	finely divided	8-9

The particle size distribution characteristics were determined by the laser analyzer "Analysette 22" manufactured by the German company Fritsch. This type of analyzer allows reliable measuring the

specified characteristics at the particle sizes in the sample from 0.3 mm to 0.3 microns.
Figures 1 and 2 show the measurement results. The graphs show 2 functions: differential and integral. The differential function reflects the distribution of particles for each class of size, the integral function sums up these indicators.
The crushed material after the particle size measurement by laser granulometry was sent to the decarbonization stage and then to leaching.

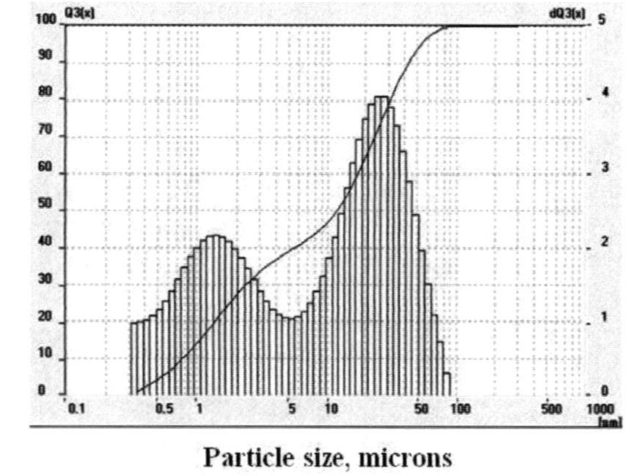

Figure 1. Granulometric composition of fine concentrate (P_{80} = 34-35 mcm).

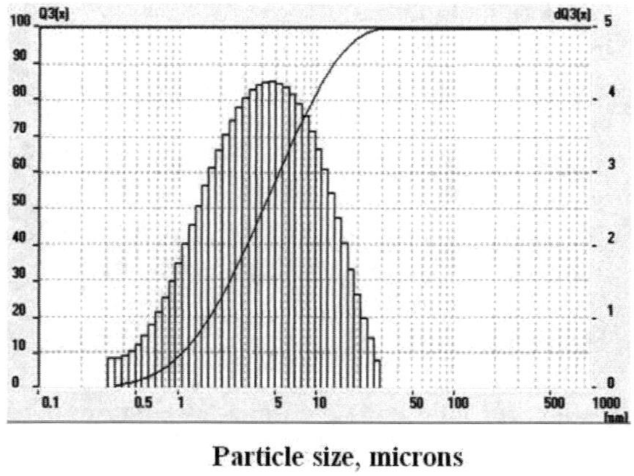

Figure 2. Granulometric composition of the crushed concentrate (P_{80} = 8-9 mcm).

Decarbonization. The raw material supplied to the autoclave leaching, as a rule, contains carbonates. A mandatory stage for persistent gold-containing concentrates is the preliminary removal of carbonate compounds from their content (the decarbonization process). Their presence in the autoclave process leads to the release of gaseous CO_2, reduces the partial pressure of oxygen and to a large extent its use. The process of decomposition of sulfides (pyrite and arsenopyrite) at relatively low parameters differs from the traditional high-temperature leaching in that during the oxidation of sulfide sulfur, part of it is oxidized to sulfate, and part is inevitably transformed into elemental sulfur.
The optimal modes of acid processing are:

- reagent – sulfuric acid (dilution 1:1);
- liquid to solid ratio on operation: 2:1;
- temperature: 60^{0}C;
- processing time 0.5 h,
- pH of the final pulp 2.0 - 2.2;
- acid consumption is ~30 kg/t.

Autoclave leaching. The prepared material (sample of crushed and decarbonized concentrate) is subjected to hydrometallurgical low-temperature processing under the following conditions:

- liquid to solid ratio – 3;
- temperature – 130^{0}C;
- partial pressure of oxygen – 1.0 MPa;
- the speed of rotation of the agitator was selected in such a way as to provide an oxygen absorption rate of 0.5 mol O_2/(l·h·ATM);
- the duration of the autoclave oxidative leaching was estimated by stopping the absorption of oxygen.

The parameters of the pulp after unloading from the autoclave are presented in table 2.

Table 2. Parameters of the pulp after autoclave leaching of low-temperature

Concentrate	Fineness	The degree of oxidation of sulfide sulfur, %	Analysis of the cake, %		
			$As_{overall}$	$Fe_{overall}$	S^0
Pyrite	conventional	90.4	0.2	8.9	7.24
Pyrite	finely divided	99.5	0.2	5.25	7.21
Pyrite-arsenopyrite	finely divided	99.2	5.5	10.0	6.58

4. Conclusion

The conducted studies show that the sample preparation of the material for technological operations plays a decisive role. Thus, the degree of sulfur oxidation in the experiment with a concentration in usual size is not high enough and is about 90%. With such a degree of decomposition of sulfides, a significant part of gold remains in the structure of the initial pyrite, which leads to its loss. Therefore, it is advisable to apply ultra-fine grinding to a class content of 10-12 microns of at least 80% in order to avoid the loss of gold with rock-forming minerals. In addition the important aspect of the autoclave technology is the careful blending of ores in order to average ores over gold and maintain an appropriate level of mineral content - acid generators (sulphides) and its consumers (carbonates).

References

[1] Sizyakov V M, Ivanik S A, Fokina S B 2012 The study of the processes of condensation and filtration of fine oxidized pulps *Obogashenie rud* **2** 24-28.

[2] Naboichenko S S, Nee L P, Shneerson Ya M, Chugaev L V 2002 *Autoclave hydrometallurgy of non-ferrous metals* (Ekaterinburg: GOU UGTU).

[3] Lapin A Yu, Bitkov G A, Shneerson Ya M 2011 Autoclave-hydrometallurgical processing of refractory gold-containing sulfide materials at low temperatures *Non-ferrous metals* **12** 39-44.

[4] Chubarov A V, Belousova N V, Drozdov S V, Maksimenko V V 2010 *Preliminary oxidation of biocaque in the technology of processing sulphide gold ores and concentrates.*

[5] Smirnov V K 1947 *Methods of research of gold ores and concentratesanch bureau tehn.inform*)Moscow: Glavspetstsvetmeta).

[6] Sizyakov V M, Ivanik S A, Boginskaya A S. Bitkov G A 2012 Dehydration of slurries after autoclave leaching of finely ground sulphide concentrates *Natural and Technical Sciences* **1** (57) 369 - 375.

[7] Lapin A Yu, Bitkov G A, Shneerson Ya M 2011 Autoclave-hydrometallurgical processing of refractory gold-containing sulfide materials at low temperatures *Non-ferrous metals* **12** 39-44.

[8] Samikhov Sh R 2006 *Processing technology of hard and poor gold ores* (Dushanbe).

[9] Chekushin V S, Oleynikova N V 2005 Processing of gold-containing ore concentrates *Proceedings of the Chelyabinsk Scientific Center* **4** (30) 94-101.

[10] Strizhko L S 2001 Metallurgy of gold and silver (Moscow: MISIS).

Methods for controlling the vibration state of technical facilities

A I Karlina[1], V E Gozbenko[2, 3], S K Kargapol'tsev[2], Yu I Karlina[1]

[1] Irkutsk National Research Technical University, 83 Lermontov Street, Irkutsk, 664074, Russia
[2] Irkutsk State Transport University, 15 Chernyshevsky Street, Irkutsk, 664074, Russia
[3] Angarsk State Technical University, 60 Chaykovskoy Street, Angarsk, 668535, Russia

E-mail: ksk@irgups.ru

Abstract. The article considers the features of the dynamics of systems with several degrees of freedom with the introduction of additional constraints. As an example, the authors consider the computational scheme of a locomotive traction engine with an axial overhung support. Different cases of the combination of perturbations were analyzed and the conditions for dynamic damping were determined. The specificity of the tasks to be solved was in the necessity to determine the conditions while choosing system parameters that would minimize the occurrence of cross-couplings. These conditions were determined and implemented through the introduction of additional inertial elements into the system structure. With an appropriate choice of system parameters, the vibration platform can be invariant with respect to the deviation to the kinematic effect xj(t). The necessary conditions that can be obtained by parallel inclusion of inertial and elastic elements in systems with low friction were considered.

1. Introduction

Transport technical devices, in terms of solving problems of protection against vibration and shock, are complex objects. The theory and practice of transport dynamics, the protection of machinery, equipment, instruments is considered in a fairly large number of works of domestic and foreign authors. At different times, various aspects of this problem were regarded, related to the refinement of mathematical models, the introduction of new constraints, including the use of external energy sources, the use of automation elements and approaches based on the methods of the theory of automatic control, including direct control using computer technology. From the consideration of individual dynamic phenomena and processes, a definite tendency emerged towards the study of the vibrational states of facilities, the formation and study of vibrational fields and methods of controlling complex dynamic states, which presupposes the further development of systemic methodological and scientific-methodological viewpoints.

The authors consider the features of the dynamics of systems with several degrees of freedom with the introduction of additional constraints. As an example, the authors consider the computational scheme of a locomotive traction engine with an axial overhung support [1, 2]. The basic computational scheme of the drive has the form of a two-dimensional model (Figure 1), which, in turn, can be interpreted by the structural diagrams (Fig. 2 and 3).

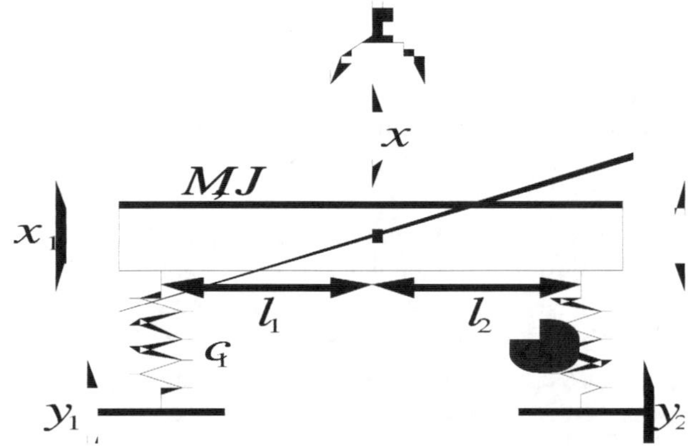

Figure 1. Basic calculated two-dimensional model.

2. Construction of the model

Additional constraints can be introduced by "connecting" the corresponding links with transfer functions *L*1 *p*2 and *L*2 *p*2 that are parallel to links *c*1 and *c*2. Necessary transfer functions with power and kinematic perturbations were subsequently determined. A number of certificates for utility models were obtained for the constructive and engineering implementations of additional constraints.

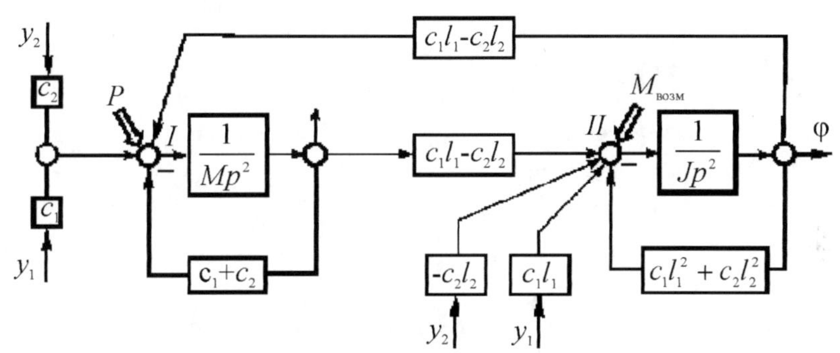

Figure 2. Structural diagram of the basic model.

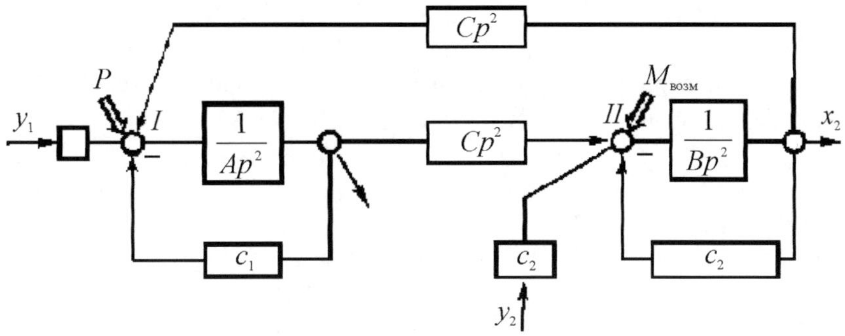

Figure 3. Structural diagram of the calculated model in the coordinate system.

$x1$, $x2$, where

$$A=\frac{Ml_2^2+J}{\left(l^1+l^2\right)^2},\ B=\frac{Ml_1^2+J}{\left(l^1+l^2\right)^2},\ C=\frac{Ml_1l_2-J}{\left(l^1+l^2\right)^2}.$$

Some approaches to the rational designing of an axial overhung support were considered (certificate No. 29504 dated May 20, 2003. Bull. No. 14 dated May 20, 2003), which was associated with the introduction of additional constraints into the interaction scheme. It was shown that the dynamic damping of oscillations significantly depended on the geometric and mass-inertial properties of the systems [3-7]. Frequency values can vary, creating certain zones to achieve oscillation damping effects. The spectrum of possibilities also depended on the nature of the introduction of additional constraints, bearing in mind the combinatorics of their distribution by degrees of mobility. In some cases, modes of dynamic oscillation damping in the coordinates might occur [1, 8].

More complex three-dimensional objects of protection and some practical developments in terms of algorithmic and software research were also considered [1, 9-16].

The problems of spatial vibration isolation and vibration protection were often encountered in practice and were characteristic, in particular, for transport systems, equipment protection, and creation of comfortable conditions for a human operator. The computational scheme of the spatial vibration protection platform is shown in Fig. 4. The dynamic properties of the system were investigated, the expressions for determining partial frequencies and relations determining the conditions for vibrations' decoupling, frequency characteristics were obtained.

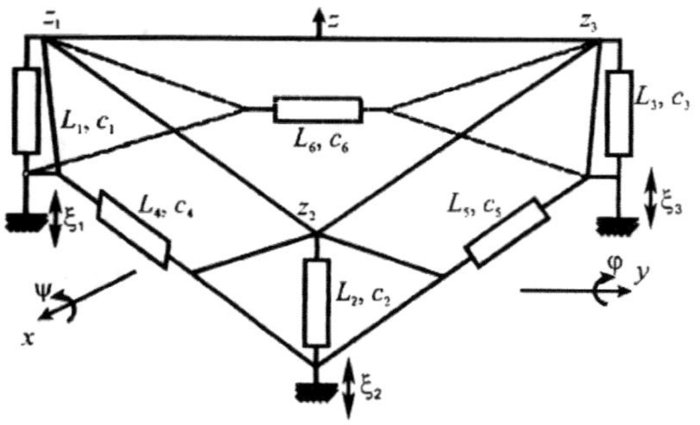

Figure 4. Computational scheme of the vibration protection platform.

The introduction of inertial elements led to the occurrence of additional modes of dynamic oscillation damping during kinematic impacts. It was shown that the conditions for dynamic damping did not depend directly on the inertial characteristics of the protected part of the system, which made it possible to use it as the filter for the protection from the vibrations of facilities with the different mass-inertia parameters. The generalized mathematical model of the system is represented by expression (1), and its structural diagram and the detailed components are shown in Figure 5, in this case:

$$A\ddot{u}+Cu=B\dot{\xi}+C\xi \tag{1}$$

where

$u=(z_1,\ z_2,z_3),\xi=(\xi_1,\xi_2,\xi_3)$; $E_{11}(p)Z_1+E_{12}(p)Z_2+E_{13}(p)Z_3=$
$=F_{11}(p)\Sigma_1+F_{12}(p)\Sigma_2+F_{13}(p)\Sigma_3$; $E_{21}(p)Z_1+E_{22}(p)Z_2+E_{23}(p)Z_3=$
$=F_{21}(p)\Sigma_1+F_{22}(p)\Sigma_2+F_{23}(p)\Sigma_3$; $E_{31}(p)Z_1+E_{32}(p)Z_2+E_{33}(p)Z_3=$
$=F_{31}(p)\Sigma_1+F_{32}(p)\Sigma_2+F_{33}(p)\Sigma_3$;

$E_{ij}(p)=a_{ij}(p)^2+c_{ij}$, $F_{ij}(p)=b_{ij}(p)^2+c_{ij}$, ($i$, j=1, 2, 3) are real polynomials, Z1, Z2, Z3 is the

Laplacian image of generalized coordinates, Σ1, Σ2, Σ3 is the Laplacian image of perturbations.

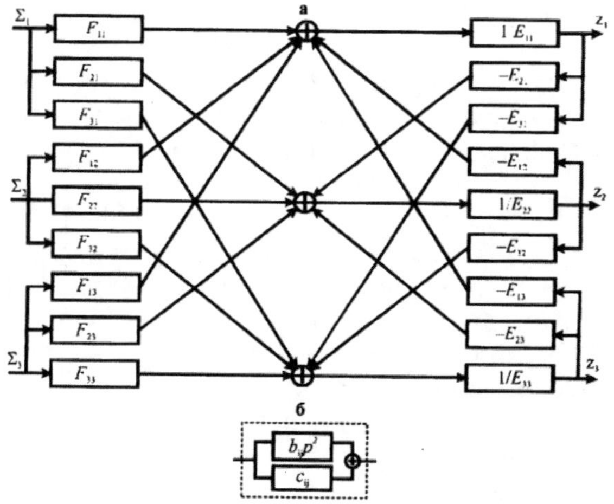

Figure 5. Generalized structural diagram (*a*) and its detailed components (*b*).

Let us note that the terms $a_{ij}p^2, b_{ij}p^2$ reflect the introduction of the additional constraints, which create the extra forces, which depend on the accelerations of relative motion.

In certain cases computational scheme can be represented by different versions. Using conditions of symmetry or proximity to them, it is possible to simplify the initial model and to determine the basic parameters of dynamic state. The expressions (2) and (3) are used for determining the partial frequencies of the system from the coordinates $x1$, $x2$ and x and φ :

with the generalized coordinates $x1, x2$

$$n_1 = \sqrt{c_1 / \left[\left(ml_2^2 + J\right) / \left(\left(l_1 + l_2\right)^2 + L_1\right)\right]}, \quad n_2 = \sqrt{c_2 / \left[\left(ml_1^2 + J\right) / \left(\left(l_1 + l_2\right)^2 + L_2\right)\right]} \qquad (2)$$

with the coordinates x and φ

$$n_1 = \sqrt{\left(c_1 + c_2\right) / \left(m + L_1 + L_2\right)}, \quad n_2 = \sqrt{\left(c_1 l_1^2 + c_2 l_2^2\right) / \left(J + L_1 l_2^2 + L_2 l_1^2\right)} \qquad (3)$$

The change in the values of partial frequencies depends on reduced mass. It is possible to note that an increase in the reduced mass leads to noticeable reduction in the natural and partial vibration frequencies. However, the authors also encountered special features: system "cutoff" effects were observed.

3. Conclusion

Different cases of the combination of perturbations were analyzed and the conditions for dynamic damping were determined. The specificity of the tasks to be solved was in the necessity to determine the conditions while choosing system parameters that would minimize the occurrence of cross-couplings. These conditions were determined and implemented through the introduction of additional inertial elements into the system structure. With an appropriate choice of system parameters, the vibration platform can be invariant with respect to the deviation to the kinematic effect *xj(t)*. The necessary conditions that can be obtained by parallel inclusion of inertial and elastic elements in systems with low friction are considered.

References

[1] Karlina A I, Gozbenko V E 2016 *Bulletin Irkutsk State Technical University* T. 20 **10** (117) pp 35-47

[2] Tsisovski T 2001 Improving the control systems of oscillations of the rolling stock of railways: a dissertation for the degree of Doctor of Technical Sciences: 05.22.07. (Moscow: Russian State Open Technical Transport University)

[3] Balanovskii A E 2016 *High Temperature* T. 54 **5** pp 627-631

[4] Balanovskiy A E 2017 *Welding International* T. 31 **6** pp 467-476

[5] Balanovskii A E, Huy V V 2017 *Letters on Materials* **7(2)** pp 175-179

[6] Balanovskii A E, Van Huy V 2018 *Journal of Friction and Wear* **39(4)** pp 311-318

[7] Medvedev S I, Nezhivlyak A E, Grechneva M V, Balanovsky A E, Ivakin V L *2015 Welding International* **29(8)** pp 643-649

[8] Balanovskii A E 2018 *High Temperature* **56(4)** pp 486-495

[9] Nikolaev V A 2003 Development of methods for analytical design of quasi-invariant systems of spring suspension of railway carriages (Omsk: Omsk State Transport University) 371 p

[10] Alekseeva L B 2018 IOP Conference Series: Earth and Environmental Science **194(2)** 022001

[11] Ershov D Y, Zlotnikov E G, Koboyankwe L E 2017 *IOP Conference Series: Earth and Environmental Science* **87(8)** 082015

[12] Zhukovskiy Y L, Korolev N A, Babanova I S, Boikov A V 2017 *IOP Conference Series: Earth and Environmental Science* **87(3)** 032055

[13] Nguyen V X, Golikov N S 2018 *Journal of Physics: Conference Series* **1015(5)** 052020

[14] Verigo M F 1988 *Dynamics of cars. Lecure notes* (Moscow: Transport Publ.) p 174

[15] Vershinsky S V, Danilov V N, Khusidov V D 1991 *Dynamics of the rail car* (Moscow: Transport Publ.) p 360

[16] Gozbenko V E, Kargapoltsev S K, Kornilov D N, Minaev N V, Karlina A I 2017 *International Journal of Applied Engineering Research* vol 12**(23)** pp 13773-13778

Method of indirect estimation of lubricating properties of petroleum products

N K Kondrasheva[1], O V Zyryanova[1], A S Ivanov[1], S M Kosolapova[1] and E V Kireeva[1]

[1]Saint Petersburg Mining University, 21 Line, Saint Petersburg, Russia, 199106

E-mail: zyryanova_olga@rambler.ru

Abstract. In this research, a new method for evaluating the lubricating properties of prophylactic agents was developed. For the method, specialized metal models and a press with a strain-gauge s-shaped gauge, using for recording shear load indicators, were designed. Using the new method, prophylactic agents based on mixtures of light and heavy gas-oil catalytic cracking and delayed coking with the addition of tar or cracking residue as depressor additives were analyzed. A research of the hydrocarbon composition of the prophylactic agent, which was correlated with the removed shear load diagrams, was also made. As a result, the dependencies of the lubricating properties of fuels with a change in their fractional composition were identified, and the most effective PAs were determined.

1. Introduction

When moistened minerals and loose materials are transported, they adhere and freeze to the metal surfaces of the dump cars and car bodies, and the pieces of material regelate. This leads to a significant increase in labor costs when unloading cars and incomplete unloading of vehicles, which leads to significant economic losses. Thus, preventing of sticking, adfreezing and regelating of loose and lumpy materials and rocks is an important task [1-3].

To solve these problems, prophylactic agents (PAs) were developed. These agents were classified into bulk, liquid and solid. The most widespread currently received liquid PAs. Their effect was based on the formation of a hydrophobic film on the surfaces of particles of materials and metal surfaces, which led to the displacement of water in a separate phase, as a result of which it formed a weak ice [4,5].

There are 4 PAs groups: oligoorganosiloxanes (silicone fluids), PAs based on hydroxyl-containing compounds, salts of inorganic and organic acids, oil-based PAs.

Methyl and ethyl siloxanes are most common in group of organosilicon fluids. They satisfy all physico-chemical and operational indicators, but have a high cost.

Hydroxyl-containing compounds are represented by products based on glycerol, alcohols, glycols and their esters, as well as bottoms from their production. In terms of environmental safety, the use of propylene glycol is most preferred.

To improve the performance properties of glycols, it is possible to add in them small amounts of inorganic salts. In addition the solutions of inorganic salts can be used as independent PAs. The most common use of a solution of $CaCl_2$, in the concentration range of 25-30% of its freezing temperature reaches -40°C. When transporting materials in conditions up to -10°C, other salts can be used: NaCl, K_2CO_3, $NaNO_3$, Na_2SO_4, and also $CO(NH_2)_2$ [6].

In the climatic conditions of the continental zones, it is preferable to use petroleum-based products, since they are able to withstand temperatures below -20°C.

The main requirements for PAs, relate to their physico-chemical and operational properties, as well as efficiency of use. First of all, they must effectively prevent adfreezing and regelation in conditions of low temperatures, be non-toxic, do not cause environmental damage, and also have multifunctional use (for various materials, minerals and ores). In addition, they must have a satisfactory viscosity for ease of spraying through the nozzles, do not have a corrosive effect and a negative impact on the materials transported [7]. Thus, the most important characteristic is the lubricity.

To assess the lubricating properties of fuels, oils and other petroleum products, the indicator is used - the wear scar diameter, which is measured using a four-ball wear test machine. With the help of this tribological analyzer, it is able to estimate the ability of substances to reduce the loads arising in the process of friction. However, to evaluate the lubricating properties of PAs, the main task of which is to prevent the adfreezing of materials to metal surfaces, this indicator is not sufficiently informative. This is due to the fact that, in addition to the lubricating properties, they must have a low pour point, a high adhesion index and an optimal chemical composition [8].

Given the diversity of PAs, it is necessary to develop a method that will allow evaluating their effectiveness, regardless of their state of aggregation or chemical composition.

Since PAs are most often used for transportation of various minerals (coal, ores, sludge), it is proposed in the methodology to simulate the conditions of transportation at low temperatures. The main cause of sticking and freezing of loose materials to the walls of cargo compartments is water contained in the material itself. During transportation, water accumulates in the lower part of the car, forming a layer of liquid phase between the car wall and the material. The strength of the freezing of the material is influenced by the chemical composition of solid phases (the presence of oxygen atoms in the structural lattice, which causes the formation of hydrogen bonds with water molecules) and the thickness of the water layer [9,10].

2. Materials and Equipment

Further, a new method for evaluating of lubricating properties of PAs was developed. For testing, a metal (steel grade St3) car model was developed and manufactured (Fig. 1). The gate for unloading was represented by a metal plate, which was subjected to processing by PA.

Figure 1. Experimental model

Figure 2. Press

The models were placed in the freezer for 24 hours. As a result of the ongoing processes of adhesion and crystallization of water on the surface of the metal and the material was frozen. Since the task of the PA is to prevent these phenomena, an assessment of their effectiveness can be carried out by extruding the plate. When the plate is shifted, the ice layer is destroyed and the material comes off the wall. Thus, it is necessary to extrude the plate and measure the force applied to shift it. This load characterizes the strength of freezing and allows indirect evaluating the lubricity of PA.

To measure the forces, a press with a DYLY-104/300 tensile-compression strain gauge was used (Fig. 2).

Initially, the sensor was connected to a laptop using the Arduino Uno programmer and the data was displayed on the screen in the Arduino program. Thus, the press required additional equipment. Then a display was connected to the programmer, as a result of which the work of the press became completely autonomous.

3. Analysis of Results and Experimental Data

Before evaluating the lubricating properties of PA, the humidity of the transported materials was measured, since it has a great influence on the degree of adfreezing and freezing [11, 12, 13, 14]. For coal, depending on whether it is pit or brown, this parameter will vary widely from 10% to 55%. The internal humidity, that is, the water molecules that make up the lattice of crystallohydrates of the mineral part, is not taken into account, since it does not affect the freezing.

In order to analyze the effect of humidity content of the tested materials on the freezing strength, a series of tests were carried out according to the previously described procedure with quartz sand (95–99.8% silicon oxide SiO_2), pit coal (86% carbon; 0.5% sulfur; 9.2% volatile matter; ash 4.3%) and urtite (ore for the production of alum slate and a number of by-products (soda, cement), containing metal oxides SiO_2 (from 40% by weight), Al_2O_3 (from 20%), Na_2O (from 10 %), as well as Fe_2O_3 + FeO, MgO, CaO, K_2O, TiO_2) [15].

The purpose of the test was to build a graphic dependence on the freezing strength of the humidity of the material. The experiments with urtite did not give a positive result, since the crushed processed material with a particle size of <40 microns participated in the tests. In the course of the experiment, it was impossible to obtain a homogeneous mixture; the applied layer of urtite was pelletized and irregularly froze. The tests with coal and quartz sand were successful, since the sizes of particles from mixed fractions were in the range of 100 μm ÷ 1 mm. The test results are shown in Figures 3 and 4.

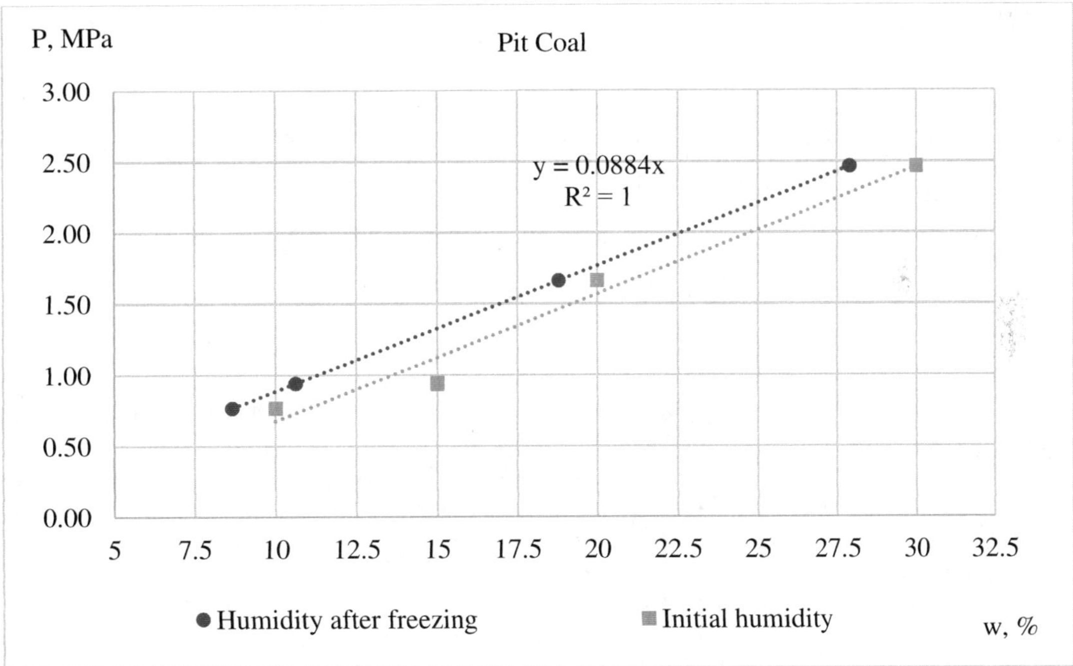

Figure 3. Dependence of shear load on coal moisture.

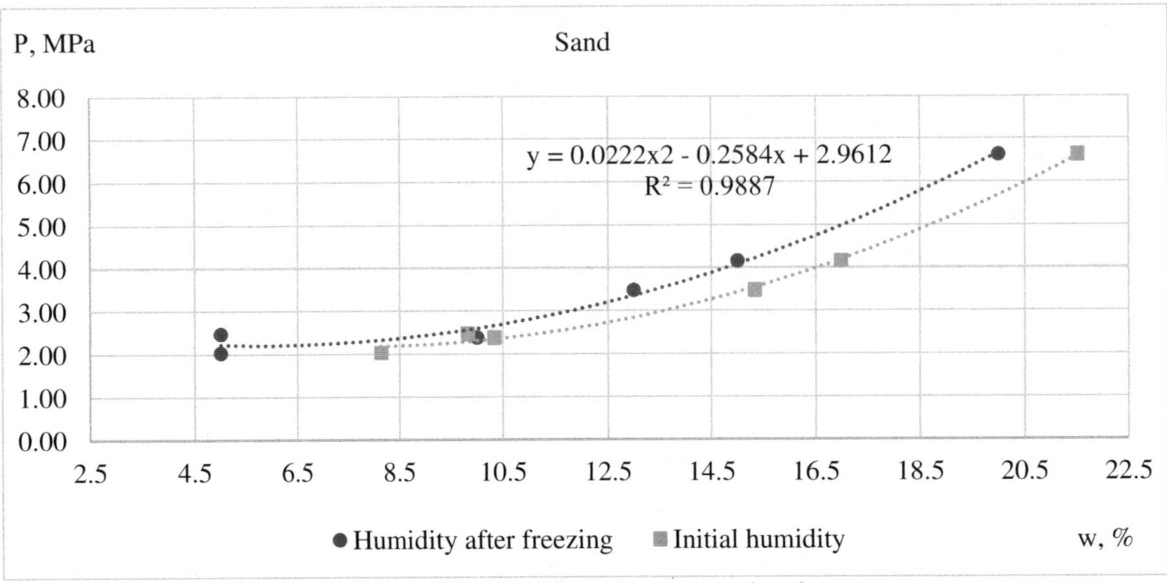

Figure 4. Dependence of shear load on sand moisture.

In the process of freezing, the humidity content of the material changed. In the study, two series of values were obtained: the initial humidity before testing (Fig. 4, black branches) and the humidity found at the end of the test after measuring the freezing strength (Fig. 4, grey branches). To measure the humidity after freezing, two weights of the material were taken from both sides of the model gate. The arithmetic average was taken as the true value. Taking into account the change in the moisture content of the material in the freezing process, more accurate graphical dependencies were obtained and empirical equations were derived using approximation.

For coal and quartz sand, linear and quadratic dependencies were derived:

$$P_{coal} = 0.0884w\,; \tag{1}$$

$$P_{sand} = 0.0222w^2 - 0.2584w + 2.9612\,; \tag{2}$$

where P_{coal} and P_{sand} - the strength of freezing of coal and sand, MPa; w - moisture content of the material, % wt.

The result obtained is explained by the presence of oxides in the structure of building sand, mainly silicon oxide (Fig. 5). Hydrogen bonds are formed between the water molecules and the oxide centers of silicon oxide in the sand and iron oxide in the steel wall of the car. Coal is almost pure carbon, so this phenomenon is not typical for it.

Based on practical data, the following humidity was taken: 15% for sand and 20% for coal.

Using a new method, PA based on light and heavy gas oil catalytic cracking and delayed coking (LGCC, HGCC, LGDC, HGDC) with the addition of tar (T) or cracking residue (CR) as depressor additives were analyzed. As a result of the tests, the following shear load values were obtained in kilograms, which were then converted into units of pressure (see Table 1).

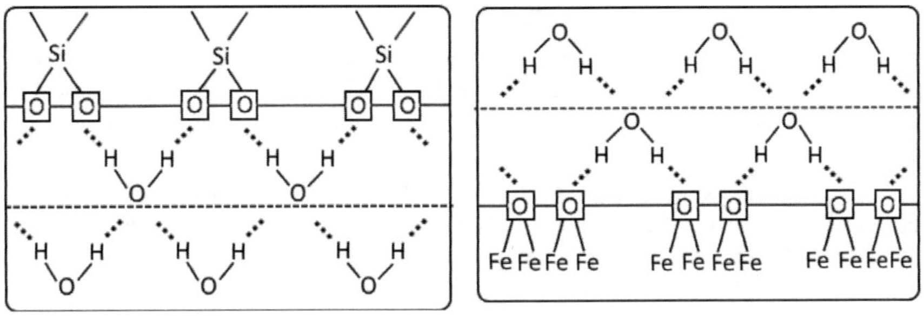

Figure 5. Hydrogen bonds in the phase section of a Sol-L-Sol.

Table 1. Dependence of the load for shear on the composition of the lubricant.

Prophylactic agent	Shear load, MPa	
	Coal with 20% moisture	Sand with 15% moisture
Without agent	3.16	7.95
Blend HGCC:LGCC+0%T	0.94	1.77
Blend HGCC:LGCC+2%T	1.46	0.88
Blend HGCC:LGCC+5%T	0.65	0.17
Blend HGCC:LGCC+10%T	1.17	5.0
Blend HGDC:LGDC+0%RC	1.72	3.84
Blend HGDC:LGDC+2%RC	0.25	5.03
Blend HGDC:LGDC+5%RC	1.24	2.46
Blend HGDC:LGDC+10%RC	1.31	5.24

4. Conclusions

According to the table, it is possible to determine the optimal PA compositions, using which a minimum shear load of the plate is required for a model filled with moistened sand or coal. The new method allows successful conducting a comparative assessment of the lubricating properties of prophylactic agents. It is also easier to use it than using a four-ball wear test machine. The method showed that mixtures based on vacuum gas-oils with the addition of heavy residues can be used as PA. To prevent sticking, adfreezing and regelation of wet loose materials, it is recommended to use preventive lubricants based on catalytic cracking gas-oil and delayed coking with a cracking residue content of 2%, or tar 5%.

In addition, the production of these mixtures will expand the range of commercial petroleum products produced at refineries, and will improve the efficiency of refining through the rational use of heavy oil refining residues.

References

The work was carried out within the framework of the state assignments 0792-2018-0002 «Rational use and deep processing of raw hydrocarbon to produce marine fuels and carbon materials».

References

[1] Oshchepkov I A and Khudonosova Z A 1999 Prevention of freezing of overburden rocks to the walls of vehicles *Mining information and analytical bulletin* **3** 75-76

[2] Koulyandin G A, Omelyanenko A V and Fedorova L L 2013 Georadar researches of structural features and water cutting of overburden rocks for coal open-cast mining *Journal of Mining Institute* **200** 49-53

[3] Gushchin A A, Ermakov A Yu and Miroshnikov A M 2016 The study of coal congealing during storage and transportation *Mining information and analytical bulletin* **4** 140-154

[4] Kondrasheva N K, Zyryanova O V, Kireeva E V and Ivkin A S 2016 Refinery byproducts in dust suppression and the prevention of rock adhesion and freezing at mines *Coke and Chemistry* **59** 338-344

[5] Korshunov G I, Kovshov V P, Kovshov S V and Yerzin A H 2014 A new chemical method of dust suppression for the storage of the rock mass *Journal of Mining Institute* **207** 116-120

[6] Gushchin A A, Ermakov A Yu and Miroshnikov A M 2017 Analyses of crystallization and consolidation of aqueous reagent solutions to prevent congelation *Mining information and analytical bulletin* **4** 403-8

[7] Vasilyev A L, Lavrenko S L and Gruszczynski M 2018 Experimental studies of rheological properties of stowing pulps *Journal of Physics: Conference Series* **1118.**

[8] Dashko R E and Lange I Y 2017 Engineering-geological aspects of negative effects assotiated with contamination of disperse soils by oil products *International Multidisciplinary Scientific GeoConference Surveying Geology and Mining Ecology Management* **17** 617-625

[9] Goldstein R V and Epifanov V P 2011 To the measurement of adhesion of ice to other materials *Bulletin of the Perm National Research Polytechnic University. Mechanics* **2** 28-41

[10] Matsumoto K and Kobayashi T 2007 Fundamental study on adhesion of ice to cooling solid surface *International Journal of Refrigeration* **30** 851–860.

[11] Adam J, Meuler J, Smith D, Varanasi K K, Mabry J M, McKinley G H and Cohen R E 2010 Relationships between Water Wettability and Ice Adhesion *ACS Applled materials & interfaces. Americal Chemical Society*

[12] Cai CZ, Li GS, Huang ZW, Shen ZH and Tian SC 2014 Rock pore structure damage due to freeze during liquid nitrogen fracturing *Arabian journal for science and engineering* **39** 9249-9257

[13] Qin L, Zhai C, Xu JZ, Liu SM, Zhong C and Yu GQ 2019 Evolution of the pore structure in coal subjected to freeze - thaw using liquid nitrogen to enhance coalbed methane extraction *Journal of petroleum science and engineering* **175** 129-139

[14] Shakhnazarov K Y, Prykhin E I and Pomeranets I B New version of Kurnakov's law about relationship between properties of binary alloys and phase diagrams balance 2017 *IOP Conference Series: Earth and Environmental Science* **87**

[15] Davidson Gp and Nye Jf 1985 A photoelastic study of ice pressure in rock cracks *Cold regions science and technology* **11** 141-153

Approximation of testing of asphalt concrete to the actual operating conditions of road surfaces

N S Kovalev[1], E N Otarova[2], VV Gladnev[1], E A oglu Sadigov[1] and P V Demidov[1]

[1]Voronezh State Agrarian University named after Emperor Peter the Great, Voronezh, Russia
[2]Air Force Academy named after professor N.E. Zhukovsky and Y.A. Gagarina, Voronezh, Russia

E-mail: NSKovalev@mail.ru, ekaterina.otarova@mail.ru, lav-78@mail.ru, elzas.sadygov@mail.ru, 79204170254@yandex.ru

Abstract. Existing test methods for asphalt concrete do not consider the actual operating conditions of road surfaces. The method, equipment and results of studies of the influence of the combined effects of vibration vacuum water saturation, ultraviolet irradiation and freezing-thawing on the structural and mechanical properties of asphalt concrete from slag materials are presented.

1. Introduction
Currently, among the improved road surfaces, asphalt concrete is the most widely used, creating maximum convenience for the movement of vehicles. These coatings are used on roads of any cargo intensity.

Asphalt concrete is one of the most complex artificial conglomerates, which is mainly due to the peculiarity of its structure, as well as the great dependence of its properties on many external factors.

The existing GOST 9128-2013 on asphalt mixes and asphalt concrete provides for the determination of compressive strength at temperatures of 20, 50, 0 °C, water saturation of samples, water resistance and water resistance during prolonged water saturation, shear resistance (by the coefficient of internal friction and adhesion during shear), crack resistance by tensile strength when split at a temperature of 0 ° C. Laboratory methods for assessing the quality indicators of asphalt concrete do not reflect the conditions of its operation in the coating, in addition, the characteristic is asphalt concrete TV is given according to separate independent indicators, and the complex influence of operational and climatic factors on the properties of asphalt concrete is practically not studied [7].

The practice of road construction shows that the use of various asphalt mixes, even fully meeting the requirements of domestic state standards and other regulatory and technical documents, does not always allow us to provide the necessary overhaul life for road asphalt concrete [2].

In recent years, many studies have appeared on the impact of operational and climatic influences on changing the properties of asphalt concrete. A significant part of these studies is devoted to modeling external influences on asphalt concrete in order to predict its operational reliability [4, 5, 8, 15].

Known methods for modeling operational and climatic effects on asphalt concrete are based on a differential approach to assessing the stability of its structure according to the results of the influence of individual factors [17, 18]. However, such an assessment does not allow to consider the simultaneous influence of a set of operational climatic and recipe factors on asphalt concrete, as is the case in actual operating conditions of road surfaces.

DIN EN 13108-1: 2006-08 also does not provide for the integrated impact of factors on asphalt concrete [16].

Thus, it would be advisable to consider the possibility of testing the properties of asphalt concrete according to the methods that most simulate the operating conditions of asphalt concrete in the upper layers of coatings.

The appropriateness of the complex impact method is based on the opinion of a number of researchers [9, 10] about the insufficiency of assessing the influence of individual factors on the change in the properties of asphalt concrete when predicting its service life and on the need to determine the patterns of change in the structure of asphalt concrete under the influence of a complex of factors.

2. Methods and materials

To approximate the nature of testing of asphalt concrete to the conditions of actual operation in road surfaces and to reduce their duration, the most appropriate test cycle seems to be the sequential accumulation of the results of the combined impact of operational and climatic factors in the same samples of asphalt concrete. We have proposed and developed a method for preparing samples of building materials for strength tests [14]. The essence of the method lies in the fact that a complex of factors acts on the same asphalt concrete sample.

For the mathematical description of the influence of a complex of factors in the study of the physical and mechanical properties of asphalt concrete, the method of mathematical planning of extreme experiments (MMPEE) was used, the essence of which is to obtain maximum information with a minimum number of experiments [3].

This method differs from multivariate regression analysis in that the researcher plans his strategy in advance, actively intervenes in the experiment, and is not a passive observer.

1) The mathematical model of the influence of a complex of factors is represented in the form of a polynomial:

$$Y = b_0 + \sum_{i=1}^{i=n} b_i x_i + \sum_{i=1}^{i=n} b_{ij} x_i x_j + \sum_{i=1}^{i=n} b_{ii} x_i^2$$

where b_o – is a free member;

b_i – is a linear coefficient;

b_{ij} – is the interaction coefficient;

b_{ii} – is the quadratic coefficient.

2) Before planning, factors are scaled to obtain the equation of a polynomial of the second degree with respect to dimensionless quantities x_i.

The transition from physical variables X_i to encoded x_i is carried out according to the formula:

$$x_i = \frac{X_i - X_i^0}{\lambda_i},$$

where $X^0 = \dfrac{X_i^в + X_i^н}{2}$ – is the middle of the range of variation of the i factor;

$\lambda = \dfrac{X_i^в - X_i^н}{2}$ – is the step of variation i factor;

$X_i^в$ and $X_i^н$ – are the upper and lower levels of variation of the i factor respectively.

As a result of encoding variables, the upper level will correspond to the value “+1”, the zero level to “0”, and the lower level to “-1”.

3) After the implementation of the experiments, the uniformity of the experiments at each point is checked by the Student t-test and the sample variances are checked according to the χ – Cochren criterion.

4) Check the significance of the coefficients by t-student test.

5) The model is checked for adequacy by the Fisher F-test.

The use of MMPEE is widespread in materials science in the selection of optimal compositions of artificial conglomerates. In our case, this method was used to study the influence of external factors on asphalt concrete [11, 12].

Ordinary asphalt concrete is destroyed by prolonged and periodic wetting, as well as by alternating freezing-thawing in a water-saturated state [1].

To model the complex of influences of external factors from all of their diversity, we took three as an example: vibrovacuum water saturation, cyclic freezing-thawing, and irradiation with artificial sunlight at temperatures up to +60 °C and periodic irrigation with water in an atmosphere containing ozone [11, 12]. We applied a 17-point three-level plan of the second order with the number of factors $k = 3$ ($N = N_1 + N_\alpha + N_0$) [3].

Hyprodornia studies have proved that the mechanism of accelerated fatigue failure of water-saturated asphalt concrete during the operation of the coating in the cyclic dynamic loading mode is caused to a certain extent by the appearance of pulsed hydrodynamic pressures in water-saturated pores [1].

Introducing the factor of vibrovacuum water saturation into the complex of actions, it was taken into account that when the car moves, depending on the speed and evenness of the coating, the frequency of vibrations transmitted by the car to the coating is from 10 to 25 Hz, and the water is filtered in the pores of the coating mainly under the influence of vacuum created by tires moving on the surface of vehicles.

The vibrovacuum water saturation unit consists of a steel cylinder 10-liter capacity, mounted on a UV-70/100 type vibration stands and an MP-10 vacuum pump.

Pre-weighed and measured samples of asphalt concrete were placed in a vacuum chamber, the inner surface of the walls of which had a rubber coating, filled it with water, and then using a vacuum pump in the chamber a residual pressure of the order of (0.133–0.199)·10 N/m^2 was generated and the vibration test rig was turned on (Fig. 1).

To simulate the effect of alternating temperatures on asphalt concrete, an apparatus was used for accelerated freezing and thawing of samples (Fig. 2).

When developing the methods, it was assumed that a high cooling and thawing rate increases the destructive effect of the influence of alternating temperatures on asphalt concrete. The installation was created based on the Gronland T 25/01 low-temperature cabinet, supplemented by an electronic device that provides automatic cooling and thawing in the temperature range from –20 to +5 ºC, and a potentiometer for graphically monitoring the magnitude and speed of temperature transitions. Temperature control was carried out by a thermocouple placed inside the control samples. The cooling rate was about 30 ºC, and the thawing rate was about 60 ºC.

To simulate the processes of thermooxidative aging of bitumen in asphalt concrete, an IP-1-3 artificial weather apparatus was used (Fig. 3), designed for accelerated testing of polymeric materials.

It is known that the most dangerous for polymeric materials is ultraviolet radiation (wavelength less than 400 nm). According to [6, 13], the share of UV radiation reaching the earth's surface accounts for no more than 3–4 % of the total flux of total solar radiation. Moreover, the share of the most effective UV radiation accounts for only 0.1 % of the total radiation. The amount of total solar radiation per year for the central regions of the European part of Russia is 2360 MJ / m^2 [13], therefore, the share of UV radiation is 2360x 0.001 = 2.36 MJ / m^2. The average value of the light energy incident on a unit area in the ultraviolet region in an artificial weather apparatus with two mercury-quartz lamps of the PRK-2 type for 1 h is 0.67 MJ / m^2.

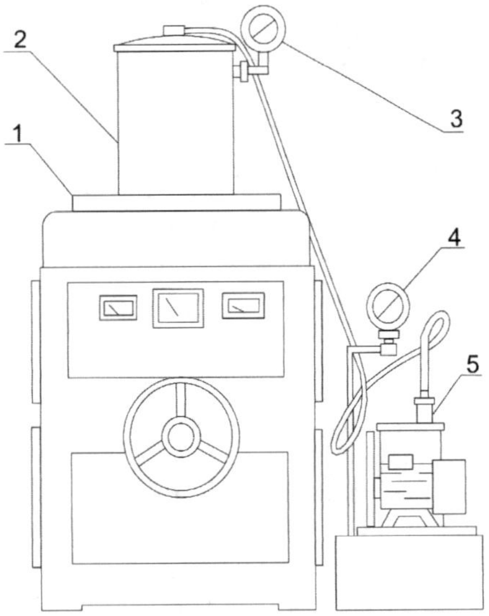

Figure 1. Installation of vibro-vacuum water saturation of samples of asphalt concrete: 1 – plate vibrostand; 2 – a vacuum chamber; 3, 4 – vacuum gauges; 5 – vacuum pump

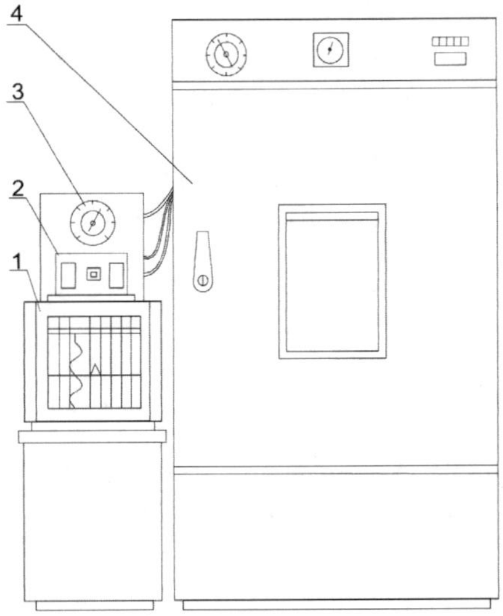

Figure 2. Installation of cyclic freezing-thawing in automatic mode: 1 – potentiometer KSP-4; 2 – automatic control unit; 3 – thermocouple switch; 4 – low-temperature cabinet "Gronland T 25/01"

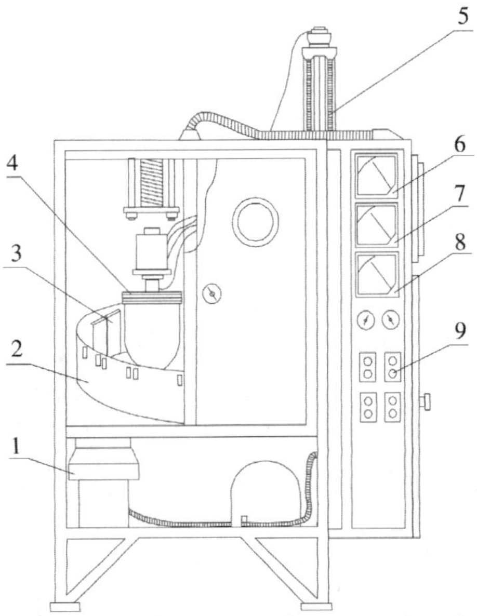

Figure 3. Artificial weather apparatus IP-1-3: 1 – fan; 2 – a rotating drum for attaching samples; 3 – sample; 4 – arc lamp; 5 – moisture meter; 6.7 – voltmeters; 8 – ammeter; 9 – start buttons

Asphalt concrete on slag materials has specific properties that are not inherent in asphalt concrete on natural stone materials. Slag materials have hydraulic binders, have a porous structure, their particle size distribution during technological operations (drying, mixing with bitumen, laying and compaction) is optimized. As a result of this, the processes of structure formation in this type of asphalt concrete occur for a long time from the moment of paving. The presence of a hydraulically active mineral component leads to the formation of a coagulation-condensation structure hardened by crystallization intergrowths at the contact points of the slag grains. This process is enhanced with prolonged water saturation [12].

We studied asphalt composition: slag sand of the Novolipetsk Metallurgical Plant – 100 %, bitumen grade BND 60/90 – 9 %. The output parameter was the adhesion at a temperature of +20 °C.

Planning conditions are given in table 1. The parameters for modeling external factors for the highway 3 of the technical category located in the Voronezh region are substantiated in [11].

Table 1. Conditions for planning an experiment for studying structural and mechanical properties of asphalt concrete from slag materials

Factors	The physical meaning of variables			Coded value of variables		
Условия	X_1 – duration of vibrational water saturation, h	X_2 – number of accelerated freeze-thaw cycles	X_3 – duration of artificial exposure, h	x_1	x_2	x_3
Upper level $X_i^в$	60	100	60	+1	+1	+1
Lower level $X_i^н$	20	24	40	-1	-1	-1
Main level X_i^0	40	62	50	0	0	0
Variation step λi	20	38	10	-	-	-

The transition from physical to coded variables is carried out according to the formulas:

$$x_1 = \frac{X_1 - 40}{20},\ x_2 = \frac{X_2 - 62}{38},\ x_3 = \frac{X_1 - 50}{10}$$

3. Results

As a result of the experiments and calculation of the planning matrix, a mathematical model was obtained, and graphs were constructed in different coordinate systems (Fig. 4):

$$C = (6{,}35 + 0{,}42\,x_1 + 0{,}55\,x_1 x_3 + 0{,}35\,x_1^2 + 0{,}89\,x_2^2 + 1{,}38\,x_3^2)\ 10^{-1}, \text{ MPa}$$

Analysis of the mathematical model and Figure 4 constructed from it made it possible to determine the influence of the studied factors on adhesion at a temperature of 20 ° C:

- of the coefficients of the variables of the first degree, only the coefficient was found to be significant for the duration of the vibrovacuum water saturation, (x1), which increases the tensile strength at split;
- the coefficients for the paired variables x1 · x2 and x3 · x2 turned out to be insignificant;
- coefficients with quadratic variables increase adhesion (the “+” sign): ultraviolet irradiation has the greatest impact, then freeze-thaw and vibro-vacuum water saturation.

The influence of factors is more clearly seen in the analysis of the data shown in Figure 4. With an increase in the duration of vibro-vacuum water saturation up to 40 h, the adhesion decreases and then increases. With an increase in the duration of ultraviolet irradiation to 50 hours, adhesion also decreases, and then an increase also occurs. With an increase in the number of accelerated freeze-thaw cycles to 62, a decrease occurs, and then – an increase in adhesion.

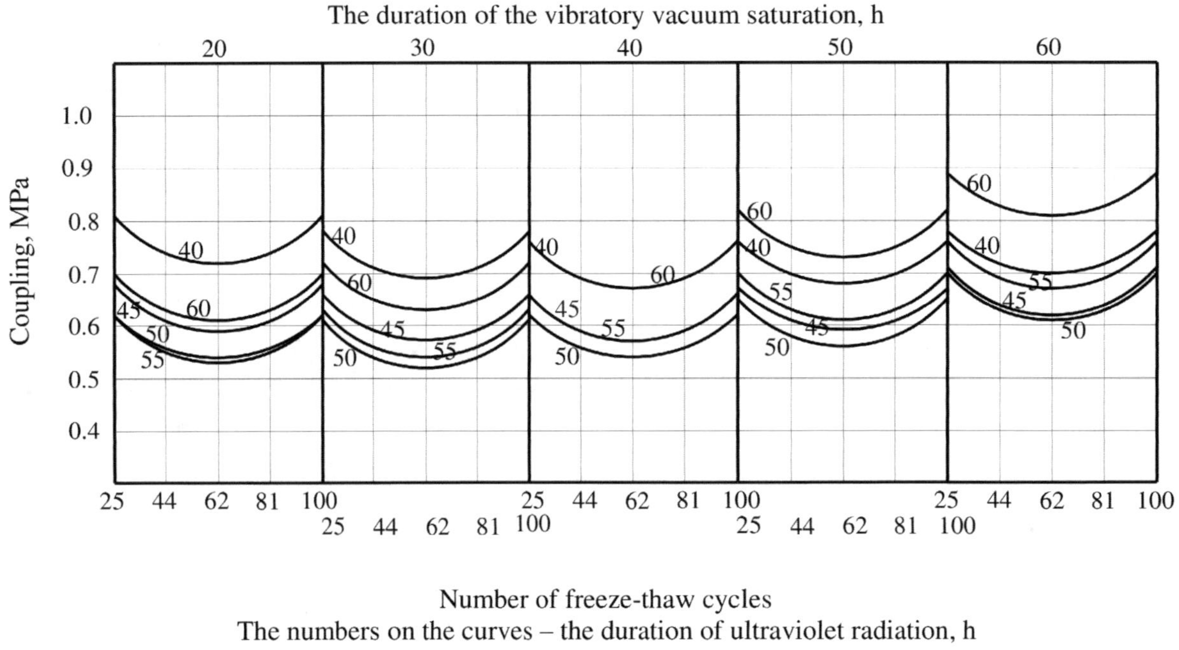

Figure 4. The influence of the combined effects of vibration vacuum water saturation, ultraviolet irradiation and accelerated freezing-thawing on adhesion at a temperature of 20 °C of asphalt concrete composition: slag sand – 100 %, bitumen grade BND 60/90 – 9 %.

Thus, under the influence of all factors, destructive processes occur in the initial stages of the test, and then constructive ones.

The results can be explained by the fact that slag sand is a hydraulically active mineral material. When exposed to water and UV irradiation at a temperature of + 60 °C, micro-steaming chambers arise [10], hydrolysis and hydration processes in slag sand are accelerated, and the coagulation structure of asphalt concrete is strengthened by crystallization intergrowths at the contact points of slag grains. Crystal hydrates continue to form under the influence of freezing and thawing [12].

4. Conclusion

1. A method and apparatus are proposed that allow modeling the complex effect of external factors on the same samples of asphalt concrete.

2. The presented modeling method will allow to more objectively characterize the work of the material of the road surface and choose the most optimal composition of asphalt concrete for specific areas of construction and operating conditions of the road.

3. Using a mathematical method for planning extreme experiments, a mathematical model has been developed that allows predicting the behavior of asphalt concrete from slag materials for operating conditions of a highway of the third technical category in the Voronezh region.

References

[1] Rudensky A V and Gorshkov I M 1973 Study of the water resistance of bitumen-mineral materials *Proceedings of Hyprodornia* P 47

[2] Styagov A S 2015 Analysis of foreign testing methods for asphalt concrete of the upper layers of the coating *Izvestia Rostov State University of Civil Engineer.* **2(20)** P 208–214

Journal of Physics: Conference Series **1384** (2019) 012021 doi:10.1088/1742-6596/1384/1/012021

[3] Nalimov V V and Chernova N A 1965 *Statistical methods for planning extreme experiments* (Moscow: Nauka) p 341
[4] Bratchun V I, Gritsuk Yu V, Bespalov V L, Pakter M K and Gubar A V 2014 Asphalt Slag Concrete on Anionic Bitumen Emulsions *Sci. and Technol. in the Road Industry* P 22–26
[5] Erofeev V T and Likomaskina M A 2016 Assessment of the durability of asphalt concrete during tests in climatic conditions with variable humidity, ultraviolet radiation and aggressive sea water *Vestnik MGSU* P 63–79
[6] GOST 16350-80 *The climate of the USSR. Zoning and statistical parameters of climatic factors for technical products* p 102
[7] GOST 9128-2013 *Mixes are asphalt-concrete, polymer-asphalt-concrete, asphalt concrete, polymer-asphalt concrete for roads and airfields* Technical conditions Enter 2014-14-11 (Moscow: Standartinform) p 50
[8] Rudenskaya I M, Rudensky A V 1984 *Organic binders for road construction* (Moscow: Transport) p 229
[9] Gezentsvei L B, Gorelyshev N V, Boguslavsky A M and Korolev I V *1993 Asphalt concrete and other bitumen-mineral materials* (All-Union Book Chamber) p 208
[10] Gorelyshev N V 1995 *Asphalt concrete and other bitumen-mineral materials* (Moscow: Mozhaysk-Terra) p. 176
[11] Kovalev N S 2014 Justification of the duration of exposure to climatic factors in the simulation of accelerated testing of asphalt concrete from slag materials *VGAU Bull.* **3(42)** P 163–171
[12] Kovalev N S 2015 *Road slag asphalt concrete* (Saarbrucken, Deutschland: LAP LAMBERT Academic Publishing GmbH & Co. KG) p 230
[13] Rusin N P 1979 *Applied Actinometry* (Gidrometizdat) p 232
[14] Sokolov B F, Samodurov S I and Kovalev N S 1978 *Method B.F. Sokolova preparation of building materials for strength tests*. Patent for invention RUS 665254, 04.01.1978.
[15] Kotlyarsky E V 2011 Frost Resistance of Asphalt Concrete *Building Materials* P 81–84
[16] DIN EN 13108-1:2006-08 *Asphaltmischgut – Mischgutanforderungen – Teil 1: Asphaltbeton* German version EN 13108-1:2006: Bituminous mixtures Material specifications. Part 1: Asphalt Concrete, P 32
[17] Leonovich I, Melnikova I and Puodžiukas V 2013 Estimation of the cracking probability in road structures by modeling of external influences *Baltic J. of Road and Bridge Engineer.* **8(4)** P 240–249
[18] Leonovich I and Melnikova I 2012 Influence of temperature on the formation of damages in asphalt concrete pavements under climatic conditions of the republic of belarus *Baltic J. of Road and Bridge Engineer.* **7(1)** P 42–47

Journal of Physics: Conference Series **1384** (2019) 012022 doi:10.1088/1742-6596/1384/1/012022

Implementation of the six sigma method in the educational process

D.A. Kremcheeva[1], **E A Kremcheev**[1]

[1] Saint-Petersburg Mining University, Vasilievski Ostrov, 21 linia, 2, Saint-Petersburg 199106, Russia

E-mail: kr.dinara@mail.ru

Abstract. The article describes the theoretical and practical aspects of the Six Sigma methodology and features of its implementation to optimize the educational process. The regulatory documents that can be used in the implementation of the method were analyzed. Particular attention was paid to the practical features of the implementation of the Six Sigma method and the development of target indicators of the educational process effectiveness. The developed system of performance targets is reliable and measurable and can be used in assessing the effectiveness of the educational process. The study also provided insights into the benefits of using the Six Sigma method in an educational institution. The goal of the study is to adapt the Six Sigma method for its use in the educational process and to analyze the main difficulties in the implementation of the method. The object of research in this article is the educational process. The analysis of scientific data, systematization was used in the article as a research method. Target performance indicators were developed by compiling and detailing information from literary sources, government standards and regulatory documents.

1. Introduction

The use of modern statistical methods can significantly improve the quality of the educational process, and improving the quality in turn makes it possible to gain competitive advantages both for a particular educational institution and in general for the education system at the state level. [1] To ensure the high quality of education, all processes must be defined, aligned with the goals and constantly improved. [2, 3]

The management of processes in an organization can be carried out by many methods of quality management, lean manufacturing, among which is the Six Sigma method. It should be noted that traditionally the Six Sigma method is used at manufacturing enterprises to reduce defects and costs, as well as for quality management in production. [4, 5, 6, 7, 8, 9] However, the educational process, like any other (including production), needs to be managed, as the controlled state of the process ensures its stability and effectiveness. In contrast to the production process, the object of management in the process of education is not a product, but a service.

It should be noted that studies that specify the implementation and application of the Six Sigma method in educational institutions are few. Most of the studies in this area are general in nature and do not reveal the practical implementation of the Six Sigma method of setting process performance targets and their confidence limits in the educational process.

2. Research methods

This article presents an overview. The following research methods were used: description, data analysis, its systematization and generalization.

As a basis for the study, materials of articles by Russian and foreign scientists were used (Zhabreva V.S., Grigorash O.V., Zolotukhina V.A., V.H. Patil and others) on the application of statistical methods of quality management in general and the method of Six Sigma in particular, as well as regulatory and legislative documents in this area: international standards for quality management, statistical management and lean manufacturing, federal educational standards.

3. Research results

The implementation and application of the Six Sigma method is based on the standard GOST R ISO 13053-2015 Statistical methods, Quantitative methods for improvement of processes "Six Sigma". However, it should be noted that this standard is aimed at product management and is not adapted for use in the educational process.

After analyzing the provisions of the standard GOST R ISO 13053-2015, it is possible can distinguish the following stages of the implementation of the Six Sigma (Figure 1). This scheme is relevant for any type of organization, including an educational institution.

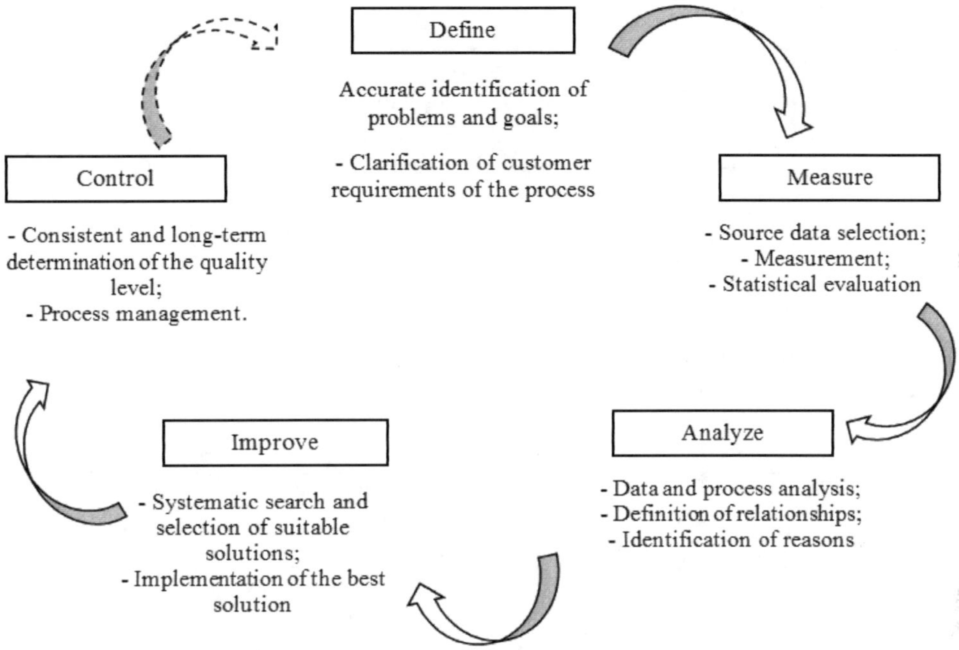

Figure 1 Stages of implementation of the Six Sigma method in the educational process

This scheme is similar to the PDCA model, which is set forth in the standard GOST R ISO 9001-2015 and provides for continuous improvement of each process in the organization. Thus, the integration of the Six Sigma method into the quality management system of an educational institution based on the process model is possible.

The main difficulties of implementing the Six Sigma method that an educational institution faces can be classified as follows:

1) Problems of correct application of the methodology and selection of performance indicators.

This is a group of problems associated with accurate identification of process targets, their confidence limits, proper measurement and analysis, accurate identification of root causes of a problem affecting the effectiveness of the educational process;

2) Resource provision problems.

The provision of the educational institution with qualified personnel (training of employees for the Six Sigma method), material support of work.

3) Organizational and managerial problems.

Staff involvement, competent planning of work on the method implementation, bringing information about the need to implement the Six Sigma method to all participants in the educational process.

If the second and third groups of problems are common to all types of enterprises implementing the Six Sigma, then the first group of problems requires specification, due to the specificity of the goals and objectives of the educational process.

It should be noted that the primary task in adapting the Six Sigma method for analyzing the educational process is the development of target performance indicators, since it is these indicators that significantly affect its quality.

Target process indicator makes it possible to assess the purpose of the educational process. The purpose of the educational process is the transfer to the student of knowledge and skills in accordance with the curriculum for the formation of the necessary competence. [10] If we consider the educational process within the framework of quality management, we can distinguish the following inputs and outputs, resources and requirements (Figure 2).

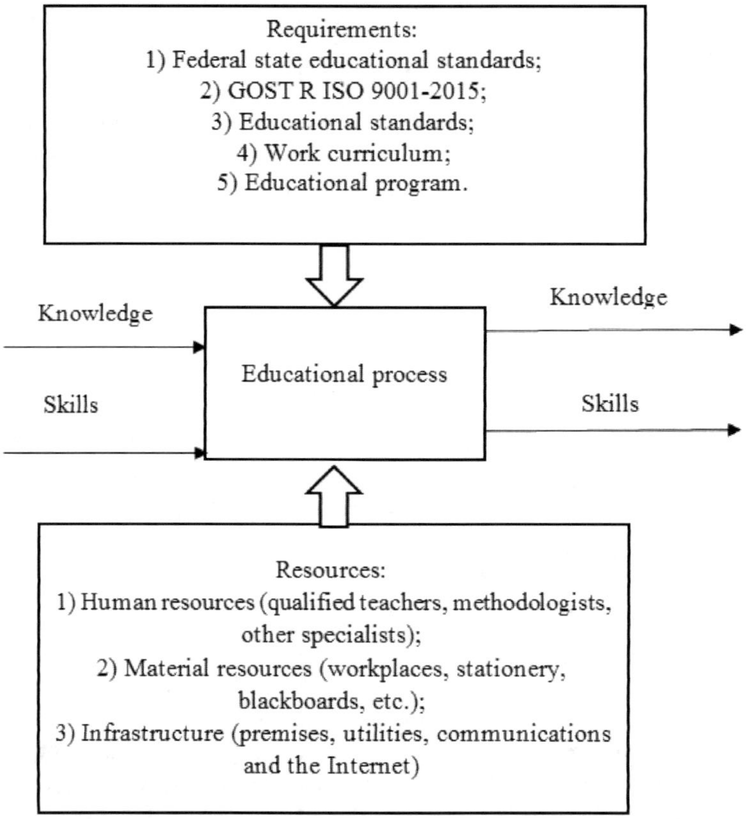

Figure 2 Simplified scheme of the educational process in the framework of quality management

Thus, the educational institution on the one hand transforms the student's knowledge and skills into deeper and more extensive ones, and on the other hand provides the student with new knowledge and skills in accordance with the requirements of the curriculum, methodological recommendations and program. However, the assessment of a student's knowledge and skills can not tell the quality level of the educational process in full, since these indicators are quite subjective and depend on the motivation and personal qualities of the consumer of educational process - a student. [11]

After analyzing a number of literary sources [10, 12, 13] and detailing the information given therein, we can single out the following system of quality indicators of the educational process (from the point of view of the process consumer) - Figure 3. It should be noted that the system of indicators can be expanded depending on the type of school, the form of education of the student, etc.

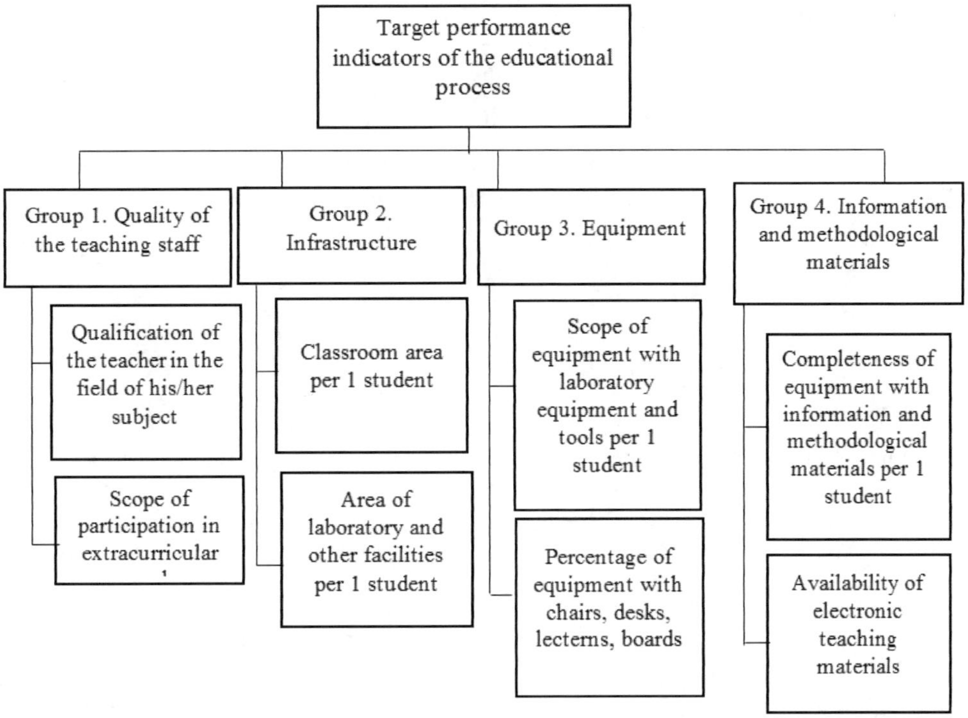

Figure 3 Target performance indicators of the educational process

These process indicators can be measured by the results of an internal audit of the quality management system, certification of specialists, inspection audit and do not depend on external conditions (progress and motivation of the student). For example, the indicator "Qualification of a teacher in the field of his/her subject" (Qt) may include such measurable components as the academic degree - AD and the title - AT of the teacher, teaching experience - TE, honored titles - HT, publication activity - PA (number of scientific articles, monographs, textbooks, etc.).

Thus, the proposed system of process goal indicators is objective and can be used as a baseline when adapting the Six Sigma method for the educational process.

Another main task is the quantitative measurement of process targets and the establishment of their permissible limits. The quality of the educational process Y is a function of the input effects of indicators X_1, ..., X_n. The magnitude of the deviation of the real value from the target ΔY is also a function of the variations in $\Delta X1$, ..., ΔXi. Reducing the variability of these deviations is the main objective of the Six Sigma method. [14] The indicators of the learning process should fall within established confidence limits, and exceeding them is just as undesirable as going beyond the minimum value. Thus, for example, increasing the number of equipment per student is a signal of inefficient use of resources, which leads to additional process costs. Figure 4 shows an example of the optimal distribution of the indicator "Amount of laboratory equipment and tools per student". Measurements should be carried out in all specialties. In this example, the boundaries are set from 1.2 to 1.8 pieces of equipment per student (since spare equipment is needed in case of a breakdown).

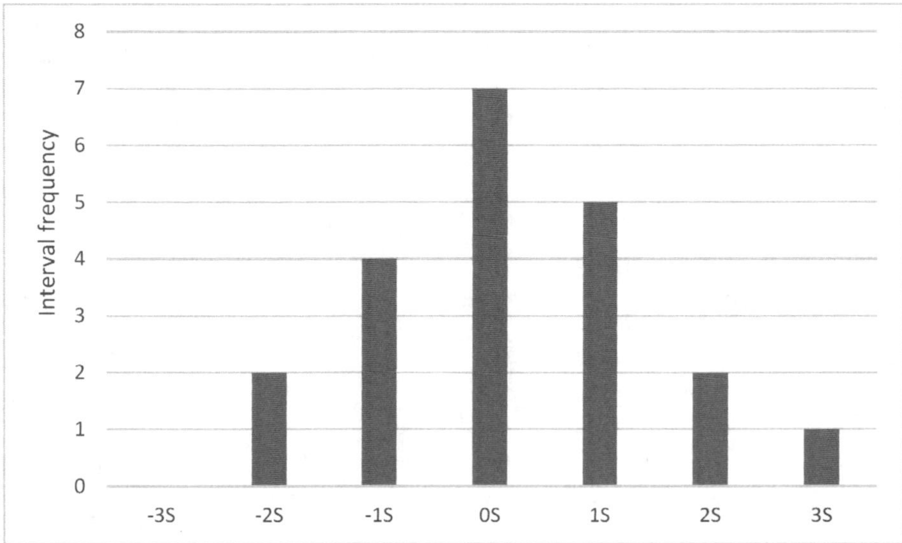

Figure 4 Distribution of the process target indicator "Amount of laboratory equipment"

This distribution is normal and shows that the indicator "Amount of laboratory equipment and tools per student" is in a controlled state (within the confidence limits). The amount of equipment in this case is optimal and does not require correction.

The boundaries of indicators can be set in a particular educational institution based on the requirements of the Federal state educational standards, professional standards for teachers of universities (Order of the Ministry of Labor of Russia No.608n dated 08.09.2015), the curriculum of a particular institution, the requirement of organization standards for a quality management system.

However, it should be noted that not all indicators are established by regulatory documents and legislative acts. Some indicators are set by the educational institution based on the analysis of statistical data, practical experience, analysis of student performance, analysis of the knowledge management system.

The implementation of innovations is always a time-consuming work, which requires special qualities from all participants and the optimal organization of process management. The importance of staff qualifications and involvement in the implementation of the Six Sigma method in the educational process should be noted. Specialists should know both the methodology for conducting the Six Sigma and the methodology for conducting the educational process, as well as, if possible, have practical skills for implementing this method in a particular enterprise. [15, 16]

4. Discussion of the research results

The article highlights the main difficulties in adapting the Six Sigma method to the educational process. The main practical task of the study is to develop targets for the educational process and their permissible boundaries.

As a result of the analysis of the inputs and outputs of the educational process, it is revealed that the input and output of the process are skills and knowledge that are transformed or expanded during the learning process before the student acquires the necessary competence. However, the quality of knowledge and skills of the process consumer - the student - depends not only on the quality of the process itself, but also on the external conditions (student's personal qualities, motivation, health status). Thus, target indicators should be measurable and objective. These indicators are proposed in the study (Figure 3), but they can be expanded or modified depending on the type of educational institution and the form of education.

The important component for the effective implementation of the Six Sigma method in an educational institution is the availability of qualified and interested professionals. Largely due to the involvement and literacy of the staff, such a tool of lean manufacturing as the Six Sigma can be implemented not only formally, but also implemented in practice.

However, until now, the Six Sigma method has not found wide application for analyzing the effectiveness of processes in educational institutions. This is due to the lack of regulatory documentation specifically in the provision of services, including educational services. The support of the method at the state level through the development of standards and recommendations for its use in education will contribute to its wider dissemination. The materials of this article can be used in practice in the development of techniques for the implementation of the Six Sigma in specific educational institutions.

5. Conclusion

The Six Sigma method is one of the tools of the lean learning system implemented in an educational institution as an element of a business strategy and is aimed at the improvement of the efficiency of the educational process.

It should be noted that the implementation of the Six Sigma method in an educational institution makes it possible to achieve the following advantages:

- To establish effective use and allocation of resources (material, administrative, personnel);
- To improve the quality of the educational process by reducing the variability of deviations of target indicators;
- To optimize the time of the learning process.

The Six Sigma method allows reducing the variability of educational process target indicators and ensuring its compliance with the requirements of concerned parties: students, teachers and the state.

References

[1] D. A. Zhevnov 2013 Innovative business development. Application of approaches "Lin" and "Six Sigma" for process optimization *Finance: Theory and practice* **1** 127 - 132.

[2] J. Antony 2014 Readiness Factors for the Lean Six Sigma journey in the higher education sector *International Journal of Productivity and Performance Management* **63 (2)** 257-264

[3] Grafe, B., Bravo, A. H., Hesse, M., Morgenstern, R., Rosin, K., Schlothauer, T., Talovina, I V, Nikiforova, V. S. (2019). Innocrush: New solutions for highly selective process chains. Paper presented at the *Innovation-Based Development of the Mineral Resources Sector: Challenges and Prospects - 11th Conference of the Russian-German Raw Materials, 2018,* 21-37.

[4] V.A. Zolotukhin 2013 Quality management at the Six Sigma level *Society. Environment. Development* **3** 212-215.

[5] Mardashov, D. V., & Vasiliev, B. U. 2016 Methodology of educational process organization using training simulator. Paper presented at the *2016 IEEE Conference on Quality Management, Transport and Information Security, Information Technologies, IT and MQ and IS 2016,* 144-146.

[6] Demenkov, P., & Verbilo, P. (2016). Methodology of prediction stress-strain state deep foundation structures of subway station's taking into account stages of its construction. Paper presented at the *Procedia Engineering, 165* 379-384.

[7] Smirnova E E, Lutskiy D S 2019 Improving the efficiency of purification in the technological cycles of limestone. ARPN Journal of Engineering and Applied Sciences 14 (12) 2306-2309

[8] Vinogradova, A.A., Leontyuk, S.M. (2018). Evaluation of current status in field of electrical voltage measurements. Journal of Physics: Conference Series, 1118(1),DOI: 10.1088/1742-6596/1118/1/012048

[9] Safiullin R.N., Afanasyev A.S., Rezchenko V.V. The Concept of Development of Monitoring Systems and Management of Intelligent Technical Complexes. Journal of Mining Institute. 2019. Vol. 237, p. 322-330.

[10] O.V. Grigorash 2013 Improving the quality management of the educational process *Higher Education in Russia* **1** 72 - 73.

[11] V.H. Patil, S.M. Kamlapur, M.L. Dhore, 2006 Six Sigma in Education: To Achieve Overall Excellence in the Field of Education In the *Proceedings of the 2006 Third International Conference on Information Technology* pp. 376-382 (St Petersburg: New Generations).

[12] E.V. Mezemtseva, E.V. Korolyuk, E.N. Astankova 2015 Assessment of the quality of educational activities of the university (for example, a branch of the Kuban State University in Tikhoretsk) *Humanities, socio-economic and social sciences* **5** 38 - 41.

[13] O.V. Grigorash 2014 On indicators for evaluating the effectiveness of the activities of universities *Polytematic electronic scientific journal of the Kuban State Agrarian University* **95** 2-14

[14] R.A. Dolzhenko 2014 Essence and assessment of the effectiveness of the use of optimization technologies "Lin" and "Six Sigma" *Bulletin of Omsk University. Series "Economy"* **1** 25-33.

[15] M. Vijaya Sunder 2016 Sunder Lean Six Sigma in higher education institutions *International Journal of Quality and Service* **8(2)** 159-178

[16] V.S. Zhabreva, T.N. Polovova 2014 Models of the implementation process of "lean learning + Six Sigma" *Bulletin of South Ural State University, Series: computer technologies, management, electronics* **3** 119-123.

Problematics and perspectives of the development of automatic control systems for concentration tables using computer simulation

Ya V Kuskova[1], O O Erokhina[2,1], A S Simakov[1]

[1] Saint Petersburg Mining University, 2, 21st Line, St.Petersburg, 199106, Russia
[2]Peter the Great St. Petersburg Polytechnic University, 29, Polytechnicheskaya, St.Petersburg, 195251, Russia

E-mail: yana.kuskova@gmail.com

Abstract. The Concentration tables are widely used for the enrichment of various materials due to a number of reasons, which include the low cost of the process and aggregates, high recovery of concentrate; environmental friendliness of the process. However, there are a number of problems associated with both the specific capacity of a single apparatus and the need to adjust the parameters of the process mode in accordance with the characteristics of the used raw materials. Currently, there are no automatic control systems for concentration tables, which is associated both with the lack of a mathematical apparatus for assessing the ore behavior during the process and with the complexity of using computer simulations for various types of materials consisting of liquid and solid systems. This article presents the prospects for the development of automatic control systems for concentration tables with regard to industry development trends.

1. Introduction

Concentration tables are widely used in mineral processing. These units are commonly used in the processing of gold, coal, silver, platinum group metals and other materials [1]. Their distribution is caused by a number of factors, which include the low cost characteristics of the units in terms of their operation, simplicity of the process and pre-commissioning activities, the absence of toxic reagents and, consequently, the ecological purity of the process.

The significant disadvantage of concentration tables is their low specific productivity, as a result of which there is a need to introduce a large number of units of this type into the production process [2]. The above mentioned aspect is the main reason for the spread of concentration tables as laboratory equipment, as well as for processes characterized by low specific productivity (including the enrichment of rare-earth and precious metals).

The enrichment on concentration tables is realized due to the gravitational / inertial forces acting on the wet material, realized due to the asymmetric reciprocating motion of the deck. In addition, flushing by a stream of water in the direction perpendicular to the direction of movement of the material is realized on the units. It allows minimizing friction of the particles, as well as effective separating them in a turbulent mode of fluid flow [3].

Material characteristics (density, particle size and shape), as well as the operation mode of the concentration table (including the parameters of the reciprocating motion of the deck, the angle of inclination of the table, the flow of water and granular material) most affect the specific performance of the concentration tables. Consequently, in order to increase in efficiency, there is a need to calibrate the mode of operation of the unit depending on the characteristics of the incoming material.

2. Prospects for the development of automated control systems for concentration tables

At present, the selection of the parameters of the operating regime of the concentration table is carried out manually on the basis of laboratory studies conducted for the average composition of particles, as well as taking into account the experience of the staff of the workshop.

One of the main problems in determining the optimal parameters of the concentration tables is related to the transfer of data due to the lack of a sufficient theoretical basis that allows making additional corrections to the process. Whereas laboratory-scale experiments examine only average ore compositions, as a result of which constant adjustment of concentration tables is necessary during their work at specific time intervals.

The correction of the operating mode of the concentration table is made manually by the staff of the workshop. The discreteness of the adjustment is often not associated with changes in the input parameters of bulk materials, but with the employment of staff, which greatly affects the unit capacity and the extraction of the target component.

The low degree of automation of the process at concentration tables is often associated with high costs for automation equipment, low efficiency of these systems due to the lack of a mathematical apparatus for characterizing the process, as well as the lack of the necessary flexibility and predictability of automated control systems that would eliminate the human factor from managing these units.

Nowadays, there are trends in the implementation of mathematical modeling of concentration tables [4-8]. There are two fundamentally different approaches in the study of concentration tables using the method of mathematical modeling. The first method explores the dependencies formed on the basis of empirical output data for the material depending on the concentration table operation mode. The data obtained is often applicable in a particular production, taking into account the use of identical concentration tables and ores similar in composition. The derived formulas frequently do not allow obtaining fundamental information in the course of the process, as a result of which the data obtained are not universal and do not allow predicting the optimal operation of the concentration table in the case of using fundamentally different ores by physicochemical properties.

Moreover, the construction of mathematical models based on experimental data is characterized by the restriction on the number of experiments. The use of a full factorial experiment to optimize the course of research introduces a complementary error in the data obtained. Concentration tables are characterized by a study that includes no more than two or three joint parameters, which ultimately affect the extraction of a valuable component, as a result of which there is no generalized view of the process of separation into this equipment.

Consequently, the similar method of mathematical modeling is reduced to the need for continuous adjustment of mathematical dependencies according to the properties of the ore used and the characteristics of the concentration table, furthermore the resulting equations are selected depending on the behavior of the pulp and do not have a physical basis.

The second approach is the use of principles of physics to describe the behavior of particles on a concentration table. In this case, there are difficulties associated with choosing the mathematical apparatus that most characterizes the work of the concentration table. Furthermore, the unit is characterized by a complex physical base due to the presence of vibration and two phases (liquid and solid). The development of additional software is impractical due to the high complexity of the described processes and the need to use powerful computing equipment. However, this approach allows estimating the trajectory of the particles, which is virtually impossible to implement on industrial and laboratory scale units in consideration of the small size of particles.

Consequently, the modeling methods of concentration tables used at the current moment do not provide a rough idea of the nature of the process. The data obtained depends either on the experiment plan drawn up by the researchers, or on the physical dependencies chosen to describe the process. Based on the above mentioned aspects, it can be concluded that it is necessary to use a fundamentally different approach for modeling concentration tables.

Currently, computer modeling of technological processes and equipment is being actively used [9-15], including these methods are beginning to be applied with respect to mineral-processing equipment. Bundled software for computer modeling is distinguished by high functionality and complexity of the principles of physics included in the program, as a result of which the results obtained are highly convergent with respect to real processes.

3. Prospects and problems of sharing methods DEM and CFD

To evaluate the pulp behavior, it is advisable to use DEM and CFD modeling methods together [16-20].

DEM-modeling (discrete element method) allows simulating bulk materials, where each element is considered separately. There are a number of problems associated with the use of DEM modeling for ore modeling. These include: the complexity of the shapes of particles and the variability of their size, the inconstancy of the composition and properties of ores, the difficulty of determining the main characteristics of materials (including dynamic and static friction, rheological properties and others) [21-25].

The problem of determining the parameters of bulk materials for further computer modeling is typical for the entire mining processing industry. The most frequently, reference data or laboratory scale experiments are used to assess the behavior of bulk materials. To achieve convergence of the results obtained on the physical and computer models, the parameters of bulk materials are selected iteratively. The foregoing significantly affects the accuracy of the measurement method.

Currently, methods for determining the parameters of bulk materials are being developed [17-21], however, there are difficulties in assessing the adequacy, predictability and scalability of models implemented using these methods, which is caused by the need to reconcile models in kind.

The use of DEM-modeling allows estimating the behavior of exclusively bulk materials, which introduces a number of measure of inaccuracies in the resulting model, namely: the lack of consideration of the influence of the regime of fluid motion on the ore behavior; the absence of segregation of particles by gravity and other.

For leveling these factors, it seems appropriate to further use the CFD method (computational fluid dynamics). The introduction of computational fluid dynamics (CFD) is necessary to assess the behavior of the liquid phase and the influence on it of the process parameters (parameters of the reciprocating motion of the deck, the angles of the table, the flow of water and bulk material) and the behavior of the granular medium environment (the so-termed two-direction integration).

It should be pointed out that the joint use of the DEM and CFD methods is quite resource-consuming and time-consuming to implement, as a result of which there is no possibility to promptly use computer simulations for adjustments to the operating mode. It is possible to use supercomputers to solve these problems. However, their implementation on the basis of enterprises is impractical from an economic point of view. As a possible option is to attract outside companies to carry out the necessary calculations, but it will also have a negative economic effect, furthermore the results obtained will characterize some averaged over the charge, which will significantly reduce their applicability to the enterprise.

Consequently, it conceivable to develop a digital twin based on the resulting model, which will greatly simplify the automation of concentration tables, as well as reduce resource costs for evaluating the behavior of the pulp depending on the variation of process parameters.

The creation of digital twins (a computer model that describes the product or process as accurately and completely as possible) is one of the main trends in engineering manufactory, automotive industry and other similar industries [26-30]. However, in consideration of the complexity of the enrichment processes, this method was not widely used. The cumulative introduction of a digital twin and a digital shadow (a computer model operating on a database obtained from inbuilt tool of the process automation) is proposed to refine the mathematical dependencies. Digital shadow will allow clarifying the results of computer modeling by means of obtaining relevant industrial data. The introduction of a digital twin and a digital shadow will solve the problem associated with varying the characteristics of the incoming charging material, which will greatly increase productivity and simplify the process of managing concentration tables.

4. Conclusion

At present, concentration tables are characterized by a low degree of automation, and they are often managed manually by the workshop staff.

The complexity of the implementation of automation systems for concentration tables is associated with a number of reasons, which include the lack of a mathematical apparatus for evaluating the behavior of the material / pulp in the enrichment conditions, and high time expenditure in terms of modeling the totality of liquid and solids.

The use of computer simulation, namely the combination of DEM and CFD methods seems to be the most promising. It will clarify the existing mathematical dependencies, greatly simplify them, which in turn will reduce the time expenditure on subsequent iterations of computer modeling, as well as develop automatic control systems with low cost estimate and high accuracy. In the time following, it is proposed to build a digital twin based on the developed computer model. The digital twin will be adjusted by using data from a digital shadow, which will optimize the process of selecting the parameters of concentration tables depending on the characteristics of the incoming charging material.

References

[1] Isaev I N 1962 *Concentrations tables* (Moscow: Gosgortekhizda)
[2] Sokur A K 2012 Overview of gravitational technologies of enrichment of coal slimes of non-flotation size *Zagagachennya korisnyh kopalin* **92(51)**
[3] Lenev L A, Kuskov V B, Vasil'ev A M 2007 Different methods of enrichment from hematite ore *Obogashchenie Rud* **1** 6-8.
[4] Bukin S L, Komarov V F and Shold R A 2013 Mathematical model of fluid motion in inter-column space of a concentration table deck *Scientific works of the Donetsk Nat. Technical University. Sir: Mining and electromechanical* **1** 47-56
[5] Yuryevich B V,. Borisovich K V 2016 Production of fuel briquettes from carbon containing materials *XVIII International Coal Preparation Congress Springer, Cham* 701-705
[6] Evdokimov S I 2013 Improving the technological schemes of jigging and concentration on the tables during the enrichment of gold-bearing raw materials *Mountain Journal* **12** 59-62
[7] Zvonarev I E, Shishlyannikov D I 2017 Information and diagnostic tools of objective control as means to improve performance of mining machines *IOP Conf. Series: Materials Science and Engineering* **177**
[8] Koptev V Yu and Kopteva A V 2018 Improving Pit Vehicle Ecology Safety **1015(5)** *Journal of Physics: Conf. Series*
[9] Coetzee C J 2017 Calibration of the discrete element method *Powder Technology* **310** 104-142
[10] Vasilyeva N V, Koteleva N I and Fedorova E R 2018 Real-time control data wrangling for development of mathematical control models of technological processes *Journal of Physics: Conference Series* **1015(3)**
[11] Sizyakov V M, Brichkin V N 2018 About the role of hydrafed calcium carboaluminates in improving the technology of complex processing of nephelines *Journal of Mining Institute* **231** 292-298
[12] Jovanović A et al. 2014 DEM/CFD analysis of granular flow in static mixers *Powder technology* **266** 240-248
[13] Fries L et al. 2011 DEM–CFD modeling of a fluidized bed spray granulator *Chemical Engineering Science* **66(11)** 2340-55
[14] Mishra B K and Tripathy A A 2010 Preliminary study of particle separation in spiral

concentrators using DEM *Int. Journal of Mineral Processing* **94(3-4)** 192-195
[15] Kopteva A V and Koptev V Yu 2018 Automatic Measuring System for oil stream paraffin deposits parameters IOP *Conf. Series: Materials Science and Engineering* **327**
[16] Boikov A V, Payor V A, Savelev R V 2018 Technical vision system for analysing the mechanical characteristics of bulk materials *Journal of Physics Conference Series* **944**
[17] Boikov A V, Savelev R V and Payor V A 2018 DEM Calibration Approach: Random Forest *Journal of Physics Conf. Series* **1118**
[18] Boikov A V, Savelev R V and Payor V A 2018 DEM Calibration Approach: Implementing Contact Model *Journal of Physics Conf. Series* **1050**
[19] Boikov A V, Savelev R V, Payor V A 2018 DEM Calibration Approach: design of experiment *Journal of Physics: Conference Series* **1015(3)**
[20] Beloglayov I I 2018 Automation experimental studies of grinding process in jaw crusher using DEM simulation *Journal of Physics: Conference Series* **1118 (10)**
[21] Rackl M and Hanley K J 2017 A methodical calibration procedure for discrete element models *Powder technology* **307** 73-83
[22] Boschert S and Rosen R 2016 Digital twin – the simulation aspect *Mechatronic Futures* 59-74
[23] Kruk M N, Semenov A S 2017 Project Risk Analysis and Management Decision-Making in Determining the Parameters of Ore Quarries Journal of industrial pollution control 33(1) 1024-1028
[24] Siziakova E V, Ivanov P V, Boikov A V 2019 Application of calcium hydrocarboaluminate for the production of coarse-graded alumina *Journal of Chemical Technology and Metallurgy* **54(1)** 200-203
[25] Xiang F, Zhi Z and Jiang G 2018 Digital Twins technolgy and its data fusion in iron and steel product life cycle *IEEE 15th Int. Conf. on Networking, Sensing and Control (ICNSC)* pp 1-5
[26] Kovalski E R, Karpov G N , Leisle A V 2018 Geomechanic models of jointed rock mass *International journal of civil engineering and technology* **9(13)**, 440-448
[27] Bazhin V Y, Kulchitskiy A A, Kadrov D N 2018 Complex control of the state of steel pins in soderberg electrolytic cells by using computer vision systems *Tsvetnye Metally* **(3)** 27-32
[28] Koptev V Yu and Kopteva A V 2016 *Mining business transportation system structure optimization* **11**
[29] Siziakova E V, Ivanov P V, Boikov A V 2019 Application of calcium hydrocarboaluminate for the production of coarse-graded alumina *Journal of Chemical Technology and Metallurgy* **54(1)** 200-203
[30] Xiang F, Zhi Z and Jiang G 2018 Digital Twins technolgy and its data fusion in iron and steel product life cycle *IEEE 15th Int. Conf. on Networking, Sensing and Control (ICNSC)* 1-5

Reduction of the recovery period of marsh ecosystem through the use of new mining technologies

V S Kuznetsov, A S Kuznetsova

Saint-Petersburg Mining University, 2, 21st Line, St Petersburg 199106, Russia

E-mail: Kuznetsov_VS@pres.spmi.ru

Abstract. The article touches upon the issue of restoring wetland ecosystems after the industrial extraction of peat. The analysis of the processes occurring in the swamp reclamation, and the ways of development of peat deposits technologies that reduce ecosystem restoration period after production.

1. Introduction

According to the International Peat Society (IPS, 1995) of peat resources in the world amount to more than 400 million hectares, of which only a little more than 305 million hectares is under construction.

Peat fuel for agriculture has long been mined in 23 countries. The largest peat reserves are concentrated in two countries: Russia - 150 million hectares and Canada - 111 million hectares. Over the years, the development of the peat industry in different countries have developed several areas of use of peat and peat products [1,19], the continuous exploitation of peat deposits has not been in vain for wetland ecosystems, and the extent of the negative impact is determined primarily by way of the development of peat deposits.

In the world there are several ways of peat mining, the main ones: milling method of production, they produce up to 95% peat, hydraulic method, excavating method [2,3,4,5,6,7]. Let us consider them in terms of impact on the ecosystem of the bolt.

Milling method for the extraction of peat layers surface [6]. In this case, the peat in the form of small crumbs going to the surface prepared for the exploitation of peat deposits. With this method, there is a maximum drainage of the peat deposits by removing most of the water drainage network is adversely increases the period of recovery after ecosystem production [17,20].

The hydraulic method of peat production, invented by Russian engineer R. Klasson, based on the use of high-pressure water jet (1.0-1.5 MPa) for the erosion of the peat deposit and turn it into a fluid hydro mass, then transported in pipelines over long distances [6,8, 22]. Technological process of sod peat by the method of hydraulic extraction consists of three stages: hydro mass production (erosion of deposits and excavation); transport and flood hydro mass on drying fields; drying and harvesting sod peat [21, 23]. Currently not applicable due to the low profitability of production, in the process of mining has contaminated nearby water bodies suspended solids after extraction of peat in this way remains a large number of pits filled with water in Figure 1.

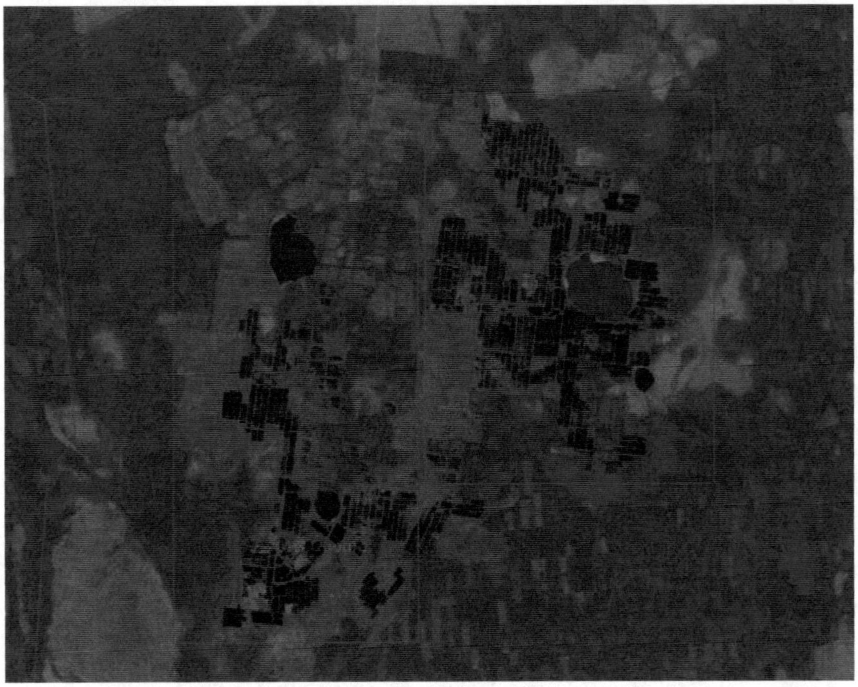

Figure 1. Space image, abandoned fields developed by hydraulic method (shown in black water flooded career)

A third way to a career is a method of extraction technology includes four basic operations: excavation of peat from the quarry across the depth of the deposit, drying peat sometimes turning it, carrying out operations, cleaning of finished products in the stacks [15]. In this process, the period of recovery bog ecosystem is the smallest, so the development of new production technology, we adhere to this method of development [9].

2. Description of the suggested technology

In the course of the scientific research carried out field experimental research as the object of the preliminary (evaluation) studies were selected peat deposits Sakkala, Priozersk district of Leningrad region, where the currently works on peat extraction were stopped.

It was found that the development of deposits excavation method involves running peatlands that have a positive impact on the local carbon footprint.

According to a comprehensive field research, peatlands with high groundwater levels are characterized by the emission of methane exceeds the amount of the absorbed carbon dioxide.

As it is known, the greenhouse potential of methane exceeds the capacity of carbon dioxide is 23 times.

Therefore, excavation of peat in the fields with low emissions of methane leads to almost total reduction of greenhouse gas emissions.

Peat deposits after peat excavation to the full depth can be quickly and easily restored as a functioning wetland. Excavation natural peat permit a progressive recovery in production areas so that only a relatively small area of peatland affected at certain times. This minimizes dust problems and by carefully monitoring the hydrological and water management will reduce the negative effects on the local water system [18].

Monitoring of air quality has confirmed that in their natural bedding peat deposits are sinks of CO_2 at high methane emissions. After peat extraction and recovery, new wetlands are carbon consumers due to more actively recovering surface vegetation. The process of extraction or recovery leads to a reduction of greenhouse gas emissions.

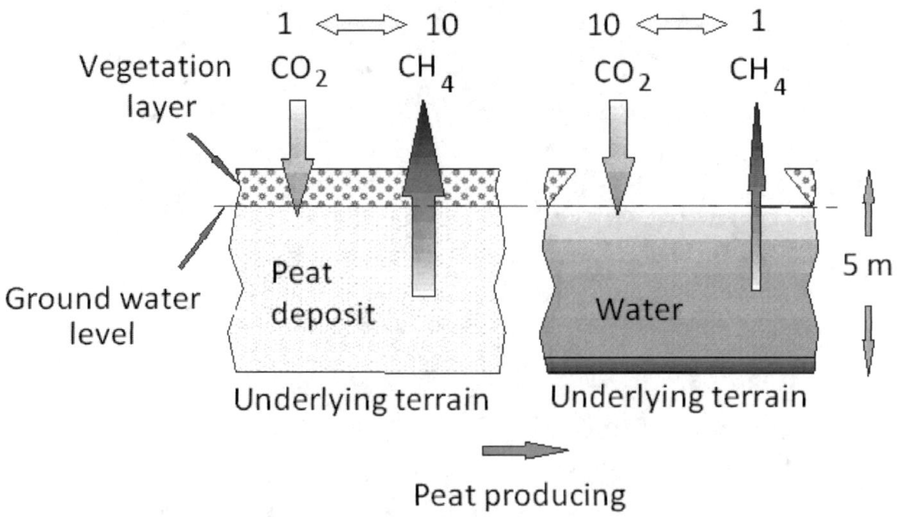

Figure 2. Schematic of greenhouse gas emission

As it is shown in Figure 2, before peat excavation - CH_4 emission more than 10 times greater than absorbed CO_2. After extraction - CO_2 absorbed 10 times greater than the emission of CH_4.

Excavator method allows virtually year-round production of various products based on peat (9-10 months of production and processing, 1 month - preventive maintenance and repairs, 1 month - a break for vacation). This reduces the overall cost and increases production efficiency.

3. Results & Discussion

The advantages of the new production of peat fuel are the following:

- Reduction of environmental impacts (dust emissions, noise and water impacts, reduction of greenhouse gas emissions);
- Expansion of the production season;
- Optimal use of weather conditions;
- Production efficiency is 20 times higher than the current production;
- Increasing the number of exploitable peat resources;
- Rapid recovery of production space absorbing carbon.

Our group of authors proposed the introduction in Russia, based on GIS technology and satellite positioning data, aimed at complete removal of peat from the field, which will reduce the human impact on the ecosystem of the bogs and lead to the speedy restoration of peatland [10,11].

Rewetting drained peatlands leads to an increase in groundwater levels, which reduces the rate of peat mineralization. However, rewetting does not always immediately lead to a reduction in greenhouse gas emission [16]. Rewetting fens often leads to increased release of methane (CH_4), whereas when the discharge of carbon dioxide (CO_2) can be high. Water level fluctuations can cause a significant increase in the release of nitrous oxide (N_2O).

Due to limited experience rewetting, long-term changes in the information available. Available data show that the natural function of the peat bog as an absorber of CO_2, recovered quickly enough. At the same time, this process can be accompanied by sufficiently high separation of methane [12].

We can distinguish three periods after rewetting with different characteristics:

1.High allocation of CH4 is accompanied by a high absorption (accumulation) of CO_2. This primary phase has an extremely negative impact on the climate.

2. Isolation CH_4 sharply reduced, while maximally absorbed CO_2. This period has several positive climate effects.
3. Poor selection of CH_4 and CO_2 absorption is weak, similar to the situation on the pristine bogs. This phase has a neutral impact on the climate.

Although drained bog (zero phase), the global warming potential is determined in addition to CO_2, and the presence of N_2O, CH_4 and CO_2 effects dominates in all types of bogs after rewetting. In phase 1 is still no clear distinction between the types of wetlands, but they are noticeably smoothed in the second phase.

Rewetting of degraded peatlands has significant benefits in any scenario. Even on the most pessimistic scenario three, rewetting will avoid greenhouse gas emissions in an amount equivalent to more than 30 million tons for hundred years [13].

4. Conclusion

Despite the fact that the qualitative changes in the dynamics of greenhouse gases from rewetting obvious need for further research in the field of long-term quantitative forecasts. This is due to the fact that peat swamp is relatively recent and long-term studies are not available. To determine the most effective methods of rewetting need comprehensive long-term study of greenhouse gas balance of peatlands.

Restrictions primary methane emissions can be achieved in the following way:

- Removing young decaying plant to water logging;
- Optimization of the water level (not deep-water logging);
- Artificial revegetation;
- Prevent the proliferation of vegetation, which decomposition results in the release of methane [14].

Whatever it is, despite all the difficulties, it is clear that the restoration of the bogs is the work carried out by trial and error, and we are only at the beginning.

References

[1] International Peatland Society: [Electronic resource]. URL: www.peatsociety.org (Reference date 02/10/2016)

[2] Antonov V J, Malkov L M, Gamayunov N I 1981 Technology Field drying peat. (Moscow: Nedra)

[3] Antonov V J, Churaev N V, Malchin S M 1957 Drying peat bricks stackable figures with surface-layered method of production of sod pea. *Proceedings of MIT* **4** 34-41

[4] Afanasiev A E , Malkov L M , Smirnov V I 1987 Technology and complex mechanization of peat deposits (Moscow: Nedra

[5] Afanasiev A E, Churaev N V 1992 Optimizing the drying process and structure formation in peat production technology. (Moscow: Nedra)

[6] Bogatov B A, Nikiforov V A 1988 Technology and complex mechanization of peat production. Publishing House "University 463

[7] Mikhailov A V 2004 *Mining Journal* **5** 20-23

[8] Patent number 6796, cl. C10F 7/04 from 16.06.1998. Systems for drying and briquetting of peat

[9] Mikhailov A, Nagornov D 2012 *Book of Abstracts of the 14th International Peat Congress*

[10] Kremcheev E. A., Nagornov, D. O. (2014). Environmentally compatible technology of peat extraction. *Life Science Journal* **11** 453-456.

[11] Patent of Russian Federation № 2470984 cl. C10F7 / 00 from 27.12.2012. The modular technological complex of peat extraction and production of fuel agglomerates

[12] Nagornov D O, Kremcheev E A 2011 Justification of the concept of modular peat enterprises in the structure of a full cycle of small energy cluster. *Collected works of the winners of the qualifying round of the national competition of research works of students, graduate students and young scientists in several interdisciplinary areas* 275-276

[13] Mikhailov A V, Kremcheev E A, Bolshunov A V, Nagorno D O 2010 *Mining informational and analytical bulletin (scientific and technical journal)* **9**
[14] Database "Peatbogs of Belarus". URL: www.peatlands.by (Reference date 01/16/2016)
[15] Nagornov D. O., Kremcheev, E. A. 2018 Framework to evaluate efficiency of peat processing using roller-disc grinding machines. Paper presented at the *Journal of Physics: Conference Series, , 1118*(1).
[16] Nagornov D O, Kremcheeva, D A 2017 *Journal of Industrial Pollution Control*
[17] Nagornov, D. O., Kremcheev, E. A., & Kremcheeva, D. A. 2019. Research of the condition of regional parts of massif at longwall mining of prone to spontaneous ignition coal seams. *International Journal of Civil Engineering and Technology* **10**(1) 876-883.
[18] Misnikov O S 2016 *Mires and Peat* **18** 1-15
[19] Misnikov O S 2006 *Theoretical Foundations of Chemical Technology* **40** 4 423-430
[20] Misnikov O S, Kopenkina L V 2018 *Gornyi Zhurnal*
[21] Mikhailov A. V., Kremcheev E. A., Nagornov D. O. 2012 Journal of Mining Institute Mechanisation of extraction of peat from undrained deposit **196** 256-260.
[22] Lebedev V, Puhova O 2017 *E3S Web of Conferences*
[23] Kremcheev E A 2018 Journal of Mining Institute, Special features of a structure of technical operations for peat excavation with stage dewatering **231** 225-234.

Journal of Physics: Conference Series **1384** (2019) 012025 doi:10.1088/1742-6596/1384/1/012025

Berkeley pit and its impact on the environment

A S Kuznetsova, A V Ivanov

Saint-Petersburg Mining University, 2, 21st Line, St Petersburg 199106, Russia

E-mail: kuznetsovaanna@hotmail.com

Abstract. The article tells about the Berkeley Pit and its environmental impact. The main chemical characteristics of mine water and water treatment systems are outlined. Possible ways of dealing with problem of ground water pollution are discussed.

1. Introduction

The Berkeley Pit was founded in 1955 in the city of Butte, Montana, USA. It was done due to the several reasons: the first one was the cost - open-pit mining of copper was cheaper than underground mining which had been used before; the second one was the personnel safety – it was safer to work on the surface. 2500 people died in the process of underground mining of copper in 1860-1976, but after changing the way of mining 6 people died.

By the moment the pit was opened, the core contained only 0.75% of copper, so, to make the mining profitable, the Berkeley Pit united the nearby mines, such as Meaderville, Dublin Gulch и McQueen. Afterwards, gradual decline in prices led to the closure of the pit in 1982 [1].

Montana Bureau of Mines and Geology developed the digital model of all mines in this area shown on figure 1. Red lines are vertical shafts, other colors are horizontal ones. The deepest shafts have the depth of 1600 meters, which is much lower than the bottom of the pit with the depth of 540 meters [5].

2. Materials and methods

The pumps were used while underground and open-pit mining to prevent flooding of the work area. When the pit was closed, the pumping stopped, which led to the filling of underground mines and the Berkeley Pit itself. Nowadays there is a lake and its water is highly acidic due to the interaction of water and sulphide rock. Water contains such metals as copper, iron, arsenic, cadmium. From the moment of stopping the pumps water level reached the level of 270 meters. Soon this level catches up with the level of groundwater in this area and dissolved contaminants will migrate farther [4].

In 1996 water drainage was redirected from the pit, water was treated from the copper and headed to the tailing pond Yankee Doodle which was built for processing plant. Due to this, filing speed reduced to 11 cubic meter per minute.

In 1998 the pumping of water from the pit started for the first time due to the opportunity of copper recovery from the water. In the process water was pumped to the treatment plant and went through concrete cells filled with iron bars. Copper replaced iron on bars, iron dissolved in water [11]. The process took about 30 minutes, then water went back to the pit. Once a week the bars were cleaned of copper, copper was dehydrated, bars went back to the process. In total 181 tons of copper was extracted and sold to pay for the building of the plant. Plant is shown on figure 2a. The process stopped in 2013 when the pit edge slipped and the pump broke.

Figure 1. Digital model of mining area

a) ***b***)

Figure 2. Water treatment plants:
a – copper extraction plant; *b* – metals extraction plant with flow of treated water back to the pit

In 2000 redirecting of water stopped. Price for copper went up and mining started again in the Continental pit near the Berkeley pit. Another water treatment plant Horseshoe Bend was opened in 2003. Plant is shown on figure 2b. That plant prevents the drainage of water to the Berkeley pit. Water treatment occurs due to the two-stage sedimentation with combination of calcium hydroxide and high-density sludge technology. Limestone, aeration and polymeric additives help in metal sedimentation. Fully automated system provides 10 times less amount of sludge. High-density sludge is provided by repetition of treatment cycle for a few times. All amount of sludge (151 cubic meters) is returned to the pit. The research shows that sludge storage in the pit may raise the pH level for the next 10-20 years which reduce the cost of water treatment. Treated water may be used in processing of ore or released in the Silver Bow Creek. In 2012 19 thousand cubic meters of water per day was treated in the plant. Project capacity of water treatment plant is 26500 cubic meters of water per day which will be reached when the treatment of water from the Berkeley pit will start.

The critical level of water for Berkeley pit is 1646 meters above sea level. This level means the beginning of water pumping from Berkeley pit and treatment of that water in Horseshoe Bend water treatment plant. Rising of water level is slower than anticipated due to the redirecting of surface discharge from the pit. In 2016 level of water was 1626 meters above sea level [6].

In 2017 quality and amount estimation of water was finished. In 2019-2021 water treatment plant Horseshoe Bend will be improved to be able to pump and treat more water. The recent research shows that it will be necessary to start pumping and treatment of water in the pit in 2023 because the critical level is reached. Level of water may be changed because of bigger amount of precipitation or pit edge slips. Last slip happened in 2013 when water level gained 18 cm at once which means a month rate. The first water treatment plant was also closed. Level of water in the Berkeley pit is checked every month, so, if necessary, it is possible to start treatment earlier or even later.

Quality of water is checked every 6 months. Water samples are taken from different depth and analyzed for dissolved contaminants. Temperature, pH level, dissolved oxygen and electrical conductivity are measured to the depth of 183 meters with increment of 3 meters. In addition, these qualities are measured at the bottom.

Till 2012 chemical composition has been changing with depth. Brown-red water at the top was the cleanest, greenish blue at the bottom was the most poisonous. At some depth chemical composition changed so fast, that the chemical border was formed. Density at the bottom was greater than at the top. Also, the pH level was higher under that border. Concentration of metals was higher at the bottom [2, 3]. Quality levels till 2012 are:

- up to 122 meters: pH=2,57, C(Fe)=0,5g/l, C(Cu)=0,07g/l;
- lower than 122 meters: pH=2,37, C(Fe)=0,9g/l, C(Cu)=0,14g/l.

Border line was at the depth of 123-132 meters.

Nowadays the chemical composition is more homogeneous. According to scientists, stratification occurred because of cleaner water intake which stayed on the top due to its lower density. Copper extraction from lower part of the pit made the quality of water better. Concentration of iron decreased by 56%, arsenic – 87%, copper – 60%. Decrease of metal concentration made the reclamation process cheaper, but much iron sedimented, so water had red colour. Also, concentration of zinc, aluminum, cadmium and manganese is still high [7, 10].

There is some research about ability of algae to reduce concentration of metals dissolved in acidic water. There are some algae in the Berkeley pit, but there are not enough nitrates for them to grow fully. It was experimentally established that algae are capable of reducing the concentration of iron and arsenic in the water if treated with additional number of nitrates [8].

In addition, some new algae species were found in the Berkeley pit which successfully adjusted to the aggressive environment of the pit. They are not only able to reduce the amount of metals in the water but also able to fight cancer cells [9].

Journal of Physics: Conference Series **1384** (2019) 012025 doi:10.1088/1742-6596/1384/1/012025

3. Conclusion

Thus, this is short description of environmental impact of one of many open-pit mines in the world and methods of dealing with the negative effects of this pit. There are several things that could be improved to better eliminate the effects on the environment, and biological ones must not be excluded.

References

[1] Bridget R Barry 2012 Toxic Tourism: Promoting the Berkeley Pit and Industrial Heritage in Butte, Montana. University of Nebraska-Lincoln

[2] Gammons C H, Duaime T E 2006 *Mine Water and the Environment* **25 (2)** 76-85

[3] Gammons C H, Madison J P *2006 Mine Water and the Environment* **25 (2)** 124-129

[4] Metesh J J 2006 *Mine Water and the Environment* **25 (2)** 108-115

[5] Montana Bureau of Mines and Geology – Berkeley Pit: [Electronic resource]. URL: www.mbmg.mtech.edu/env/env-berkeley.asp. (Reference date 08/10/2016)

[6] PitWatch: [Electronic resource]. URL: www.pitwatch.org/. (Reference date 08/10/2016)

[7] Stierle A A, Stierle D B 2005 *Studies in Natural Products Chemistry* **32** 1123-1175

[8] Stierle A A, Stierle D B, Kelly K 2006 *The Journal of organic chemistry* **71 (14)** 5357-5360

[9] The Berkeley Pit: URL: www.atlasobscura.com/places/berkeley-pit. [10] Voropanova L A 2015 *Journal of Mining Institute* **214** 28-32

[11] Zhmurova V V, Nemchinova N V 2018 *Journal of Mining Institute* **233** 506-511

Application of electronic systems of georadical surface sensing

D V Limanskiy[1], **R B Kuskildin** [2]

[1] St. Petersburg State Public Institution «Directorate of transport construction», 8, Neyslotsky lane, St. Petersburg, 194044, Russia
[2] St. Petersburg Mining University, 2, 21-line VO, St. Petersburg, 199106, Russia

E-mail: saturn.sun@mail.ru

Abstract. This article discusses the use of a complex of GPR near-surface sounding and the possibility of using GPR in determining the location and spatial parameters of the outbreaks of coal and coal dumps with the aim of quickly making management decisions to minimize the negative impact on the environment.

1. Introduction

The burning of rock mass in the waste dumps leads to the release into the environment of more than two dozen harmful and hazardous substances. In particular, it is carbon dioxide, sulfur dioxide, sulfuric anhydride, nitrous oxide, hydrogen sulfide, sulfuric acid, sulfur, ammonia, cyanide, etc. Given the fact that most of the waste dumps are located in residential areas, their extinguishing is required, and some cases and complete disassembly [1].

In order to comprehensively improve the ecological and geodynamic situation of urban areas, it is rational to make a complete disassembly of extinguished waste heaps with simultaneous hydraulic laying of the worked-out space under the promising areas with empty rocks from the nearest dumps. The results of research conducted in recent years, suggest the possibility of using in coal mines hardening bookmarks on a cementless basis using advanced pumping equipment [2-4].

The results of monitoring the thermal state of extinguished waste dumps show that a recurrence of fire occurs quite often [5]. In this regard, for the operational control and timely localization of sources of ignition of waste dumps, various methods of thermal control of waste dumps are used: remote sensing from unmanned aerial vehicles, thermal imaging of dumps using multi-zone temperature sensors, etc. However such monitoring methods can only give an idea about the start of the ignition process, and as a consequence - to be a signal for making operational decisions to minimize risks from primary or secondary fire of coal mass. To effectively design work for the suppression, reclamation or complete disassembly of dumps of rock mass, it is necessary to have information about the depth of the location of the source of fire, its length and spatial parameters.

This article describes the method of using the electronic complex of geophysical equipment for remote determination of the spatial parameters of the sources of ignition of coal and coal dumps.

2. Materials and methods

When conducting surveys with GPR, the method of vertical traveltime is usually used [6]. In this case, the GPR moves along the profile with a constant separation between the receiving and transmitting antennas, and at each point of the profile a signal is recorded consisting of a probe pulse and secondary waves (Fig. 1). Any subsurface discontinuities associated with changes in the thickness of the rock mass will give a response in the wave structure of the received signal.

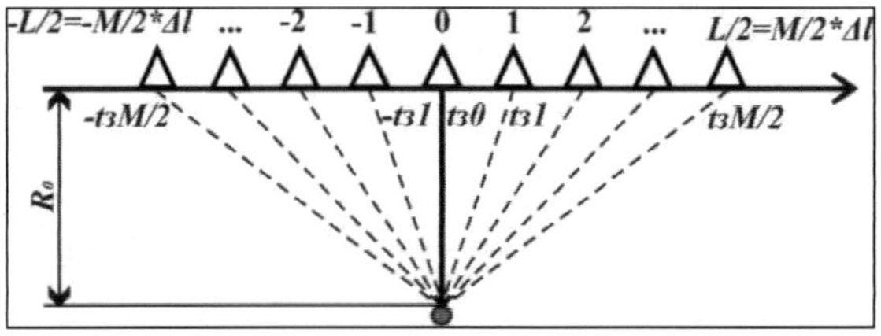

Figure 1. Scheme of movement of the GPR antennas along the route of the waste dump

The most important parameters characterizing the possibility of using the GPR method in various environments are the specific attenuation and the propagation speed of electromagnetic waves in the medium, which are determined by its electrical properties. Specific attenuation determines the depth of sensing. And knowledge of the propagation velocity of radio waves is necessary to recalculate the time delay of the reflected pulse in the distance to the reflecting boundary. Figure 2 shows the propagation of radio waves in the layers of moldboard mass.

3. Field surveys of volumetric parameters of the outbreaks of coal and coal dumps

The most important parameters characterizing the possibility of using the georadiolocation method in various environments are the specific attenuation and propagation velocity of electromagnetic waves in the environment, which are determined by its electrical properties. Specific attenuation determines the depth of sounding. The value of the propagation velocity of radio waves is necessary for recalculating the time delay of the reflected pulse in the distance parameter to the reflecting boundary (Fig. 2).

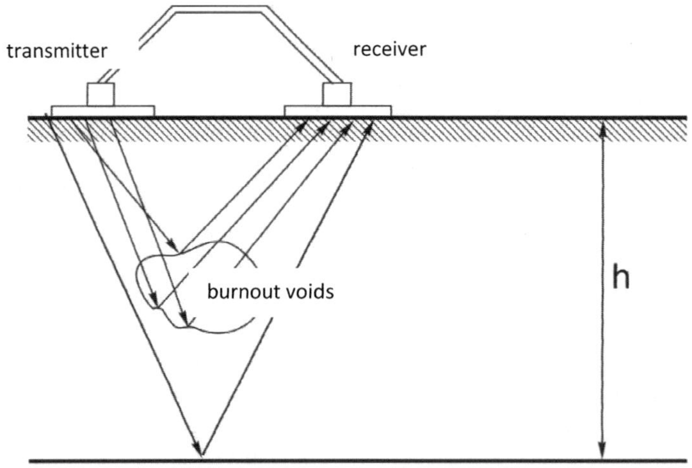

Figure 2. Propagation scheme of radio waves in the decompacted layers of dump rocks

The determination of combustion zones using the GPR method is based on studying the effect of temperature on the dielectric constant and conductivity [7].

The dielectric constant ε shows how many times the field is weakened by the dielectric, quantitatively characterizing the dielectric's property of polarizing in an electric field. The value of the relative dielectric constant of a substance, which characterizes the degree of its polarizability, is determined by the mechanisms of polarization. However, the value largely depends on the state of aggregation of a substance, since during transitions from one state to another, the density of a substance, its viscosity and isotropy significantly change.

The dielectric constant characterizes the speed of passage of a pulse in a medium and in solids can take on very different numerical values in accordance with the variety of structural features of a solid dielectric. In solid dielectrics, all types of polarization are possible.

The main parameters necessary to ensure the correct technology for surveying and interpreting GPR data for most rocks are presented in Table 1 [8].

Table 1. Main parameters for the interpretation of GPR data

Parameter	Interpretation
Dielectric and magnetic permeability	$\mu \approx 1$
Phase velocity	$V_{\Phi} = \frac{1}{R_e\sqrt{\varepsilon}}$
Medium wavelength	$\lambda = \frac{\lambda_0}{R_e\sqrt{\varepsilon}}$
Specific attenuation	$\Gamma = 8{,}68\alpha = \frac{54{,}6}{\lambda ln\sqrt{\varepsilon}}$, $\alpha = \frac{\omega}{cln\sqrt{\varepsilon}}$
The degree of attenuation of the wave at a distance z in dB	$L = 20lg\left[\frac{E_0}{E_z}\right] = 8{,}68\alpha z$

The inspections of areas of burning sites of coal and rock dumps were conducted near the town of Shakhty, Rostov Region. The purpose of these studies was to determine the depth parameters of the source of ignition of coal and coal dumps, as well as zones of decompaction of rocks caused b y burnout of the rock mass [9].

The surveys were carried out according to the scheme shown in Figure 3 in compliance with safety requirements when working in the immediate vicinity of fire sources.

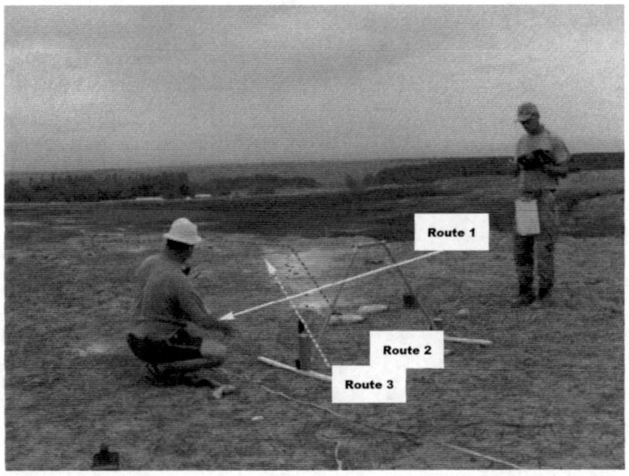

Figure 3. Propagation scheme of radio waves in the decompacted layers of dump rocks

4. Results

The difference in phase velocity for different frequencies determines the variance. The probe pulse in a medium with dispersion changes its shape due to phase distortion of the shape.

The approximate values of the electrical characteristics of some soils and rocks at a field frequency of 100 MHz used to operate in the study area are given in Tables 2 and 3 [10].

Journal of Physics: Conference Series **1384** (2019) 012026 doi:10.1088/1742-6596/1384/1/012026

Table 2. Dielectric values of materials in the frequency range recommended for measurements

Material	Dielectric	Conduction	Specific	Phase velocity
Air	1	0	0	300
Fresh water	81	10^{-3}	0,18	33
Sea water	81	4	330	15
Sandy soil is dry	2,6	$1,4*10^{-4}$	0,14	190
Sandy soil moist	25	$6,9*10^{-3}$	2,3	60
Dry loam	2,5	$1,1*10^{-4}$	0,11	190
Wet loam	19	$2,1*10^{-2}$	7,9	69
Clay soil is dry	2,4	$2,7*10^{-4}$	0,28	190
Clay soil moist	15	$5*10^{-2}$	20	74
Basalt wet	8	10^{-2}	5,6	110
Granite wet	7	10^{-3}	0,62	110
Clay slate moist	7	10^{-1}	45	83
Sandstone wet	6	$4*10^{-2}$	24	110
Limestone moist	8	$2,5*10^{-2}$	14	110
Iron	1	10^{6}	$1,7*10^{7}$	-

As the moisture content of the soil increases, the specific dielectric constant also increases. The electrical conductivity of the soil increases.

Table 3. Dependence of the basic electromagnetic parameters of soils on humidity

Material	Dielectric constant	Conduction
Road construction	5…10	0,0002…0,00002
Rock	4…10	0,01…0,0000
Clay	4…16	0,05…0,0002
Loam	2,5…19	0,021…0,00011
Sand	3…25	0,007…0,00002
Peat wet	50…78	0,002…0,001
Moraine	9…25	0,01…0,0001
Il	9…23	0,001…0,0001
Metal	1…2	1000000
Ice	3…4	0,001
Water	80…81	0,002…0,001
Air	1	0
Road construction	5…10	0,0002…0,00002
Rock	4…10	0,01…0,0000

The minimum values of the dielectric constant correspond to the dry state of the material, the maximum - to the wet [11].

With an increase in the water content in the examined layers, the parameters of the specific dielectric constant also increase, and the electrical conductivity also increases.

When examining the zones of dumps intended for thermal imaging, the values of the dielectric constant of the corresponding rocks in the non-wetted state are taken, since in the field of examination the surface is exposed to heating and drying under the thermal influence of the combustion centers.

Journal of Physics: Conference Series **1384** (2019) 012026 doi:10.1088/1742-6596/1384/1/012026

Figures 4-6 show the results of the traversing of the profiles of tracks 1-3 by the GROT 12E receiver and the 9 kV transmitter using antennas - 1 m.

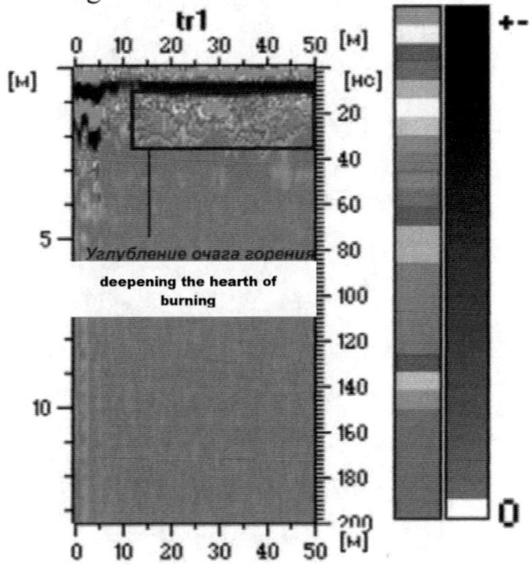

Figure 4. Radarogram profile on route 1

The results obtained from the survey of route 2 are presented in Figure 6 and largely repeat the results obtained from route 1. There is a clear change in the dielectric constant parameter of rocks as the source of combustion approaches the outer surface of the blade, as well as the difference in the areas of decompaction of rocks caused by the burning out of rock volumes [12-13].

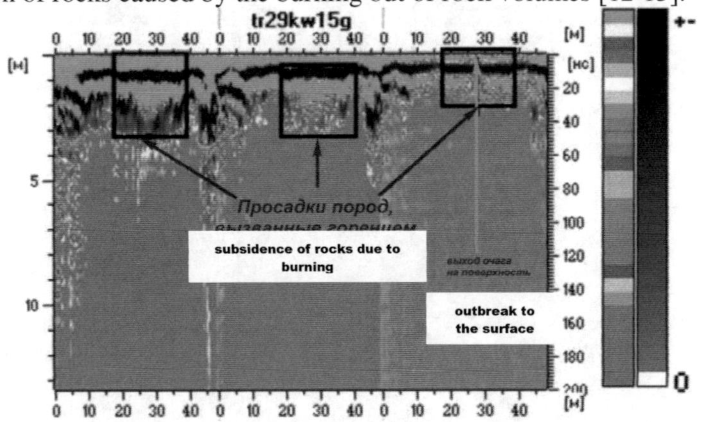

Figure 5. Radarogram profile on route 2

When examining the territory along the route section 3 (Figure 6), a zone of burning of maximum intensity was identified. In this area, the temperature of the surface rocks was higher than the ambient temperature, which was clearly determined without the use of special devices. On the radarogram this zone is highlighted as a separate section.

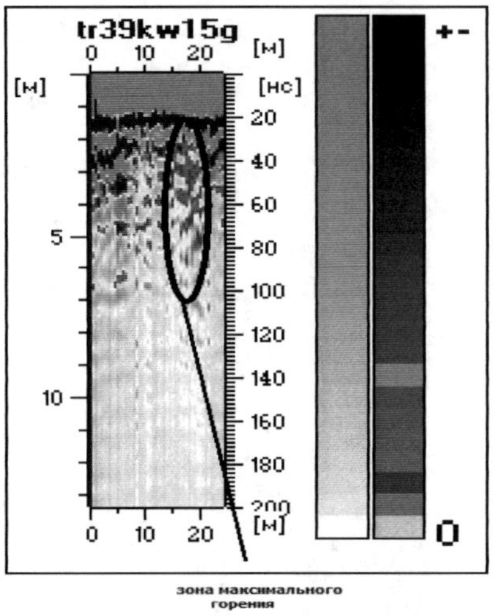

Figure 6. Radarogram profile on route 3

maximum burning area

To confirm the obtained results, the passage on route 3 using antennas 2 m, GROT 12E receiver, 9 kV transmitter was repeated. The results of the site survey are presented in Figure 7.

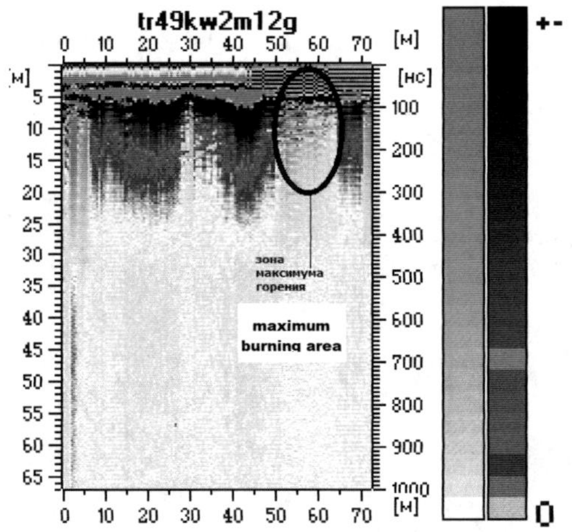

Figure 7. Radarogram profile on route 3 using 2 m antennas

5. Conclusion

Burnout zones of rocks in coal dumps are clearly visible on radarograms due to the appearance of softening and voids in the dump mass, formed as a result of the combustion process. In addition, this method allowed tracing the trajectory of the spread of combustion along the length of the survey routes.

The obtained results allow concluding about the consistency of this method to obtain additional information used in planning activities for the suppression and recultivation of coal dumps located in residential areas.

Further study of the temperature dependence of the dielectric constant, as well as the application of the above-described method of remote sensing of zones of softening and burnt voids in combination with the use of additional thermal control methods will allow building three-dimensional models of high accuracy before starting the work of extinguishing and reclamation. This will reduce labor costs and the cost of work, making their planning more accessible and taking measures to improve the environmental situation of mining towns and settlements more accessible.

References

[1] Zaidenvarg V E, Navitniy A M, Semikobyl Ya G, 2002, Scientific foundations of the complex ecological area of the closed mines. - Moscow: Rosinformugol. 226 p
[2] Liss N.Yu., Kremcheev, E.A., Afanasiadi, K.I. 2017 *Gornyi Zhurnal* **10** p 19-26
[3] Vasilyeva M.A., Volchikhina A.A., 2018 *Journal of Physics: Conference Series* **1118(1)** 012047
[4] Technical report on the results of mountain-ecological monitoring at the mining allotment of the mined mines of Eastern Donbass for the fourth quarter of 2011.
[5] Krasavin A P, Navitniy A M, Kaplunov Yu V, 2010, Environmental monitoring in the liquidation of mines and sections. - Perm, OJSC "MNIIECOTEK". 315 p.
[6] Saraev A, Simakov A, Shlykov A, Tezkan B, 2017 *Journal of applied geophysics* **146** 228-237
[7] Limanskiy D V., Vasilyeva M A., *Journal of Physics: Conference Series* **1118(1)** 012023
[8] Ikeda K, Steiner M, Thompson G 2017 *Weather and forecasting* **32(3)** 949-967
[9] Limanskiy A, Vasilyeva M, 2016 *Ecological Engineering* **91** 41-43
[10] Zibrov G V, Zakusilov V P, Buryak Z A, 2017 *Research journal of pharmaceutical biological and chemical sciences* **8(2)** 2439–2443
[11] Slukovskii Z I 2015 *Russian meteorology and hydrology* **40(6)** 420–426
[12] Tskhai A, Shirokova S, Konev D, Koshelev K, Tskhai T 1995 *Modelling and management of sustainable basin-scale water resource systems* **231** 263–270

Method of determining the axial profile and diameter in various micro-holes' cross-sections after electrical discharge machining

A M Loiko[1], A F Bojko[1] and A I Shestakov[1]

[1]Belgorod State Technological University named after V.G. Shukhov, 46, Kostyukova str., Belgorod, 308012, Russia

E-mail: lam.bel@mail.ru

Abstract. The paper presents a new method for the assessment of the accuracy of micro-holes obtained by electrical discharge machining, which was called the "blades method" by the authors. This method allows estimating to a high precision the errors in size and shape (taper, ovality) of the holes in the cross-sections of different heights and the axial longitudinal section of the hole. The application of this method is especially effective in studies of the blind micro-holes, as well as non-cylindrical holes, the measurement of which is especially difficult. The developed method was tested in the study of hole profiles processed at different pulse energy. A technological method for EDM of the holes with the help of two modes is proposed, which reduces the main hole shape errors. The application of the developed method allowed increasing significantly the efficiency of studying the electrical discharge machining micro-holes accuracy.

1. Introduction

As the practice of modern mechanical engineering production shows, for the processing of precision micro-holes, the application of the method of electrical discharge machining (EDM) is especially effective. With the development of mechanical engineering, there is a complication of products with micro-holes and an increase in the basic requirements for the accuracy of their manufacture. In this regard, studies aimed at reducing errors arising from EDM processing and increasing the accuracy are currently actual [1–6].

The most common errors of the micro-holes shapes are: in cross sections – ovality (deviation from circularity), in the longitudinal axial section - taper (deviation from cylindricity). The ovality of the hole in the cross-sections occurs due to a deviation from parallelism of the geometric axis of the tool electrode relative to the direction of its supply. The taper of the hole in the longitudinal axial section (increasing the hole's diameter from the lower cross sections to the upper) is formed due to physicochemical processes in the lateral interelectrode gap: by additional electrical discharges going through the processing products during their evacuation from the interelectrode gap, as well as electrochemical anodic dissolution of the hole surface material [7–12].

In order to research the main shape errors that occur during EDM of micro-holes, in this paper a method for determining the axial profile and hole diameter in various cross-sections, which the authors called the "blades method", was developed and applied in practical studies.

2. Equipment and materials

Experimental research was performed on a Russian EDM machine 04EP-10MF2 with a short bipolar pulses transistor generator of the nanosecond range. The processing was carried out by tungsten tools electrode with a diameter of 0.25 mm. The workpiece is made up of multiple tightly pressed blades made of stainless steel 65H13 (Russian designation, analogue of AISI 425 Mod) with a total thickness up to 3.8 mm. During processing, as the working fluid was used the tap water.

In this work, the size measurements and shape errors of the obtained holes were measured using the Levenhuk D70L digital microscope, previously calibrated to determine the sizes with the help of the Olympus micrometer object, which is a glass plate with a micrometer scale. In order to process the obtained digital images of microholes, the Levenhuk ToupView software package was used.

3. Results and Discussion

The technological scheme of the developed method is presented in Figure 1, a. The sample in which the micro-hole is machining, presents a pile that consists of many *N* blades *3*, tightly pressed against each other by means of a screw *4*, an insert *5* and metal clips *6*. The thickness of one blade is about 0.1 mm. The processed sample is fixed in the device *1*, adjusted on the table *2* of the EDM machine. After machining holes in the pile of blades with an tool electrode *7*, it is disassembled and, successively, starting from the upper blade, the obtained holes are measured using a digital microscope (figure 1, b).

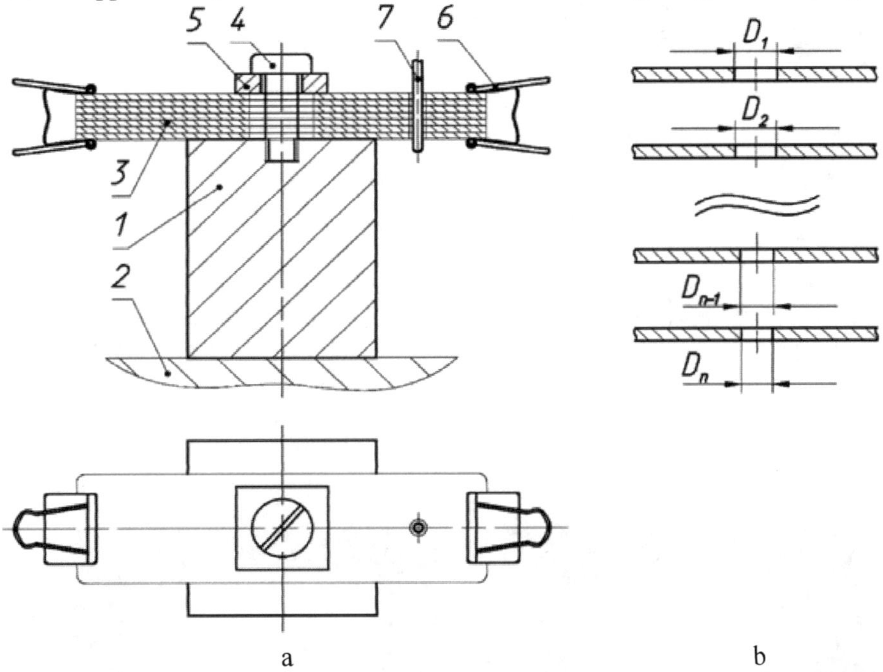

Figure 1. Method of determining the axial profile and diameter in various micro-holes' cross-sections: a – processing scheme; b – hole measurement; 1 – device; 2 – machine table; 3 – workpieces; 4 – screw; 5 – insert; 6 – metal clips; 7 – tool electrode.

While using this method and workpiece construction, it seems possible to determine the diameter in many *N* sections of the hole, to establish the nature and regular relationships of changes in the diameter (taper) of the hole, as well as its deviation from circularity (ovality), depending on the depth of the cross-section micro-hole, to find a mathematical model of changes in the hole diameter in depth. The application of this method is especially effective in research of blind micro-holes, as well as non-cylindrical holes, the measurement of which is especially difficult.

Based on the results of research of 10 holes profiles with a diameter of 0.25 mm and a depth of 3.8 mm processed using the developed method, it was concluded that the most significant breakdown and the greatest ovality are formed in the upper sections of the hole. At the entrance of the hole, the ovality and average taper increase slightly, because as the tool electrode enters the workpiece, it is affected by the resulting force of electric discharges, which can deviate the axis of the non-rigid tool electrode from the feed axis within the backlash value. Even a slight deviation of the tool electrode axis from the feed direction during entry can lead to a significant increase in the breakdown and ovality of the hole.

In connection with this observation, there was carried out an experiment where the dual-mode EDM of hole was used. A small part of the hole (about 10% of the total depth) for stabilizing the process of entering the tool electrode into the workpiece was processed in the "soft" mode with low pulse energy E_{min}, then the main part of the hole was machined in a significantly more "rough" high-performance mode with high pulse energy E_{norm} (figure 2).

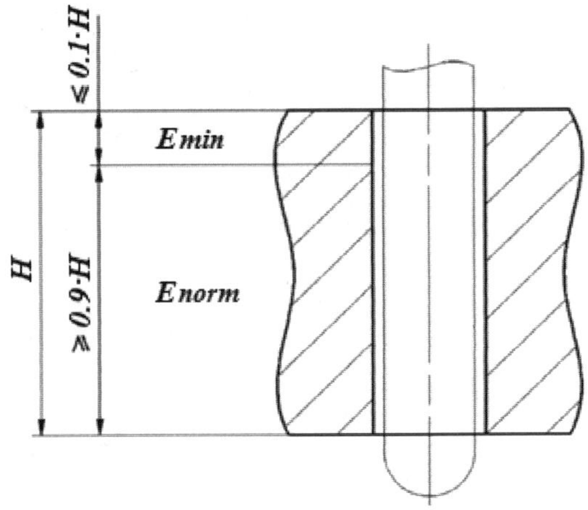

Figure 2. The micro-hole dual-mode EDM scheme.

Within the experimental research, two groups consisting of four holes were processed. The EDM of the first group was performed in a constant high-performance mode with a pulse energy E_{norm} = 0.353 mJ. The EDM of the second group was carried out using two modes: to a depth of 0.3 mm, the processing was performed in the "soft" mode with low energy E_{min} = 0.021 mJ, then the mode was switched to the E_{norm} = 0.353 mJ, where the rest part of the hole was processed. Processing was carried out at a constant pulse frequency f = 44 kHz. To stabilize the processing, the tool electrode was given vibration with a frequency of 380 Hz and an amplitude of 0.01 mm.

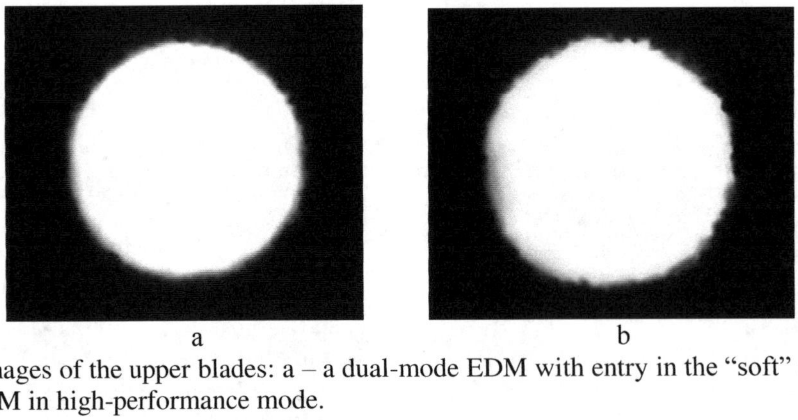

a b

Figure 3. Images of the upper blades: a – a dual-mode EDM with entry in the "soft" mode; b – with standard EDM in high-performance mode.

According to the results of the experiment, the two-mode EDM of the small holes showed a significant decrease in the breakdown (by 17.5%) and taper (by 10.5%) of the holes. Figure 3 shows the images of the holes in the upper blades, that were obtained with the two considered processing options. It is clear that with a dual-mode EDM there are no surface defects on the input edges of the holes, and that is an important parameter for ensuring the machined holes quality.

4. Conclusion

Based on the research results, a new method for the assessment of the EDM micro-holes accuracy was developed. This method allows estimating to a high precision the errors in size and shape (taper, ovality) of the holes with different cross-section heights and with the axial longitudinal section of the hole. The application of this method is especially effective in research of blind micro-holes, as well as non-cylindrical holes, the measurement of which is especially difficult. The application of this method allows finding a mathematical model for changing the diameter of the hole depending on the cross-section depth.

The developed method was used in an experimental research of the hole profiles, that were obtained by EDM. It was found that at the input of the holes, the ovality and average taper slightly increase. In this regard, an experiment was carried out using dual-mode hole machining with the entering of the tool electrode into the workpiece in the "soft" mode with low pulse energy. According to the results of the experiment, the two-mode EDM of small holes showed a significant decrease in the breakdown and taper of the holes, as well as the absence of surface defects of the input edges of the holes.

Application of the developed method of measurements allowed increasing significantly the efficiency of studying the EDM micro-holes accuracy.

5. Acknowledgments

The article was prepared within development program of the Flagship Regional University on the basis of Belgorod State Technological University named after V.G. Shukhov, using equipment of High Technology Center at BSTU named after V.G. Shukhov.

References

[1] Jahan M P, Rahman M and Wong Y S 2011 *International Journal of Machine Tools & Manufacture* **51** 837–858
[2] Puzacheva E I, Bojko A F and Macpherson A V 2014 *Research Journal of Applied Sciences* **9(22)** 691-695
[3] Boyko A F, Loyko A M and Shestakov A I 2016 *Bulletin of BSTU named after. V. G. Shukhov* **11** 128–131
[4] Saushkin B P 2004 *Elektronnaya Obrabotka Materialov* **1** 4–17
[5] Pham D T, Dimov S S, Bigot S, Ivanov A and Popov K 2004 *Journal of Materials Processing Technology* **149** 50–57
[6] D'Urso G, Maccarini G and Ravasio C 2014 *International Journal of Advanced Manufacturing Technology* **72** 1287–1298
[7] Davim J P 2013 *Nontraditional Machining Processes* (London: Springer-Verlag) p 232
[8] Chung D K, Shin H S, Park M S and Chu C N 2011 *International Journal of Precision Engineering and Manufacturing* **2** 195–201
[9] Nguyen M D, Rahman M and Wong Y S 2012 *International Journal of Advanced Manufacturing Technology* **58** 533–544
[10] Gosavi V Y and Phafat N G 2016 *International Journal of Materials Science and Engineering* **1** 35–45
[11] Jahan M P, Rahman M, Wong Y S and Fuhua L 2009 *Proceedings of the International Mechanical Engineering* **224** 795–814
[12] Tai T Y, Masuzawa T and Lee H T 2007 *Materials Transactions* **2** 205–210

Low-cost sorption technologies for large-tonnage wastewater treatment in mining industries

V A Matveeva, M A Chukaeva, I P Sverchkov

Saint Petersburg Mining University, 2, 21st Line, Saint Petersburg, 199106, Russia

E-mail: poveriya@mail.ru

Abstract. The current state of water bodies in the zone of influence of the largest mining complexes in Russia – JSC "Apatit" and JSC "Kovdorsky GOK" was assessed on the basis of the conducted monitoring studies. The review of the natural and artificial materials with the high sorption capacity was carried out on the purpose of elimination of the revealed sources of hydrochemical anomalies. The paper presents the laboratory test results of sorbents depending on the methods of their modification, composition and properties of purified water. The introduction of developed sorbents into the wastewater treatment systems will significantly reduce the negative impact of the mining enterprises on surface waters, thus it will improve the environmental situation in the studied areas and the life quality of the local community.

1. Introduction

The development of human activity is accompanied by the involvement of new materials in the production process. In turn, to obtain them, it is necessary to increase the mining and processing volume of minerals, which is leads to the expansion of search geography and to the alienation of new territories. As a result, today there are several serious problems, including environmental ones, in the mining and processing industries. Modern mining and processing technologies of minerals are still remaining multi-waste in some cases, which leads to contamination of all components of the environment. The annual extraction of all kinds of raw materials in the world is about 25 billion tons; overall, less than 1.5 billion tons of them are used as finished products. The remaining part is stored in the waste, to accommodate those more than 10 thousand hectares of land are annually alienated [1, 2].

As a result of long-term waste storage, not only soils but also natural waters and the atmosphere are polluted, which adversely affects on humans health. The negative impact persists for 150 - 200 years or more even after the termination of field development and conservation of mine workings, due to the hypergene transformation of rock formations.

Anthropogenic waters are one of the most dangerous types of mining waste due to their high migration capacity. Mine waters and sewage from the tailings of processing plants pose a greater threat to water resources, located in the areas of mining and processing enterprises. A huge amount of pure water, which is taken from surface water bodies, is used at each technological process to obtain marketable products from ores of various metals. Processed water is saturated with various pollutants, including dangerous synthetic reagents and components of the extracted ore, during ore processing. Significant volumes of waste waters, which are generated as a result of minerals extraction and processing, their multicomponent composition and the uneven inflow, currently make it impossible to solve the problem of sewage treatment using standard methods at the majority of mining enterprises. This fact is confirmed by the increasing total amount of sewage discharge. Thus, according to the Federal Agency for water resources, as a result of mining, the total volume of sewage in 2018 in the Russian Federation was 1394 million m^3, including 996 million m^3 of polluted wastewater.

The long-term irrational use of mineral resources has led to a shortage of drinking water in a number of mining regions.

The zones of influence of such enterprises as JSC "Apatit" and JSC "Kovdorsky GOK" are vivid examples of mining areas with unfavorable ecological situation connected with the formation of extensive hydrochemical halos and streams of heavy metals pollution. The first of them is JSC "Apatit" - the world's largest manufacturer of high-grade phosphate raw materials and nepheline concentrate. The activity of this enterprise is annually accompanied by a discharge of 50 million m^3 mine waters. As a result, a high-contrast hydrochemical anomaly in the basin of the lake Imandra was formed, in which the molybdenum content is 100 times exceeds the maximum allowable concentration for fishery waters (MAC) [3, 4].

The second of them is JSC "Kovdorsky GOK" - manufacturer of iron ore, apatite and baddeleyite concentrates. Water pollution in the Kovdor river basin is caused by the annual discharge of 50 million m^3 of insufficiently treated industrial wastewater into water bodies. 20 million m^3 of sewage, in which a manganese (II) concentration 90 times exceeds the MAC, are come from the tailing pond [5].

The studied objects - mining and processing plants - are located on the territory of the Kola Peninsula in the Northern part of the Russian Federation, which aggravates the environmental problem due to poor stability and low self-cleaning ability of the investigated ecosystems. At the same time, the main pollutants of wastewater belong to heavy metals and have a toxic effect on hydrobionts and the ecosystem as a whole. In addition, these pollutants can accumulate in sediments, becoming a source of secondary pollution.

Taking into account the above, the search for effective and safe sewage purification technologies from heavy metal is one of the priority areas in the field of environmental protection. The choice of the optimal sewage treatment method is determined by many factors, including the concentration and form of the element in the wastewater. The selection of the sewage treatment scheme depends on the amount and composition of wastewater, the possibility and economic feasibility of extracting impurities, the requirements for the quality of purified water and so on. Nowadays, combined treatment methods based on mechanical, chemical, physico-chemical and biological methods are used for large-tonnage sewage purification from heavy metals.

2. Materials and methods

Nowadays, several methods for purification of manganese-containing wastewater have been developed and are actively implemented at industrial facilities. The five main methods of wastewater treatment from manganese are ion-exchange, reagent methods, oxidizing filters, aeration under pressure with subsequent filtration, chemical oxidation with subsequent deposition and filtration, sorption.

Numerous works and developments are devoted to solving the problem of molybdenum-containing wastewater treatment. The main methods of wastewater treatment from molybdenum ions are ultrafiltration, electrodialysis, reagent methods, sorption and ion exchange.

Taking into account the specifics of the sewage formation at the mining enterprises, typical methods are required a significant refinement and additional field and laboratory studies.

A detailed literature review showed that nowadays methods of waste water purification from molybdenum and manganese based on sorption are most widespread. Their application is characterized by high productivity, as well as the possibility of selective treatment of multicomponent wastewater.

Sorbents are divided into carbon and non-carbon. The first are active coals, peat and other materials, those are obtaining mainly by the processing of various organic materials. At the same time, non-carbon sorbents prepared from natural and artificial materials (clay rocks, zeolites and etc.) are increasingly used. The wide using of such sorbents is due to their selectivity, relatively low cost, availability and cation exchange properties of some of them. The relevant trend is using production waste as cheap raw materials to obtain sorbents and, at the same time, recycling of accumulating by-products.

Despite this, the widespread use of sorption methods for industrial wastewater treatment from heavy metals is limited for a number of reasons. The most important of them is insufficient sorption capacity of the material in its original form in relation to heavy metals and lack of reliable regeneration methods of sorbents [6, 7].

Among the materials with high sorption activity in relation to manganese are montmorillonite, limestone and dolomite [8, 9]. However, these sorption materials are not in demand for the industrial sewages purification for the following reasons: insufficient cleaning efficiency, lack of selective extraction of heavy metals, limited raw material base for the sorbents production.

The review of sorption materials providing sewages purification from molybdenum showed that their effective work is observed in the pH range from 1 to 7 pH units. In addition, the existing sorbents are practically not used on an industrial scale due to the high cost and limited amount of feedstock, which includes bentonite clay, activated bone charcoal, aluminum oxide, bean skin, peat and others [10].

Thus, the search and study of sorbents on the basis of natural materials and production waste, capable to provide an increase of industrial wastewater treatment efficiency, is an urgent problem.

In order to solve the current environmental problem related to wastewater treatment from manganese and molybdenum, the sorption activity of natural materials and industrial waste was assessed in the laboratory.

Vermiculite from Kovdor Deposit, located in close proximity to sources of formation of hydro-chemical anomalies, was chosen as a potential sorbent for treatment of the considered sewage. It has a high selectivity to large cations of alkaline, alkaline earth and heavy metals, which indicated the prospects of its application in wastewater treatment of manganese and molybdenum. The study of vermiculite as a sorbent began with the analysis of its composition and properties. The chemical composition of Kovdorsky vermiculite was determined by the method of atomic emission spectrometry with inductively coupled plasma on the analyzer ICPE-9000. The results of the analysis are presented in table 1.

Table 1. Chemical composition of Kovdor vermiculite concentrate

Components	Content, %
SiO_2	40.0
Fe_2O_3	12.5
Al_2O_3	12.1
MgO	29.7
TiO_2	0.9
CaO	2.3
K_2O	1.8

As it can be seen from table 1, the main components of vermiculite are Si, Fe, Al, Mg. Data of x-ray structural analysis showed that vermiculite belongs to water aluminosilicates of complex chemical composition.

The parameters of the porous structure of vermiculite were studied using the method of mercury penetration porosimetry, which allowed obtaining a qualitative and quantitative assessment of the vermiculite pores, as well as calculating the specific surface area (table 2). The value of the specific surface of vermiculite studied samples amounted to 700-800 m²/g.

Table 2. Parameters of the porous structure of Kovdor Deposit vermiculite

		Varieties of pores					
		Transitional pores			Micropores		
Sorbent	V_s, cm³/g	V_{tr}, cm³/g	S_{tr}, m²/g	R_{tr}, Å	V_{mic}, cm³/g	τ, Å	V_{mic}/V_s
Kovdor Deposit vermiculite	0.18	0.03	12	90	0.15	5.6	0.83

*Note: V_s is the limiting adsorption volume; V_{tr} is the volume of transitional pores; S_{tr} is the specific surface area of transient pores; R_{tr} is the radius of the transitional pores; V_{mic} is the volume of micropores; τ is the thickness of the micropores.

The results of the comprehensive analysis showed that the vermiculite refers to a microporous adsorbent with changing pore sizes in the process of adsorption. In addition to the primary microporosity, due to the crystalline structure, vermiculite has secondary, mainly, transitional pores formed by gaps between the contacting particles. At the same time, the surface and volume of secondary pores in vermiculite is much smaller than the surface and volume of primary pores.

The evaluation of sorption properties of vermiculite in relation to the studied cations was carried out on model solutions under laboratory conditions using high-tech equipment of the Environmental laboratory of the Mining University. The analysis of metals residual concentrations in solutions by atomic absorption method on Shimadzu AA-7000 spectrometer showed that after contact with a potential sorbent under static conditions the degree of manganese (II) extraction was 51 %, molybdenum sorption from the solution was not observed. In this regard, there was a need to develop and evaluate possible methods of vermiculite modification in order to increase its sorption capacity with respect to manganese (II), as well as to continue the search for a sorbent capable of extracting molybdenum ions from aqueous solutions.

To solve the problem of surface water protection in Apatitsky district, a detailed study of the mineral resource base of the Kola Peninsula was carried out. Clays were chosen among the available and common natural sorbents. During the laboratory experiment, the maximum efficiency of model solution purification equal to 69% was achieved with Pechenga Deposit clay. A characteristic feature of this clay is an increased iron content, which according to chemical analysis reaches 22% in terms of Fe_2O_3.

It was established, due to the applying of modern instrumental methods of research, that the chemisorption process, based on the formation of chemical compounds $FeMoO_4$ and $Fe_2(MoO_4)_3$ and their subsequent sorption on developed surfaces, is an essential factor in the purification of molybdenum-containing wastewaters. However, using clays did not allow achieving the required standard values of molybdenum concentrations in purified solutions. In this regard, the next stage of laboratory research was the study of sorption abilities of iron-containing waste, which included sulfuric acid production slag and metalworking waste. The degree of molybdenum extraction from the model solution reached 98-99 % in case of using each of the waste types. Despite this, further work with slag was impractical due to desorption of a number of toxic components (Zn, Mn, Fe, Cu) from the waste during the purification process. Thus, the sewage treatment using metal processing waste (steel chips), providing efficiency up to 99 % has become the most promising method.

The development of the preparation technology of the adsorbent and assessment of its work depending on the composition and properties of the treated water has become the main objectives of further laboratory research.

3. Results

Given that the properties of vermiculite predetermine its possible grinding by impact and cutting, at the initial stage, the method of mechanical activation of vermiculite was chosen, which will allow showing the maximum sorption activity of the sorbent with respect to the extracted manganese ions. To grind the material by impact, jaw and hammer crushers were used; layer-by-layer disintegration was carried out in a rotary crusher with a cutting-grinding nozzle. In order to increase the specific surface area of vermiculite concentrate, heat treatment was also used, after which vermiculite is able to swell. Based on the above, parts of the vermiculite test portions were incubated in a muffle furnace at temperatures of 300 ^{0}C and 600 ^{0}C for two hours. Laboratory tests showed that high efficiency of purification of manganese-containing water, reaching 88 %, was observed in the vermiculite disintegration with the predominant use of cutting and shear forces without additional heat treatment. However, the results were obtained using samples of size 1-2 mm. It is known that the grinding of the sorption material leads to an increase in the contact area with the purified water and, as a result, significantly increases the amount of adsorption of pollutants from the solution. The study of the influence of the vermiculite samples size on the purification efficiency showed that the maximum permissible concentration (0.01 mg/dm^3) in the model solution with the initial manganese content of 1 mg/dm^3 can be achieved in the case of using a layer-by-layer crushed sorbent with a size of 0.5-1.0 mm and less. Further, the adsorption values of the solutes on the solid sorbent were obtained experimentally. The adsorption value was 2.1 mg/g for vermiculite with a size of 0.5-1 mm. In General, the results of laboratory tests showed that for a smaller fraction the adsorption value is much higher than the adsorption value for large fractions at the same equilibrium concentrations. This is due to an increase in the specific surface area of the sorbent and, possibly, opening more micropores. The study of sorption kinetics showed that the rate of sorption of manganese on vermiculite is maximum in the first minutes of contact. First of all, the equilibrium was occurred in the system with the participation of vermiculite with a size of 0.125-0.315 mm and amounted to 20 minutes, with the participation of vermiculite with a size of 0.315-0.5 mm and 0.5-1 mm – 30 minutes.

Further, the influence of the composition and properties of the purified solution on the vermiculite sorption properties was studied. The limiting parameters of the solution were: the concentration of manganese in the solution, the temperature of the solution and the value of the hydrogen index.

Model solutions with concentrations ranging from 0.5 to 5 mg/dm^3 were prepared to assess the degree of manganese extraction by vermiculite depending on the initial concentration. Contact of the sorbent with the solution was carried out under static conditions in the mode of mixing at a constant temperature of 20 ^{0}C. The obtained experimental data showed that the degree of manganese cations extraction with an increase in the concentration of solutions decreases. This is due to the fact that the ion strength of the solutions increases with the increase in the concentration of the initial solution and the activity of manganese cations decreases. Consequently, the number of free manganese ions in the solution becomes smaller, and therefore the degree of cation extraction is reduced.

The effect of temperature on sorption from aqueous solutions is far from clear. It is known that the penetration of substances into the pores depends on their kinetic energy in case of the sorption of these substances on microporous sorbents, the size of which is close to the effective size of the pores. At a sufficient temperature, the sorbate molecules penetrate into the pore and are sorbed; otherwise, there is only a slight absorption on the surface of the meso- and macropores. In other words, the sorption capacity increases with temperature. This phenomenon is called "activated adsorption". At the same time, physical sorption, like any exothermic process, generally deteriorates with increasing temperature. Therefore, the total externally recorded manifestation of these two phenomena can have a maximum at a certain temperature. The experiment was conducted to confirm the proposed assumptions. The results of the research allow making a clear conclusion that with the decrease in temperature by 5 ^{0}C, the degree of manganese extraction from the model solution falls by 3 %. The peak of sorption on the selected temperature range of the model solution 5-20 ^{0}C was not recorded.

One of the most important factors in the sorption extraction of ions from the solution is the value of the acidity of the medium, which affects both the form in which the ion under study is in the solution and the state of the ionogenic groups. In most cases, the decrease in acidity causes an increase in the adsorption of pollutants from wastewater. However, it was found that acidic solutions destroy the crystal lattice of vermiculite. The destruction of the lattice occurs unevenly and leads to a decrease in the exchange capacity of vermiculite. The study of the influence of pH on the processes of manganese cation sorption was carried out in the range from 3 to 10 units. At pH = 3, there was a sharp decrease in the degree of manganese extraction from the solution, by almost 30 %, due to the partial destruction of vermiculite in an acidic medium. In all other solutions, the purification efficiency was not less than 95%.

As a result of studies purification efficiency up to 95-99% of manganese-containing waters of the Kovdor Deposit by the using of the modified vermiculite has been proven.

As it was noted earlier, the waste, represented by metal chips, has proven to be available for use as a sorbent for the molybdenum extraction from wastewater. Due to the fact that the main role in the purification process is played by iron oxides, which have a developed specific surface area and increased reactivity, chip samples were treated with distilled water and kept in the air for 24 hours for the intensification of the corrosion processes. Further, the analyses were carried out on a Mossbauer spectrometer MC1104Em with the 57Co (Cr) source in order to determine the exact composition of corrosion products of iron-containing metal waste

According to the results of Mossbauer spectrometry the content of Fe_2O_3 is 32.2 %, and Fe_3O_4 - 67.8 %. At the same time, the assessment of the efficiency of model solutions purification with each of the corrosion products separately showed that the main role in the molybdenum-containing waters treatment belongs to Fe_2O_3.

It should be noted that the intensification of the corrosion process, which is directly related to the subsequent rate of molybdenum extraction from the solution, is influenced by such parameters as the temperature of the purified solution, its acidity level, and the content of dissolved oxygen. A series of experimental studies have shown that:

- with an increase in temperature by 5 0C, in the studied temperature range (5 0C – 25 0C), the efficiency of model solutions purification from molybdenum increases by 5- 10 % on average, as the rate of chemical reactions, the diffusion rate and the solubility of corrosion products increases
- cleaning efficiency is inversely proportional to the pH of the purified solution. So at pH = 9, the amount of molybdenum ions extracted from the solution is 3 times lower than at pH = 3, which is due to the high corrosion rate of metal waste in an acidic medium;
- with an increase in the content of dissolved oxygen 2.5 times the efficiency of cleaning the model solution increases by 2 times. This is due to the intensification of the process of steel chips corrosion, as well as to the formation of mainly iron oxide (III), which showed greater efficiency of cleaning compared with iron oxide (II).

The study of the chemisorption process of molybdenum from aqueous solutions showed that steel chips ion exchange capacity is limited only to the passivation process of iron. This process occurring in the formation of iron molybdate on the steel chips surface with a specific surface area less than 15 cm^2/g. The obtained results allow concluding about the possibility of purification of JSC "Apatit" mining waters up to the standards values with the chemical adsorbent represented corroded metal steel chips application.

4. Conclusion

The results of the complex scientific and practical research pointed to the prospects of using of the sorbents based on natural materials and industrial waste in the purification process of multi-tonnage sewage from heavy metals, which will make it possible:

to increase the efficiency of sewage purification in relation to manganese and molybdenum up to 99 %;

- to develop environmentally efficient and cost-effective sorption technologies for wastewater treatment using the obtained sorbents;
- to reduce the risk of secondary pollution of purified water by using safe methods to increase the sorption activity of the sorption materials under study and the absence of their chemical treatment;
- significantly reduce the cost of manganese-containing wastewater treatment due to using vermiculite as a sorbent, which is a natural mineral available in the study area with the deposit located in close proximity to the mining facility;
- to implement waste-free technology at the industrial facility as a result of using of metal production waste for the molybdenum-containing waters purification;
- to improve the environmental situation by preventing further formation of hydrochemical anomalies;
- to improve the quality of life of the population in mining areas.

5. Acknowledgments

The reported study was carried out according to the state target № 5.12850.2018/8.9 of the Ministry of Education and Science of the Russian Federation.

References

[1] Kuznetsov V S and Petrov D S 2017 Assessing the environmental condition of minor rivers in urban areas *Journal of Ecological Engineering* **18(6)** 110-114

[2] Kondrasheva N K, Nazarenko M Y and Saltykova S N 2018 Efciency of applying oil shale and ash-shale wastes for water treatment from organic pollutants *Bulletin of the Tomsk Polytechnic University, Geo Assets Engineering* **327(9)** 95-103

[3] Isakov A E and Chukaeva M A 2016 Ecological and geochemical peculiarities of surface water transformation in the area of the enterprise JSC "Apatit" impact *International Journal of Ecology Development* **31(2)** 90-98

[4] Limanskiy A V and Vasilyeva M A 2016 Using of low-grade heat mine water as a renewable source of energy in coal-mining regions *Ecological Engineering* **91** 41-43

[5] Pashkevich M A, Matveeva V A and Danilov A S 2019 Migration of pollutants from the mining waste disposal territories on the Kola Peninsula *Gornyi Zhurnal* **1** 17-21

[6] Sulimova M A, Sizyakov V M, Litvinova T E and Vasilyev V V 2016 On possibility of the use of metallurgical production wastes as a sorbent in the industrial water cycle *Chernye Metally* **8** 43-49

[7] Tryakina A S 2017 Development of sustainable water treatment technology using scientifically based calculated indexes of source water quality indicators *Journal of mining institute* **227** 608-612

[8] Levkin N D, Afanaseva N N, Malikov A A and Fisher W L 2014 Purification of waste waters by natural sorbents *News of Tula state University* **4** 37-41

[9] Pimneva L A and Koroleva M N 2015 Sorption treatment of industrial wastewater from manganese and nickel ions *International journal of experimental education* **3(1)** 83-84

[10] Alekseenko V A, Maximovich N G and Alekseenko A V 2017 Geochemical Barriers for Soil Protection in Mining Areas *Assessment, Restoration and Reclamation of Mining Influenced Soils* **1** 255-274

Using the LABVIEW software package to automate production and research tasks

A V Melnikova[1]

[1]Saint-Petersburg Joint-Stock Company WERTEKS, 27A, 24 Line, Saint-Petersburg, 199106, Russia

E-mail: mlnkval77@gmail.com

Abstract. Currently, improving the quality indicators of mass production and consumer goods is impossible without the use of modern software systems, means and methods of automation. The main systems capable of satisfying the tasks of modern enterprises and scientific organizations include Scada-systems, which provide the necessary control accuracy and versatility in conditions of low costs for additional staff training. Consider the prospects for improving the technical and economic indicators of metallurgical and chemical enterprises through the introduction of automation, on the example of the operation of concentration plants. Obviously, the main tool to increase the economic efficiency of production is to reduce various types of costs. Among the main cost items, fuel, feedstock and chemical reagents are distinguished, however, the undisputed leader is the cost of electricity, the work of the mills needed to grind the feedstock. On the other hand, the problem of timely and quick diagnosis of the internal components of the mills and the ability to choose the optimal mode of operation are also relevant tasks. These reasons significantly affect such process indicators as the quality of the crushed raw materials, the energy efficiency of the grinding process, reducing the cost of the production cycle, etc. To control the processes of grinding, loading of raw materials, monitoring the filling of the drum, the use of vibroacoustic primary transducers installed on the housing of the axle bearing of the drum mill is promising.

1. Introduction

LabVIEW (Laboratory Virtual Instrumentation Engineering Workbench) is a major National Instruments product. It is used in data collection and processing systems, as well as for the management of technical objects and technological processes. Ideologically, LabVIEW is very close to SCADA-systems, but in contrast to them, it is more focused on solving problems not so much in the field of automated process control systems (APCS), but in the field of an automated research system. However, the functionality of this product seems to be broader.

National Instruments' primary focus was measurement and automation tools. Although the LabView complex is closely related to National Instruments hardware, it is nonetheless not related to specific hardware and can be used with devices running Windows, Linux, MacOS, etc., which makes the environment universal and can be used on most enterprises and in research projects.

Automation of technological processes is a necessary element of improving the production of modern enterprises to increase the efficiency of processes in all areas of industry [1, 2], where a large number of primary converters (sensors) and various devices are used, the metrological characteristics and indications of which must be continuously monitored. Their characteristics are greatly influenced by the input signal power parameters, the increase of which is an urgent automation task [3, 4].

Due to the features of the LabView software package, it can be used in production processes, not only for automation, but also for quality control of materials. For example, laboratory methods for studying the tribological characteristics of metals from works [5–11] based on the use of the LabView software package are used at enterprises for sampling and also monitoring production processes in the production of standardized products. The method consists in measuring the tribological characteristics of metal-oil systems using a laboratory bench, improved by a software package based on LabView, which allows you to speed up and simplify work at times. An advanced installation can be used at any enterprise by unskilled personnel, since LabView has a graphical interface and intuitive controls.

At the enterprises of the mineral resource complex, the stage of grinding the source material to alarge extent affects the subsequent stages of processing, affecting the quality of the finished product. The methods of automatic regulation and manual regulation by the grinding complex used at existing enrichment factories, preserved at some plants, do not guarantee the stability of the process, and are not able to ensure the constancy of the parameters of the products obtained. Automation of the grinding stage allows you to comply with the required process criteria, including taking into account the variability factor of the properties of the feedstock, and quickly carry out readjustment of the equipment used. As an example, which allows production to respond quickly to constantly changing properties of the feedstock, which entails the adjustment of individual operations of the process chain, automated control over the volume of filling, for example, grinding mills, can be used.

The grinding process is carried out using mills containing, in addition to the charge, solids made of materials such as steel or aluminum. Grinding bodies and crushed material significantly affect the integrity of the lining, intensifying wear, as a result of which a set of measures is used to increase the life of the mills: the use of combined linings, the use of DEM modeling to predict places of increased wear, and others [12-15].

The process of quality control of the mills is a non-trivial task. This is primarily due to the different duration of the grinding processes, which depends on the properties of the feedstock and the required qualitative characteristics of the material at the end of the process. Also, parameters of grinding media, a protective inner coating of the drum, occupancy rate, etc., have a great influence. Maximum productivity is achieved by optimizing the entire set of parameters, including taking into account the number of grinding elements, since their disadvantage is characterized by an insufficient degree of grinding of the material due to the "filtering" of grinding elements to the walls.

The parameters of the crushed bulk material significantly affect subsequent redistribution, which is explained by the fact that if the required fineness was not obtained at the grinding stage, the required value of mineral disclosure was not achieved, and, therefore, it is necessary to make additional adjustments to the parameters of the production process at the following stages of production.

Therefore, the competent organization of the grinding process: monitoring the degree of filling of the equipment, the state of interior decoration, the number and distribution of grinding media, can reduce the energy consumption of the enterprise, the number of unexpected stops of the equipment, increase the service life of the lining of the mill and the unit as a whole, thereby increasing its efficiency.

2. Proposed Methodology

To solve certain production problems, it is necessary to use various types of mills. The grinding process is so important in the further enrichment of raw materials, since it is in it that minerals are discovered due to the destruction of bonds between useful components and waste rock. Therefore, the goal of this process can be considered to be optimal for further enrichment procedures of dimensional parameters or size required by the customer, as, for example, when sorting coal.

There are five of the most common automated control methods for grinding process parameters. The first method is based on the fact that it is necessary to adhere to the required percentage ratio between the flow rate of the water used in the mill and the noise indicators in the zone where slag formation occurs. This is due to an automated device that analyzes the supplied signals and changes, if necessary, the amount of water supplied. The second method is to automatically adjust the mill loading level based on an analysis of the vibrational characteristics of the process. The influence of changes in the viscosity of the crushed product on the flow of water formed the basis of the third method. The fourth method involves the study and adjustment of the composition of the resulting waste after the grinding process using special expensive equipment for express analysis. After receiving the data, the program can make a download adjustment. The fifth option is based on the relationship between the noise emitted by the grinding equipment and the required amount of feedstock loaded. The efficiency of the last disassembled method can be greatly negatively affected by units located in the immediate vicinity of the sensor mounting point, which registers the received data.

The most rational option for adjusting the grinding process indicators is to study the volumetric filling of the grinding equipment, the amount of raw material loaded, the number of grinding media used for the subsequent automated process of loading the raw materials and grinding media.

In most enterprises, processes, especially those associated with heat treatment, play a significant role in the processing of heat transfer processes, due to the fact that modern science and technology require a reliable forecast of such processes, the experimental study of which in laboratory or field conditions is very difficult and expensive, and in some cases, it is simply impossible [16]. This requires considerable calculations and labor, but the use of Scada-systems and LabView complexes allows you to speed up the calculations, thereby reducing the risks of production errors that affect the quality of products, which is the main goal of modern production. For the effective application of the calculated data of heat transfer processes, it is necessary to provide the necessary accuracy, without which it is impossible to achieve the required metrological parameters, which affects the built-in product quality [17].

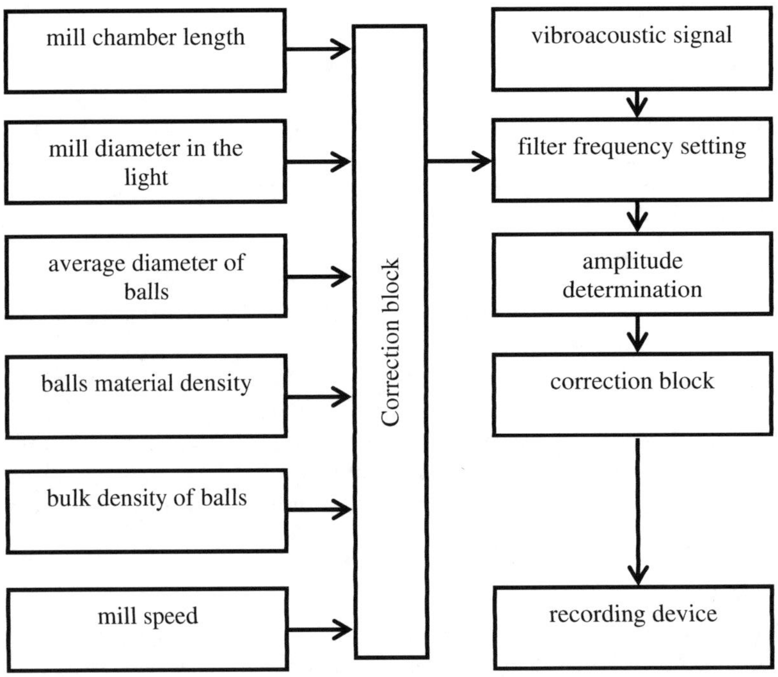

Fig. 1. Block diagram of a device for automatic control of the degree of filling of a ball mill with ground material

3. Results and Discussion

A block diagram of a device that implements a method for automatically controlling the degree of filling of a ball mill with crushed material [18] is presented in Figure 2.

The considered option [18, 19] of automated control of the volumetric filling of grinding equipment and the size of grinding bodies is based on the study of vibroacoustic vibrations of the mill body at a characteristic frequency, which is the disturbing frequency of the shock pulses of grinding bodies against the mill drum body.

The frequency of shock pulses is significantly affected by such characteristics as the size and composition of grinding bodies, the frequency of rotation of the drum, the diameter of the mill, etc.

The period of revolution of the grinding bodies of the ball load can be found from the expression (1):

$$T_{shz} = \frac{T_m}{Ц_n} = \frac{\phi}{n(1-K^2)}, \tag{1}$$

where $Ц_н = \frac{1-K^2}{\varphi}$, $K = \frac{R_2}{R_1}$, R_1 - the radius of the grind equipment, m; R_2 - internal radius of the load, m; φ - fill factor of grind in equipment; n - rotational speed of the mill drum, 1/s; T_{shz} - period of circulation of grinding bodies, s; T_m - rotation period of the grinding equipment case, s; $Ц_n$ - the circulation coefficient of the load of grinding equipment.

The number of grinding media concentrated at the upper level can be calculated using the following formula:

$$M = \frac{\pi R_1^2 L \phi \rho_{sh}}{m_{sh}}, \tag{2}$$

where M - is the number of grinding media at the upper level; L - chamber length, m; ρ_{sh} - bulk density of grinding media, kg / m^3; m_{sh} - the mass of the grinding body of medium diameter, kg.

Knowing the period of circulation of the grinding media and the number of grinding media at the upper level, it is possible to determine the frequency of collisions of the outer layer of grinding media against the mill body:

$$f = \frac{M}{T_{shz}} = 0.1nL\frac{\rho_{sh}}{\rho_m}\frac{(2R_1 - d_{sh})}{d_{sh}^2}, \tag{3}$$

where d_{sh} - is the average diameter of the grinding media, m; ρ_m - steel density (material density of grinding media), kg / m^3.

The presented expression (3) determines the characteristic frequency of disturbing impulses of the impact of grinding media on the inner surface of the drum of grinding media (maximum acoustic noise of the mill).

Taking $a = 0{,}1\rho_{sh}/\rho_m$, you can adjust (3), which determines the frequency of impact of grinding media on the mill drum housing:

$$f = anL\frac{(2R_1 - d_{sh})}{d_{sh}^2}. \tag{4}$$

By changing the expression (4), the relation is obtained for determining the diameter of the grinding media:

$$d_{sh} = \frac{1}{2}\left(\sqrt{\frac{aLn}{f}\left(\frac{aLn}{f} + 4D\right)} - \frac{aLn}{f}\right). \tag{5}$$

With constant monitoring of the collision frequency of grinding media, the frequency of rotation of the grinding equipment in automatic mode, you can get the average size of the grinding media.

4. Conclusion

Using LabView software package helps to achieve the necessary results, such as automation of production, acceleration of production processes, ensuring the required metrological characteristics and specified accuracy, as well as quality control of products and raw materials used. As shown in the article, these tasks are relevant, since at present there are more and more new methods for using the LabView software package, using examples of enterprises and research organizations.

The method considered is based on the study of vibro-acoustic signals of grinding media, therefore, if an object with large dimensional parameters gets into the grinding process, then an analysis of the oscillations, namely, sharp amplitude jumps, will allow us to fix and evaluate the complications that have arisen. Therefore, it will be possible to correctly assess the level of workload of the units, automatically carry out its control, and quickly respond to the ingress of foreign bodies into the crushed material, preventing equipment breakdown, thereby increasing its service life and reducing operating costs, to improve the quality of the finished product.

References

[1] Fedorova E R, Vinogradova A A 2018 *IOP Conference Series: Materials Science and Engineering* **327(2)**.
[2] Kremcheeva D, Kremcheev E. 2018 *Journal of Mechanical Engineering Research and Developments* **41(1)** 151-155.
[3] Zhambalova S T, Vinogradova AA 2018 IOP *Conference Series: Materials Science and Engineering* **327(2)**.
[4] Vinogradova A A, Silivanov M O, Leontyuk S M 2018 *Journal of Physics: Conference Series* **1118 (1)**.
[5] Syrkov A G, Silivanov M O, Kushchenko A N 2016 *Journal of Physics: Conference Series* **729(1)**.
[6] Nazarenko M Yu, Kondrasheva N K, Saltykova S N 2018 *Coke and Chemistry* **61(5)** 184-187
[7] Syrkov A G, Kabirov V R, Silivanov M O 2017 *Journal of Physics: Conference Series* **872(1)**.
[8] Nazarenko M Yu., Kondrasheva N K, Saltykova S N 2018 *Coke and Chemistry* **60(2)** 86-89
[9] Syrkov A G, Sychev. M, SilivanovM O, Rozhkova N N 2018 *Glass Physics and Chemistry* **44(5)** 474-479.
[10] Nazarenko M Yu, Kondrasheva N K, Saltykova S N 2017 *Tsvetnye Metally* **7** 29-33
[11] Syrkov A G, Silivanov M O, Kabirov V R, Pleskunov I V 2018 IOP Conference Series: Materials Science and Engineering **387(1)**.
[12] Boikov A V, Savelev R V, Payor V A 2018 *Journal of Physics Conference Series* **1118**.
[13] Boikov A V, Savelev R V, Payor V A 2018 *Journal of Physics Conference Series* **1050**.
[14] Boikov A V, Savelev R V, Payor V A 2018 *Journal of Physics Conference Series* **1015**.
[15] Boikov A V, Payor V A, Savelev R V 2018 *Journal of Physics Conference Series* **944**.
[16] Bryukhanov M A, Tsvetkov N V, Vinogradova A A 2018 *IOP Conference Series: Earth and Environmental Science* **194(6)**.
[17] Vinogradova A A Leontyuk S M 2018 *Journal of Physics: Conference Series* **1118(1)**.
[18] Kremcheev E A, Kremcheeva D A 2017 *Journal of Industrial Pollution Control* **33(1)** 787-791
[19] Nagornov D O, Kremcheev E A 2018 *Journal of Physics: Conference Series* **1118(1)**.

Method of reliability and speed increase of work of automatic input of a reserve on the basis of the microprocessor terminal

T E Minakova

Department of Electric Power and Electromechanics SPMU

Abstract. The new operation algorithm of the automatic input device of a reserve (AIR) together with the system of relay protection of power transformers of substation is offered. The increase in reliability and speed of the automatic input of a reserve is provided due to use of signs and algorithms of identification of emergencies and collaboration of a system of relay protection and automatic equipment of substation. The sign of identification of level of undervoltage on tires of the distributing device with automatic input of a reserve and also type of the relay protection which switched-off the equipment in emergency operation is entered. The formulas for calculation of parameters at which operation of AIR is provided are offered. The proof of efficiency of the offered signs and algorithms is executed on the basis of the analysis of oscillograms of parameters of emergency operation. The program realization of the offered algorithm on the basis of the microprocessor terminal of relay protection Sepam 1000+ is presented.

1. Introduction

The microprocessor devices of relay protection (MDRP) [1] became the main devices of relay protection in electrical networks of the industrial enterprises and city electrical networks.

The advantages of MDRP are rather known: these are small dimensions, wide range of protective characteristics, functions of measurement and archiving of electrical quantities at the time of protection operation. Microprocessor devices continuously carry out self-checking, there is a possibility of remote control and expeditious change of values of settings of protection [2, 3].

Except these obvious advantages of MDRP it is important to specify the following features having crucial value for operation.

1. Due to the gadgets and technologies used in activity and household clever decisions, the relay protection executed on the basis of the microprocessor becomes more convenient and intuitively clear for quick and technicians at service and operation, than devices on semiconductor and electromechanical relays [4].

2. Relay protection on the basis of the microprocessor equipment allows using signs and algorithms of identification of emergencies which were theoretically proved, but did not receive practical application [5, 6].

Therefore, the development and improvement of new means of relay protection and automatic equipment of electrical networks and power supply systems is relevant [7]. So far insufficiently effective are devices of local automatic equipment [8]. In this regard expedient is use of MDRP [9] for widespread devices of the automatic input of a reserve (AIR) and automatic repeated switch-on (ARSO).

2. Materials and methods

The standard scheme of two-transformer substation of 110 kV with distributing devices of the low voltage of 6-10 kV assumes separate work of sections (systems) of tires and the AIR installation on the section Q5 switch (fig. 1) [10, 11].

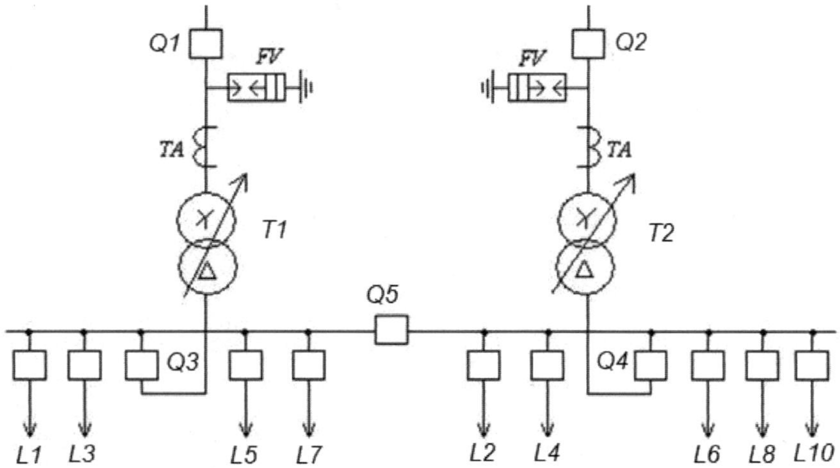

Figure 1. Scheme of two-transformer substation of 110/10 kV

For such scheme the following options of emergency operation are possible:

1. SC on the feeding transformer, shutdown of the transformer the main protection - longitudinal differential, without endurance of time.
2. SC on the feeding transformer, without shutdown of the transformer high-speed protection.
3. SC on tires – is started by MCP with endurance of time or reserve protection against external overcurrents.
4. SC on the departing lines. In this case SC has to be disconnected by the corresponding protection of the departing line, as a rule, with relative selectivity, and with endurance of time.

AIR has to fulfill in the 1st and 2nd cases, including the stand-by power supply after shutdown of the main (damaged) transformer [12, 13]. In the first emergency operation shutdown of the transformer differential protection can be a signal to operation of AIR. In this case the response time of AIR can build up with a selectivity step from the response time of high-speed protection of the transformer and be accepted approximately $t_{o.p.}$ = 0.3 s. The signal of operation of high-speed protection of the transformer can be identification sign of need of turning on of the section switch (permission of operation of AIR) after confirmation signals about presence of voltage on a reserve source (the second transformer), strain monitoring of parallel inputs and inclusion of The Departing Line mode.

In the second case as criterion of existence of damage to the transformer (that is after the lead-in switch) critical undervoltage at least can serve in one of phases at simultaneous increase in current of a phase (fig. 2). Previously the setting on voltage can be set, building up from undervoltage at external SC (that is on buses or the departing lines). In practice voltage in such mode is set within U_{min} = (0,2 – 0,4) $\cdot U_{nom}$=$U_{C.Z.1}$. In case of SC in a transformer chain voltage on tires of secondary voltage will be lower as there will be no feed of voltage from the power source (power transformer). Therefore for an initial setting it is possible to offer formula (1):

$$U_{o.p.2} = (0{,}2 - 0{,}4) \cdot U_{nom} / K_N. \quad (1)$$

where K_N – reliability coefficient.

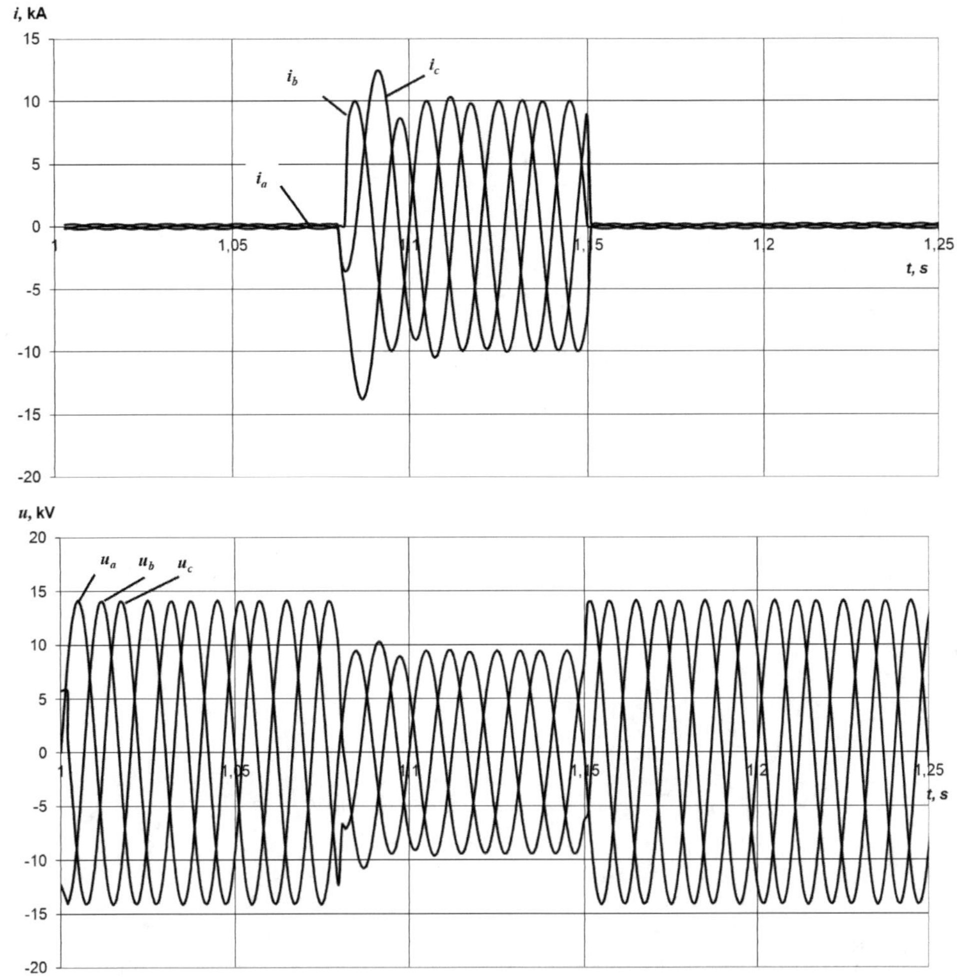

Figure 2. Two-phase SC on buses RU of 10 kV

In this case (that is to turning on of the section switch) the signal of the interlock contact of the section switch (at shutdown) can be a signal to operation of AIR. Actuation time of AIR in this case will be $t_{o.p} = t_{off}$ input off, and this time is minimum possible as inclusion on parallel operation of the damaged main source and a reserve source inadmissibly.

In the 3rd and 4th options of succession of events and operation of AIR should be blocked as at SC on buses and not disconnected SC on the main power supply departing lines is not damaged and does not demand connection of a reserve source. AIR should be prohibited even in case of shutdown of SC on accession and buses the switch on input of the transformer (main power supply). Identification criterion of such option of succession of events is smaller undervoltage in comparison with voltage of $U_{o.p.}$ as in these cases voltage is supported by the EMF of loading of 6 (10) kV, especially motive loading (fig. 2).

3. Results

The implementation of this algorithm can be considered on the example of MDRP Sepam of the second generation having flexible logic that allows setting the sequence of actions demanded for the specific protected object [9].

The described algorithm is implemented by the scheme submitted in fig. 3.

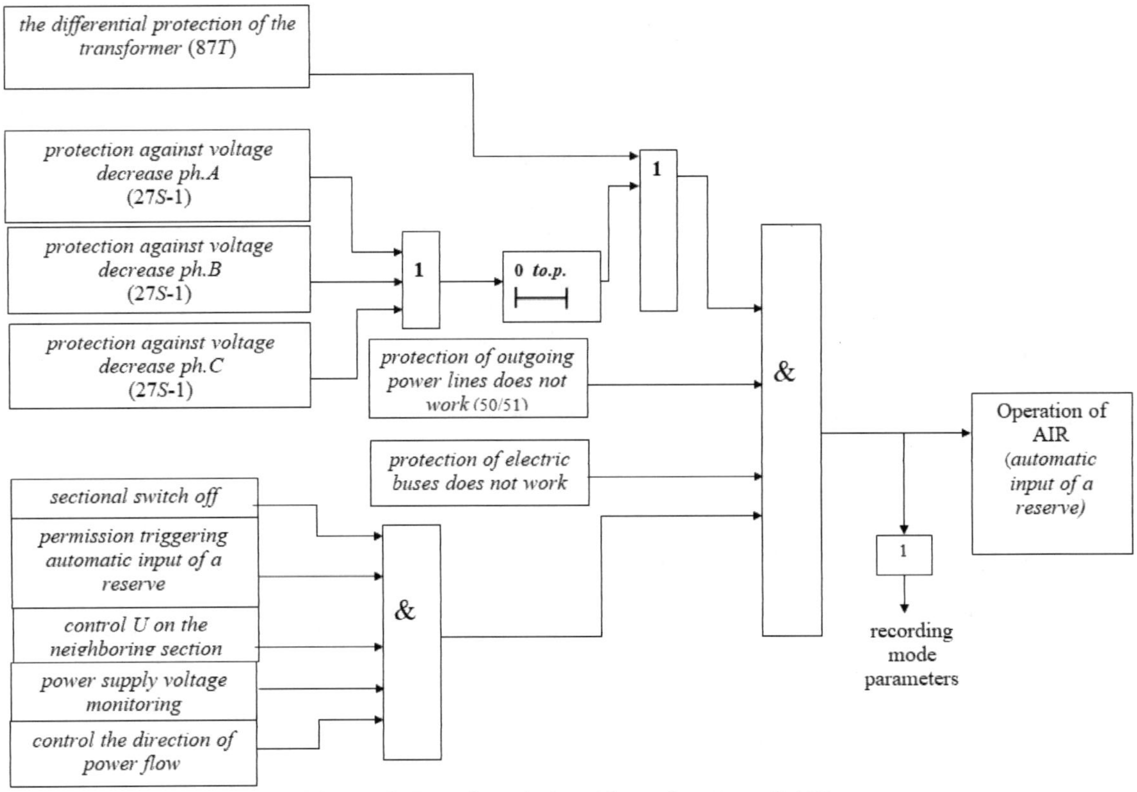

Figure 3. Developed algorithm of action of AIR

4. Conclusion

MDRP allow using difficult and multicriteria signs and algorithms of identification of faults and by that to provide the selection action of relay protection. Transition to similar algorithms will allow reducing the number of excessive operations of systems of relay protection and automatic equipment, to increase sensitivity and high-speed performance of systems of relay protection. The offered algorithm of action of AIR will allow reducing time of a break of power supply of consumers that leads to increase in reliability of power supply of consumers. Transition from simple calculations of settings of protection to accounting of interference and interaction of different types of protection and automatic equipment is the beginning of transition to the self-adaptive smart systems of the relay protection which is not demanding constant control from service staff [14, 15] and providing more reliable power supply, the minimum time of a break of supply of electricity and the maximum sensitivity and high-speed performance.

References

[1] Kostin V N and Minakova T E 2017 Protective and control relays as coal-mine power-supply acs subsystem *IOP Conf. Ser.: Earth and Environmental Sci.* **032017**

[2] Belsky A A and Korolyov I A 2018 Thermal oil recovery method using self-contained windelectric sets *J. of Physics* 3

[3] Abramovich B N and Sychev Y A 2017 The structure selection of hybrid correction device for centralized and distributed electrical networks *11th International IEEE Scientific and Technical Conference "Dynamics of Systems, Mechanisms and Machines", Dynamics 2017 Proc.* pp 1–6

[4] Minakov V F, Lobanov O S, Makarchuk T A, Minakova T E, Leonova N M 2017 Dynamic management model of innovations generations *Proceedings of 2017 20th IEEE Int. Conf. on Soft Computing and Measurements* pp 849–52

[5] Gulkov Y V and Ustinov D A 2017 Electric power quality distortions influencing oil processing plant parameters and characteristics *Int. J. of Applied Engineer. Res.* **12(9)** 1127–30

[6] Belsky A and Dobush V S 2017 Autonomous electrical heating facility supplied by wind turbine for elimination of oil wellbore paraffin deposits *Int. Conf. on Industrial Engineer., Applications and Manufacturing* pp 8076256

[7] Ustinov D A and Baburin S V 2016 Synthesis procedure of the power supply systems topology at mineral resource enterprises based on logical-probabilistic assessments *Int. J. of Applied Engineer. Res.* **11(9)** 6402–6

[8] Kostin V N, Minakova T E and Kopteva A V 2018 Urban substations transformers allowed loading *Proc. of the 2018 IEEE Conf. of Rus. Young Res. in Electrical and Electronic Engineer., ElConRus* pp 692–5

[9] Minakov V F, Minakova T E, Galstyan A Sh and Shiyanova A A 2015 Time constant of innovation effects doubling *Mediterranean J. of Social Sci.* **6** 36

[10] Abramovich B N, Sychev Y A and Prokhorova V B 2017 Electrical complex of combined power supply on the base of renewables and hybrid correction device *Conf. of Open Innovation Association, FRUCT-2017* pp 3–9

[11] Minakov V F, Minakova T E, Lobanov O S, Kostin V N and Makarchuk T.A. 2017 The Expansion of Time Series Innovations in a Series of Sigmoid *Int. J. of Appl. Business and Economic Res.* **15** 311–9

[12] Lobanov O S and Minakov V F 2018 The hierarchy model of telecommunication needs in geoinformation resources *18 Int. Multidisciplinary Scientific GeoConf. SGEM* pp 781–8

[13] Minakov V F, Lobanov O S, Minakova T E, Makarchuk T A and Kostin V N 2017 The law of diminishing marginal productivity in the model of pure discounted income of innovations *Int. J. of Economic Res.* **14(14)** 435–41

[14] Minakov V F 2016 Smart – innovations in the industry 4.0. *Intellectual and information technologies in formation of digital society* pp 69–73

[15] Bozhkov M I and Kostin V N 2016 Technocenological approach to managing power supplying substation loads *Int. J. of Appl. Engineer. Res.* **11(9)** 6736–9

Peat and mineral anticaking agents for powdered nitrile-butadiene rubbers

O S Misnikov

Tver State Technical University, 22, A. Nikitin Street, 170026, Tver, Russia

E-mail: oleg.misnikov@gmail.com

Abstract. The article discusses caking of loose materials and main methods of its prevention. The main characteristics of nitrile-butadiene rubber are outlined. It is possible to obtain hydrophobic-modifying additives during thermochemical degradation of organic matter peat. The author presents a method for analyzing anticaking agents based on peat and mineral compositions. The comparative analysis of the effectiveness of anticaking agents made of kaolin and chalk is made. It is established that the treatment of powdered nitrile butadiene rubber with a hydrophobically modified chalk makes it possible to reduce its ability to cake by 22.5 times during long-term storage compared with the control sample.

1. Introduction

Organic, mineral, amorphous and crystalline substances belonging to the class of high-molecular compounds are widely used in various areas of industrial production. Amorphous polymers that possess high-elasticity properties – elastomers are of particular significance. Compared to other polymers they have a lower glass transition temperature. Natural and synthetic elastomers – rubbers – possess elasticity, hydrophobicity and electrical insulation properties [1].

Currently, the range of rubber products has more than 36,000 items with the separation of raw materials for their manufacture into general- and special-purpose rubbers. Particularly, nitrile-butadiene rubbers (NBR) are the products of polymerization of butadiene and acrylonitrile. NBR are one the most spread special-purpose rubbers that are currently produced in about 20 countries.

Nitrile butadiene rubbers and their vulcanizates are highly resistant to the exposure of mineral oils and fats. Oil and benzine resistance of such rubbers increases along with the increase of content of acrylic acid nitrile in them. NBR vulcanizates compare favorably to other rubbers vulcanizates because of their higher resistance to heat aging and abrasion. Rubber products based on NBR can be used for a long time at 120°C with air exposure and at 150°C in different oils [1]. Therefore, they are used in the manufacture of various rubber technical products: rings, bushings, hoses, cuffs, etc. Rubber based on them is used as anti-corrosion materials for internal coatings of devices operating in aggressive environments.

Nitrile-butadiene rubbers are produced in briquetted, granulated or powder form. Nitrile-butadiene rubbers are stored in polyethylene, paper or tissue bags.

The powder form of rubber allows using of modern loose material dispensers on industrial production lines. This increases the accuracy of dispensing of components and consequently increases the quality of end product. In addition, production costs are reduced due to partial automation of the production cycle. However, the storage of rubber in a powder form results in particles sticking together due to complex interaction of different forces. The process of caking is aggravated by the storage of bags with powdered rubber under pressure.

Caking refers to the ability of loose materials to lose their flowability during long-term storage due to the interactions between their particles. The main causes of caking include [2, 3, 13]:

• appearance of capillary forces due to an increase in the moisture content of the loose material;

• formation of crystallization bridges between particles as a result of the growth of crystals from substances dissolved in water;

• occurrence of autoadhesion interactions at the interface of two or more particles.

Capillary forces and formation of crystallization bridges in the «moistening – drying» cycle are crucial for disperse soluble systems, for example, mineral fertilizers.

In the case of rubbers that possess hydrophobic properties, autoadhesion forces are the main factor of caking. Autoadhesion is considered to be a diffusion process. It depends on the temperature and can occur both with pressure and without it. Pressure has a positive effect on the process of autoadhesion due to the increase in the contact area. The increase in temperature weakens the intermolecular bonds and intensifies thermal motion of the segments of rubber macromolecules. This facilitates the mutual penetration of areas of the hydrocarbon chains of two or more particles that are in direct contact.

In order to avoid caking of powdered rubbers and other loose materials, they are occasionally dried or stirred. In addition, methods of dry or wet surface treatment of particles or granules of loose bodies are used to avoid caking. The main task of such treatment is to prevent or minimize the moistening of the particles and their contact with each other during storage and transportation.

Powdered kaolin, chalk, kieselguhr, talc and other materials – inert fine dispersed systems – are used as anticaking agents of dry surface treatment (or dusting) for powdered rubber. Such additives, with the exception of talc, do not possess hydrophobic properties. In the case of mineral fertilizers, they are used to absorb liquid or gaseous moisture, compensating for the hydration of the water-soluble matter. It is obvious that, as a result of the particles surface treatment of hydrophobic rubber with a hygroscopic substance, an upper layer of hydrophilic mineral carrier is formed between these particles. In relation to such a layer, undesirable caking factors described above may occur. However, inert finely dispersed systems that do not initially possess hydrophobic properties can be modified with various water-repellent additives. Such a composition can prevent the main causes of caking of powdered rubber and a number of other bulk materials.

Today there is a relatively large number of hydrophobic additives that are obtained from products and wastes of oil-processing. Organosilicon liquids (or silicone oils) as well as some products of wood chemical processing are better water-repellant agents. Analysis of the results of scientific research on the thermochemical processing of peat, carried out in Russia in the twentieth century and the beginning of the 21st century [5–8], made it possible to develop a new direction in the use of compounds extracted from its organic matter. Their physical and chemical properties allow hydrophobic modification of a wide range of mineral dispersed materials. Despite the fact that the main part of the organic matter of peat consists of hydrophilic components [9-11], it also contains bitumens consisting of waxes, resins, oils and paraffins. Bitumens obtained from peat are hydrophobic components of peat organic matter [1, 3, 12]. It is known that organic matter of peat undergoes chemical composition change during its thermal processing. The character of such changes depends on the amount and time of thermal exposure. As a result of thermochemical conversion, the organic matter of peat decomposes into individual components, which transit from the class of hydrophilic into the category of hydrophobic matters. Moreover, these are of all aggregative states (solid, liquid, gaseous). Thus, one of the main tasks is to obtain and apply them to the treated surface. This method was implemented upon obtaining of model hydrophobically-modified cement [12]. Currently, it is analyzed whether it is possible to replicate it to other types of mineral dispersed materials. In particular, this paper proposes the treatment of hydrophilic materials – chalk and kaolin – for their subsequent use as anticaking agents for NBR according to the developed method.

2. Materials and methods

The approach is based on the fact that chalk and kaolin are part of the basic composition of a number of rubber mixtures used for specific operating conditions: roll coatings in the textile industry, for shock absorbers with benzine and oil resistance, for various protective coatings, etc. Moreover, their amount in the compositions can vary in a wide range of concentrations.

The method of hydrophobic treatment of kaolin and chalk was as follows. Peat hydrophobic semi-finished product obtained from cotton-grass-sphagnum peat with decomposition degree of 25-30% was introduced to the mineral material. To increase the particle dispersion and homogeneity of their mixing, as well as to create conditions for the formation of primary adhesion contacts, the mixture was placed in a ball mill. An additional grinding of the mixture was carried out in the mill for 5 minutes with a speed of 56-60 rpm. The choice of parameters was made based on the design features of the mill, as well as the optimal quality characteristics of the organomineral mixture.

The obtained two-component composition was activated in a laboratory reactor installation according to the method described in detail in the previous papers [4, 12]. During activation, a protective water-repellent film coating is formed on the surface of the mineral particles from bitumen compounds released during heat treatment. The presence of the film does not adversely affect the flowability of the powder.

The hydrophobic composition was used for the treatment of powdered NBR with particles size lower than 1 mm. The concentration of the components was: 90% of NBR and 10% of the anticaking additive. To model the process of storage under pressure, the control and experimental samples were poured into a special device (Fig. 1). A 1 kg weight was placed on the cylinder head, which created additional pressure on the material.

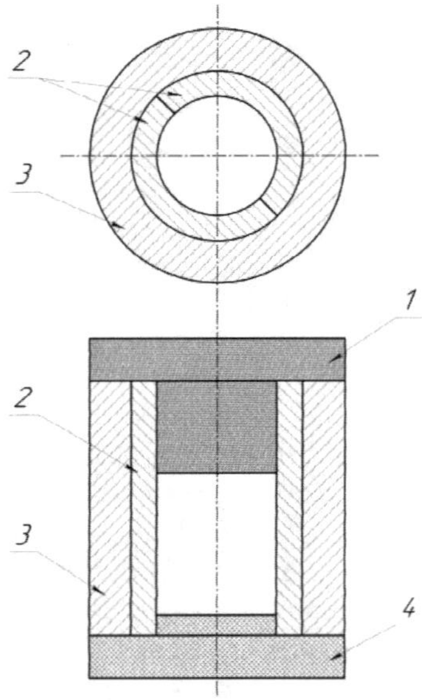

Figure 1. Scheme of the installation for determining of caking during the storage of dispersed materials

The experiments of caking of the samples were carried out in seven and thirty days of storage under a pressure of 14.2 KPa, a temperature of 20-25°C and a relative humidity of 70-80%.

Upon expiration of the storage period, the contents of the cylinders (compacted NBR, Fig. 2) with the respective groups of samples were removed from the internal detachable device and subjected to pulsed destruction on a specially designed experimental setup (Fig. 3) using the following procedure.

Figure 2. Appearance of the experimental sample of rubber located in a two-piece cylinder

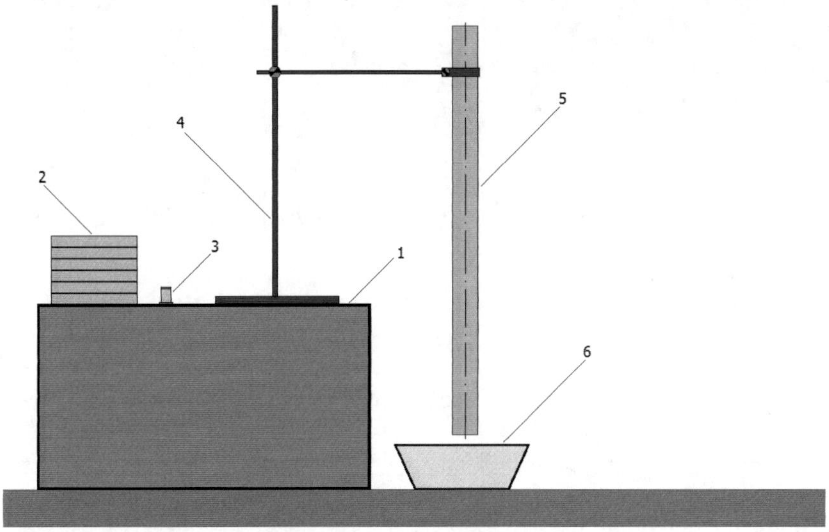

Figure 3. Experimental setup for determining caking of samples of loose materials: 1 - stand, 2 - set of sieves, 3 - experimental sample, 4 - tripod; 5 - hollow tube; 6 - crystallizer

A test sample was put to the plastic tube fixed in the laboratory tripod, the top part of which was at a height of 150 cm.

Moving under the action of gravity down the tube with the acceleration of free fall, it fell into the crystallizer and partially or completely destroyed. Next, the contents of the crystallizer were subjected to sieve analysis using sieves with round cells of dimensions 10, 7, 5, 3, 2, 1 mm. The purpose of the analysis was to determine the mass of material in loose and aggregated states. In addition, the degree of aggregation of particles.

3. Results & Discussion

The experiments allowed establishing the following change in the quality characteristics during storage of nitrile-butadiene rubbers (Table 1). In the first seven days, the proportion of caked mass (the oversize fraction on a sieve with a mesh size of d = 1 mm) in the control sample was 57.27%. It is characteristic that its main part (39.05%) remained on a sieve with a mesh size of d = 10 mm. In experimental samples, the best indicators were obtained by treating NBR with hydrophobically-modified chalk – 1.82%. When hydrophobic kaolin was used as an anticaking agent, the caking rate was 13.8 times higher – 25.19%. It is noteworthy that relatively large agglomerates (d > 5 mm) did not form when using hydrophobically modified chalk in seven days. With the increase in the storage time up to 30 days, the trends identified above persist, and the quality indicators deteriorate significantly – in the control sample, the amount of caked mass is 96.55%. Moreover, the majority are agglomerates with a size greater than 10 mm (Figure 4, a). In the samples treated with hydrophobically-modified chalk and kaolin the caking rates were 11.03% and 96.43%, respectively. Moreover, the main part of the sample, treated with chalk, almost completely retained its flowability (Figure 4, b). The substantial part of the material (~ 89%) passed through the sieve with a 1 mm mesh.

a)   ***b***)

Figure 4. Appearance of samples of NBR destroyed after 30 days storage:
a - control; *b* - treated with hydrophobically-modified chalk

Table 1. Comparative sieve analysis of the control and experimental samples of powdered nitrile-butadiene rubber after storage

Sieve residue, mm	Fraction content, g / %		
	Control sample of NBR	Experimental sample of NBR treated with hydrophobically-modified	
		chalk	kaolin
	7 days of storage		
$d > 10$	11.78 / 39.05	0 / 0	0.94 / 3.72
$7 < d < 10$	0.92 / 3.05	0 / 0	0.39 / 1.54
$5 < d < 7$	0.54 / 1.78	0 / 0	0.16 / 0.64
$3 < d < 5$	0.75 / 2.48	0 / 0	0.3 / 1.19
$1 < d < 3$	3.29 / 10.91	0.44 / 1.82	4.57 / 18.10
$d < 1$	12.89 / 42.73	23.80 / 98.18	18.89 / 74.81
Total weight, g	30.16 / 100	24.24 / 100	25.25 / 100
	30 days of storage		
$d > 10$	28.06 / 94.12	0.5 / 2.14	20.18 / 92.36
$7 < d < 10$	0.13 / 0.43	0.1 / 0.43	0.11 / 0.51
$5 < d < 7$	0.09 / 0.30	0.1 / 0.43	0.25 / 1.14
$3 < d < 5$	0.13 / 0.43	0.09 / 0.38	0.13 / 0.59
$1 < d < 3$	0.38 / 1.27	1.79 / 7.65	0.4 / 1.83
$d < 1$	1.03 / 3.45	20.81 / 88.97	0.78 / 3.57
Total weight, g	29.81 / 100	23.39 / 100	21.85 / 100

Thus, the choice of chalk as a mineral carrier for the production of hydrophobic anticaking agents makes it possible to fully implement the principle of forming on its surface solid bitumen nano-films extracted from organic matter of peat during thermochemical exposure [4, 12]. In this regard, it was necessary to establish how the anti-agglomeration properties of natural chalk change after its hydrophobic treatment.

For this purpose, a series of experiments with NBR treated with powders of same dispersion from natural and hydrophobically-modified chalk. The results of the experiments (Table 2) demonstrate that the natural chalk, which is a typical hydrophilic material, solves the issues of caking decreasing only for a relatively short time of storage.

Table 2. Efficiency analysis of the anticaking effect of fine natural and hydrophobically-modified chalk during storage of nitrile butadiene rubber

	Sieve residue, mm	Fraction mass of NBR on the sieve, %		
		samples		
		control	Treated with chalk	Treated with hydrophobically-modified chalk
7 days of storage	$d > 10$	68.34	0.00	0.00
	$7 < d < 10$	3.79	0.00	0.34
	$5 < d < 7$	1.54	0.00	0.26
	$3 < d < 5$	2.42	0.12	0.52
	$1 < d < 3$	9.38	4.22	2.88
	$d < 1$	14.54	95.66	96.01
30 days of storage	$d > 10$	92.95	58.21	2.14
	$7 < d < 10$	0.56	0.00	0.43
	$5 < d < 7$	0.37	0.00	0.43
	$3 < d < 5$	0.56	0.96	0.38
	$1 < d < 3$	1.61	7.07	7.65
	$d < 1$	3.95	33.76	88.97

The experiment time that did not change the tendency to cake for the samples treated both with chalk and hydrophobically-modified chalk was seven days. With a longer storage time (30 days), a sufficiently high caking ability, which is approximately 63% of the control sample (data from the analysis of the fraction above 10 mm), is observed in the nitrile-butadiene rubber treated with chalk. Most likely this is due to the sorption of water vapor from the environment, which eventually begins to condense and moisten the chalk particles during storage. In addition, with the adhesion of rubber particles, moisture condensation occurs on the chalk particles with the subsequent formation of rather large lumps.

In rubber treated with hydrophobic-modified chalk, the caking ability was only 2.3% of the indicator of the control sample. Moreover, the vast majority of it (practically 89%) turned out to be in the undersize product when sifting on a sieve with a mesh of less than 1 mm.

4. Conclusion

Thus, it is experimentally determined that hydrophobic processing of chalk with bitumen compound of peat increases its efficiency as an anticaking agent for powdered nitrile-butadiene rubbers by 2.6 times compared standard chalk and by 22.5 times compared to the control sample. Moreover, a lower price of chalk as a raw material compared to kaolin will increase the economic efficiency of its use in technological processes of obtaining of different types of industrial-rubber products.

References

[1] Misnikov O S, Korolev I O 2017 *Polymer Science. Series D.* **10 (3)** 255-259.

[2] Afanas'ev A E, Boltushkin A N 1996 *Colloid Journal of the Russian Academy of Sciences: Kolloidnyi Zhurnal,* **58 (2)** 139-144

[3] Afanas'ev A E, Efremov A S 2011 *Theoretical Foundations of Chemical Engineering* **45 (1)** 120-126

[4] Misnikov O S 2016 *Mires and Peat* **18** 1-15

[5] Kosivtsov Y Y, Chalov K V, Lugovoy Y V, Sulman E M, Stepacheva A A 2015 *Chemical Engineering Transactions* **45** 667-672

[6] Navosha Y Y, Prokhorov S G, Strigutskiy V P, Tomson A E 1992 *Eurasian Soil Science* **24**(5) 1-6

[7] Chalov K V, Lugovoy Y V, Kosivtsov Y Y, Sulman M G, Sulman E M, Matveeva V G, Stepacheva A A, *Bulletin of Chemical Reaction Engineering and Catalysis*, **11**, 3, 330-338 (2016)

[8] Gamayunov N I, Maslennikov B I, *Eurasian Soil Science*, **24**, 7, 32-37 (1992)

[9] Kremcheev E A, Ivanov A V 2016 *Water and Ecology*, **2** 48-57.

[10] Nagornov, D. O., & Kremcheev, E. A. 2018 Framework to evaluate efficiency of peat processing using roller-disc grinding machines. Paper presented at the *Journal of Physics: Conference Series* **1118**(1).

[11] Nagornov, D. O., Kremcheeva, D. A. 2017 Assessment of gravity component of moisture flow in raw peat in conditions of convective and radiative-convective heat input. *Journal of Industrial Pollution Control* **33**(1) 749-752.

[12] Misnikov O S 2006 *Theoretical Foundations of Chemical Technology* **40**(4) 423-430

[13] Vasilyeva M.A. 2016 Justification of the choice matrix material of the magnetoactive elastomer for working camera-channel peristaltic unit *Materials Science Forum* **870** 13-19

Influence of the shape of quartz sand particles factor on single particle erosion damage

M I Nasyrova[1], P A Kulakov[1]

[1] Ufa State Petroleum Technological University, 1, Kosmonavtov St., Ufa, 450044, Russia

E-mail: nasyrova.marina@mail.ru

Abstract. The paper presents data on the influence of aerodynamic factor of quartz sand particles on a single particle erosion damage as a part of investigation of gas-abrasive wear of field equipment steel elements caused by solid particles impact in conditions of sand production of wells. The studies are mainly focused on calculating geometric characteristics of the crater obtained as a result of a single particle impact, such as maximum penetration depth and equivalent diameter of the crater, affected by the shape of the sand grains. The results have shown the significance of experimental and numerical data, including sphericity, roundness and shape factor of particles determination towards further calculation of particles impact parameters.

1. Introduction

The transportation of oil and gas with large quantities of solid particles can cause repetitive impacts of particles with the inner wall of a range of field equipment elements, for example, high values of erosion rate are often observed for gas-gathering system elements – pipelines, elbows, chokes, blind tees, T-junctions, mesuring devices which are locally installed in the pipelines [1].

The erosion of steel elements caused by solid particles is one of several forms of material degradation processes, most commonly classified as wear. The erosion wear of materials is a complex phenomenon including a range of specific processes happening simultaneously. The prediction of maximum erosion rate of steels is quite a complicated task involving a huge quantity of calculation parameters such as chemical, physical, mechanical properties of materials, flow regimes, interaction of a pair of materials specific features and others [2].

In industrial field most of the equipment elements are thin-walled parts, so the calculation of wall thinning rate and residual resource of equipment and single elements is of particular interest.

In oil and gas industry, the particulate erosion is one of the main reasons for premature wear of equipment, which leads to expensive repairs, loss of production time and temporary decrease of efficiency.

Calculating and prognosing erosion rate is an extensive industrial problem of current interest, and at the same time it's a complicated question that doesn't fully have the answer nowadays.

There is a huge range of parameters that affect the absolute erosion rate value – particles impact parameters (velocity and angle), properties of the target material, size, shape and properties of the solid particles etc.

The studies are mainly focused on exploring geometrical characteristics of the solid particles as one of potentially most dangerous factors on the way to predict wall thinning due to erosion wear.

This paper presents the investigation of different shapes of quartz sand particles. The work includes calculation of shape factor of sand grains according to the particles laboratory size and shape analysis, and penetration rate for each particle in order to find out the most dangerous variant.

At the present work the most promising method to calculate the parameters listed above is usage of CAD and CAE technologies as an addition to laboratory analysis.

2. Problem overview

The effect of particles properties, size and shape on the mechanisms of material removal during single impacts under different valocity value has been widely studied as a model of solid particles erosion [2-9]. At low impact velocities sphere shaped particles formed circular depressions at impact angle of 90 degrees [2]. Plastic deformation was indicated by the formation of an slightly raised ridge of material around the crater edge. A detailed study of the craters revealed a limited amount of mass loss from the crater recess by the combination of melting and tearing due to particle surface adhesion [9].Extensive mass loss was observed at sufficiently high impact velocties which is explained by the fact inertia of the crater edge induced fracture aling the plane of concentrated shear [7].

Brittle sphere shaped erodent particles (glass beads) fractured at the impact angle of 90 degrees at high velocities [10]. As the studies showed, angular brittle particles also fragmented at the same impact angle. Non-brittle particles produces craters at decreased impact velocities, which mirrored impact particles features [11].

All the listed papers are pioneers in the study of particles shape influence on the erosion rate and are supposed to be fundumental towards developing newest models describing different types of solid particles behavior in appropriate conditions.

One of the most burning problems about developing a calculation process including particles shape factor is a range of empirical coefficients that are hard or unable to get from experimental data.

According to the newest studies in single particle erosion effects, geometric properties of the crater are calculated according to the critical impact angle of the particles and the corresponding critical sliding friction force of the particle [12]. There is also a semi-mechanistic model developed for predicting maximum erosion and relating it to measurable mechanical properties such as hardness and fracture toughness [13]. These model showed a good agreement without involving of a huge amount of empirical coefficients.

Most of the newest calculation methods are based on finite element method (FEM) and computational fluid dynamics (CFD) and have a really good agreement with experiments in comparison.

3. Calculation details

As it was mentioned above, CFD-method is concidered to be one of the most full techniques to obtain the operating parameters of particles dynamic movement in gas or liquid field. The attractiveness of CFD-methods is connected not only with simplicity of usage, but also with the level of accuracy of the evaluated results and an opportunity to combine fluid dynamics and structural analysis methods. In this paper a single sand grain motion model is obtained with the help of coupling two major methods: modelling dynamics of the particles in the gas flow field and modelling impact dynamics of the particles at the moment of particles's impingement into the flat steel target sample.

It is possible to predict a trajectory of the discrete phase particle by integrating the force balance of the particle, which is written in Lagrangian reference frame. This force balance equates the particle inertia with the forces acting on the particle, and can be written (for the x direction in Cartesian coordinates) as:

$$\frac{du_p}{dt} = F_D\left(u - u_p\right) + \frac{g_x(\rho_p - \rho)}{\rho_p} + F_x, \tag{1}$$

where $F_D\left(u - u_p\right)$ is the drag force per unit particle mass and

$$F_D = \frac{18\mu}{\rho_p d_p^2}\frac{C_D Re}{24} \tag{2}$$

Here, u is the fluid phase velocity, u_p is the particle velocity, μ is the molecular viscosity of the fluid, ρ is the fluid density, ρ_p is the density of the particle and d_p is the particle diameter. Re is the relative Reynold number which is defined as

$$Re \equiv \frac{\rho d_p |u_p - u|}{\mu} \tag{3}$$

The drag coefficient C_D can be taken from

$$C_D = \frac{24}{Re}\left(1 + b_1 Re^{b_2}\right) + \frac{b_3 Re}{b_4 + Re} \tag{4}$$

where

$$b_1 = \exp(2{,}3288 - 6{,}4581\varphi + 2{,}4486\varphi^2)$$
$$b_2 = 0{,}0964 + 0{,}5565\varphi$$
$$b_3 = \exp(4{,}90 - 13{,}8944\varphi + 18{,}4222\varphi^2 - 10{,}2599\varphi^3)$$
$$b_4 = \exp(1{,}4681 + 12{,}2584\varphi - 20{,}7322\varphi^2 + 15{,}8855\varphi^3) \tag{5 – 8}$$

which is taken from Haider and Levenspiel [14]. The shape factor φ is defined as

$$\varphi = \frac{s}{S} \tag{9}$$

where s is the surface area of a sphere having the same volume as the particle, and S is the actual surface area of the particle.

4. Experimental data

According to the analysis of mechanical impurities samples selected from a range of sand productive wells (Fig.1, a) the quantity of quartz sand is supposed to be prevalent. Since quartz sand appears to be the most aggressive abrasive agent according to the particles hardness analysis (Fig.1, b), quartz sand particles are chosen as discrete phase for modelling.

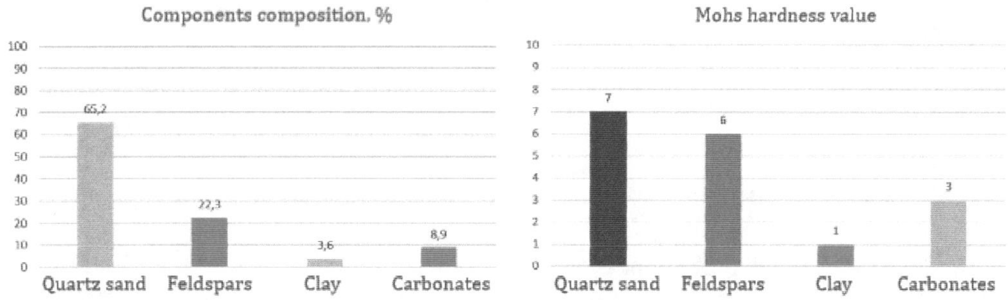

a b

Figure 1. Solid particles component analysis result obtained from sand productive wells:
a – Components composition, %; b – particles Mohs hardness value

According to the laboratory analysis of quartz sand grains fractions size, the smallest grain is supposed to be about 45 μm in it's equivalent diameter, the biggest one – about 500 μm (Fig.2).

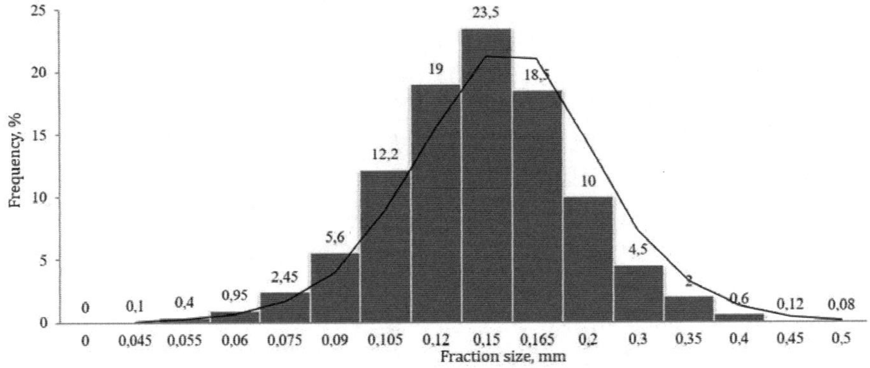

Figure 2. Size of quartz sand fractions

For further modelling mean diameter of the quartz sand particles equals 150 μm as the most common size for solid particle met in the examined wells.

For the samples obtained from a range of sand productive wells particles geometric characteristics were evaluated. Particle shape was determined visually from microscopic images by comparison with samples from the Crumbien-Schloss diagram [15]. According to the diagram there are two common parameters by which particles geometry is determined – roundness and sphericity coefficients. There are mean values of roundness and sphericity of the examined quartz sand particles in table 1 for each of the well.

Table 1. Roundness and sphericity coefficients for quartz sand particles

Well no.	Mean roundness coefficient	Mean sphericity coefficient
1	0,4	0,3
2	0,3	0,5
3	0,6	0,4
4	0,4	0,7
5	0,5	0,5
6	0,3	0,4
7	0,6	0,8
8	0,7	0,4

Sand grains parameters like roundness and sphericity are necessary in order to determine the mean geometry of the sand grains for further impact parameters calculation. Despite the fact these characteristics are common in industrial field to understand the particles geometry, the data is not enough to determine the sand grains shape in visual representaion for further penetration depth calculation, thus according to the information from table 1 3d-models of particles have been created automaticly (Fig. 3).

As it is seen from Figure 3, most of the particles represent angular geometry for the sand grains. Taking into account previous research [16] angular particles are much more aggressive to the target material than rounded particles comparing total mass loss for both of the cases.

Since a 3d-model is a full geometry representation for each of the sand grains, calculation of surface area for each of the particle is automatic. Data on the surface area for the particles and for the equivalent volume spheres are shown in table 2.

According to (9) shape factor is calculated due to the known surface area values.

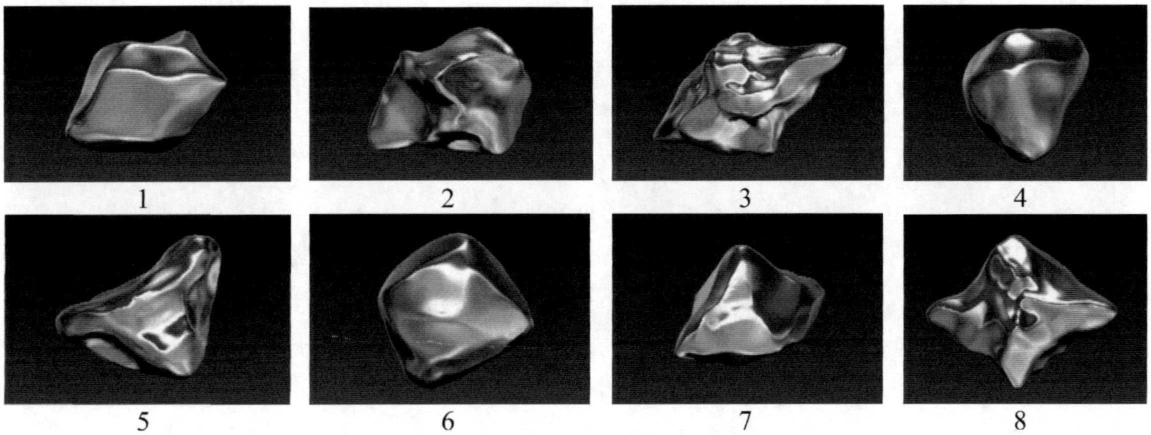

Figure 3. 3d-models of quartz sand grains generated randomly from roundness and sphericity coefficients data

Table 2. Shape Factor data

Particle no.	Particle surface area, mm^2	Body volume, mm^3	Equivalent volume sphere surface area, mm^2	Shape Factor, φ	Shape Factor Mean Value
1	0,0508047	0,0009043	0,0452162	0,89	
2	0,1106665	0,0002571	0,0907465	0,82	
3	0,1139832	0,0023509	0,0854874	0,75	
4	0,0736023	0,0017663	0,0706582	0,96	0,85
5	0,0918318	0,0018379	0,0725471	0,79	
6	0,0873741	0,0021436	0,0803841	0,92	
7	0,1105326	0,0028048	0,0961634	0,87	
8	0,1240693	0,0030521	0,1017368	0,82	

With the Use of CFD-method a single particle motion in gas flow filed has been simulated for different shaped grains.

Taking into account so called near wall effects, the area around the whole particle is supposed to be the boundary layer, so the size of the mesh elements around the particles are generated by adding the inflation method with elements growth rate coefficient 1,4 (Fig.4).

Particles are tracked in a gas flow field which has properties of a real gas. Absolute impact velocity of the grains is 10 m/s. In addition to the basic boundary conditions, gravity force is also taken into account. As a result of the calculation there are streamline contours for each of the particle in the gas flow field (Fig. 5).

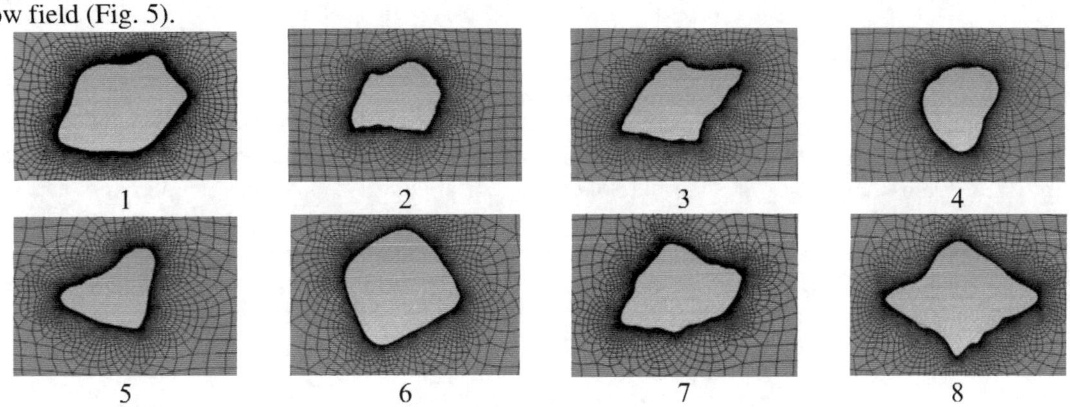

Figure 4. Boundary layer mesh for the particles

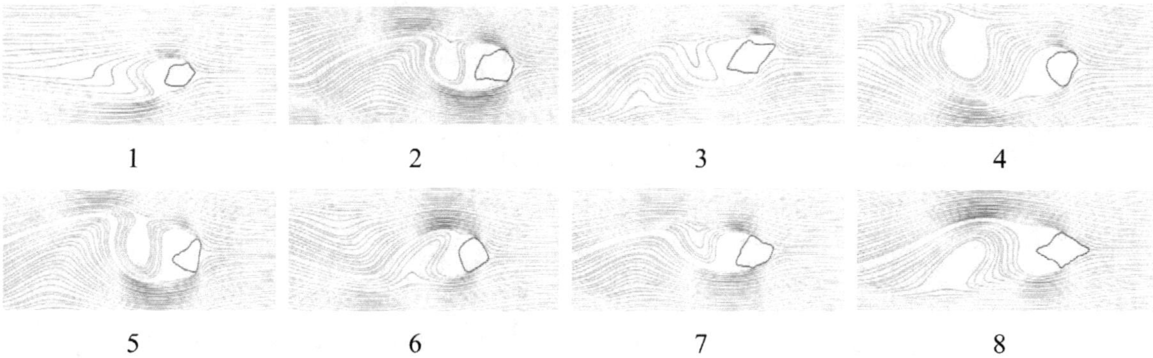

Figure 5. Influence of shape factor of the particles on the streamline contours

The process of indentation of the particles into the target material (steel density is 7850 kg/m^3) was calculated at the impact angle of 90 degrees. As far as sliding angles (0-30 degrees) represent a much more complicated mechanism including frictional force components, this aspect will be fully described in the next research devoted to complex single particle erosion impact.

Typical single particle indentation depth values for each of the particles are shown in table 3. Maximum penetration depth was analysed for different shapes of the grains.

Table 3. Geometric characteristics of the crater for the different shaped grains after the impact

Particle no.	Maximum Penetration Depth, μm	Approximate particle crater diameter after the impact, μm
1	24	20,5
2	22	15
3	56	12,5
4	12	24
5	10	24,5
6	28	13,5
7	48	14
8	76	12

As it is seen from Table 3 maximum penetration rate is observed for the most angular particle of all. It is also necessary to take into account the orientation of the particle in space and additional forces, including particle rotational law, since the geometry of the crater after the particle impact directly depends by which side the particle hits the wall during the indentation process.

5. Conclusion

The main purpose of this paper was to study the dependence of geometric features of the quartz sand particles on the penetration characteristics caused by a single particle impact.

After conducted studies of sand grains shape factor on the penetration depth analysis, it is possible to conclude that the material removal mechanism appears to be the function of the particles' shape. Quartz sand grains having lower shape factor (or higher angularity) produces deep craters and most possibly generates higher mass loss as compared to particles having higher shape factor.

To sum up, penetration depth obtained by a single angular solid particle impact, is significantly high. Numerical modelling shows the importance of a single particle impact simulation, in contrast to calculating total erosion rate, to determine the mechanism of abrasive wear for different shapes of particles.

References

[1] Mansouri A. 2016 *A Combined CFD-experimental Method for Developing an Erosion Equation for Both Gas-Sand and Liquid-Sand Flows* (Ph.D. Dissertation) University of Tulsa.
[2] Rickerby D G, Macmillan N H 1979 *Proc. 5th Int. Conf. on Erosion by Solid and Liquid Impact* Cavendish Laboratory: Cambridge 29-1
[3] Hutchings L M, Winter R E 1974 *Wear* **27** 121 - 128
[4] Hutchings L M, Winter R E 1974 *Wear* **29** 181 - 194
[5] Christman T, Shewmon P G 1979 *Wear* **52** 57 - 70
[6] Christman T, Shewmon P G 1979 *Wear* **54** 145 - 155
[7] Quadir T, Shewmon P G 1981 *Metall Trans* **12** 1163 – 1170
[8] Gulden M E 1979 *Proc 5th Int. Conf. on Erosion by Solid and Liquid Impact* Cambridge: Cavendish Laboratory
[9] Brown R, Edington J W 1982 *Wear* **77** 347 – 353
[10] Tilly G R, Sage W 1970 *Wear* **16** 447 – 465
[11] Ives L K, Ruff A W 1978 *Wear* **46** 149 – 162
[12] Cheng J, Zhang N, Wei L, Mi H, Dou Y 2019 A New Phenomenological Model for Single Particle Erosion of Plastic Materials
[13] Ben-Ami Y, Uzi A, Levy 2016 Modelling the Particles Impingement Angle to Produce Maximum Erosion *Powder Technology* **301** 1032 – 1043
[14] Haider A, Levenspiel O 1989 Drag Coefficient and Terminal Velocity of Spherical and Non-Spherical Particles *Powder Technology* **58** (1) 63-79.
[15] GOST R 54571-2011. Magnezial-quartz proppants. Specifications
[16] Oka Y I, Okamura K, Yoshida T 2005 Practical Estimation of Erosion Damage Caused by Solid Particle Impact *Wear* **259** 1-6 95 – 101.

Numerical simulation of a draining vessel

M I Nasyrova and P A Kulakov

Ufa State Petroleum Technological University, 1, Kosmonavtov St., Ufa, 450044, Russia

E-mail: nasyrova.marina@mail.ru

Abstract. In this paper the models of a theoretical tank and a real size draining vessel are described and numerically simulated. The main idea is to demonstrate the Toricelli's law theorem for fluid dynamics for ideal liquid both for a theoretical vertical cylindrical tank and for a different geometry horizontal vessel of the same volume as the tank and compare the total time of liquid leaving the space of both of the reservoirs. The simulation is a multiphase problem containing two phases, which are water and air, is based on the VOF (volume of fluid) method discrete elements method (DEM) for simulating the sand bed formation for outlet nozzle capacity evaluation. As a result there are two-phase (gas-liquid) models of an "ideal" tank and an industrial vessel numerically calculated for determining and comparing the total time of water draining, and an extra model showing the dependence of sediment layer thickness on the outlet nozzle capacity.

1. Introduction

In oil industry there are many different types of equipment used for storage of liquids and gases. A cylindrical storage vessel is one of the most popular type of fluid storage equipment and it is widely used for storage due to being less expensive than other types of storage tanks, like a sphere, for example [1].

As industrial vessels are usually filled with a definite amount of liquid, they experience hydrostatic pressure. Hydrostatic pressure is the pressure exerted by a fluid at a given point within the fluid, due to the force of gravity, and increases in proportion to depth measured from the surface because of the increasing weight of fluid exerting downward force from above [2]. An industrial vessel or a tank experiences hydrostatic pressure either during a hydro testing or during keeping the liquid product inside.

Since a typical vessel represents a thin-walled construction, it is limited to stand the exact pressure value for an exact time period. Thus, there is a necessity to drain the liquid phase out of the apparatus either after the definite time of the exploitation or after a hydro testing.

A drain system that is connected directly to pressure vessels is called a "pressure" or "closed" drain system. A drain system that collects liquids that spill on the ground is an "atmospheric" , "gravity" , or "open" drain. The liquid in a closed drain system must be assumed to contain dissolved gases that flash in the drain system and can become a hazard if not handled properly. In addition, it must be assumed that a closed drain valve could be left open by accident. Once the liquid has drained out of the vessel, a large amount of gas will flow out of the vessel into the closed drain system (gas blowy) and will have to be handled safely [3]. Thus, closed drain systems should always be routed to a pressure vessel and should never be connected to an open drain system.

Due to the fact every kind of action during exploitation of industrial apparatus must be strictly planned and controlled, calculating the exact time of draining a tank or a vessel is a particular interest in this paper. This is basically reached by knowing or calculating the product outlet nozzle's passage capacity, that's why this value has the most important meaning in the entire draining calculation process. Knowing the exact time of draining a vessel gives an opportunity to schedule the work plan, correct the calendar schedule and even predict a range of risks caused by unexpected time loss.

Another problem is connected with separation of sediments from liquid caused by long settling of liquid in the vessel. The bottom sediment can include sand, wax or paraffin-like substance that make the draining process difficult due to clogging the outlet nozzle.

The newest calculation methods are based on finite element method (FEM). In oil industry there are many different types of equipment used for storage of liquid and gases. A cylindrical storage vessel is one of the most popular type of fluid storage equipment and it's widely used for storage due to being less expensive than other types of storage tanks, like a sphere, for example [1].

As industrial vessels are usually filled with a definite amount of liquid, they experience hydrostatic pressure. Hydrostatic pressure is the pressure exerted by a fluid at a given point within the fluid, due to the force of gravity, and increases in proportion to depth measured from the surface because of the increasing weight of fluid exerting downward force from above [2]. An industrial vessel or a tank experiences hydrostatic pressure either during a hydro testing or during keeping the liquid product inside.

Since a typical vessel represents a thin-walled construction, it's limited to stand the exact pressure value for an exact time period. Thus, there is a necessity to drain the liquid phase out of the apparatus either after the definite time of exploitation or after a hydro testing.

A drain system that is connected directly to pressure vessels is called a "pressure" or "closed" drain system. A drain system that collects liquids that spill on the ground is an "atmospheric," "gravity," or "open" drain. The liquid in a closed drain system must be assumed to contain dissolved gases that flash in the drain system and can become a hazard if not handled properly. In addition, it must be assumed that a closed drain valve could be left open by accident. Once the liquid has drained out of the vessel, a large amount of gas will flow out of the vessel into the closed drain system (gas blowy) and will have to be handled safely [3]. Thus, closed drain systems should always be routed to a pressure vessel and should never be connected to an open drain system.

Due to the fact every kind of action during exploitation of industrial apparatus must be strictly planned and controlled, calculating the exact time of draining a tank or a vessel is a particular interest in this paper. This is basically reached by knowing or calculating the product outlet nozzle's passage capacity, that is why this value has the most important meaning in the entire draining calculation process. Knowing the exact time of draining a vessel gives an opportunity to schedule the work plan, correct the calendar schedule and even predict a range of risks caused by unexpected time loss.

2. Calculation details

2.1 Brief concept of axisymmetric and non-axisymmetric loading of thin-walled shells

A shell of revolution is called axisymmetric if it is loaded with loads uniformly distributed about its axis. Examples of axisymmetric shells are spherical, cylindrical, conical and elliptical shells of revolution, loaded with internal gas or liquid pressure [4].

A simple example of an axisymmetric shell is a vertical cylindrical storage tank.

In a horizontal apparatus filled with water, each point of the cross section experiences different pressure and, therefore, this shell is no longer considered to be axisymmetric.

A hydrostatic pressure distribution diagram for cylindrical axisymmetric and non-axisymmetric shells respectively are shown in figure 1.

It is important to consider whether the shell is axisymmetric or not, since the type of the shell depends on both the distribution of hydrostatic pressure and, consequently, the stresses in the metal of the shell, and the rate of flow of fluid from the shell, depending on the location of the drainage hole, which is one the main objectives affecting the calculation result in this paper.

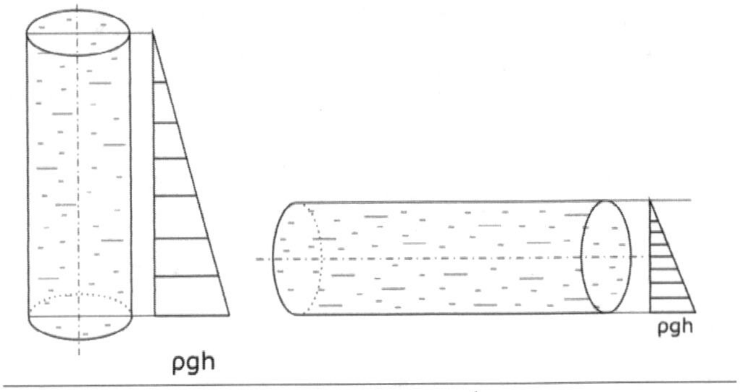

Figure 1. Hydrostatic pressure distribution diagram for a typical example of cylindrical axisymmetric and nonaxissymmetric shells respectively: a – vertical (axissymmetric) shell, b – horizontal (non-axissymmetric) shell

2.2 Volume of fluid (VOF) method

The volume-of-fluid (VOF) is one of the most widely spread and used method in computational fluid dynamics (CFD) nowadays.

The VOF model allows modelling two or even more immiscible fluids by tracking the volume fraction of each of the fluids throughout the domain and solving a set of volume fraction, momentum, energy and additional scalar equations, including the main equation which is so called fraction function C [5–8].

The fraction function represents the integral of fluid characteristic function in the control volume, which is the volume of a computational grid cell. The volume fraction of the entire fluid domain is tracked through every cell of the grid, however all the fluid domains share the same momentum equations. If the computational cell is empty, not filled with the fluid, the fraction function value is zero (C = 0); if the cell is 100%-full with the fluid, the fraction function value is 1 (C = 1). So if two different fluids are separated by a fluid interface, the fraction function in the interface cell ranges like $0 < C < 1$ [9–11].

In VOF method the evolution of the m-th fluid in the whole system is governed by solving the transport equation

$$\frac{\partial C_m}{\partial t} + u\nabla C_m = 0, \tag{1}$$

where u id the velocity. Function C_m is the volume fraction of the m-th fluid in the system [12].

The volume of the fluids is considered to be constant, thus the properties of the fluids can be calculated using the volume fraction function in the computational cell [13]. For instance, density on the fluid can be defined as

$$\rho = \sum_{m=1}^{n} \rho_m C_{m,} \tag{2}$$

The main advantage of using the VOF method is the fact the calculations require minimum time and memory storage. Moreover there is an opportunity to deal with difficult non-linear tasks and track the fluid cells in more complicated models connected with multiphase problems [14–15].

However, due to the problem of superfluous diffusion in the transport equation there is a question about calculating and tracking the fluid-fluid interface because of so called smudging the interface area between two immiscible fluids [16].

2.3 Toricelli's law
Studying a theoretical model of a vertical cylindrical tank, full of uncompressible (ideal) fluid, the instantenious flow out of the nozzle on the bottom of the tank is calculated using an equation

$$Q = \frac{dV}{dt}, \tag{3}$$

which basically means the instantenious flow of the fluid equals the change of the volume of fluid in time [17–18].

According to the Toricelli's law the velocity of the fluid flow through a hole of a thin wall, located on a definite h height over the surface, is the velocity of a body freefalling from a h height [19]

$$v = \sqrt{2gh} \tag{4}$$

where g is gravitational acceleration.
Fluid flow rate can be calculated as

$$Q = av \tag{5}$$

where a is the cross-sectional area of the nozzle, m^2.

Since the main interest in this paper is calculation the time of totally emptying the tank, there is the final equation for calculation the total time of fluid leaving the inner space on the tank based on the Toricelli's law [20–21]

$$T = \frac{R^2}{r^2}\sqrt{\frac{2H}{g}}, \tag{6}$$

where R is the cylindrical tank inner radius, r is the nozzle inner radius and H is the total fluid height inside the tank.

2.4 Concept of nozzle capacity
The calculation of the nozzle capacity is based on a number of so-called calculated models representing certain assumptions that simplify the actual physical picture of the flow.

The calculation is based on the calculated model, and then amendments are made to the result, taking into account the imperfection of the model and the additional influence of various factors [22].

All calculations of the nozzle bandwidth are based on the ideal nozzle model.

According to this model, the flow through the nozzle is firstly calculated for adiabatic (without heat exchange with the environment) and isentropic (without hydraulic friction loss) flow through the nozzle. Then correction factors (for example, additional friction, local or hydrostatic pressure loss) are applied to the calculated nozzle bandwidth.

Since the models in this paper are considered to be close to ideal, the nozzle bandwidth can be substituted with volume flow rate

$$Q = \frac{V}{t} \tag{7}$$

where V is the volume of fluid passing through the flow cross-section during time t, m^3.

Or the nozzle bandwidth for the current problem also can be calculated according to (5).

3. Boundary conditions

Both of the reservoirs are of the same volume which is 3.85 m^3. The tank is a vertical cylindrical reservoir with both a flat top and a flat bottom. The vessel is a horizontal apparatus consisting of a cylindrical shell and two elliptical dishes on both of the ends. It should be noticed that the nozzles filled with the fluid are not included in the total fluid volume. Geometry parameters of the tank and the vessel are shown in Figure 2.

The tank is supposed to be a typical cylindrical axisymmetric shell, so it is considered that the pressure value at any cross-sectional area of the shell is constant. It also should be mentioned that the nozzle (the orifice) for water draining is located symmetrically in the middle of the bottom of the tank.

The vessel represents a more complicated task since it's not only a non-axisymmetric shell model, but also includes a couple of nozzles for water draining and for air entering, that are located on the sides of the vessel. Comparing to the tank the complexity of the vessel model is explained by the absence of a free surface on top – this element is actually substituted by a nozzle for air entering. Moreover the vessel construction includes a couple of elliptical dishes that makes it more complex to calculate analytically.

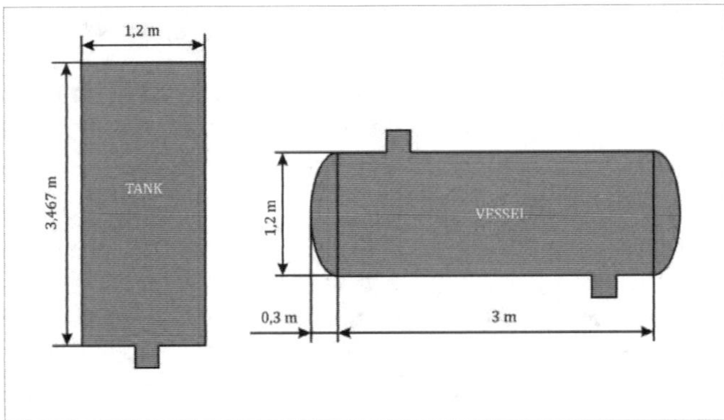

Figure 2. Geometry parameters of the tank and the vessel

The nozzle inner diameter is 200 mm both for the tank and for the vessel for the numerical calculation.

The numerical calculation for both of the reservoirs is a pressure-based task, based on the VOF method with two eulerian phases, where air is considered to be the primary phase and water is the secondary one. Implicit body force formulation is enabled in the task.

Both of the fluids (water and air) are set as ideal fluids. Surface tension between water and are is 0.072 H/m. The entire volume both inside the tank and the vessel are initialized as water, so the volume fraction function by the beginning of the calculation equals 1 for the water and 0 for the air phase for the whole computational grid area.

Operating conditions are normal, so operating pressure is atmospheric which is 1 atm.

Boundary conditions for numerical calculation for both of the reservoirs include gravitational acceleration which equals 9.81 m/s.

4. Calculation results discussion

According to the research work (6) for the tank there are values of fluid velocity and total time of fully emptying the tank depending on the nozzle inner diameter in Table 1.

Table 1. Values of fluid velocity and total time of fully emptying the tank depending on the nozzle inner diameter

Nozzle inner diameter, mm	Total draining time, s
200	30.27
150	53.80
100	121.07

The maximum velocity magnitude is detected at the highest liquid level which is 3.467 m.

Volume fraction of water phase at the beginning and at the end of the tank and the vessel draining respectively are shown in figure 3.

As the result it took 31.5 s to totally empty the tank and 41.93 s for the vessel.

Comparing the numerical calculation time (31.5 s) to the theoretically calculated value (30.27 s) it should be mentioned that the result is almost the same. The slight difference might be caused by the rough geometry for numerical simulation.

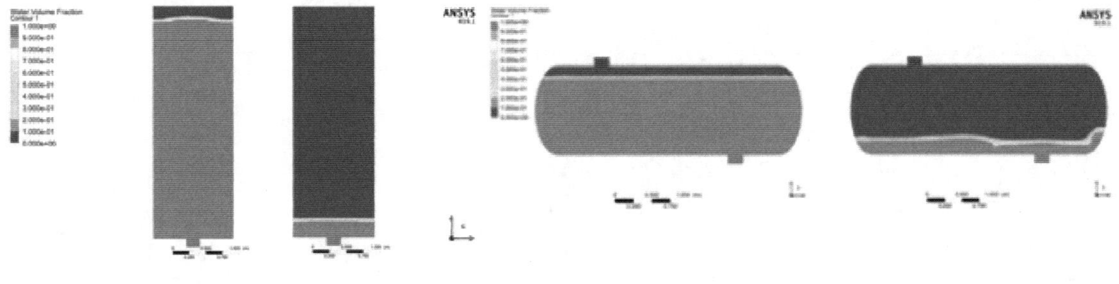

a b

Figure 3. Volume fraction of water phase at the beginning and at the end of the tank and the vessel draining respectively: a – model of a tank draining; b – model of a vessel draining

Talking about the result for the vessel, it took about 30 % more time to empty the space inside the reservoir, which is caused by:

- a much lower height of the liquid as the initial parameter, and as a consequence, lower hydrostatic pressure for emptying the tank comparing to the tank;
- non-standard geometry of the reservoir;
- asymmetrical nozzle location, so the weight of the water wasn’t concentrated evenly in the nozzle area due to the hydrostatic pressure.

There is also another problem including nozzle capacity evaluation due to impurities sedimentation on the bottom of the vessel, representing the third discrete phase. The sand is granular material with mean particles size 150 mkm. The calculation of sedimentation process was stopped until the height of the sand layer reached value of 10 mm.

Bottom sediment distribution by time residence (hours) is shown in figure 4.

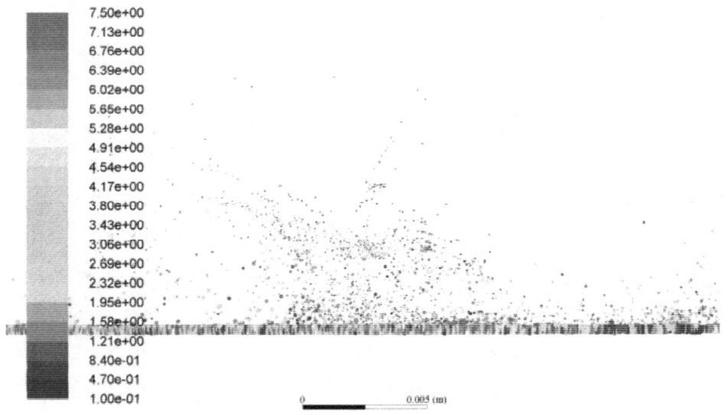

Figure 4. Sediment forming a sand bed on the bottom of the vessel causing a draining problem

As it is seen from fig.4 draining process discrete phase forms a thick layer on the bottom of the vessel, mainly caused by gravity force. Draining process of the vessel was calculated taking into account the sand bed presence.

According to the research work (7) the bottom nozzle capacity was calculated. Dependence of the nozzle capacity (for a nozzle with 100 mm inner diameter) on the sand bed layer thickness in shown in table 2. It is evident that clogging up the nozzle cross sectional area with sediment increases the total time of the draining process for the vessel.

According to the calculation results in Table 2 it is evident that the sediments layer growth causes less nozzle capacity and at a critical value of the sand bed thickness (approximately 200 mm for this case) there is no way for fluid to leave the space of the vessel due to clogging up the liquid outlet, so in this case the nozzle capacity is supposed to equal zero.

Table 2. Dependence of nozzle capacity and total draining time on the sediment layer thickness for the vessel

Sediment layer thickness, mm	Nozzle Capacity, x10^{-4} m^3/s	Total draining time, s
1	318	121,07
5	277	138,50
10	213	180,78
50	154	250,30
100	80	480,65
200	0	–

5. Conclusion

To sum up, the time calculation methods overview and experiment was made in order to demonstrate different ways of solving the problem of industrial vessels total time emptying, which are one of the most important aspects in projects planning, risks and time loss prediction. Using numerical calculation methods allows not only saving time while planning the process, but also creating a range of ways for solving the same problem.

Comparing the numerical calculation time (31.5 s) to the theoretically calculated value (30.27 s) it should be mentioned that the result is almost the same. The slight difference is caused by near wall effects calculation moments in the numerical model.

Talking about the result for the vessel, it took about 30 % more time to empty the space inside the reservoir, which is caused by:

- non-uniform vessel cross section hydrostatic pressure distribution;
- a much lower height of the liquid as the initial parameter;
- a different shape of the reservoir;
- asymmetrical nozzle location, so the weight of the water was not concentrated evenly in the nozzle area due to the hydrostatic pressure.

According to the results it should be noticed that the theoretical method of calculating the total time for draining a reservoir could be applied including appropriate correction factors since the real result depends on the geometry imperfection, real liquid compressibility, additional forces (frictional, inertial etc.) and others.

Taking into account nozzle capacity calculation it is possible to conclude sediment formation is a burning industrial problem causing not only clogging up the outlet nozzles, but also a range of unexpected troubles. One of the ways to cope with sedimentation is to maintain the homogeneity of the fluid inside the vessel by constant phases mixing process.

Despite the calculations being considered to be close to ideal models, numerical calculation provides quite accurate information due to a more complex system of initial and boundary conditions for the experiment and a possibility to solve the problem using discrete phase method to reach a closer to real model result.

References

[1] Couper J R, Fair J R, Penney W Roy and Walas S M 2004 *Chemical process equipment. Solution and design* 2nd ed pp 655–75

[2] Nakayama Y and Boucher R F 1999 *Introduction to fluid dynamics* pp 30–8

[3] Tanaka R, Uddin M A and Kato Y 2018 Flow characteristics related to liquid/liquid mixing pattern in an impeller-stirred vessel *ISIJ Int* pp 620–66

[4] Wagner W and Gruttmann F 1994 A simple finite rotation formulation for composite shell elements *Engineering Computations* 145–76

[5] Xu W and YangY-R 2011 Numerical simulation of multiphase flow based on VOF (Sichuan Daxue Xuebao, Gongcheng Kexue Ban) pp 294–6

[6] Smirnov B M and Stephen Berry R 2015 Growth of bubbles in liquid *Chem. Cent. J.* 1–8
[7] Wang J, Doussin J F, Perrier S, Perraudin E, Katrib Y, Pangui E and Picquet-Varrault B 2011 Design of a new multi-phase experimental simulation chamber for atmospheric photosmog, aerosol and cloud chemistry research *Atmos. Meas. Tech.* 2465–94
[8] Boisson N and Malin M R 1996 Numerical prediction of two-phase flow in bubble columns *Int. J. Numer. Methods Fluids* 1289–310
[9] Waltrich P J, Falcone G and Barbosa J R Jr 2011 Performance of vertical transient two-phase flow models applied to liquid loading in gas wells *Proc. SPE Annu Tech. Conf. Exhib* pp 3568–82
[10] Lee Y K, Ahn K H and Lee S J 2014 Local shear stress and its correlation with local volume fraction in concentrated non-Brownian suspensions: Lattice Boltzmann simulation *Phys. Rev. E Stat. Nonlinear Soft Matter Phys.* **062317**
[11] Sun X, Smith T R, Kim S, Ishii M and Uhle J 2003 Interfacial structure of air-water two-phase flow in a relatively large pipe *Exp. Fluids* 206–219
[12] Pfleger D, Gomes S, Gilbert N and Wagner H-G 1999 Hydrodynamic simulations of laboratory scale bubble columns fundamental studies of the Eulerian-Eulerian modelling approach *Chem. Eng. Sci.* 5091–5099
[13] Asad Sangabi F and Talebbeydokhti N 2011 Numerical modeling of two phase flow in Morning-Glory spillways *Dam Maint. Rehabil. Proc. Int. Congr. Dam Maint. Rehabil* pp 529–35
[14] Marder M, Chen C-H and Patzek T 2015 Simple models of the hydrofracture process *Phys. Rev. E Stat. Nonlinear Soft Matter Phys.* **062408**
[15] Al Jibory M W, Al Turaihi R S and Al Jibory H N 2018 An experimental and numerical study for two phase flow (water-air) in rectangular ducts with compound turbulators *IOP Conf. Ser. Mater. Sci. Eng.* **012049**
[16] Tanaka R, Uddin M A and Kato Y 2018 Flow characteristics related to liquid/liquid mixing pattern in an impeller-stirred vessel *ISIJ Int.* 620–6
[17] Swanepoel K J 1999 The fermat-toricelli problem and a balancing game *J. Geom.* 23–24
[18] Yetilmezsoy K 2017 IMECE—Implementation of mathematical, experimental, and computer-based education: A special application of fluid mechanics for civil and environmental engineering students *Comput Appl Eng Educ* 833–860
[19] Thevenin D and Janiga G 2014 *Fluid dynamics for engineers* pp 68–9
[20] Otto J 2018 *Toricelli's law for large holes* pp 5–7
[21] Bistafa Sylvio R 2015 First Theoretical constructions to the fluid mechanics problem of the discharge *Advances if historical studies* 172–99
[22] Korelshtein L B 2016 Calculation of safety values bandwidth in the dfaft of the new State Standard *Calculations and design* 40–3

The main types of jigging machines structures

S A Nebogin[1], V A Ershov[1], V O Gorovoy[1], A S Govorkov[1], A S Suhanov[1]

[1] Irkutsk National Research Technical University, 83 Lermontov Street, Irkutsk, 664074, Russia

E-mail: s.a.nebo@yandex.ru

Abstract. The article presents the results of the study and analysis of the main structures of depositors. Identified the strengths and weaknesses of individual models. Structurally, the jigs are divided into two groups: movable and immovable sieve. It is believed that jigging machines with movable sieve offer better performance, require less water, but outcomes for them are worse than the jigging machine with a fixed sieve. Given the technical characteristics of the diaphragm jigging machines. Illuminated the depositors technical characteristics with a conical pul-sators. Most designs have a limited height undersize (running) part, which ensures machines compactness and ease of use in various condi-tions. A positive feature of the machine is the consistency of the diaphragm stroke, providing a "hard" mode pulsation environment.

1. Introduction

The numerous jigging machines designs development influenced the their application conditions variety. In the enrichment of ores and sands, diaphragm jigging machines and machines with a cone pulsator have become predominant, allowing the process to be carried out in a high-frequency oscillation mode, meeting the requirement of compactness for use in cramped conditions, for example, on dredges (table 1). In diaphragm jigging machines, water relative to the sieve moves in two directions-up and down, in pulsators – only up.

There are about 90 designs of jigging machines that differ from each other in their intended purpose, the principle of drive operation and unloading devices, the number of allocated products and other characteristics [1-8]. As the main defining feature for the classification of jigging machines, the drive type with the decoding of an additional difference, for example, the location of the drive in relation to the jigging sieve, is adopted. According to this classification, all known jigging machines are merged into seven classes: air-operated; with diaphragm actuator; with cone pulsator; with piston actuator; with movable sieve; with blade drive and hydraulic pulsator. Here are diaphragm jigging machine and jigging machine with cone pulsators.

Diaphragm jigging machines (Fig.1) – the most common design used in the enrichment of dispersed ore minerals. This is due to the relative simplicity of the device, its compactness, the ability to operate in the high-frequency mode and regulate them within the required limits [4, 9-14]. Aperture in various machines types arranged horizontally side jigging separators, horizontal and inclined under the sieve, vertically in the partition wall between adjacent treads or in the housing jig walls. Most designs have a limited height undersize (running) part, which ensures machines compactness and ease of use in various conditions. A positive feature of the machine is the constancy of the diaphragm stroke, which provides a "hard" mode of medium pulsations.

2. Jigging machines structures

In domestic practice, the enrichment of ores of non-ferrous and rare metals have spread machines with vertical diaphragms located in the side (NIM-5) or end (MOD-4 and MOD-2P) walls of machines below the jigging sieves. This class striking representative machine is jigging machine MOD-4 (Fig.2) used for enrichment of ores of size up to 30 mm. The machine body consists of four cameras that pairs constitute two step of the machine, separated by a threshold. The circular diaphragm configuration is located in the end walls of the each four chambers. They are connected to the body flanges the with flexible cuffs and are driven in oscillating motion from the drive through a system of longitudinal rods and conical traverse. Each pair of diaphragms is equipped with an individual drive. Screens have sieve slit size 3×22 mm. Sieve is installed with a slight slope in the direction of enriching material. Sieve fixed between the under-size and oversize frames (stencils). The over-lattice frame has cells for uniform laying on an artificial bed layer sieve. Jigging is carried out in the high-frequency mode [8, 15-18]. The heavy product (concen-trate) is unloaded through an artificial bed and released in the lower part of the sublattice chambers through unloading devices with replaceable rubber nozzles. The light product (tails) is discharged at the end of the jigging sieve through a vertical drain threshold of adjustable height [19-22]. According to tech-nological parameters jigging machine MOD-4 is on a par with the machine MOD-3 (Fig.3 and 4).

Jigging machine MOD-2P has two stages, separated by a vertical threshold. Pulsations in each stage are created by the vertical diaphragm of the end arrangement. An essential feature of the machine is the abil-ity to work with an increased thickness of the enriched material layer, which increases the specific capac-ity twice as compared to the performance of the MOD-4. To remove heavy products from the sublattice chambers, hydroelevators are used, feeding the material to the dehydrating device to separate part of the water and feed it under the sieve of the jigging machine for reuse [17, 23]. The machine is more often used on dredges for primary enrichment of efels.

Jigging machine 2-OVM-1 with a vertical diaphragm in the partition between the two stages is designed for the enrichment of small ore classes (from 0.1 to 3 mm) in the mode of high vibration numbers (up to 850 Qty/min). Jigging sieves in both stages are equipped with grids for artificial bed grains [24-26].

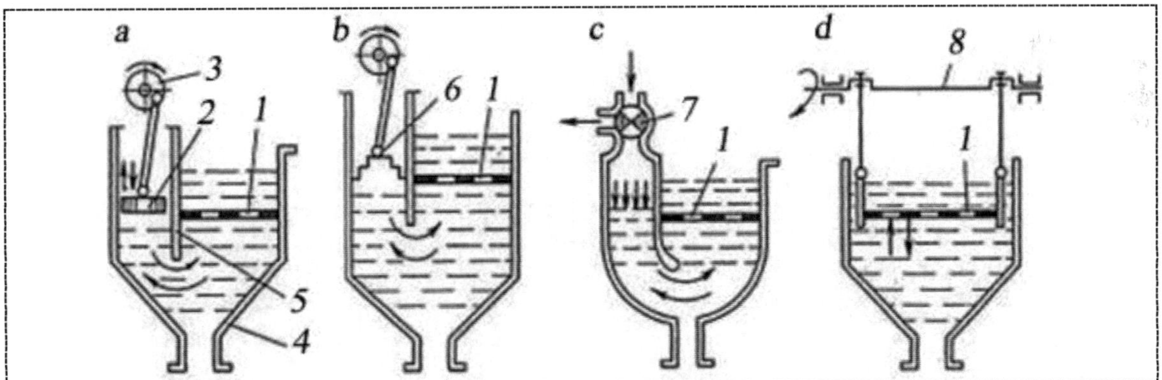

Fig. 1. Scheme jigging machines with diaphragm actuator: a) lateral diaphragm; b) with sublattice arrangement of the diaphragm; c) with a vertical diaphragm in the bulkhead; d) with vertical diaphragms in the outer walls

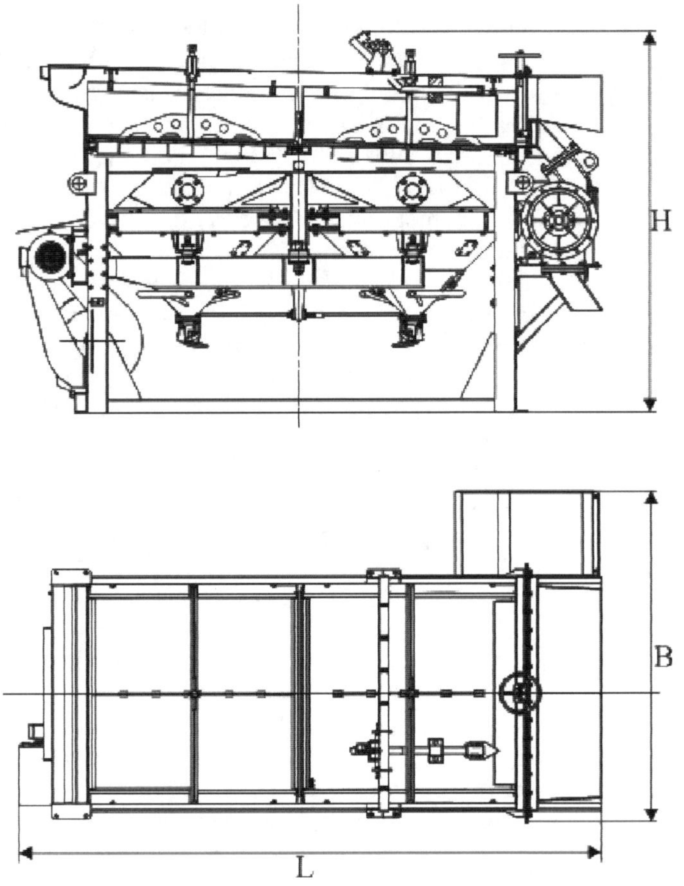

Fig. 2. Jigging machine MOD-4

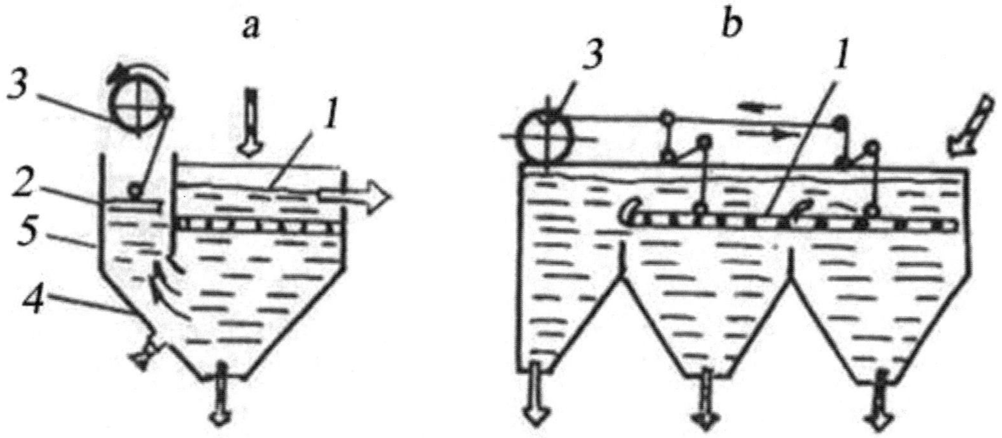

Fig. 3. Scheme jigging machines with cone pulsator: a) with conical bottoms-pulsators; b) with pneumatic cones

At the end of each stage there is a vertical, height-adjustable threshold. Vibrations are communicated to the dia-phragm from the drive mechanism through a hollow longitudinal rod, which in some versions of the ma-chine is used for the intake of sublattice water.

Heavy products (concentrate) are collected in pyramidal sublattice chambers and periodically dis-charged through a cork crane or other locking device (table 1).

Table 1. Technical characteristics of diaphragm jigging machines

Indicator	MOD-4	MOD-2P	2-OVM-1
Capacity, m3 / h	up to 17	26	up to 4
Maximum size of processed ore, mm	up to 30	up to 30	up to 3
Number of cameras	4	2	3
Working area of sieves, m2	4	2	0.18
Chamber length, mm	1060	1060	300
Traverse stroke, mm	up to 75	up to 75	0÷16
Maximum ripple amplitude. mm	37	37	20
Number of pulsations per minute.	130÷350	197	400÷850
Electric motor power, kW	2x2.2	2.8	0.6
Overall dimensions, mm:	3196 2432 1650	3230 1520 2950	1220 646 1000
Length	2600	2016	240

Our NIM-5 is the lateral location of the vertical aperture and a separate actuator for the communi-cation of high frequency fluctuations of the jigging sieve [22-24, 27-30].

To machines with cone pulsators (Fig.3) include the jig with a tapered bottom - a pulsator and less known design Mehanobr with pneumopuncture. They are characterized by simplicity and compactness, adjustable number of vibrations of the liquid.

Table 2. Technical characteristics of depositors with a conical pulsators

Indicator	Conical bottom-pulsator			Pneumatic cone
	MOD-1	MOD-2	MOD-3	MOBK-8S
Effectiveness, m^3/h	6	10	14	70÷120
Sieve width, mm	700	1000	1000	2000
Screening area, m^2	1	2	3	8
Number of pulsations per min.	130÷350	130÷236	164÷348	57÷71
Air pressure, кH/m^2	-	-	-	3.5÷4.0
Air consumption, m^3/h	-	-	-	до 4000
Electric engine power, kW	1.1	2.2	2×2.2	2.2
Overall dimensions, mm:				
length	1850	2450	3750	5610
width	1000	1260	1260	3270
height	1805	2000	2000	4140
Weight, kg	891	1460	2445	19000

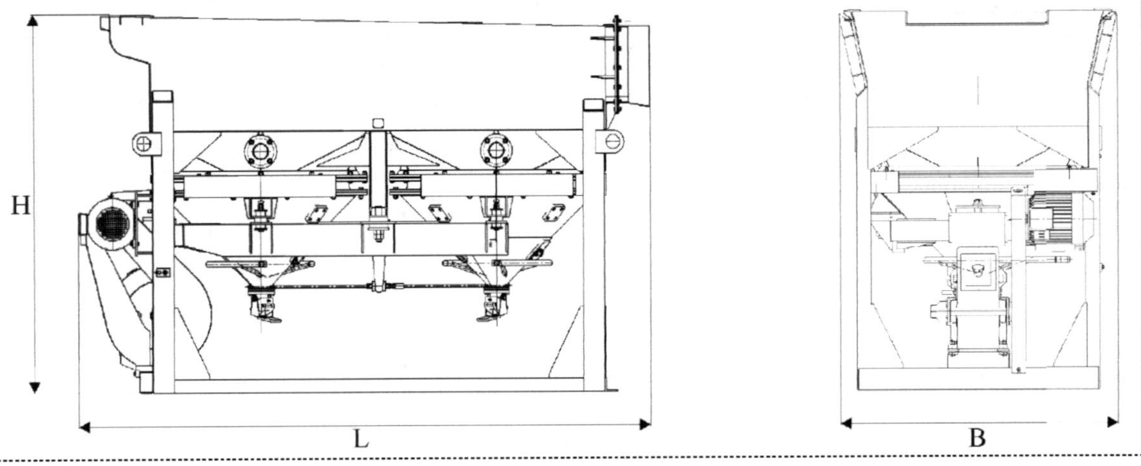

Fig. 4. Jigging machine MOD-3

3. Conclusion

Jigging machines MOD-1, MOD-2 (47B-OT) and MOD-3 (48B-OT) (table 2) designed for enrichment of ores and sands up to 15 mm. Machines MOD-1 and MOD-2 have two, and MOD-3 – three steps, separated by transverse partitions with a vertical threshold. Designs of all three standard sizes are similar, their device can be considered on an example of the MOD-3 machine (Fig.4 and tab.2). In the welding rectangular case of sheet steel jigging sieve is fixed between the undersize and oversize frames, ledges (in each successive step below the previous one) [25-26]. Each step in the lower part passes into a cone, to which a movable conical bottom with an unloading device at the top is attached through a flexible cuff to release a heavy product (concentrate). The bottom is fixed to the balance frame, which balances the static loads and transmits the reciprocating motion from the drive to the bottoms. In machines MOD-1 and MOD-2 drive and balancer frame common to the two bottoms, in the car MOD-3 one frame with the drive is connected to the first two bottoms (in the direction of the material). The third bottom has a separate drive and is attached to its balance frame.

References

[1] Greenberg I ., Ragozin A V, Tenygin A Y 2006 SPb.: MANEB pp 144-145

[2] Kondratiev V V, Govorkov A S, Kolosov A D, Gorovoy V O, Karlina A I 2017 International Journal of Applied Engineering Research 12 22 12373-12377

[3] Berezin A I, Rodnov O O, Mazubowski I V, Stont P D, Klykov V A RF patent № 2204629

[4] Kondrat'ev V, Govorkov A, Lavrent'eva M, Sysoev I, Karlina A I 2016 International Journal of Applied Engineering Research 11 19 9979-9983

[5] Suslov K V, Stepanov V S, Solonina N N 2013 *IEEE International Symposium on Electromagnetic Compatibility* pp 841-845

[6] Stotzer M, Styczynski Z, Gronstedt P, Buchholz B M, Suslov K V, Glaunsinger W 2012 *IEEE Power and Energy Society General Meeting* 6345090

[7] Suslov K, Solonina N, Stepanov V 2015 Proceedings - 2015 International Symposium on Smart Electric Distribution Systems and Technologies, EDST **7315218** pp 260-264

[8] Nguyen AV 2003 Int. J. Miner. Process 68 pp 167–182

[9] Sivtsov A V, Sheshukov O Y, Tsymbalist M M, Nekrasov I V, Makhnutin A V, Egiazar'yan D K, Orlov P P 2018 *Russian Metallurgy (Metally)* Issue 12 pp 1108-1113

[10] Nekrasov I V, Sheshukov O Yu, Sivtsov A V, Tsymbalist M M, Metelkin A A 2016 Steel in Translation Vol. 46 **6** pp 435-442

[11] Sivtsov A V, Sheshukov O Y, Tsymbalist M M, Nekrasov I V, Egiazar'yan D K 2015 *Metallurgist* Vol. 59 **5-6** pp 380-385

[12] Nekrasov I V, Sheshukov O Y, Sivtsov A V, Tsymbalist M M, Egiazar'yan D K, Metelkin A A 2015 *Metallurgist* Vol. 59 **3-4** pp 300-304
[13] Sheshukov O Y, Nekrasov I V, Sivtsov A V, Tsymbalist M M, Egiazar'yan D K, Metelkin A A 2014 *Steel in Translation*Vol. 44 Issue 1 pp 43-46
[14] Sivtsov A V, Tsimbalist M M, Fadeev V I, Ostrovskii Y I, Veselovskii I A, Afanas'ev V I 2012 *Russian Metallurgy (Metally)* Vol. 2012 Issue 12 pp 1036-1040
[15] Ivanchik N N, Balanovsky A E, Shtayger M G, Sysoev I A, Karlina A I 2018 *IOP Conference Series: Materials Science and Engineering* vol 411 012035
[16] Yolkin K S, Yolkin D K, Shtayger M G, Kolosov A D, Ivanov N A 2018 *IOP Conference Series: Materials Science and Engineering* vol 411 012028
[17] Kondratiev V V, Nebogin S A, Sysoev I A, Gorovoy V O, Karlina A I 2017 *International Journal of Applied Engineering Research* T. 12 **22** pp 12809-12813
[18] Nemarov A, Lebedev N, Kondrat'ev V, Kornyakov M, Karlina A 2016 *International Journal of Applied Engineering Research* T. 11 **20** pp 10222-10226
[19] Karlina A I, Kondratev V V, Kolosov A D, Balanovskiy A E, Ivanov N A 2019 IOP Conference Series: Materials Science and Engineering 012183
[20] Kondratiev V V, Karlina A I, Guseva E A, Konstantinova M V, Kleshnin A A 2018 IOP Conference Series: Materials Science and Engineering 042068
[21] Rassokhin A S, Ponomarev A N, Figovsky O L 2018 *Engineering and Construction Journal* **3 (79)** pp 132-139
[22] Ponomarev A N, Rassokhin A S 2016 *Engineering and Construction Journal* **8 (68)** pp 45-57
[23] Yolkin K S, Yolkin D K, Nemarov A A, Sysoev I A, Karlina A I 2018 IOP Conference Series: Materials Science and Engineering 012029
[24] Kondrat'ev V V, Ershov V A, Shakhrai S G, Ivanov N A, Karlina A I 2016 *Metallurgist* T. 60 **7-8** pp 877-882
[25] Vdovin K N, Tochilkin V V, Tochilkin V V 2016 Refractories and Industrial Ceramics 57(3) pp 221-223
[26] Zhukovskiy Y, Koteleva N 2018 *Journal of Physics: Conference Series* **1050(1)** 012106
[27] Alekseeva L B 2018 *IOP Conference Series: Earth and Environmental Science* **194(2)** 022001
[28] Ershov D Y, Zlotnikov E G, Koboyankwe L E 2017 *IOP Conference Series: Earth and Environmental Science* **87(8)** 082015
[29] Vasilev Y, Nevskaya M 2017 *International Multidisciplinary Scientific GeoConference Surveying Geology and Mining Ecology Management, SGEM* **17(53)** pp 419-426
[30] Nokhrina O I, Rozhikhina I D, Khodosov I E 2016 *Steel in Translation* Vol. 46 **10** pp 686-691

Study of the dynamics and analysis of the effect of the position of the vibration motor to the oscillation of vibrating screen

Van Xo Nguyen[1], Khac Linh Nguyen [2], Gia Nghiem Dinh[3]

[1]Ha Noi University of Mining and Geology, Ha Noi, Vietnam
[2] Saint-Petersburg Mining University, Saint-Petersburg, Russia
[3] Ministry of science and technology, Ha Noi, Vietnam

E-mail: nguyenvanxo@humg.edu.vn and khaclinhhumg@gmail.com

Abstract. A vibrating screen is a device for sorting materials, which is widely used in screening plants. The working ability of a vibrating screen relies on various parameters, such as dynamic parameters, technological parameters, and so forth. If it is necessary to find a reasonable set of parameters in order to enhance the machine's performance, it is essential to set up the vibration equation of the vibrating screen. This is also a part of the research content that the article mentions. The next step is to establish the equation of motion of random point-P on the screen box, conduct the simulation, and discuss the effect of the position of the vibratrion motor on the oscillation of vibrating screens. The result reveals that the farther the point is from the focus of vibrating screen, the higher the effect of the vibration motor placement on the amplitude of vibration will be. The points are symmetrical with each other across the focus tend to create the rule of change is opposite. When the vibration motor is moving in the positive direction of the horizontal direction, the vertical oscillation amplitude of the points lying on the upper sieve will increase gradually, while the bottom points will decrease progressively. When the vibration motor is moving in the negative direction of the horizontal direction the change rule is opposite to the case of positive direction; if the vibration motor is 0.3 meters away from the focus of screen, the vertical vibration amplitude of the points on the sieve reaches the maximum value. The research results are of considerable significance to reasonably design parameters for vibrating screen working well..

1. Introduction

A vibrating screen is a sorting device which is used to separate bulk lump material into different sized particles by screen mesh [1]. Ther structure and working principle of vibrating screen as shown in figure 1, including vibration motor (2) with a screen mesh (3) placed on springs (4) and on a screen frame to install vibration motor (1). When the vibration motor (1) is operated, it will create a vibrating force that causes the screen frame to oscillate and implements the screening process [2].

Journal of Physics: Conference Series **1384** (2019) 012035 doi:10.1088/1742-6596/1384/1/012035

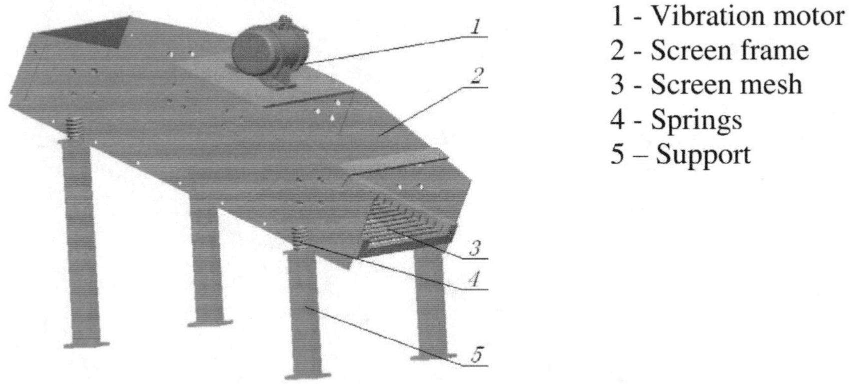

Figure 1. The structure of a vibrating screen

Currently, vibrating screens are commonly used in a range of screening plants [3, 4, 5], and there have also been tons of studies on vibrating screen in the world [6-8], However, how to improve the efficiency of the vibrating screen has been still a challenging question because the productivity, performance, and useful life of vibration screens depend on a lot of factors such as material properties (moisture content, mud content, particle size, density), dynamics of vibrating screen, technological parameters of the machine, etc. [9, 10, 11, 12]. In which the dynamic parameters in the working process plays a crucial role, that significantly affect the working ability of the device [13, 15-18]. Therefore, in this study, the author is going to dig deeper into the dynamic research of the vibrating screen and establish the oscillation equation of the vibrating screen, the equation of motion of random point on screen box, then applying Matlab software to build up the relationship between the influence of the position of vibration motor and the oscillation of the vibrating screen [14].Materials and methods

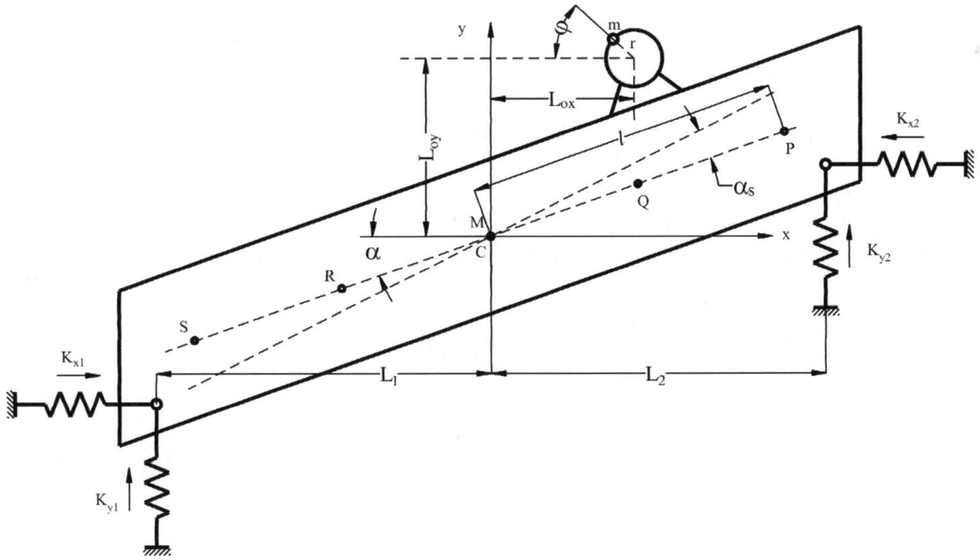

Figure 2. Dynamic model of vibrating screen

2. Dynamics of the vibrating screen

2.1. Establishing the dynamic equation of vibrating screen

From the working principle and structure of the vibrating screen, as it is shown in figure 1, by dint of using some simplified rules like concentrated volume and concentrated hardness, the author built the dynamic model of vibrating screen as shown in figure 2.

Where,

M- The total mass of screen refers to the focus C, kg;
m- The mass of the eccentric plate, kg;
r- The eccentric radius, mm;
k_{x1}, k_{x2}, k_{y1}, k_{y2} - The spring hardness in the x and y directions, N/m;
L_1, L_2 - The distance from the focus of screen to the springs in the x direction, mm;
L_{ox}, L_{oy} -The distance from the focus of screen to the center of vibration motor, mm;
α - Initial tilt angle of the screen, degree;
φ - Angle of rotation around the axis of eccentric plate, $\varphi = \omega.t$ (rad);
ω - Angular frequency of eccentric plate, rad/s;
α_s - Position angle of the screen while oscillating, rad;

According to the moving theorem of an object in the plane, the general dynamic equation of the vibrating screen is expressed in formula 1 [11]

$$\left.\begin{aligned} M\frac{d^2\vec{r}}{dt^2} &= \sum \vec{F}^{(e)} \\ J\frac{d^2\vec{\varphi}}{dt^2} &= \sum M(\vec{F}^{(e)}) \end{aligned}\right\} \quad (1)$$

Based on the dynamic model of the vibrating screen (figure 2) and some transformations, we find the dynamic differential equation of vibrating screen as follows:

$$\left.\begin{aligned} M\ddot{x} &= A_o\cos\omega t - f_x\dot{x} - (k_{x1}+k_{x2})x \\ M\ddot{y} &= A_o\sin\omega t - f_x\dot{y} - (k_{y1}+k_{y2})y - (k_{y1}L_1 - k_{y2}L_2)\alpha_s \\ J\ddot{\alpha}_s &= A_o(L_{oy}\cos\omega t - L_{ox}\sin\omega t) - f_{\alpha s}\dot{\alpha}_s \\ &-(k_{y1}L_1^2 + k_{y_2}L_2^2)\alpha_s - (k_{x1}L_1^2 + k_{x2}L_2^2)\alpha_s\sin^2\alpha - (k_{y1}L_1 - k_{y2}L_2)y \end{aligned}\right\} \quad (2)$$

Where F^e is the main force caused by the eccentric plate.

In order for the oscillation of screen from the nonlinear to linear, we take $k_{y1}L_1 = k_{y2}L_2$ then formula (2) can be presented as follows:

$$\left.\begin{aligned} (M+m)\ddot{x}_c + f_x\dot{x}_c + (k_{x1}+k_{x2})x_c &= A_o\cos\omega t \\ (M+m)\ddot{y}_c + f_x\dot{y}_c + (k_{y1}+k_{y2})y_c &= A_o\sin\omega t \\ J\ddot{\alpha}_s + f_{\alpha s}\dot{\alpha}_s + (k_{y1}L_1^2 + k_{y_2}L_2^2)\alpha_s &= A_o(L_{oy}\cos\omega t - L_{ox}\sin\omega t) \end{aligned}\right\} \quad (3)$$

Let $\xi_1 = \dfrac{f_x}{2\omega_1(M+m)}$, $\xi_2 = \dfrac{f_y}{2\omega_2(M+m)}$, $\xi_3 = \dfrac{f_{\alpha s}}{2\omega_3 J}$;

$$\omega_1 = \sqrt{\frac{k_{x1}+k_{x2}}{M+m}},\ \omega_2 = \sqrt{\frac{k_{y1}+k_{y2}}{M+m}},\ \omega_3 = \sqrt{\frac{k_{y1}L_1^2 + k_{y2}L_2^2}{J}}$$

Then formula (3) becomes:

Journal of Physics: Conference Series **1384** (2019) 012035 doi:10.1088/1742-6596/1384/1/012035

$$\left.\begin{aligned}
&\ddot{x}_c + 2\xi_1\omega_1\dot{x}_c + \omega_1^2 x_c = \frac{A_o}{M+m}\cos\omega t \\
&\ddot{y}_c + 2\xi_2\omega_2\dot{y}_c + \omega_2^2 y_c = \frac{A_o}{M+m}\sin\omega t \\
&\ddot{\alpha}_s + 2\xi_3\omega_3\dot{\alpha}_s + \omega_3^2\alpha_s = \frac{A_o}{J_c}(L_{oy}\cos\omega t - L_{ox}\sin\omega t)
\end{aligned}\right\} \quad (4)$$

Where,

ω_1, ω_2 - the natural frequency of the screen in x and y directions;

ω_3 - the natural shaking frequency in α_s direction.

The oscillation of the screen is a function of the oscillator frequency, at steady state:

$$\left.\begin{aligned}
&x_c = X\cos(\omega t - \phi_1) \\
&y_c = Y\sin(\omega t - \phi_2) \\
&\alpha_s = \alpha_{01}\cos(\omega t - \phi_3) + \alpha_{02}\sin(\omega t - \phi_4)
\end{aligned}\right\} \quad (5)$$

Substituting E_q (5) into (4), yielding:

$$\left.\begin{aligned}
&X = \frac{A_o / \left[(M+m)\omega_1^2\right]}{\sqrt{[1-(\frac{\omega}{\omega_1})^2]^2 + [2\xi_1\frac{\omega}{\omega_1}]^2}} \qquad \phi_1 = \arctan\frac{2\xi_1\frac{\omega}{\omega_1}}{1-(\frac{\omega}{\omega_1})^2} \\
&Y = \frac{A_o / \left[(M+m)\omega_2^2\right]}{\sqrt{[1-(\frac{\omega}{\omega_2})^2]^2 + [2\xi_2\frac{\omega}{\omega_2}]^2}} \qquad \phi_2 = \arctan\frac{2\xi_2\frac{\omega}{\omega_2}}{1-(\frac{\omega}{\omega_2})^2} \\
&\alpha_{01} = \frac{A_o L_{oy} / (J\omega_3^2)}{\sqrt{[1-(\frac{\omega}{\omega_3})^2]^2 + [2\xi_3\frac{\omega}{\omega_3}]^2}} \qquad \phi_3 = \arctan\frac{2\xi_3\frac{\omega}{\omega_3}}{1-(\frac{\omega}{\omega_3})^2} \\
&\alpha_{02} = \frac{-A_o L_{ox} / (J\omega_3^2)}{\sqrt{[1-(\frac{\omega}{\omega_3})^2]^2 + [2\xi_3\frac{\omega}{\omega_3}]^2}} \qquad \phi_4 = \phi_3
\end{aligned}\right\} \quad (6)$$

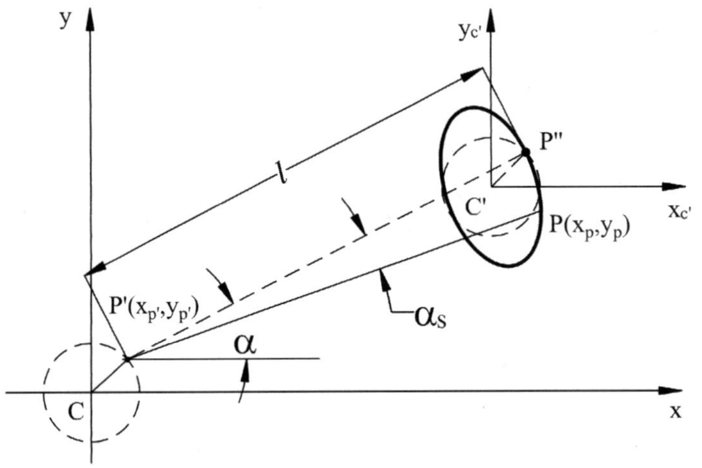

Figure 3. Movement of random point-P on the screen surface

2.2. Establishing equations of motion of a random point on the screen surface

Figure 3 shows that point C represents the initial position of the focus of screen. After a period of work the screen has a tendency to move constantly, then the focus C will have an approximate orbit of the circle centered at C with a radius as CP '. If we neglect the shake of the screen, the trajectory of point P will be an approximation of the circle centered at C ' with a radius as C'P' (P is a random point on screen surface, P is about *l* focus away from C); If the sway factor of the screen is included, the trajectory of point P will be in an ellipse form. Hence, the position of point P can be expressed as formula (7).

$$\begin{cases} x_P = x_{C'} + l\alpha_s \sin\alpha \\ y_P = y_{C'} - l\alpha_s \cos\alpha \end{cases} \tag{7}$$

Since point C ' and C have similar motion trajectories, then:

$$\begin{cases} x_{C'} = X\cos(\omega t - \phi_1) \\ y_{C'} = Y\sin(\omega t - \phi_2) \\ \alpha_s = \alpha_{01}\cos(\omega t - \phi_3) + \alpha_{02}\sin(\omega t - \phi_4) \end{cases} \tag{8}$$

Substituting E_q (8) into (7), yielding:

$$\left.\begin{aligned} x_P &= A_1\cos(\omega t + \psi_1) \\ y_P &= A_2\cos(\omega t + \psi_2) \end{aligned}\right\} \tag{9}$$

Where,

$$\left.\begin{aligned} A_1 &= \sqrt{a_1^{\,2} + b_1^{\,2}} \\ A_2 &= \sqrt{a_2^{\,2} + b_2^{\,2}} \end{aligned}\right\} \tag{10}$$

$$\left.\begin{aligned} a_1 &= X\cos\phi_1 + l\alpha_{01}\sin\alpha\cos\phi_3 - l\alpha_{02}\sin\alpha\sin\phi_4 \\ b_1 &= X\sin\phi_1 + l\alpha_{01}\sin\alpha\sin\phi_3 + l\alpha_{02}\sin\alpha\cos\phi_4 \\ a_2 &= -Y\sin\phi_2 - l\alpha_{01}\cos\alpha\cos\phi_3 + l\alpha_{02}\cos\alpha\sin\phi_4 \\ b_2 &= Y\cos\phi_2 - l\alpha_{01}\cos\alpha\sin\phi_3 - l\alpha_{02}\cos\alpha\cos\phi_4 \end{aligned}\right\} \tag{11}$$

If $a_1 \ge 0$ then $\psi_1 = \arctan(-\frac{b_1}{a_1})$; If $a_1 < 0$ then $\psi_1 = \arctan(-\frac{b_1}{a_1}) + \pi$

$$\text{If } a_2 \geq 0 \text{ then } \psi_2 = \arctan(-\frac{b_2}{a_2})\text{; If } a_2 < 0 \text{ then } \psi_2 = \arctan(-\frac{b_2}{a_2}) + \pi$$

3. Effect of the position of vibration motor on the oscillation of screen

In order to study the influence of the position of the vibration motor on the oscillation of screen, we would choose four points (S, R, Q, P) on the screen surface, which will be the research object as shown in figure 1.

Deriving from formula (9), (5) and (6) we consider L_{ox} and L_{oy} as variables and take the values of $L_{ox} = -0.5 \div 0.5$m; $L_{oy} = -0.5 \div 0.5$m. Other parameters have the following values: $l_{PC} = l_{SC} = 0,3$m; $l_{QC} = l_{RC} = 0,3$m; L = 1.2m; B = 0.6m; M = 228.6kg; m = 30.4 kg; J=17.8kg.m^2; $f_x = 0.01$; $f_y = 0.005$; $f_{\alpha s} = 0.02$; $\alpha = 25^o$; $A_o = 2500$N; $\omega = 104.7$ rad/s; $L_1=L_2 = 0.5$m; $k_{x1} = k_{x2} = 100000$N/m; $k_{y1} = k_{y2} = 50000$N/m. Using Matlab to develop the correlation between the vibration motor position and the amplitude of vibration of points P, Q, R, S as shown in figure 4 and figure 5.

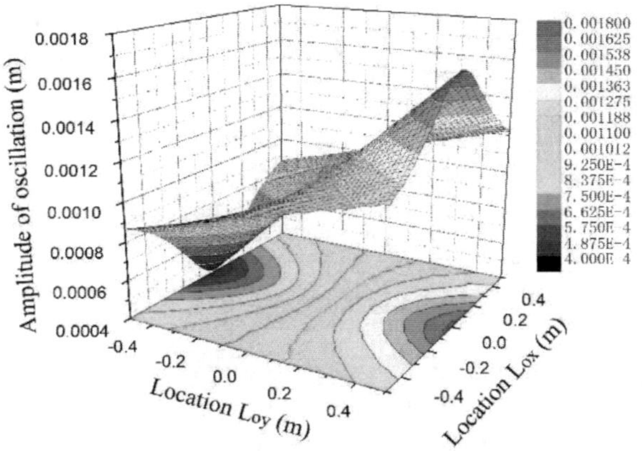

(a) Correlation between L_{ox}, L_{oy} and the amplitude of oscillation of point Q in the x direction

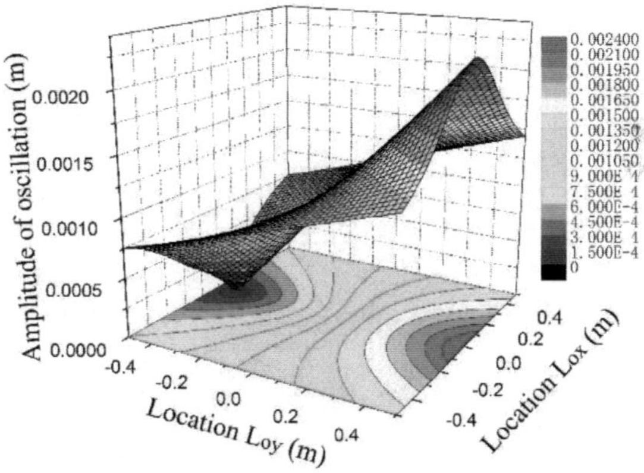

(b) Correlation between L_{ox}, L_{oy} and the amplitude of oscillation of point P in the x direction

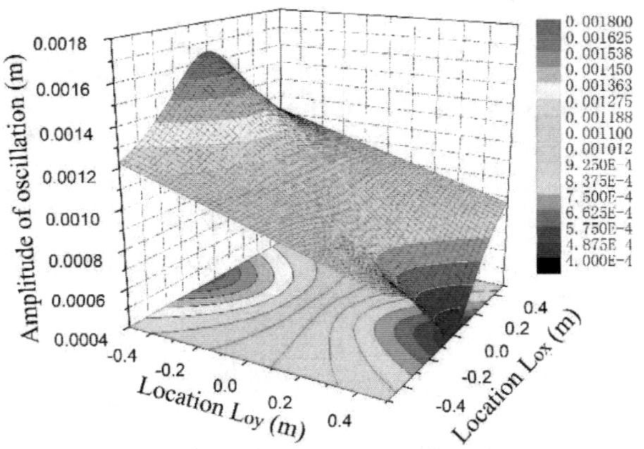

(c) Correlation between L_{ox}, L_{oy} and the amplitude of oscillation of point R in the x direction

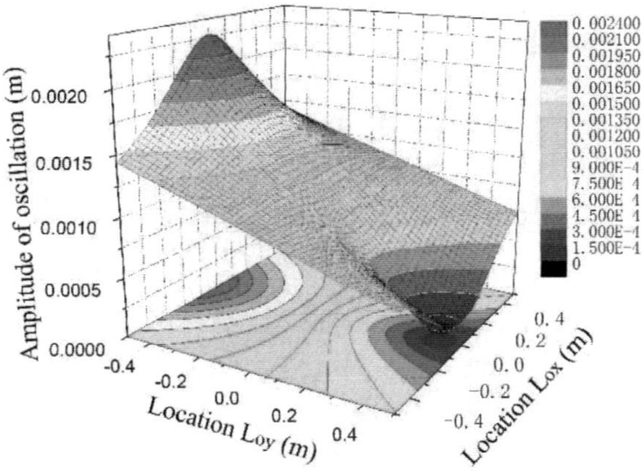

(d) Correlation between L_{ox}, L_{oy} and the amplitude of oscillation of point S in the x direction

Figure 4. Influence of the position change of excitation force the x-orientation amplitude of vibration

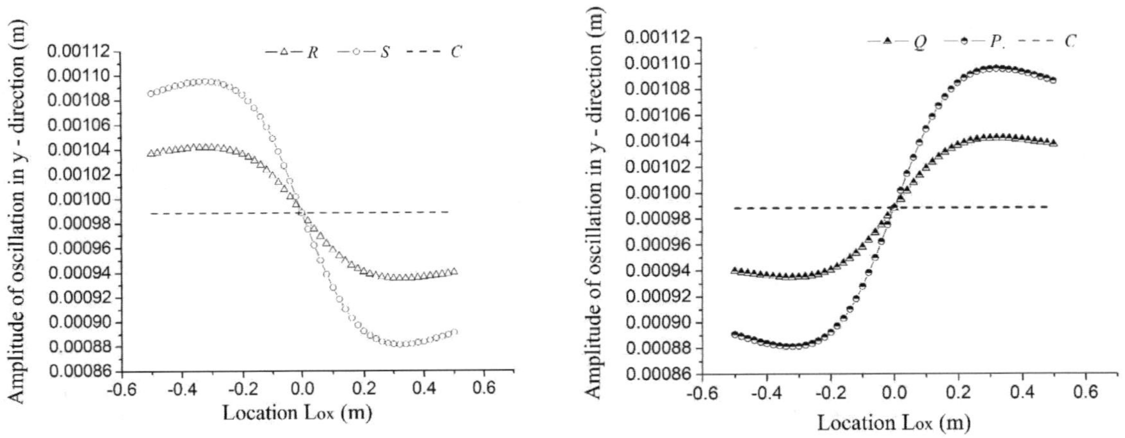

Figure 5. Influence of the position change on excitation force the y-orientation amplitude of vibration

From the results in figure 4, it is evident that:

- Figure 4a and figure 4b are graphs showing the characteristics of the two points Q and P above the screen surface above the centre of focus C. From the results we can see that the points on the screen surface in the x direction are farther away from the focus of screen, the effect of the amplitude on the position of vibration motor will be higher, and their rules of change are basically the same.
- Figure 4a and figure 4c are graphs showing the characteristics of the two points on the screen surface (Q, R) that are symmetrically identical through the focus of screen C. This indicates that the distribution of the amplitude of the two points is entirely opposite.
- Figure 4b and figure 4d are graphs showing the characteristics of the two points on the sieve (P, S) that are symmetric through the focus of screen C. The outcomes also reflect the same features as mentioned above (points Q and R). As a result, it can be affirmed that any two points are symmetrical through the focus; the rule of change follows the above rule.

From the results of figure 5, leading to:

- When the vibration motor is moving in the positive direction of the x-axis, the amplitude of the oscillation in the y-direction of the points on the screen surface above (Q, P) will increase gradually, while the points lying on the screen surface at the bottom (R, S) will decrease progressively. When L_{ox} approximately equals 0.3m, the amplitude of vibration in the y-direction of the points on screen suface approaches the maximum values (the point one on the above screen surface equals 0.0011m; the point on the bottom screen surface is 0.00088m) then go up and go down gradually;
- When the vibration motor is moving in the negative direction of the x-axis, the rule of change of the points on the screen is utterly opposite to the case where the vibration motor is moving in the positive direction of the x-axis;
- The farther the position of the points on the screen to the focus is, the greater the effect of changing the location of the vibration motor on the amplitude of vibration will be, (e.g., when $L_{ox} > 0$ then $A_{Py} > A_{Qy}$; A_{Py}, A_{Qy} is the oscillation amplitude of point P, Q according to the y-axis).

4. Conclusion

According to the working principle, the structure of the vibrating screens and some simplified rules, the author has built a dynamic model and vibration equation of vibrating screen, established the equation of motion of random point-P on screen box and conducted the simulation. Subsequently, the analysis demonstrates the effect of the position of the vibrating screen on the oscillation of vibrating screens. These results o might help scientists in their design of vibrating screens.

References

[1] Zhang Wen Bin, Zhang Long Quan 2009 Study on the probability of screening of agricultural materials in wind screen cleaning equipment *Agricultural mechanization research* **8** 50-53.

[2] Nguyen Van Xo 2018 Study kinetics, dynamics and parametric technology to improve the working efficiency of linear vibrating screen *Earth sciences and natural resources for sustainable development* 44-48.

[3] Gabov V V, Zadkov D A, Nguyen Khac Linh 2019 Features of elementary burst formation during cutting coals and isotropic materials with reference cutting tool of mining machines *Journal of Mining Institute* **236** 153.

[4] Wang Yong 2009 The future development trend of vibratory screening machinery *China powder industry* **4** 54.

[5] N S Golikov, I P Timofeev 2018 Determination of capacity of single-toggle jaw crusher, taking into account parameters of kinematics of its working mechanism *IOP Conference Series: Journal of Physics* **1015** 052008. DOI: 10.1088/1742-6596/1015/5/052008.

[6] Cao Shu Qian, Zhang Wen De, Xiao Long Xiang 2001 *Vibration structure modal analysis – theory, experiment and application* (Tianjin: Tianjin university publishing house).

[7] Wang Ranfeng, Yao Haisheng, Xiong Shibo 2003 Research on large-scale vibrating screen dynamic parameters based on test modal analysis technology In *Proceedings of the international symposium on test and measurement* vol. 1 763-766.

[8] A I Chahande, J S Arora 1994 Optimization of large structures subjected to dynamic loads with the multiplier method *International journal for numerical methods in engineering* **37** 413-430.

[9] Wen Bang Chun, Liu Feng, Li Jie 1989 *Vibrating screen vibrating feeder design and modification of vibrating conveyor* (Beijing: Chemical press).

[10] Yan Feng, Liu Ying Sheng 1995 *Screening machinery* (Beijing: Coal industry press).

[11] Nguyen Van Xo, N.S. Golikov 2018 Analysis of material particle motion and optimizing parameters of vibration of two-mass GZS vibratory feeder *IOP Conference Series: Journal of Physics* **1015** 052020. DOI: 10.1088/1742-6596/1015/5/052020.

[12] L V Fedorova, Yu S Ivanova, M V Voronina 2017 Improvement of Threaded Joint Reliability by Means of Electromechanical Processing *Journal of Mining Institute* **226** 456-461.

[13] R M Grice and R J Pinnington 2000 Method for the vibration analysis of built-up structures. Part II: analysis of the plate-stiffened beam using a combination of finite element analysis and analytical impedances *Journal of sound and vibration* **230**(4) 851-875.

[14] Nguyen Van Xo, N.S. Golikov Numerical simulation of fretting fatigue for endless rope winch drum *Advances in Engineering Research* **177** 543-548.

[15] S V Sorokin 2002 Analysis of vibrations and energy flows in sandwich plates bearing concentrated masses and spring-like inclusions in heavy fluid-loading conditions *Journal of Sound and Vibration* **253**(2) 485-505.

[16] J K Dong, A B Yu, I Brake 2009 DEM simulation of particle flow on a multi-deck banana screen. Minerals Engineering **22**(11) 910-920.

[17] P W Cleary, M D Sinnott, R D Morrison 2009 Separation performance of double deck banana screens–Part 1: Flow and separation for different accelerations *Minerals Engineering* **22**(14) 1218-1229.

[18] P W Cleary, M D Sinnott, R D Morrison 2009 Separation performance of double deck banana screens–Part 2:Quantitative predictions *Minerals Engineering* **22**(14) 1230–1244.

Study of the influence of structure and parameters of loading and transporting devices of a cleaning combine on the efficiency of coal loading

Nguyen Khac Linh[1], V V Gabov[1], Doan Van Giap[2] and Pham Van Tien[2]

[1]Saint-Petersburg Mining University, St. Petersburg, Russia
[2]Ha Noi University of Mining and Geology, Ha Noi, Vietnam

E-mail: khaclinhhumg@gmail.com

Abstract. The article analyzes the factors that influence the process of unloading coal from the zone of its separation by cutters from the massif and loading the destroyed rock mass to the face conveyor by screw executive bodies of shearers. The modeling of the process of loading coal with a shearer to the face conveyor was carried out in order to improve the quality of loading. When modeling, the program EDEM 3D was used. As indicators of the effectiveness of the process adopted: productivity and specific energy consumption. Simulation of the process of loading coal with a shearer was carried out for the following options: the cross-sectional area of the loading window of the auger varied from 0.55 to 1.00 m^2; the gap between the screw and conveyor is from 250 to 450 mm; combine feed speed – from 2 to 6 m/min; the angles of winding the blades were taken from 16° to 24°. The purpose of research is to find a comprehensive technical solution that provides resistance to the movement of coal flow, coal circulation in the stream, achieving the completeness of cleaning the soil and loading coal on the conveyor, the formation of a rational section of the coal flow on the conveyor.

1. Introduction

Currently, in underground coal mining from shallow formations of medium power, narrow-capture shearers with screw actuators are prevailing, which are distinguished by the combination of mass destruction operations, coal unloading from the destruction zone and its loading onto the conveyor [1–4], simplicity of design and reliability functioning.

The theory of the function of screw executive organs is based on the main principles of the theory of screw conveyors. Methodological and normative materials for determining the geometric and structural parameters of screw executive organs on shearers and their operating modes are based on the results of experimental studies of the irreal samples, models and on the results of their wide testing in production conditions.

Many experts note that with an increase in the intensity of the mining process, the disadvantages of screw executive organs of shearers manifests itself more significantly: soil cleaning by a lagging screw worsens, circulation and concomitant grinding of coal in the stream increase, the formation of fine dust, including volatile, hazardous by sanitary standards, is intensified and explosion, and increases specific energy consumption. Therefore, improving the process of unloading coal lagging screw executive organs, from the destruction zone and loading it onto the downhole conveyor is an urgent task.

The main features of the process under consideration as an object and subject of research are determined by external and internal factors, their influence on the loading process (Fig. 1). External factors include mining and geological conditions, strength parameters of the developed formation and technological parameters of the treatment works.

2. Materials and Method

A significant influence on the process under study is exerted by internal factors, which include (Fig. 1) the geometric parameters of screw executive organs, the parameters and operating modes of shearers and shearer mechanized complexes in general. The geometrical parameters include [1, 2, 5–9]: reduced diameter of the actuator – $D_{ш}$, diameter of the screw hub – $d_{ш}$, pitch of the blades winding– S,thickness of the screw blades – $\delta_{ш}$, number of visits of the screw blades – N_3, interlobe filling factor augers' paces – $K_з$, parameters of the arrangement of incisors, sectional area of the loading window – S_o, hedistance between the screw and the conveyor – L, the shape and dimensions of the loading space, the height of the conveyor – H_K, etc.The operational parameters of shearers and treatment mechanized complexes include: feed rate – $v_п$ and cutting – v_p, convey or chain speed – $v_к$. The values of these parameters depend, as a rule, on the external conditions, the organization of work in the complex mechanized face, the established operating modes of the shearer and canvary, in certain ranges: for example, during designer operation.

The values of some parameters characterizing these factors change deterministically with the mining of the extraction column and can be represented by the average (expected) values, others have a random nature of their formation, and can be represented by dispersion, standard deviation, coefficient of variation and spectrum.

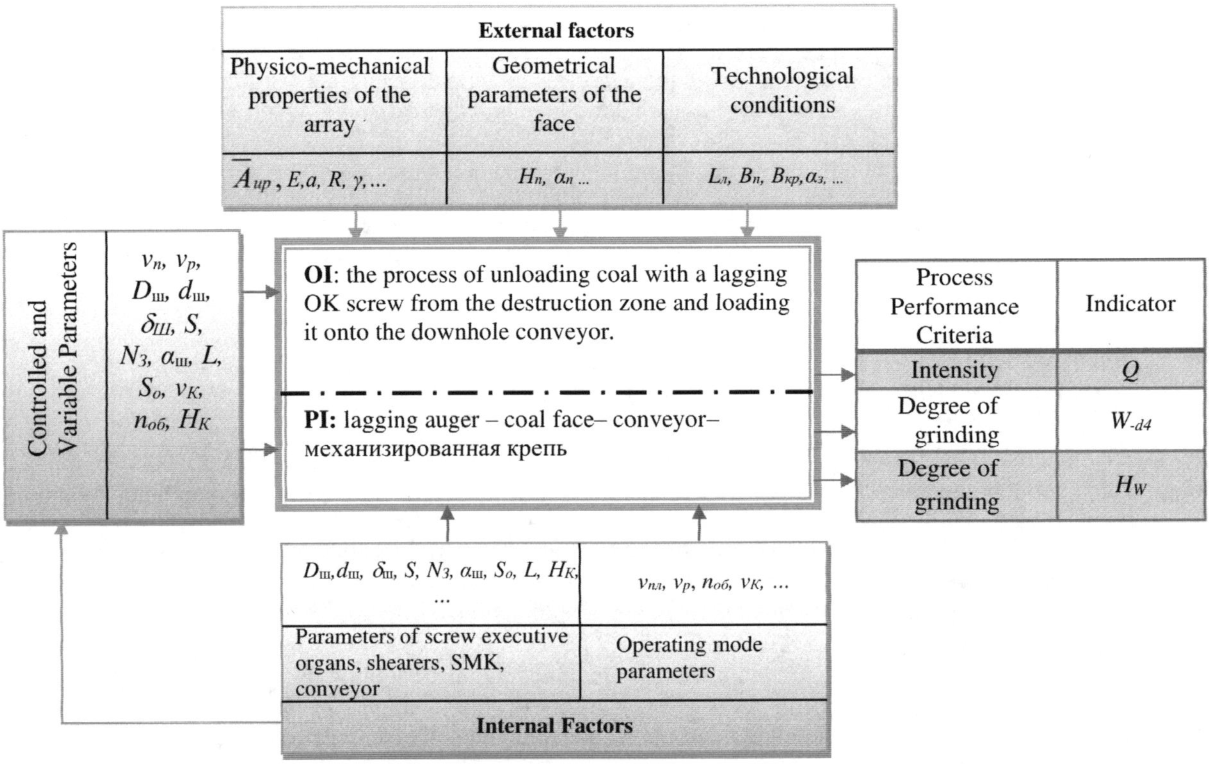

Figure 1. Factors affecting the loading process of screw executive organs

The components of the loading process as a whole in its different series-conjugated zones (Fig. 2): separation of coal from the array –1, unloading of the destroyed mass by the auger blades from the destruction zone – 2, passive movement of the destroyed mass in the spatial zone between the auger and the conveyor side – 3 and the formation of the cargo flow on the conveyor – 4 are different in their mechanical essence, the number of factors influencing them and the nature of their influence on these processes. Therefore, these processes should be attributed to complex, multifactorial and analyze them in order to identify features and areas of improvement should be separately in each of the four spatial zones in which the processes under consideration.

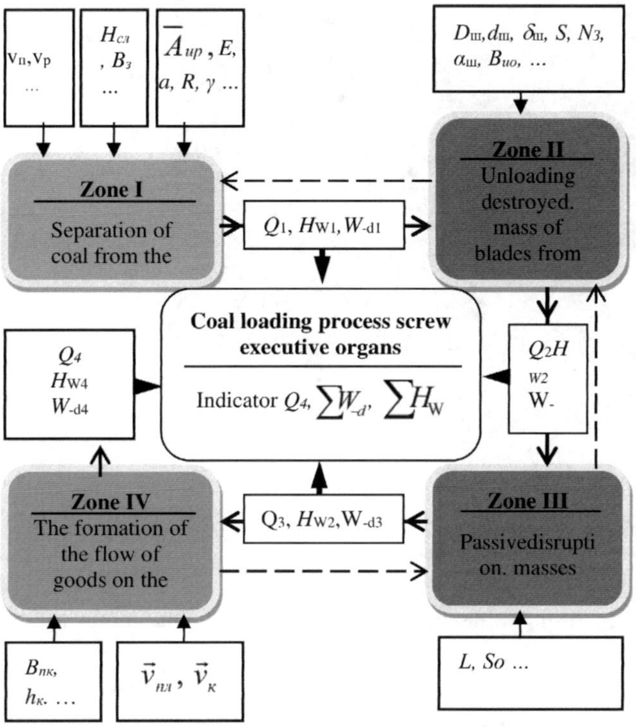

Zone	Coalloadingefficiencyindicator
I	$Q_1 = v_n \cdot B_з \cdot H_{пл} \cdot \lambda$ (1) $W_{-d1} = 100 \cdot \left[1 - \exp\left(-\frac{k_м}{m_u^2} d_c^{m_u}\right)\right]$ (2) $H_{w1} = \frac{P}{60 \cdot B_з \cdot H_{пл} \cdot v_n}$ (3)
II	$H_{w2} = 17{,}6.10^{-6} \frac{M}{(S_{сл} - S_1)\frac{v_n}{n_{об}}}$ (4) $Q_{ШН} = [\frac{\pi}{4}(D^2_{Ш.ПР} - d^2_{Ш.ПР}) - \frac{(D_{Ш.ПР} - d_{Ш.ПР}) \cdot \delta_ш \cdot N_з}{2 \sin \alpha_ш}] \cdot S \cdot n_{об} \cdot K_з \cdot c$ (5)
III	The process is not formalized
IV	The process is not formalized

Figure 2. Structural diagram of the processes of loading coal onto a conveyor with a screw

The degree of influence of a factor can be significant or insignificant depending on a combination of conditions and factors, which determines the need for further comprehensive study of the patterns of the loading process, taking into account the influence of the most significant factors.

From the analysis of factors affecting the process of loading coal by screw executive bodies onto the conveyor, it follows:

- each of the processes, by their physical and mechanical nature, is complex, and by the number of influencing factors, it is multifactorial, therefore a comprehensive research method has been adopted, including analysis, theoretical research and modeling;
- the process of passive movement of the destroyed mass in the area between the auger and the side of the conveyor and the process of forming the flow of cargo on the conveyor are most poorly studied.

3. Results and Discussion

The most important parameters affecting the efficiency of coal loading include the geometric, structural, and operational parameters of the auger and handling devices. It is these parameters that can change with the improvement of the process of loading coal onto the conveyor. The most effective method for studying such complex and multifactorial processes is modeling.

Modeling is carried out in order to assess the efficiency of the process of loading coal onto the conveyor with a shearer with changing parameters: the dimensions of the cross-sectional area of the loading window, the gap between the screw and conveyor, the feed speed of the shearer and the angles of winding the blades. In the simulation, the EDEM 3D program was used [10–12]. The following indicators were taken as indicators of process efficiency: productivity (Q_K), specific energy consumption (H_w) and particle size distribution (W_{-d}) for evaluating the secondary grinding of rock mass during loading.

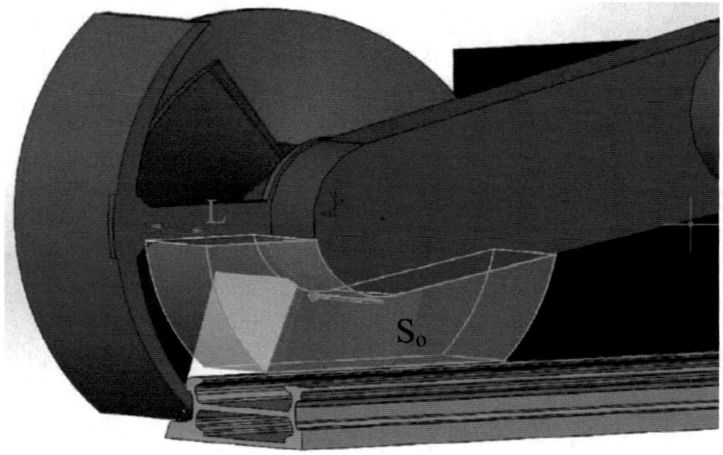

Figure 3. Loading window of the shearer

Taking into account the practice of using auger actuators with a diameter of up to 1800 mm, the ranges of change [13] for parameters were adopted for modeling: the cross-sectional area of the loading window of the screws from 0.55 to 1.00 m^2; the gap between the screw and conveyor from 250 to 450 mm; combine feed speeds from 2 to 6 m / min; the angle of winding of the auger blades 16°, 18°, 20°, 22°, 24°.

Designation	Managed Parameters	Units	Value levels				
			1	2	3	4	5
So	Sectional area of loading window	m^2	0,55	0,69	0,81	0,91	1
L	Loadingwindowlength	mm	250	300	350	400	450
v	Feed rate	m/min	2	3	4	5	6
α	screwwindingangle	degree	16	18	20	22	24

During the simulation, the following values of design parameters were adopted [2, 4, 14]: screw diameter – 1800 mm; hub diameter – 600 mm; the angles of winding the auger blades α = 20 °; the number of blades – 3; blade thickness – 50 mm; working width – 800 mm; conveyor bead h_k = 350 mm; conveyor pan width 800 mm; gearbox wall height 350 mm; screw rotation speed – 50 rpm; and values of parameters characterizing the dispersed mass of coal: average particle diameter – 30 mm; coefficient of friction of particles in the dispersed mass – 0.8; the coefficient of friction of particles on the metal is 0.6, the angle of repose of the material in the bulk ρ = 350.

The performance of loading coal on to the convey or with a lagging screw executive body during modeling was determined by the formula:

$$Q_K = \frac{M}{t} \text{ kg / s, or } Q_K = 3,6 \cdot \frac{M}{t} \text{ t / h,} \quad (6)$$

where M is the mass of coal loaded on to the conveyor, kg and t is the time, s.

The simulation results of the process of loading coal onto the conveyor are shown in Figures 4–8.

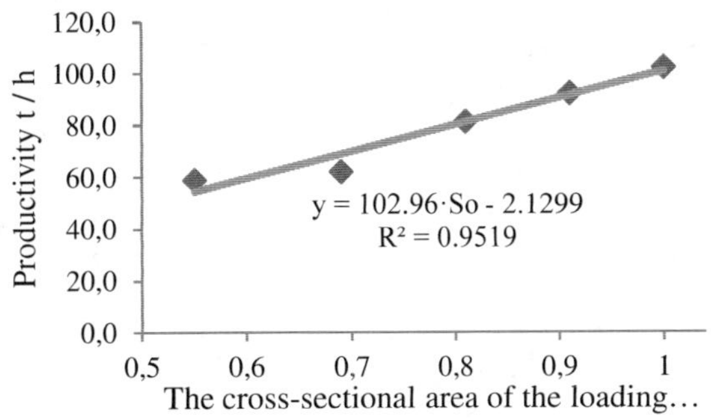

Figure 4. Dependence of loading performance on the cross-sectional area of the loading window

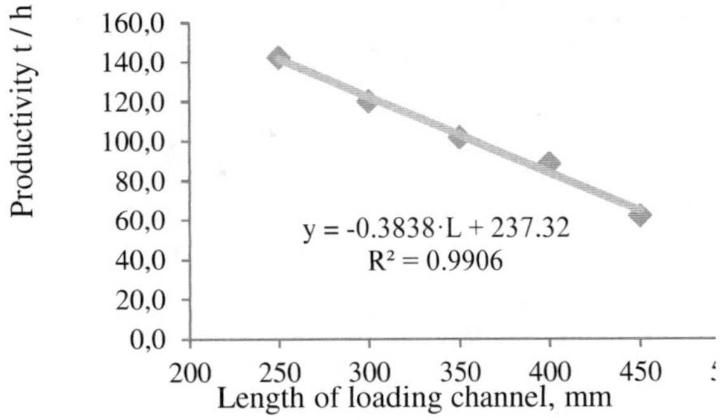

Figure 5. Dependence of loading performance on the length of the loading window

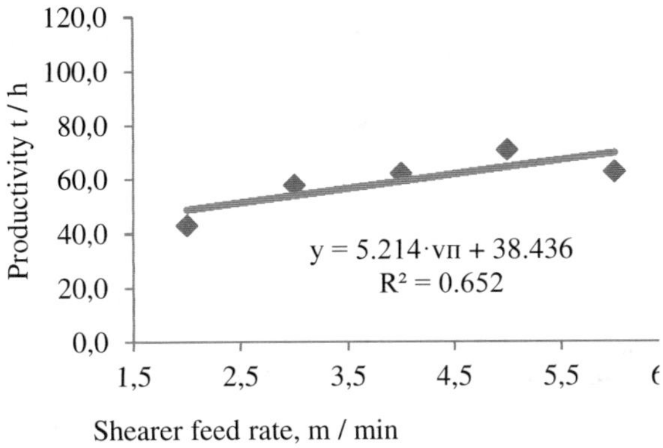

Figure 6. Dependence of loading performance on feed rate

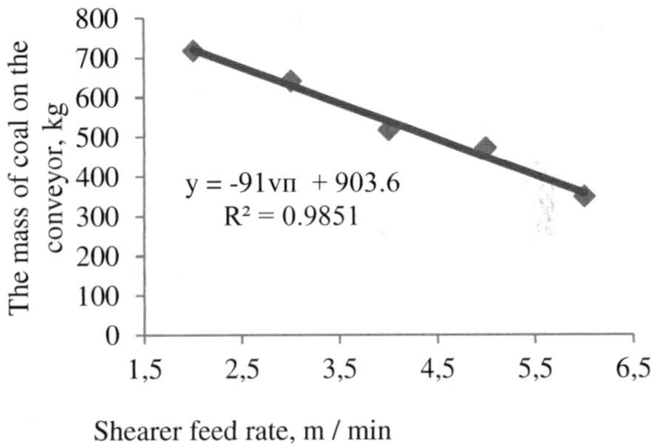

Figure 7. the relationship of the volume of coal loading onto the conveyor over time

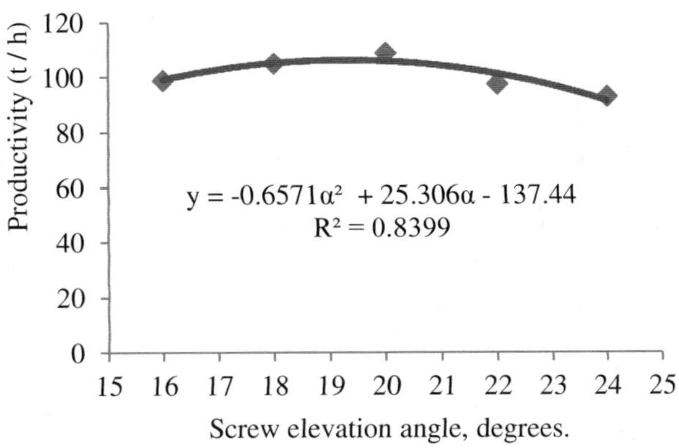

Figure 8. Dependence of loading performance from the angle of winding of the auger blades

The discussion of the results:

An increase in the cross-sectional area of the So loading window leads to a significant increase in the intensity (productivity) of the coal loading process (Figure 4).

The greater the distance (L) from the screw actuator to the conveyor, which is due to the layout features of the units of the mating equipment, the greater the resistance to the movement of the coal stream during its loading onto the conveyor. From figure 3 it follows that the shorter the path of movement (L) of the destroyed mass to the conveyor, the higher the efficiency of the loading process. However, in modern designs of shearers and conveyors, this distance (L) is 300 mm or more. Reducing this gap between the screw and the conveyor is difficult due to the design features of their layout.

From the analysis and comparison and analysis of the simulation results (Fig. 6), it follows that with an increase in the feed rate of the shearer, the intensity of the loading process (productivity) increases, and at the same time, the amount of coal (Fig. 7) remaining on the soil increases. Therefore, when choosing a harvester operating mode (feed rate), it is necessary to consider the values of the process indicators: coal breaking capacity, loading capacity, torque and specific energy consumption.

The process of loading coal onto a conveyor can be represented by a parabolic dependence of the loading intensity on the angle of winding of the auger blades in the range of their actual values of 16°–24° (Fig. 7). In this case, the maximum performance corresponds to an angle of 19.2^{o}.

4. Conclusion

Based on the foregoing, we can draw the following conclusions:

- the process in each local spatial zone differs from others in physical and mechanical nature, relates to complex and multifactorial;
- increasing the efficiency of the loading process can only be achieved by a comprehensive technical solution that provides a reduction in resistance to the movement of the coal stream, a decrease in the circulation of coal in the stream, an increase in the completeness of the soil cleaning and the formation of a rational section of the coal stream on the conveyor;
- an increase in the cross-sectional area of the loading window from 0.55 to 1.00 m^2 leads to an increase in loading performance by 1.73 times, an increase in the gap between the screw and conveyor from 250 to 450 mm reduces the loading performance by 2.28 times due to intermittent resistance in the passive zone; a three-fold increase in feed rate (from 2 to 6 m / min) increases loading performance by only 1.45 times;
- the maximum productivity of loading coal on the conveyor corresponds to the angle of winding of the screw blades of 19.2°.

References

[1] Pozin E Z, Melamec V Z, Ton V V 1984 *Coal destruction with excavation machinery* (Moscow: Nedra)

[2] Peng S S 2006 *Longwall Mining*

[3] Gorbatov P A, Ptrushkin G V 2006 *Mining machines for excavation* (Donetsk: DonTU).

[4] L V Fedorova, Yu S Ivanova, M V Voronina 2017 Improvement of Threaded Joint Reliability by Means of Electromechanical Processing *Journal of Mining Institute* **226** 456-461.

[5] Khakimova R I, Malkova R I 1984 Backwall configuration's influence on strain distribution and deformation on the bed selvedge Sovershenstvovanie teknologii gornykh rabot **227** 90-94.

[6] Eiderman B A 1984 Theory of the traffic organization on drag-type conveyer (Moscow: Nauka).

[7] Kuidong Gao Kuidong 2016 Particle movement behavior in drum coal loading process by discrete element method *EJGE* **21** 163-173

[8] Zagrivniy E A, Basin G G 2016 External dynamics of the mining machinery *Bulletin of Mining university* **217** 140-149.

[10] Gabov V V, Zadkov D A, Nguyen Khac Linh 2019 Features of elementary burst formation during cutting coals and isotropic materials with reference cutting tool of mining machines *Journal of Mining Institute* **236** 153.

[11] Nguyen Khac Linh, Gabov V V и Zadkov D A 2018 Improvement of drum shearer coal loading performance *EURASIAN MINING* **2** 22-25.

[12] Nguyen K L, Gabov V V, Zadkov D A, Le T B 2018 Justification of process of loading coal onto face conveyors by auger heads of shearer-loader machines *IOP Conference Series:*

Materials Science and Engineering. DOI: 327 042132 doi:10.1088/1757-899X/327/4/042132.

[13] Gospodarczyk P 2016 Modeling and simulation of coal loading by cutting drum in flat seams *Archives of Mining Sciences* **61** 365-379.2.

[14] Ayhan M, Eyyuboglu E M 2006 Comparison of globoid and cylindrical shearer drums' loading performance *The Journal of The South African Institute of Mining and Metallurgy* **106** 51-56.

[15] Grundke K et al 2015 *Advances in Colloid and Interface Science* **222** 350–76

Determination of nanocharacteristics of strength based on acoustic emission diagnostics

V V Nosov[1,2,3], E V Grigoriev[1], I A Pavlenko[1], E R Gilyazetdinov[1]

[1] Saint-Petersburg Mining University, 2, 21st Line, St. Petersburg, 199106, Russia
[2] Peter the Great St.Petersburg Polytechnic University, 29, Polytechnicheskaya, St. Petersburg, 195251, Russia
[3] Corresponding author, e-mail: nosovvv@list.ru

Abstract. During the period of production and exploiting of materials, complex physicochemical processes of interaction of their components with each other and with the environment take place. Identifying the mechanisms of this interaction will help to understand its various aspects and to optimize the technological processes of manufacturing materials with desired properties. The solution of this problem can be based on the interpretation of the results of acoustic emission (AE) tests from the standpoint of a multilevel model of the time dependences of the parameters of AE heterogeneous materials.

1. Introduction

Non-destructive testing of the strength of structural materials is based on the connection of the test results with their strength characteristics. The heterogeneity of the strength properties of different zones of the material makes this relationship ambiguous, introducing uncertainty in the test results. The solution of the problem relates to the necessity of formulating the concepts of heterogeneity and evaluating its quantitative characteristics, that it why in this work we consider the model of the strength heterogeneity of the material and the method for determining its parameters.

Non-uniform material is a material with non-uniform physical properties or heterogeneous material (is a material which consists of many structural elements) or "a certain mathematical model, described using material functions that are discontinuous in coordinates (for example, coordinates of elastic moduli as a function of coordinates) or defining relations (for example, ratios of stresses and strains)"[1-4]. Defects of the material and complexity of the structure lead to heterogeneity of their structural, stress-strain and strength states, behavior uncertainty and the need to increase the safety margin which is not always possible. The most heterogeneous area of a material is its surface, on which various technological damages (scale, various surface defects) accumulate and increased technological and operational stresses occur. Heterogeneity indicators are: differences in the shape, size, coordinates of structural elements, intensities and scales of the processes of their destruction or deformation, values of acting or destructive stresses, deformations, deformation energies, etc., and the scatter of these indicators forms structural, spatial, kinetic, large-scale, force, deformation or energy criteria of heterogeneity (figure 1).

In particular, the structural criterion of material heterogeneity is associated with the parameters of the distribution of the number of defects in size, large-scale - with the spatial level of destruction (nano-, micro-, meso-, macro-, etc.), power - with the variation of the calculated and appropriate stresses, kinetic - with a change in the intensity of material restructuring processes. Spatial heterogeneity, which manifests itself in the dispersion of the coordinates of acts of destruction, is a prerequisite for the localization of destruction when the critical concentration of microcracks is reached, which limits the object from complete disintegration into microelements. The energy inhomogeneity of the fracture process is manifested in the variation of the ratio of the energy of destruction of structural elements released during fracture and dissipated in the material, the kinetic heterogeneity of the first stage of cracking, the completion of which is determined by the strongest of the destroyed structural elements and decreases in the intensity of its destruction. In cases where heterogeneity is controlled and actively used to optimize properties, directed to the management of the structure, composition and properties of the material, the material is called composite [4].

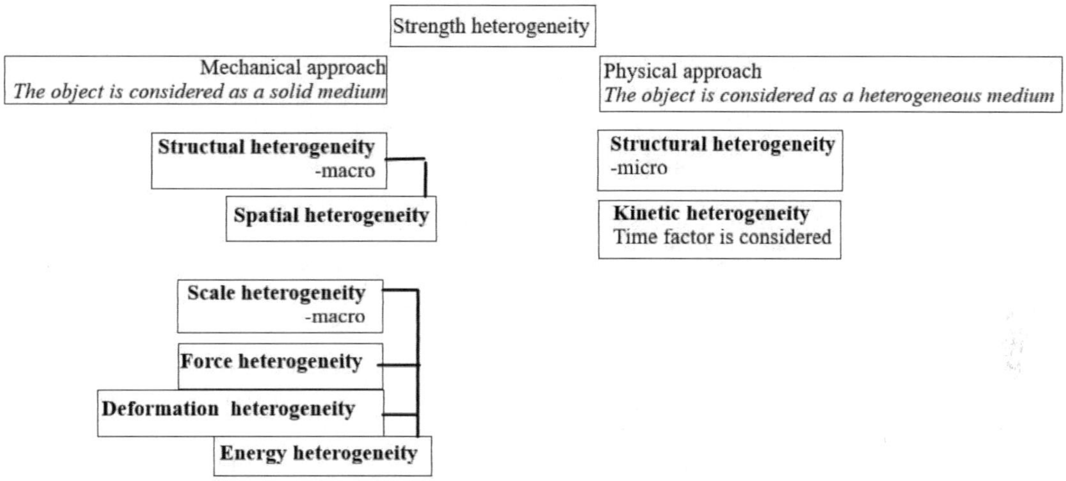

Figure 1. Components of the strength of the heterogeneity of the material.

Different types and degrees of heterogeneity affect the properties of structural materials and the predictability of their behavior in different ways. Energy inhomogeneity is manifested by the spread of pulsed emissions of accumulated energy in the form of acoustic radiation. It is considered [5] that the amplitude of the emitted pulse will depend on the amount of previously stored energy and on its distribution to the formation of a new surface and conversion to acoustic radiation. "In the article [6], this provision was formulated as a new system-forming mechanism for the destruction of a heterogeneous system in the process of fine disintegration based on acoustic emission studies and scanning electron microscopy revealed stages with “sufficient” and “suppressed” dissipative properties. The criticality of a state that limits a resource is determined by the dynamic relationship between the released energy in the discrete act of structural adjustment and dissipative properties of the nearest environment or the system as a whole.

The uncertainty of the behavior of inhomogeneous materials and objects made from them causes the need of the additional diagnostic of effects on the test object. In particular, the production tests carried out under conditions of growing load are substantially inhomogeneous, both new commissioned and welded structures that have served a long period of time, focus on the nature of the time dependences of the AE parameters during reloading [7-11]. The absence of signals under load is less than the original, called the Kaiser effect (figure 2), and the decaying or stable nature of the AE is interpreted as a non-hazardous state (section AB, figure 2), when AE signals appear long before the initial sample load value is reached (Filisiti effect) or accumulate with increasing activity are interpreted as signs of the presence of dangerous defects (FD section, figure 2); AE signals recorded before full discharge (Elber effect) indicate the presence of cracks ins (table 1). However, repeated loading is not always acceptable, and this creates the need to increase the informativeness of the results of registration of AE primary loading signals.

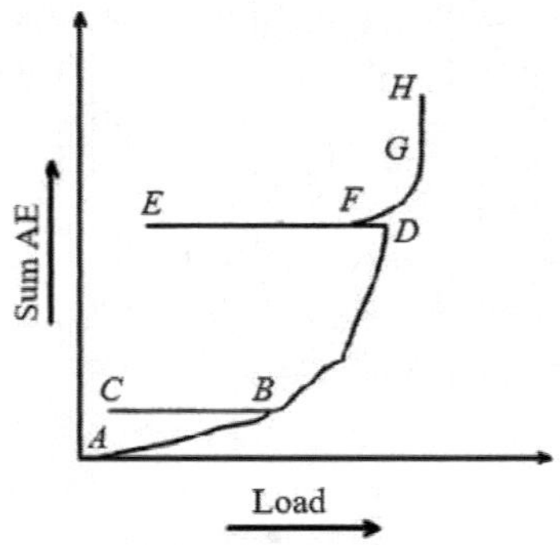

Figure 2. Graphic image of the Kaiser effect observed when testing samples and industrial facilities.

Table 1. The relationship of the state of the structure of the material with the types of strength heterogeneity, stages of destruction and diagnostic AE signs of these stages

Structure state	Stages of destruction	Types of strength heterogeneity			Diagnostic AE signs
		Spatioal	Kinetic	Energy	
Destructive (weak)	Delocalized fine inhomogeneous	++[a]	++	++	Fall of AE activity and AE amplitude before final destruction, DRT[d] variation, Kaiser effect
Without hub	Delocalized fine inhomogeneous	+[b]	+	+	Drop of activity, amplitude of AE, variation of DRT[d], Kaiser effect
	Delocalized fine homogeneous	+	–[c]	–	DRT variations, the Felicity effect, the ability to assess the concentration-kinetic strength AE parameters
With hub	Localized fine inhomogeneous	–	+	+	Drop in activity, AE amplitudes, DRT[d] invariant, Kaiser effect
	Localized fine homogeneous	–	–	–	Invariant DRT[d], the Felicity effect, the ability to assess the concentration-kinetic strength AE indicators
Hub Development	Crack formation and growth	–	+	+	Increasing the spread of amplitudes, duration of pauses, the ability to assess the concentration-kinetic strength AE parameters, the Elber effect
	Plastic destruction	–	–	+	Invariant DRT[d], increase overlap ratio

[a] «++» - increased heterogeneity;
[b] «+» - a significant heterogeneity;
[c] «-» - insignificant heterogeneity;
[d] DRT is the difference between the arrival times of AE signals on the registration channels.

Solving the problem is possible on the basis of modeling the processes determining the working capacity and physically grounded formulation of the prognostic criterion for their heterogeneity, developing a non-destructive method for estimating heterogeneity indicators, described in the framework of the information-kinetic approach [3,12]. This article presents the rationale for the relationship of AE parameters with indicators of the strength heterogeneity of the material of test objects, formulated the prerequisites for predicting the resource while eliminating the need for repeated diagnostic loading.

2. Research methods

The most representative characteristic of strength is the time to failure; therefore, the degree of strength heterogeneity of a structural material should be characterized by the spread of rupture times of the structural elements constituting a heterogeneous material, and its evaluation should be made by determining the time dependences of the AE associated with the moment of rupture of the micromechanical model. For the analysis of the adequacy and disclosure of the physical essence of this assessment, its results are compared with the presence in the material of various shapes of defects or the surface area of thermally untreated welds, which is the most defective and overstressed area. The description of the fracture process that determines the strength and the analysis of experimental data, the formulation of the criterion and indicators of strength heterogeneity, the development of a method for their quantitative evaluation is made from the standpoint of the micromechanical fracture model, temporal dependencies of the AE parameters and the use of simulation computer modeling.

The research methodology consisted in the experimental determination of the influence of various technological and operational factors on the values of the coefficients included in the model AE parameters. The proposed model of time dependence (on time t) of the number N_Σ of pulses of AE materials has the following general form:

$$N_\Sigma(t)=V\iiint_{\Delta t,f,u}\Phi(\Delta t,f,u)dudfd\Delta tC_0\int_{\mu}^{\mu+\Delta\omega}\psi(\omega)\left\{1-\exp\left[\frac{-\int_0^t dt'}{\theta(U_0,\omega(t'))}\right]\right\}d\omega \qquad (1)$$

where $\theta(U_0,\omega(t'))=\tau_0\exp\left(\frac{U_0-\gamma\sigma(t')}{KT}\right)$ is Zhurkov's formula.

Every parameter of the model (1) has its specific physical nature and depends on distinct factors what allow revealing mechanisms of impact of these factors on material's features:

- Parameter $V\iiint_{\Delta t,f,u}\Phi(\Delta t,f,u)dudfd\Delta tC_0$ where V is controlled volume of material, $\Phi(\Delta t,f,u)$ is AE signals' density function of pauses' duration Δt, frequency f and amplitude u, C_0 is structural elements concentration in material, characterizes amount of AE sources which are literally structural elements which can be "heard" by AE equipment during the process of destruction;
- parameter U_0 (activation energy of destruction process of molecular links) does not depend on state of material structure and is defined through characteristics of interatomic interaction (chemical ties) of structural element;
- $\omega=\gamma\sigma/KT$ is parameter, characterizing decrease of activation energy of destruction process, and being a strength characteristic of structural microelements;
- parameter γ (activation volume) is characteristic of molecular nanostructure of material. Parameters γ and ω are faintly sensitive parameter to its chemical nature.
- correspondence of the variables of $\psi(\omega)$ function characterizes the degree of heterogeneity of material's mechanical state at a molecular level;

(figure 3) and non-uniform (figure 4) destruction stages.

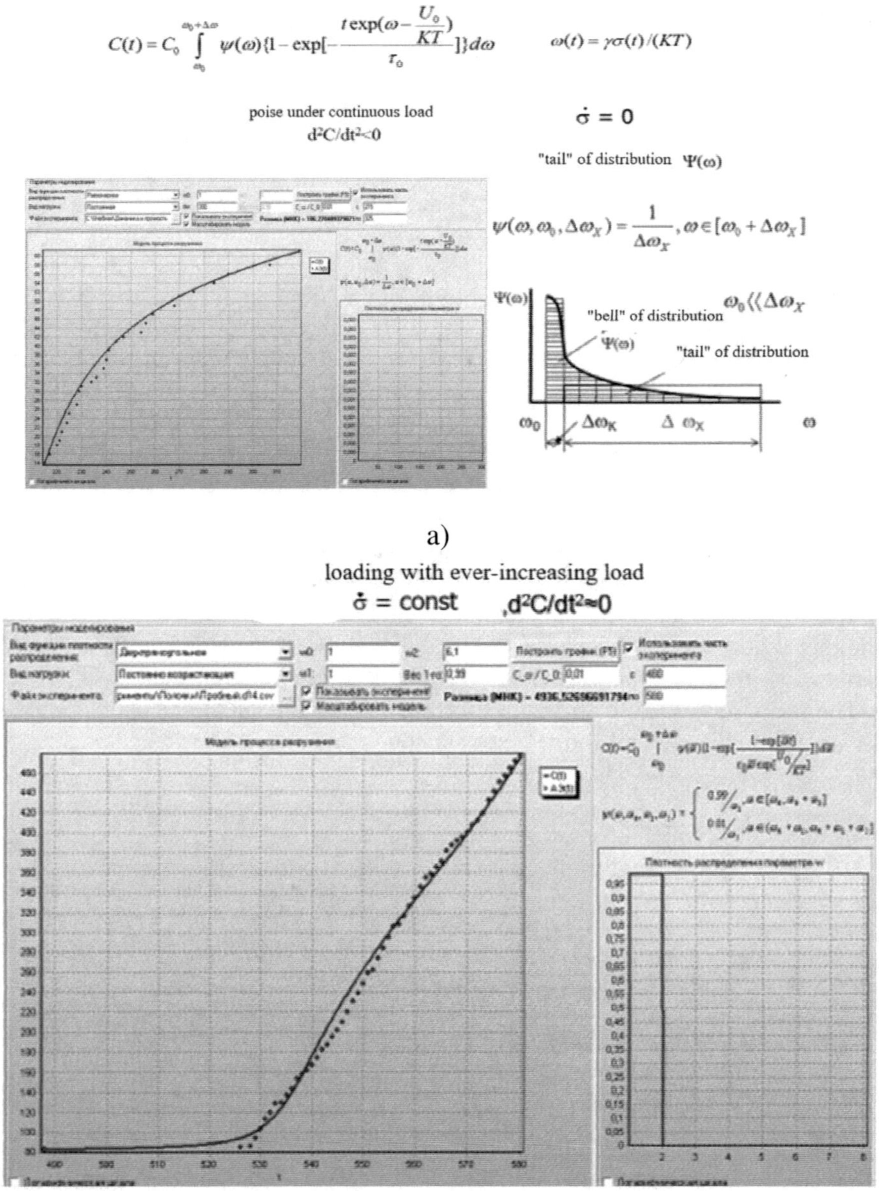

b)

Figure 3. Modeling of the stage of heterogeneous destruction at holding (a) and uniform loading (b) of the material. $\dot{\sigma}$ - average stress growth rate.

poise under continuous load $\dot{\sigma} = 0$

$d^2C/dt^2 \approx 0$

$\omega(t) = \omega = const$

"bell" of distribution $\Psi(\omega)$

$$\psi(\omega, \omega_0, \Delta\omega_K) = \frac{1}{\Delta\omega_K}, \omega \in [\omega_0 + \Delta\omega_K]$$

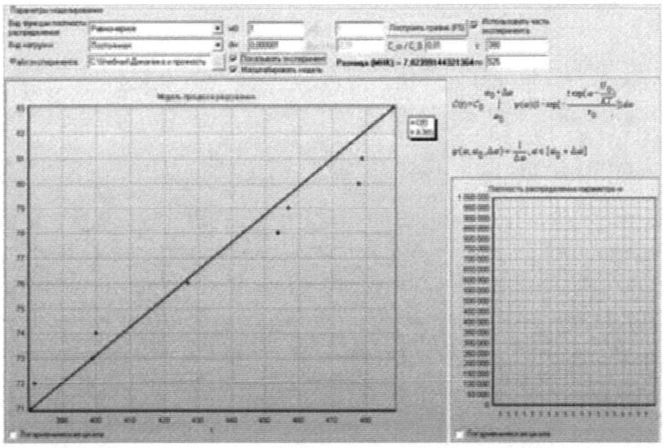

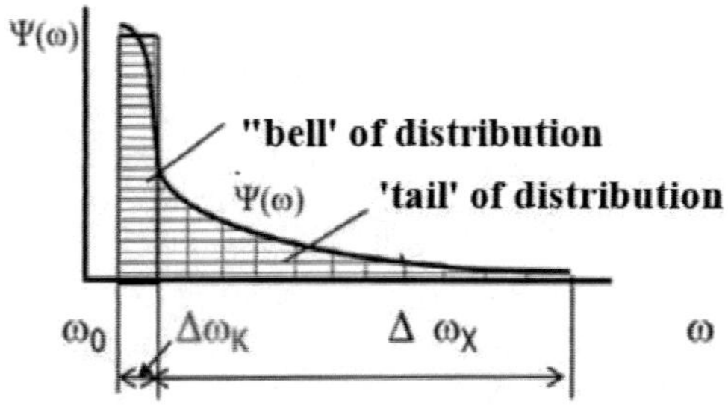

$$N_\Sigma(t,\sigma) = k_{AE}C_o\, t \exp\omega/\{\tau_o \exp[U_o/(KT)]\},\ N'_\Sigma(t,\sigma) = k_{AE}C_O \exp\omega/\{\tau_O \exp[U_O/(KT)]\}$$

при $k_{AE}=const$

$$d\ln N'_\Sigma/d\sigma = Y_{AE} = d\omega/d\sigma = \gamma/(KT)$$

a)

uniform loading with a constant increase in stress

$\dot{\sigma} = const$ $d^2\ln C/dt^2 \approx 0$

'bell" of distribution $\Psi(\omega)$

$$\psi(\omega, \omega_0, \Delta\omega_K) = \frac{1}{\Delta\omega_K}, \omega \in [\omega_0 + \Delta\omega_K]$$

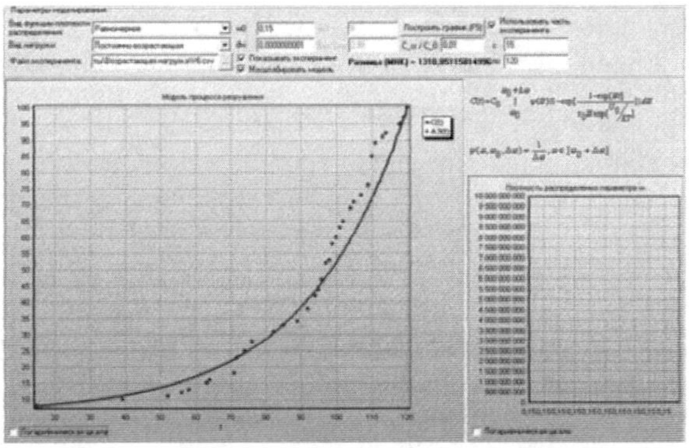

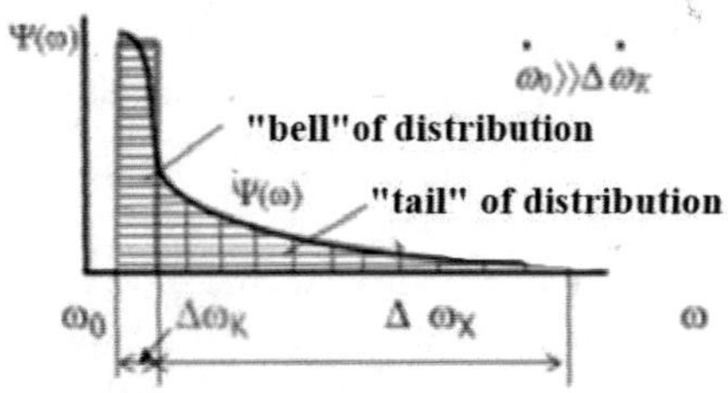

$$N_\Sigma(t) = k_{AE}C_oKT\exp[(\gamma\dot{\sigma}t - U_o)/(KT)]/(\tau_o\gamma\dot{\sigma}),\ N'_\Sigma(t) = k_{AE}C_o\exp[(\gamma\dot{\sigma}t - Uo)/(KT)]/\tau_o$$

where $k_{AE}=const$

$$d\ln N_\Sigma/d\sigma = d\ln N'_\Sigma/d\sigma = Y_{AE} = d\omega/d\sigma = \gamma/(KT)$$

$$d\ln N_\Sigma/dt = X_{AE} = \gamma\dot{\sigma}/(KT)$$

b)

Figure 4. Simulation of the stage of homogeneous destruction at holding (a) and uniform loading (b) of the material. $\dot{\sigma}$ - average stress growth rate.

There are could be used the following types of function modeling $\Psi(\omega)$:

- logarithmic-normal allocation

$$\Psi(\omega,\mu,\sigma_3)=\frac{1}{(2\pi)^{1/2}\sigma_3\omega}\exp\left[-\frac{1}{(2\sigma_3)^2}(\ln(\omega)-\mu)^2\right],$$

where μ, σ_3 are parameters of allocation;

- two-rectangular with scales 0,99÷0,999 and 0,01÷0,001

$$\Psi(\omega,\omega_0,\omega_1,\omega_2)=\begin{cases}0{,}99/\omega_1, \omega\in[\omega_0,\omega_0+\omega_1];\\ 0{,}01/\omega_2, \omega\in[\omega_0+\omega_1,\omega_0+\omega_1+\omega_2].\end{cases}$$

The estimation of strength inhomogeneity is carried out on the basis of fine-dispersed breaking into the first stage into homogeneous

During a non-uniform stage, the least durable elements of the “loosened” area of material with high dissipative properties are exposed to destruction, which are destroyed after the first loading and, due to their small number, are completely eliminated from the destruction process and no longer manifest themselves upon repeated loading. Homogeneous destruction is less intense, it proceeds in “cramped” conditions in an area with suppressed dissipative properties, however, it begins to dominate with time, without decreasing its intensity when re-loading due to the large number of structural elements of approximately the same strength. Modeled from these positions, the behavior in time of AE parameters re-loaded with an increase in the load of samples (figure 2) is a symbiosis of two idealized variants of the destruction process shown in figure 5.

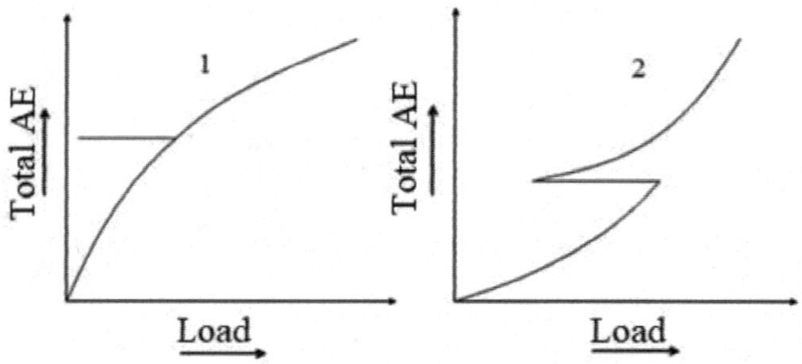

Figure 5. Behavior of AE parameters in idealized variants of the process of destruction:
1 - kinetically inhomogeneous process of destruction, 2 - kinetically homogeneous process of destruction

Thus, the degree of heterogeneity of destruction carries information about the state of the object and its resource. By virtue of the priority of the non-uniform stage, the fact of its registration indicates remoteness before the completion of the first stage and the non-hazardous state of the object, and the course of the stage of uniform destruction, on the contrary, indicates the approach to the end of the first stage and the formation of a macro-crack. It is possible to estimate the degree of approximation by determining the ratio of the parameters of the function (ω). Thus with $\omega_x/\omega_k=\omega_2/\omega_1\leq1$, $\omega_2/\omega_0\leq1$ (figure 4) the process of destruction has a homogeneous character, since the length of the “tail” of the function $\psi(\omega)$ will be relatively small, which means that the number of the least durable structural elements, the destruction of which is intense during the primary and not significant during repeated loading, is extremely small; otherwise, non-uniform destruction of the loosened zone of the material with developed dissipative properties, suggesting the manifestation of the Kaiser effect during repeated loading and the non-hazardous state of the test object [9,10].

Journal of Physics: Conference Series **1384** (2019) 012037 doi:10.1088/1742-6596/1384/1/012037

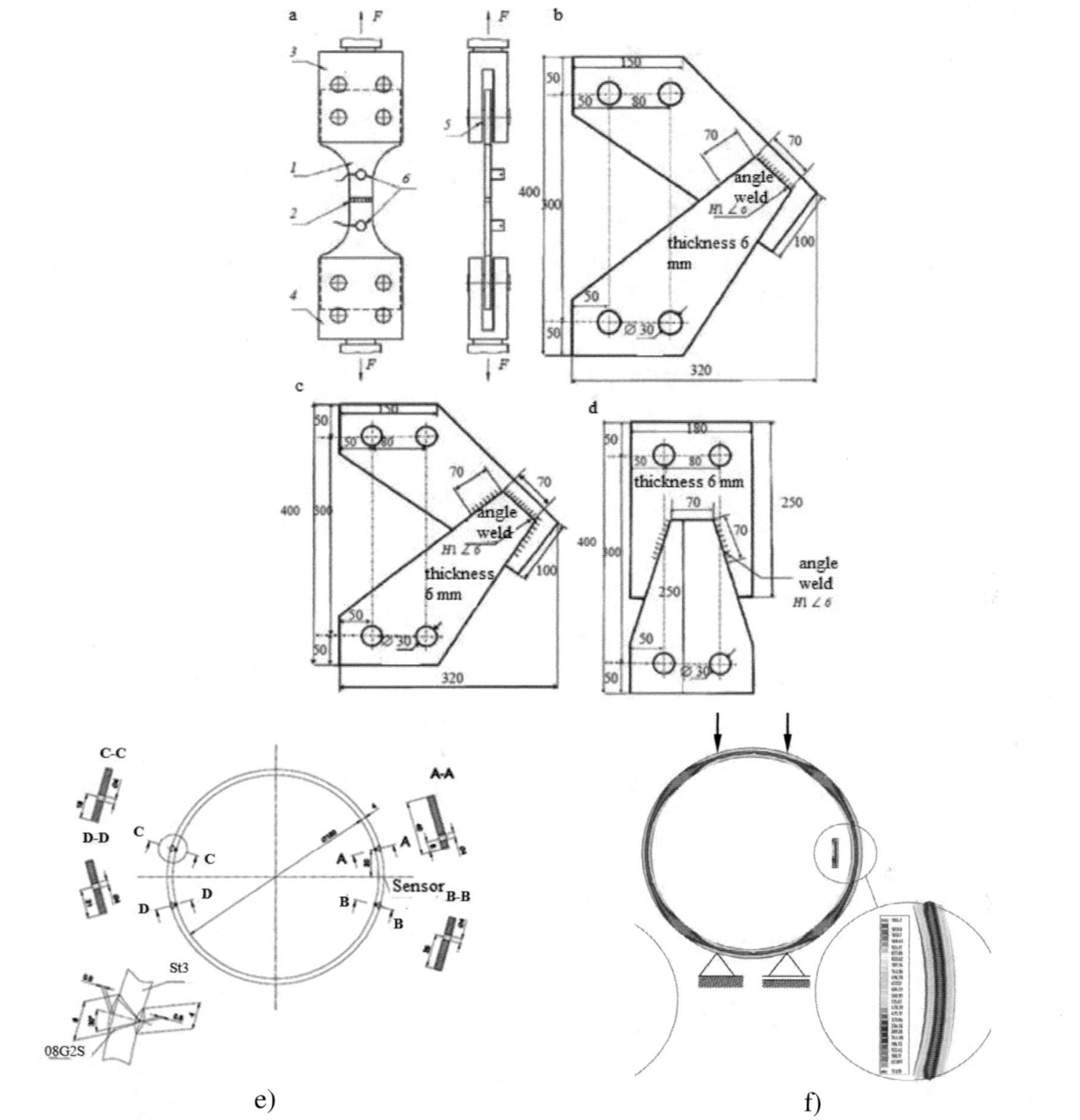

Figure 6. Test pieces with various shapes, types of loading and degree of imperfection: a) butt and b), c), d) -lap-welded joints 1- sample, 2-weld, 3-top grip loading device, 4-bottom grip loading device, 5-finger, 6- TAE; e,f) ring welded samples.

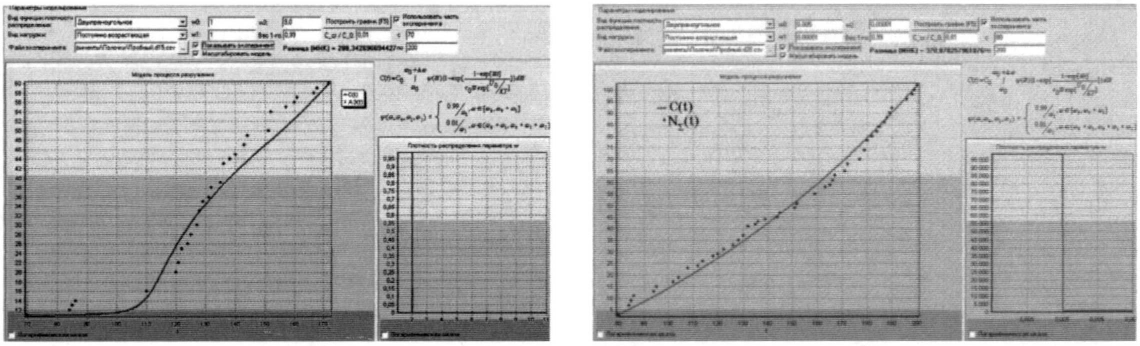

a) b)

Figure 7. Results of simulation of microcrack formation and registration of AE of a butt-welded joint sample: a) -with two rounded side cuts at the stage of elastic deformation, two-rectangular distribution (ω); ω_2/ω_1> 1; ω_2/ω_0> 1; ω_1/ω_0 = 1, non-uniform destruction; b) - with two “sharpened” side cuts, two-rectangular distribution ψ (ω), ω_1/ω_0 <1, ω_2/ω_0 <1, ω_2/ω_1 = 1, homogeneous failure

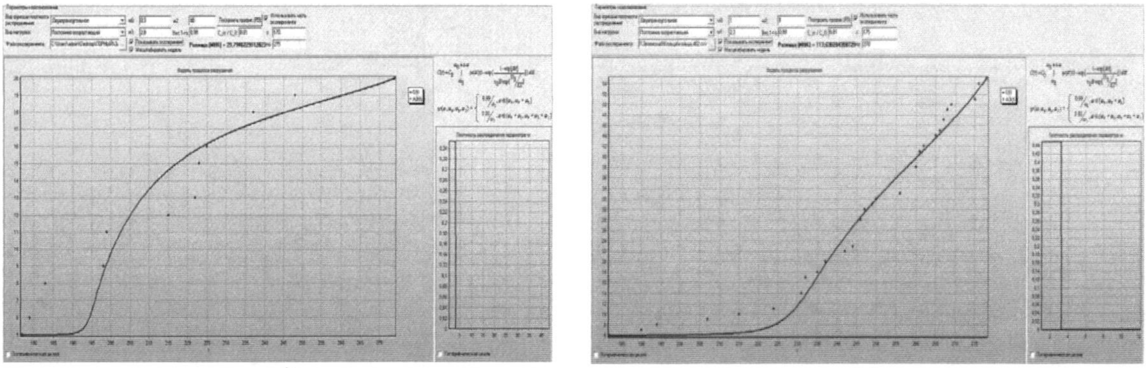

a) b)

Figure 8. Results of the simulation microcrack and registration of AE: a) defect-free ring sample: two-rectangular distribution ψ (ω); ω_2/ω_1 > 10; ω_2/ω_0> 10; ω_1/ω_0 > 10, highly heterogeneous fracture; b) an annular sample with 2 non-through holes made outside: a two-rectangular distribution ψ (ω); ω_2/ω_1 > 1; ω_2/ω_0 > 1; ω_1/ω_0 > 1, non-uniform destruction.

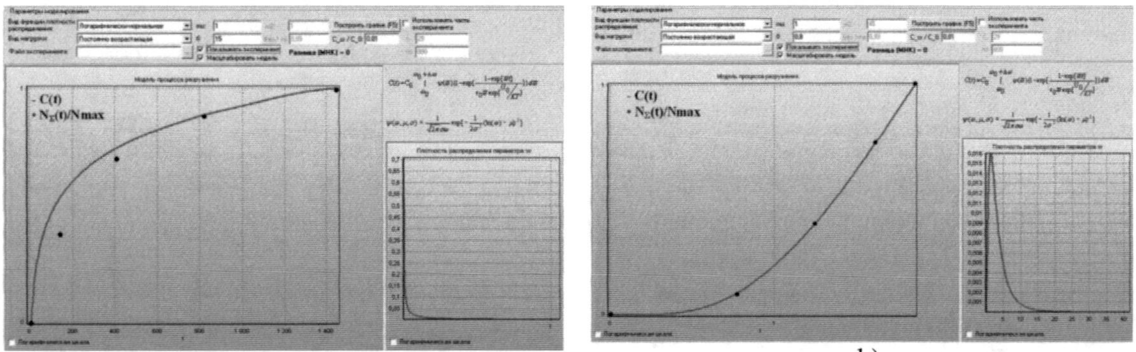

a) b)

Figure 9. Comparison of the results of registration of the AE of a cement stone sample: a) with an unformed highly heterogeneous structure (daily age) σ_3 / μ > 1; b) structured sample of cement stone (the age of the sample is 132 days) σ_3 / μ < 1.

3. Results and discussion

The results of both their own and third-party research were analyzed. The studies were carried out on welded steel samples (figure 6), registration of AE signals using an automated diagnostic acoustic emission system SDAE-16, described in [3,4]. It was established that for samples with "rounded" defects, the ratio $\omega_x/\omega_k = \omega_2/\omega_1 > 1$, $\sigma_z > \mu$ (σ_z, μ are parameters of the log-normal distribution of the function ψ (ω), figure. 7a, 8b), that is, heterogeneous failure is observed in samples made with "sharp" stress concentrators $\omega_2/\omega_1 < 1$ (figure.7b, 9b), that is, homogeneous failure is observed. Samples with increased heterogeneity and a structure that was not formed in the process are characterized by the ratios $\omega_2/\omega_1 > 10$, $\sigma_z > 10\mu$ (figure 8a, 9a).

For samples with "acute" defects, the kinetically inhomogeneous fracture section is short or completely absent; upon repeated loading of such samples, the fracture intensity and AE activity does not change (figure 10).

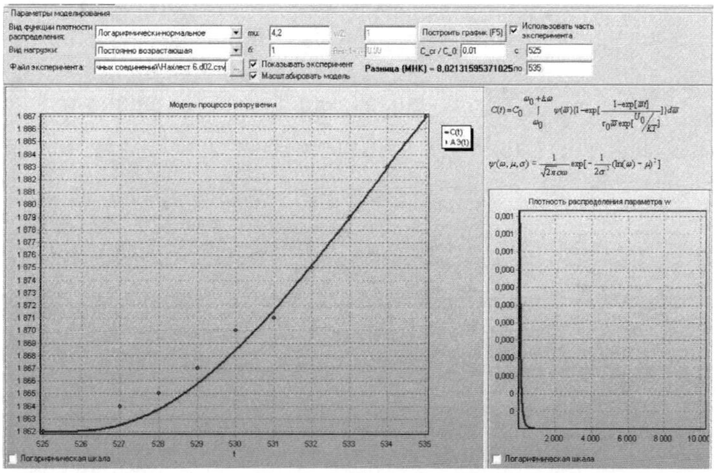

a)

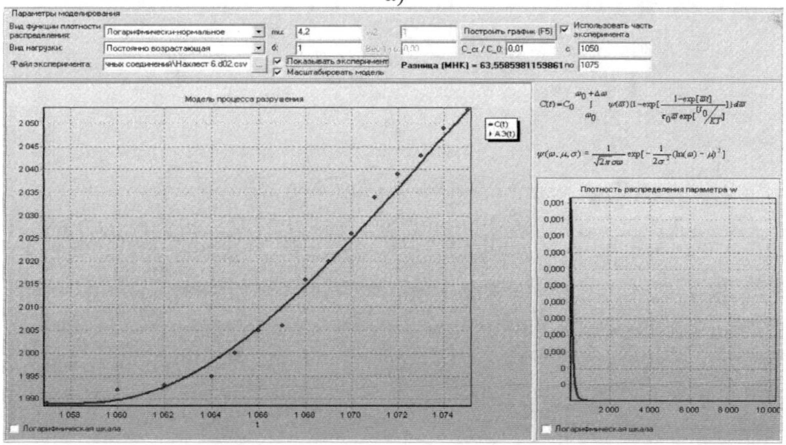

b)

Figure 10. Primary (a) and repeated (b) loading of sample No.6 without holes with a crack with a load of 44-45 kN. Homogeneous destruction ($\mu > \sigma z$).

The physical meaning of the indicated ratios of the parameters of the function ψ (ω) is also revealed by the comparison with other indices of heterogeneity, and, in particular, according to the results of processing experimental data obtained during AE testing of welded samples of various degrees of surface layer processing, where according to the research work [12] geometrically heterogeneous elements of heterogeneity. In particular, there was a good correlation of the ratio ω_2/ω_1 with the removed surface area of overlap welds (table 2) and ring welded samples (table 3) [13]. The results of the study indicate a relationship between the parameters of the function parameters (ω) and the area of the most structurally inhomogeneous region of samples of welded joints.

Table 2. Correlation of the ratio of the parameters of the density distribution function ψ (ω) with the area of the removed surface of the samples of overlap welded joints (figure 6b, c, d)

Sample type	Sample number	Defect Type(mm)	Surface removed(mm^2)	ω_2/ω_1
front seam	1	-	0.000	2.833
	2	2 holes Ø6	56.520	3.000
	3	4 holes Ø6	113.040	24.000
front seam and 2 flank	4	1 hole Ø6	28.260	4.667
	5	-	0.000	3.938
	6	6 holes Ø6	169.560	31.429
	7	12 holes Ø6	339.120	242.857
	8	3 holes Ø6	84.780	66.667
	9	3 holes Ø6	84.780	10.667
The correlation coefficient of the ratio ω_2/ω_1 with the area of defects			0.918	

Table 3. Correlation of the ratio of the parameters of the density distribution function ψ (ω) of ring samples (figure 6 d)

Sample number	Defects(mm)	σ_3/μ	ω_2/ω_1	ω_1/ω_0	Area A of the removed surface of the thermally untreated seam, (mm^2)	Maximum stress near defects σ_{max}, (MPa)
5	2 blind holes inside: Ø4 and Ø3	0.92	0.875	0.89	19.6	268
4	2 blind holes outside: Ø2.4 and Ø3.2; flaw 1mm	2.1	4.1	2.2	9.48	247
1	2 through holes Ø4 (burrs)	3.56	6.45	3.1	25	259
3	2 non-through holes: inside Ø3.5 and outside Ø3	3.375	6	2	16.7	266
2	Without defects	12	14.29	9.3	0	188
Correlation coefficient with σ_{max} values		-0.95	-0.89	-0.97		
The correlation coefficient with the values A		-0.75	-0.68	-0.76		

A significant decrease in the strength inhomogeneity is observed under cyclic loading of cast structures. During the primary loading of the first cycle, it is maximum, the number of AE pulses is maximum and decreases with increasing load of the primary loading, while at subsequent loading cycles the correlation between the number of AE pulses and the load increases significantly [14-18].

4. Conclusion

Thus, the proposed criterion of strength inhomogeneity of the material, representing the ratio of the parameters of the micromechanical model of temporal dependencies of acoustic emission parameters and determined by means of simulation computer modeling, is consistent with the criteria of structural and energy heterogeneity, which confirms the adequacy of the criterion. The values of the criterion make it possible to identify various stages of destruction and diagnose the state of the object during the primary loading without the need to use a repeated one, which causes the high diagnostic value of the criterion, making it testable and informative.

References

[1] Gilyarov L V 2011 *Physics of the solid state* **53(4)** pp 707-710
[2] Gushchin A N and Pachurin G V 2005 *Modern high technologies* **9** pp 53-54
[3] Nosov V V and Eltchaninov G S 2011 *Russian journal of nondestructive testing* **47(12)** pp 824-833
[4] Nosov V V 2014 *Russian journal of nondestructive testing* **50(12)** pp 719–729
[5] Ivanov V I and Belov V M 1981 *Acoustic emission control of welding and welded joints* (Moscow: Mashinostroenie)
[6] Leksovsky A M, Baskin B L and Yakushev P N 2016 Correlated non-linear interaction of micro / meso defects - a new systemically forming mechanism for the destruction of a heterogeneous system when it is deformed under conditions of constrained deformation *The fourth tectonophysical conference at the Institute of Physics of Physics, RAS. Tectonophysics and current issues of Earth sciences* (vol 1) ed Rebetsky Y L (Moscow: Institute of Physics of the Earth, Russian Academy of Sciences) pp 282-289
[7] PB 03-593-03 2003 *The rules for the organization and conduct of acoustic emission monitoring of vessels, apparatus, boilers and process pipelines* (Moscow: Gosgortekhnadzor of Russia) p 50
[8] Parkhomenko I V, Yaremenko M A and Zhbanov Y V 1990 Application of the AE method for testing alloyed steel at low temperatures *Technical diagnostics and nondestructive testing* **1** pp 68-71
[9] Tsarev I K, Tuikin O R and Kuznetsov G P 1984 Acoustic-emission control of metal structures of bridge cranes *Acoustic emission of materials and structures* **2** pp 87 - 88
[10] Kuzmin A N and Filippov S Y 2008 *NDT World* **2** pp 46-48
[11] Gongtian S and WuZhanwen 2010 Study on the Spectrum of Acoustic Signs of Bridge Crane *10th European conference on non-destructive testing* **52(3)** pp 144-147
[12] Nosov V V 2016 *Russian journal of nondestructive testing* **52(7)** pp 386–399
[13] Khromchenko F A 2002 *Resource welds steam lines* (Moscow: Mashinostroenie)
[14] Zelensky N A 2017 Substantiation of the method of non-destructive testing of the strength of structural elements of deep-sea structures based on the use of the phenomenon of acoustic emission (St. Petersburg: St. Petersburg Mining University) p 174
[15] Stepanova L N, Bobrov K V, Kanifadin K V and Chernova V V 2014 *Deformation and fracture of materials* **6** pp 41–47
[16] Bobrov A L and Stepanova L N 2013 Study of the parameters of acoustic emission signals in the control of cast parts of a freight car truck *Defektoskopiya* **12** pp 61–68
[17] Plotnikov D G, Sokolov S A, Borovkov A I and Mikhailov A A 2015 *St. Petersburg state polytechnic university journal of engineering science and technology* **1(214)** 186-193
[18] Leontev A, Aleshin M, Klyavin O and Borovkov A 2018 *MATEC web of conferences* **148** 1–4

Information and kinetic approach to the acoustic emission diagnostics of technical facilities

V V Nosov[1,2,3], E V Grigoriev[1], I A Pavlenko[1], E R Gilyazetdinov[1].

[1] Saint-Petersburg Mining University, 2, 21st Line,, St. Petersburg, 199106, Russia
[2] Peter the Great St.Petersburg Polytechnic University, 29, Polytechnicheskaya, St. Petersburg, 195251, Russia
[3] Corresponding author, e-mail: nosovvv@list.ru

Abstract. The analysis of trends in the development of resource assessment methods is given, a multi-level information-kinetic approach to acoustic emission diagnostics is described and its methodological relationship with the traditional experimental method of resource assessment is shown. The informative multi-model diagnostic parameters associated with various scale level strength characteristics of structural materials and resource technical objects are given. The examples of resource assessment of welded joints and rolling bearings are considered.

1. Introduction

The reasonable assessment of residual life is a necessary component of a set of preventive measures that ensure the reliability and safety of operation. Long-term prediction of residual life is made on the basis of an assessment of their pressure-strain-state, parameters of fatigue curves, hypotheses of damage summation-up, methods of fracture mechanics and non-destructive control [1-10]. The variety of defects, types of loading, the complexity of operation and the behavior of hazardous industrial facilities leads to forecast uncertainty, which implies the need for destructive testing of reference samples with various concentrators [5-8, 24-28].

The traditional destructive approach to the assessment of the resource has already covered most of the construction materials, due to it a good database of empirical data has been created, but the problem of taking into account the influence of the defectiveness of a specific product operating under unique industrial conditions remains. The creation of a sample of this model, identical to the object, is a very difficult task, and in connection with the use of space-geometry-geometric systems in the framework of traditional concepts of mechanics. The criteria for the similarity of defects weakly correlating with the parameters of the accumulation of damage and retaliation, the traditional methods for solving the problem do not provide the expected result. The search for the possibility to remove the uncertainty with non-destructive observation methods of the process of growth of damage is distinguished among other methods of acoustic emission as the most promising method. However, to evaluate a resource under industrial conditions of significant noise and the influence of interference, the AE method also requires revision [11, 12].

The solution of the long-term acoustic emission forecasting of the residual life of hazardous technical objects seems possible based on the information-kinetic approach developed by the author of the article [13, 14, 18-23], which focuses on the information of AE signals optimization of control, allowing formulating criteria for metrological and strength identity (similarity) of objects and control results, to create operational methods for assessing their strength characteristics tick and residual resource prediction. The purpose of the work was to uncover the connection between the information-kinetic approach and current trends in the development of methods for assessing the resource of hazardous technical objects and the traditional empirical approach, as well as justifying its pre-so-so rationality on the basis of comparison with existing ones.

2. Idea of problem solution

The analysis of modern approaches to the development of methods of acoustic emission diagnostics of the state of technical objects (figure 1) allows dividing them into:

- statistical, based on comparison with the permissible values of the primary parameters of the AE (number of pulses, total amplitude, energy of AE signals, etc.) or their distribution parameters, various statistical invariants and variations, regression, spectral and wavelet analysis of the results of registration of AE signals;
- physical, based on the laws of mechanics or micro fracture mechanisms, the kinetic concept of strength, the principles of information optimization of AE control technologies and physical models of AE parameters.

The increase in the sensitivity, the increase in the amount of primary information about AE signals, and automation of control, modernization of the instrumental part of the method led to the domination of a statistical approach in AE diagnostics. However, weak noise immunity, short-term prediction and laboriousness of learning statistical methods have sharpened the understanding of the need for fundamental methodological development, optimizing the search for the connection of a wide range of primary parameters of AE control with performance teria to abilities, indicators of reliability and resource. In this regard, the increasing interest of AE control specialists is attracted by physical approaches.

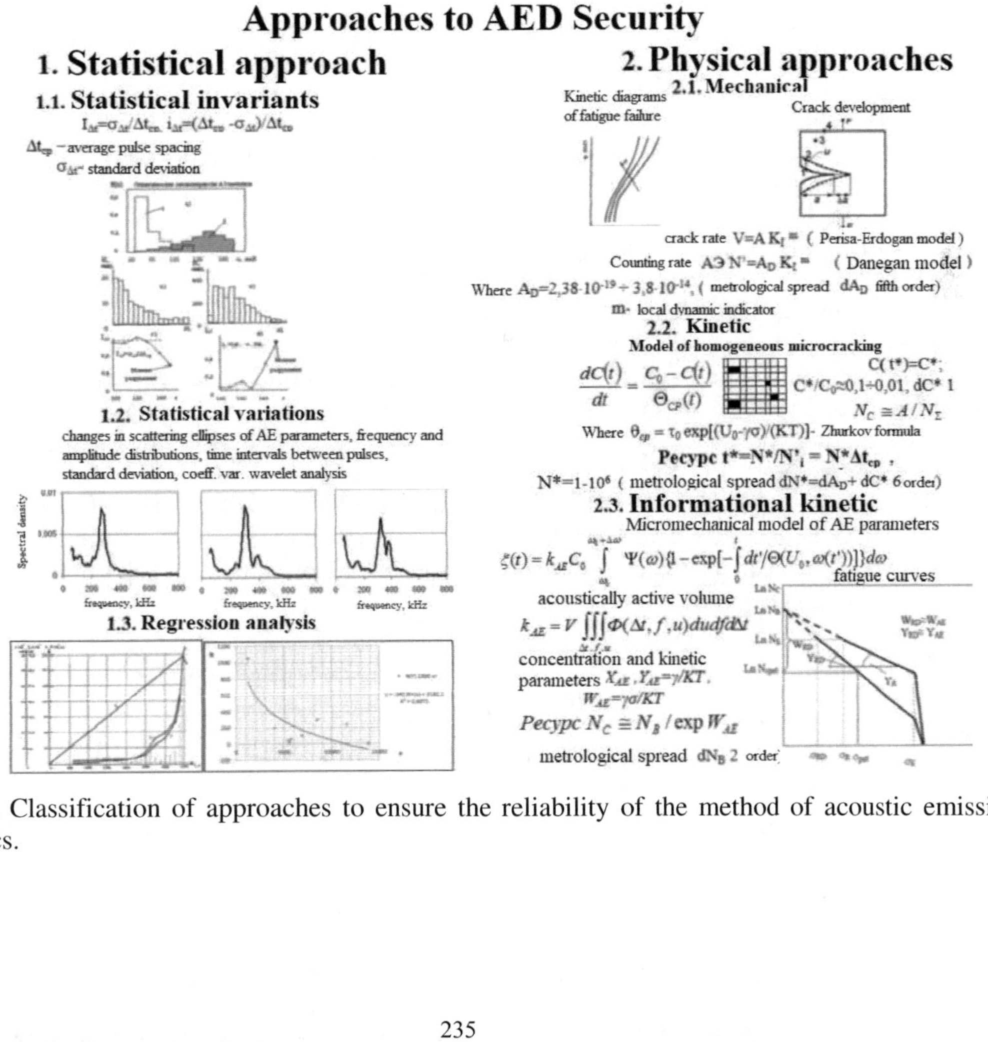

Figure 1. Classification of approaches to ensure the reliability of the method of acoustic emission diagnostics.

Focusing on the parameters of the process of the development of cracks, the mechanical approach (section 2.1, figure 1) is most widely used [11, 12]. However, it is short-term, since it does not consider the object until the formation of a crack and is not always informative in the conditions of strength heterogeneity of the material and metrological heterogeneity of the control. The kinetic approach [15] (section 2.2, figure 1) is more informative, since it divides the destruction process into two stages, adding to the crack growth stage the previous stage of fine destruction, considers the hierarchy of the process and uses a steady concentration sign of destruction. Based on the hypothesis of linear summation of damages and stability of the critical value of relative concentration of microcracks ($C^*/C_0 \approx 10^{-2}$ in section 2.2, figure 1), the approach allows predicting the resource according to the results of observation of the first, longest stage long before the formation of a crack or fragmentation of the material, which makes it long-term. The tendency of the further development of the approach has been outlined through the modeling of durability heterogeneity and the separation of the first stage of the destruction process into stages with differing dissipative properties and the results of the competition of acts of destruction and relaxation of the material [16]. However, the absence of specific quantitative methods for estimating the duration of the stages and the high uncertainty of the parameters in the resource estimation formulas (section 2.2, figure 1) that depends on interference makes the diagnosis method no longer possible. sufficiently informative with respect to the resource. The information-kinetic approach (section 3.3, figure 1) is associated with the modeling of the strength inhomogeneity of the material and the metrological non-rational nature of control, specific methods of identifying stages stages of dispersed fracture, information filtering of AE signals and complex analysis of information obtained on various scale levels of information on the strength properties of the material necessary to assess the resource of the control object.

The effectiveness of the approach is shown on products made of composite materials, building structures, metallurgical billets, various types of welded joints, pressure vessels, structures of hoisting-and-carrying machines, and cast metal structures of rolling stock elements. The requirements and criteria for the optimization of AE control have been formulated, which have become the basis of new ways to assess the performance and predict the life of hazardous technical facilities [13].

Considering the dependence of the characteristics of strength, resource, parameters of the process of destruction and AE materials on the result of competition simultaneously occurring in a heterogeneous material processes of destruction and plastic deformation of structural elements [17], the resource of the majority of long loaded materials, structures and structures according to the approach is determined by the elastic process of microcracking. Reference representatives of the strength properties of the product are its structural elements, selected from their total number by kinetic and temporal similarity criteria using AE control. The resource object technical assessment is based on the experimental determination of the moment and intensity of elastic uniform destruction of representative structural elements by recording the AE signal (figure 2). Informative about the moment of destruction is the time t of this registration, information about the number of destroyed structural elements carries the parameter ξ, which can be the number N_Σ of recorded pulses of a discrete AE, the total score N of AE, the relative total amplitude or a combination of these parameters.

The time dependence $\xi(t)$ under the conditions of the strength inhomogeneity of the material has the form of a multilevel model:

$$\xi(t) = k_{AE} C_0 \int_{\omega_0}^{\omega_0+\Delta\omega} \Psi(\omega) \left\{ 1 - \exp\left[-\int_0^t dt' / \Theta(U_0, \omega(t')) \right] \right\} d\omega \qquad (1)$$

where k_{AE} is the acoustic emission coefficient (AEC) responsible for the "metrological" concentration conversion C scattered in the material gaps in the structural elements of the scale level preceding the formation of the main crack and fragmentation of the material into the ξ AE parameter that has the meaning of the acoustically active volume of the material influenced by noise and factors destabilizing the relationship between C and ξ. Its value

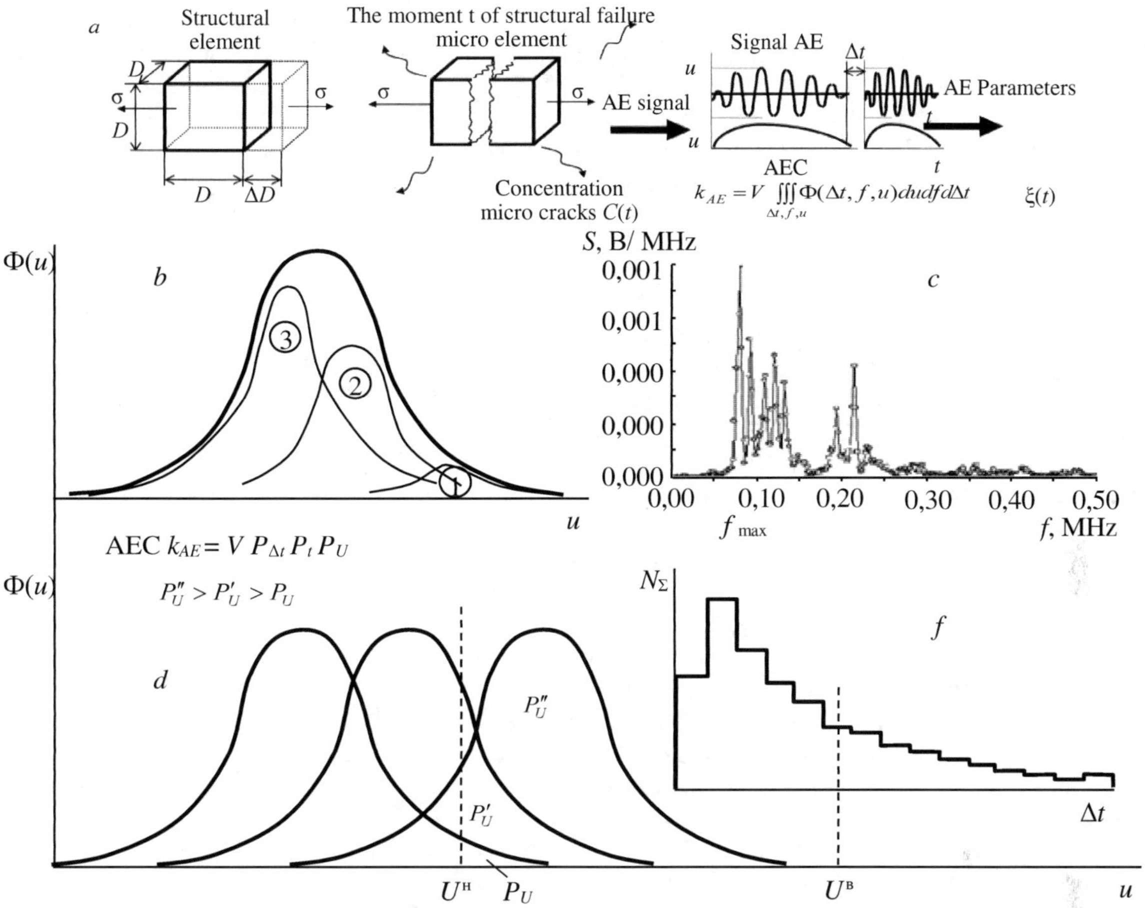

Figure 2. Model of the AE source and interpretation of the value that includes the path factors of the AE signal of the acoustic emission coefficient: a is model of the AE signal source and factors of the AE recording, σ^* strength limit, E is modulus of elasticity, D is characteristic size of the structural element; b is amplitude distribution of AE signals; c is the frequency distribution of AE signals, d is the probability of recording AE signals at a given amplitude range; d is distribution of AE signals by the duration of pauses; 1, 2, 3 are distributions from the destruction of structural elements of the weld material (1), the normalization zone and the base metal of the welded joint (2), the elements of the weakened welded joint zone (3); P_U, P'_U, P''_U, are probabilities of the amplitude of the AE signal entering the recorded amplitude range [$U^н$; $U^в$] with a uniform, exponential and with the presence of a maximum in the observed amplitude distribution of AE signals, respectively; $P_{\Delta t}$, P_f are the probabilities of signal registration in a given time and frequency range, respectively.

$$k_{AE} = V \iiint_{\Delta t,f,u} F(\Delta t,f,u)\,du\,df\,d\Delta t \qquad (2)$$

where V is the controlled volume of the material, the macroscale value; F(Δt,f,u) is the density function of the distribution of AE signals over the duration of the pauses Δt, frequency f, and amplitude u; C_0 is the initial concentration of structural elements, as a rule, on a micrometric scale; θ is the time of microelement destruction, given by Zhurkov formula

$$\Theta(t) = \tau_0 \exp\left[\left(U_0 - \gamma\sigma(t)\right) / KT\right]$$

where U_0 is the activation energy of the destruction process, γ is the activation volume; K is the Boltzmann constant; T is the absolute temperature; $\omega(t) = \gamma\sigma(t) / KT$ is the dimensionless parameter of the strength state of the structural microcell of the material of the object, depending on the structure and time-varying tensile stresses on the microcell σ (t); $\Psi(\omega)$ is the density distribution function of the parameter ω simulating the strength inhomogeneity over the structural elements of the monitored volume V of the material. Different types of the function $\Psi(\omega)$ were considered, and in particular:

• log-normal distribution with density function

$$\Psi(\omega, \mu, \sigma_3) = \frac{1}{(2\pi)^{1/2}\sigma_3\omega} \exp\left[-\frac{1}{(2\sigma_3)^2}(\ln(\omega) - \mu)^2\right] \quad (3)$$

• two rectangular with weights of 0.99÷0.999 and 0.01÷0.001

$$\Psi(\omega, \omega_0, \omega_1, \omega_2) = \begin{cases} 0{,}99 / \omega_1, \omega \in [\omega_0, \omega_0 + \omega_1]; \\ 0{,}01 / \omega_2, \omega \in [\omega_0 + \omega_1, \omega_0 + \omega_1 + \omega_2]. \end{cases} \quad (4)$$

ω_0, ω_1 , ω_2 are the scattering boundaries of the values of the parameter ω, respectively; σ_3, μ are the parameters of the function $\Psi(\omega)$.

The ratio of the parameters of the function $\Psi(\omega)$ included in formulas (3), (4) characterizes the degree of heterogeneity of the strength state of the material, and their change over time is the plastic rearrangement of its structure [18]; parameters (1) τ_0 and U_0 are the most conservative, determined by the characteristics of the interatomic interaction of the structural element and do not depend on the state of the structure, which gives them a valuable property of universality, which reduces uncertainty when predicting a resource. The values of the parameter $\gamma \approx 10^{-26} \div 10^{-29}$ m^3 (activation volume) are characteristics of the nanostructure of the material, which is weakly sensitive to its chemical nature. Together with the stresses σ, the parameter γ reflects the strength individuality of the structural element.

The integral entering into formula (2) is given the meaning of the probability of recording AE signals coming from an AE source into the range of frequencies f recorded by the instrumentation, amplitudes u of AE signals and time intervals Δt (duration of pauses) between them. The type and behavior of the distributions determining the F(Δt, f, u) function are shown in figure 2. Stability (2) determines the conditions of metrological similarity C and ξ, that is, the metrological homogeneity of the AE control, and the experimental separation of fine disintegration into heterogeneous and homogeneous stages by the kinetic feature from (1) allows selecting an informative fragment of the AE signal flow. The adequacy of the model (1) to the process of microcracking of welded joints is shown in figure 3. The nanometric level of AE fracture monitoring is currently not possible due to the complexity of creating the corresponding microwave AE converter.

The time dependence of the number of impulses of AE at the stage of the most informative homogeneous destruction, recorded, for example, with uniform diagnostic loading of the test object with a constant rate of growth of stress, will have an exponential form:

$$N_{\Sigma_{homog}}(t) = k_{AE} C_0 KT \exp\left[\left(\gamma \dot{\sigma} t - U_0\right) / KT\right] / (\tau_0 \gamma \dot{\sigma})$$

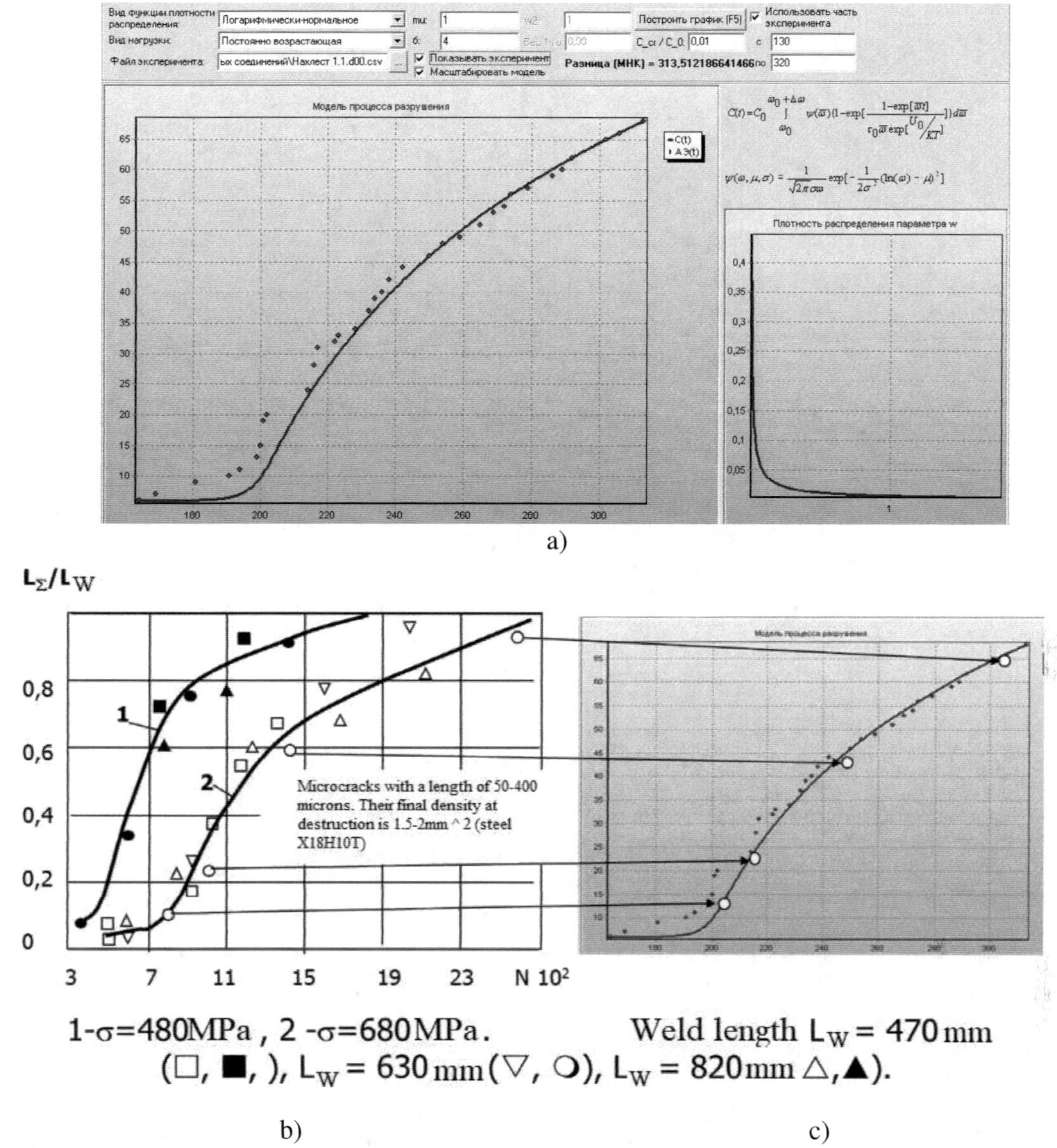

Figure 3. Assessment of the adequacy of the model. a) Comparison of computer simulation results (solid line) with the results of recording the number of AE pulses (chart points) under loading of overlapping welded joint; b) dependence of the total length of cracks L_Σ in the weld on the number of loads N for different values of nominal stresses σ and the length of the weld *Lw* (according to [2]); c) comparison of the results of simulation modeling and microscopic study of the time dependence of the concentration of microcracks.

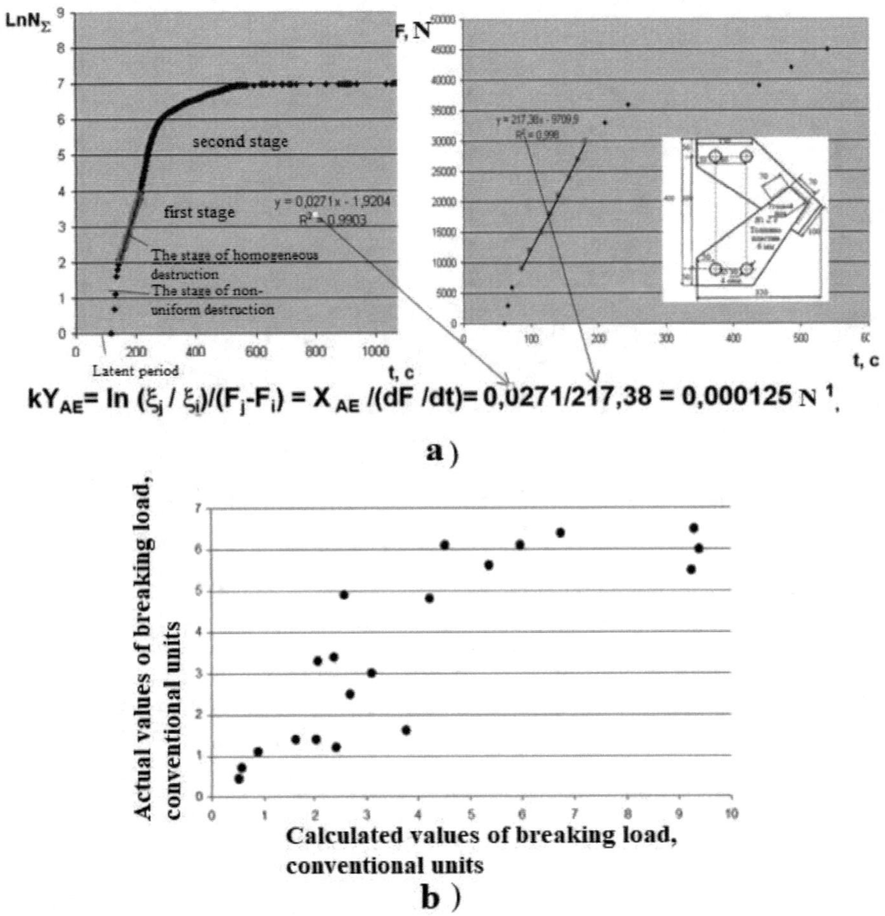

Figure 4. Algorithm for assessing the AE-index of strength kY_{AE} defect-free welded steel sample type 2, made of steel St3 (a) and the results of the resource assessment of samples of welded joints (b).

The informative concentration-kinetic AE information derived from it, which are free from AEC and therefore are resistant to additive interference and destabilizing factors, are presented in Table. 1. An example of a non-destructive determination of these parameters when loading samples of welded joints is shown in figure 4a, and their relationship with the life and fatigue curve indices for samples of overlap welded joints are shown in figure 4b, figure 1, section 2.3. The values of breaking load, calculated through kY_{AE} by the formula:

$$F_p = \frac{U_0}{KT} + \ln\left(\frac{\tau_0 C^*}{C_0 F_{p'} \cdot kY_{AE}}\right) / (kY_{AE}) \tag{5}$$

where F_p' is the load growth rate under loading, $C^*/C_0=10^{-2}$, satisfactorily agree with experimentally determined load values corresponding to the moment of crack occurrence with uniform diagnostic loading of samples of welded joints of various types, shapes and sizes with various defects. The average value of the activation energy U_{0av} is 113.6 J/mole, the confidence interval of values (109.7; 117.6) with a probability of 0.99.

According to figure 4, we have:

$X_{AE}=d\ln N_{\Sigma}/dt=0.0271\ s^{-1}$,

dF/dt=217..38 N/s,

$kY_{AE}=X_{AE}/(dF/dt)=0.000125\ N^{-1}$

The value of the coefficient k is determined by the geometric characteristics of the sample cross section or the ratio of the nominal values of the maximum normal stresses in the weld and the load on the sample. With the load on the sample F = 1 N, we have σ = 8225 Pa = 0.008225 MPa, therefore

$$Y_{AE}=\frac{d\ln N_{\Sigma_{homog}}}{d\sigma}=\frac{\gamma}{KT}=\frac{kYAEF}{\sigma}=0.000125\cdot\frac{1}{0.008225}=0.015MPa^{-1} \quad (6)$$

This value coincides with the value of the parameter $Y_R = \gamma/(KT)$ fatigue curves figure 5a., described by the same expression as Y_{AE}, provided that the fatigue curves are expressed by Zhurkov formula.

Sample activation volume $\gamma = Y_{AE}\,KT = 1.5 \cdot 10^{-8} \cdot 1.38 \cdot 10^{-23} \cdot 300 = 6.21 \cdot 10^{-29}\ m^3 = 6.21 \cdot 10^{-2}\ nm^3$ has a nano-scale dimension, and is consistent with the values obtained in [18] for AE measurements and deformations.

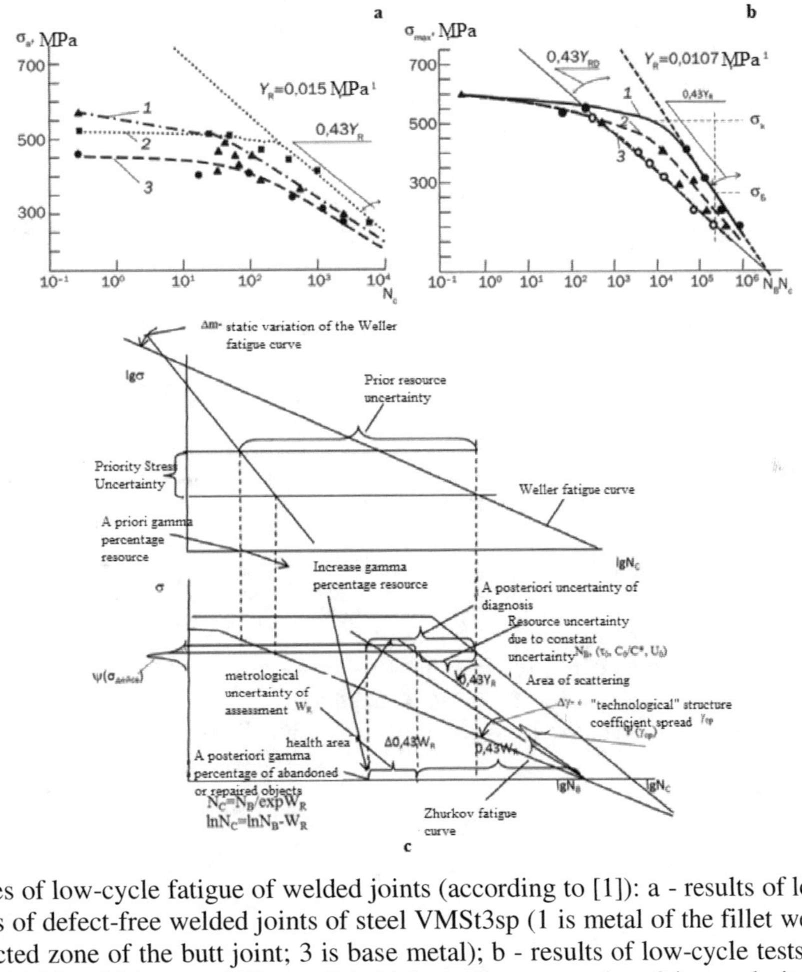

Figure 5. Curves of low-cycle fatigue of welded joints (according to [1]): a - results of low-cycle tests of various zones of defect-free welded joints of steel VMSt3sp (1 is metal of the fillet weld; 2 is metal of the heat-affected zone of the butt joint; 3 is base metal); b - results of low-cycle tests of butt joints of 10HSND steel with a thickness of 20 mm (1 is high-quality connection; 2 is angularity of 8 mm at a length of 1 m; 3 is lack of penetration of 4 mm.); c - illustration of the increase in reliability due to the use of concentration-kinetic indicators $W_{AE} = W_R$ and $Y_{AE} = Y_R$

Table 1. Multi-model multi-level concentration-kinetic AE-strength indicators, resistant to the effects of interference and destabilizing factors AE control

AE indicator	Micromodel	Nano model	Macromodel
$X_{AE}(s^{-1})$	$d\ln\xi/dt$	$\gamma\dot{\sigma}/KT$	-
$Y_{AE}(Pa^{-1})$	$d\ln\xi/d\sigma$	γ/KT	$d\ln N_c/d\sigma^a$
$kY_{AE}(N^{-1})^b$	$d\ln\xi/dF$	$k\gamma/KT^b$	$d\ln N_c/dF^a$
W_{AE}	$d\ln\xi/dK_n{}^c$	$\omega = \dfrac{\gamma\sigma}{KT}$	$\ln N_B - \ln N_w{}^a$

[a] N_C, N_B, N_w are the slave parameters of the material fatigue curve (figure 3).
[b] $k = \sigma/F$ is the proportionality coefficient between the load F and the nominal stresses σ
[c] Kn is the load factor (ratio of the diagnostic load to the working one).

The universality of the value of N_B, as cut off by extrapolating the coordinate system of the fatigue curve to the abscissa of the coordinate system (figure 5 b,c), is explained by the stability of the values entering the formula describing the fatigue curve. At constant temperature T and period τ_{cycle} cyclic loading the number of N_C cycles to failure:

$$N_C = \frac{\theta}{\tau_{cycle}} = \frac{\tau_0}{\tau_{cycle}}\exp\left[\left(U_0 - \gamma\sigma\right)/KT\right]$$

with $\sigma = 0$, the value $N_C = N_B$, where $\lg N_B = \lg(\tau_0/\tau_{cycle}) + 0.43U_0/(KT)$.

Expression mapping $Y_{AE} = d\ln N_{\Sigma homogeneity}/d\sigma = \gamma/KT$ and $Y_R = -d\ln N_C/d\sigma = \gamma/KT$, as well as the numerical values of the YAE (6) and YR parameters (figure 5a) reveals their identity ($Y_{AE} = Y_R = 0.015\ MPa^{-1}$) and suggests the inverse proportionality of the number of N_C cycles to failure and the number of N_Σ of AE impulses registered at the stage of homogeneous destruction

$$N_C \cong \frac{A}{N_{\Sigma homog}}$$

This demonstrates the validity of the hypothesis of linear summation of damages at the stage of homogeneous destruction, which makes it possible to carry out a forecast on the basis of linear extrapolation.

Resource of the control object

$$N_C \cong N_B/\exp W_{AE}$$

A graphical interpretation of the relationship of concentration-kinetic strength indicators with a resource shown in figure. 1, section 2.3 and the diagnostic features of the stages of the destruction process and the generalized formulas for resource estimation in Table 2.

Table 2. Diagnostic features of the stages of the destruction process and the resource assessment formulas

Stage	Stage of destruction	Diagnostic sign of the destruction stage	Resource evaluation formula (T-moment diagnosis)
I	Delocalized, fine, no heterogeneous	$d^2\xi/dt^2 < 0$ at $\sigma = 0$; $d^2\ln\xi/dt^2 < 0$ at $\sigma = 0$; $dk_{AE}/dt < 0$ ($dP_U/dt < 0$); $\omega_2/\omega_1 > 1$, $\omega_2/\omega_0 > 1$; $\sigma_3 > \mu$; DRT = var	$\tau^* = (1 \div 10)T$
I	Delocalized, fine, heterogeneous	$d^2\xi/dt^2 = 0$ at σ = const; $d^2\ln\xi/dt^2 = 0$ at σ = const; $dk_{AE}/dt = 0$; $\omega_2/\omega_1 < 1$, $\omega_2/\omega_0 < 1$; $\sigma_3 < \mu$; DRT = var	Time to localization $\tau^* = f(Y_{AE})$ or $\tau^* = f(W_{AE})$
I	Localized, fine, no heterogeneous	$d^2\xi/dt^2 < 0$ at $\sigma = 0$; $d^2\ln\xi/dt^2 < 0$ at $\sigma = 0$; $dk_{AE}/dt<0$ ($dP_U/dt<0$); $\omega_2/\omega_1>1$, $\omega_2/\omega_0>1$; $\sigma_3 > \mu$; DRT = invar	$\tau^* = (0,1 \div 0,5)T$
I	Localized, fine, homogeneous	$d^2\xi/dt^2 = 0$ at σ = const; $d^2\ln\xi/dt^2 = 0$ at σ = const; $dk_{AE}/dt = 0$; $\omega_1/\omega_0 < 1$, $\omega_2/\omega_0 < 1$; $\sigma_3 < \mu$; DRT = invar	Time before growth hub $\tau^* = f(Y_{AE})$ $\tau^* = f(W_{AE})$
II	Formation and crack growth	$d^2\xi/dt^2 > 0$ at σ = const; $d^2\ln\xi/dt^2 > 0$ at σ = const; $dk_{AE}/dt > 0$ ($dP_U/dt < 0$); $\omega_1/\omega_0 > 1$, $\omega_2/\omega_0 > 1$; $\sigma_3 > \mu$; DRT $\approx$ invar	$\tau^* = (0,01 \div 0,1)T$
II	Plastic destruction	$d^2\xi/dt^2 < 0$ at σ = const; $d^2\ln\xi/dt^2 < 0$ at σ = const; $dk_{AE}/dt < 0$ ($dP_{\Delta t}/dt < 0$); $\omega_1/\omega_0 < 1$, $\omega_2/\omega_0<1$; $\sigma_3 < \mu$; DRT $\approx$ invar	$\tau^* = (0,01 \div 0,1)T$

[a] DRT-difference of arrival times of the AE signal to the AE converter.

Figure 4c shows a graphical illustration of the increase in reliability due to the assessment of concentration-kinetic parameters $W_{AE} = W_R$ and $Y_{AE} = Y_R$.

3. Discussion of results in scientific and applied aspects

Based on a multi-level model of time dependence of AE parameters (1), the information-kinetic approach develops a traditional experimental approach to resource assessment, based on destructive tests of similar working reference samples, construction of fatigue curves and the linear summation hypothesis of damage, differing from it by the type of body samples, the degree of damage to the object and the type of control of their destruction, indicators of strength characteristics, criteria for heterogeneity and strength similarity. The essence of development reduced to the implementation of the following subject-methodical procedures that reduce the uncertainty in the assessment of the resource:

1. The transition from the visual control of the number of cycles to the destruction of standard samples with conditionally identical defects to the automatization of acoustic emission control of the time of destruction of the preset microelements of the material of an industrial object in order to ensure the testability of the damage accumulation process;
2. The transition from the strength characteristics of the strength of standard macro samples to the concentration-kinetic indicators of strength and strength heterogeneity of the material of a technical object, which made it possible to provide informative control;
3. Replacing the spatial-geometric criteria of strength similarity (shapes, sizes, geometric parameters of samples, defects, cracks, etc.) with time (time to failure, parameters of density distribution functions with its associated strength nanocharacteristics), which made it possible to

select AE signals from representative structural elements material through sound filtering of the flow of AE signals from the object of diagnosis by kinetic and statistical features;

4. Identification of parameters of material fatigue curves and time dependencies of AE parameters, which allowed justifying the choice of stable strength characteristics of the material and to ensure the possibility of using the information base of fatigue tests of standard samples.

4. Example of practical use

The practical application of the approach in this paper considered when diagnosing the condition of the outer ring of a rolling bearing. The results of the registration of AE, obtained by its static loading on the stand, presented in figure 6.

For a bearing with a defect grid of 3x3, according to the AE control, we have the number of AE pulses recorded from 145 to 150 seconds equal to 5 and 54, and the change in the diagnostic load during this time occurred at 10,000 N. Then, with the load on the bearing R_{work} = 40,000 N we have

$$W_{AE}=\ln(54/5)/(1/4)=9.52$$

The constant N_B is calculated from the bearing fatigue curve (C/P) n = Lh, ppm, converted to an exponential form.

One turn the bearing will withstand at a load of P = 100 C, ln1 = 0.

With a load equal to the dynamic load capacity P = C, the number of revolutions of the bearing L = 106, ln106 = 13.82, from the proportions of the exponential fatigue curve [(100C-C) / C = $\ln 10^6$ / ($\ln N_B$-$\ln 10^6$) we have

$\ln N_B = \ln 10^6 (1+1/99)$= 13.82 ·1.01=13.953

Durability of the bearing in revolutions

$L=N_B/\exp W_{AE} = \exp(1{,}01\ln 10^6)/\exp 9{,}52 = \exp(13{,}953-9{,}52)=\exp 4{,}43=82{,}8$ turnover)

For P_{work}= 4000 H then $W_{AE} = \ln (54/5)/ (1/0{,}4)= 0.952$

Durability in revs

$L=N_B/\exp W_{AE}=\exp(1{,}01\ln 10^6)/\exp 0.952$= 425631 turnover

5. Conclusion

Thus, by the example of welded structures and rolling bearings, the possibility of an effective assessment of their life based on the information-kinetic approach to acoustic emission diagnostics has been shown. As a methodological basis, a multilevel model of temporal dependencies of AE parameters was taken, describing their behavior under conditions of strength and metrological heterogeneity, as a prognostic one, that is, informationally related to the resource and parameters of fatigue curves, the final stage of uniform elastic fine-dispersed destruction is considered heterogeneous material in the form of destruction of representative structural elements. The criteria of metrological and strength similarity arising from the model make it possible to organize the selection of informative AE signals by filtering their flow according to kinetic and statistical features, to justify the choice of valuable multi-level diagnostic AE strength indicators. The proximity of information-kinetic and traditional destructive approaches to resource assessment allows combining the results of AE control with a wide base of accumulated data of fatigue tests, ensuring a reduction of uncertainty in the resource assessment of hazardous technical objects. The effect of applying an approach that allows separating the influence of material of factors of various scale levels (acoustically active and activation volumes) on AE is especially apparent when diagnosing objects exposed to thermochemical effects (hydrogen absorption, corrosion, heat treatment, etc.), when the change in AE activity does not correlates with changes in the degree of danger source [17].

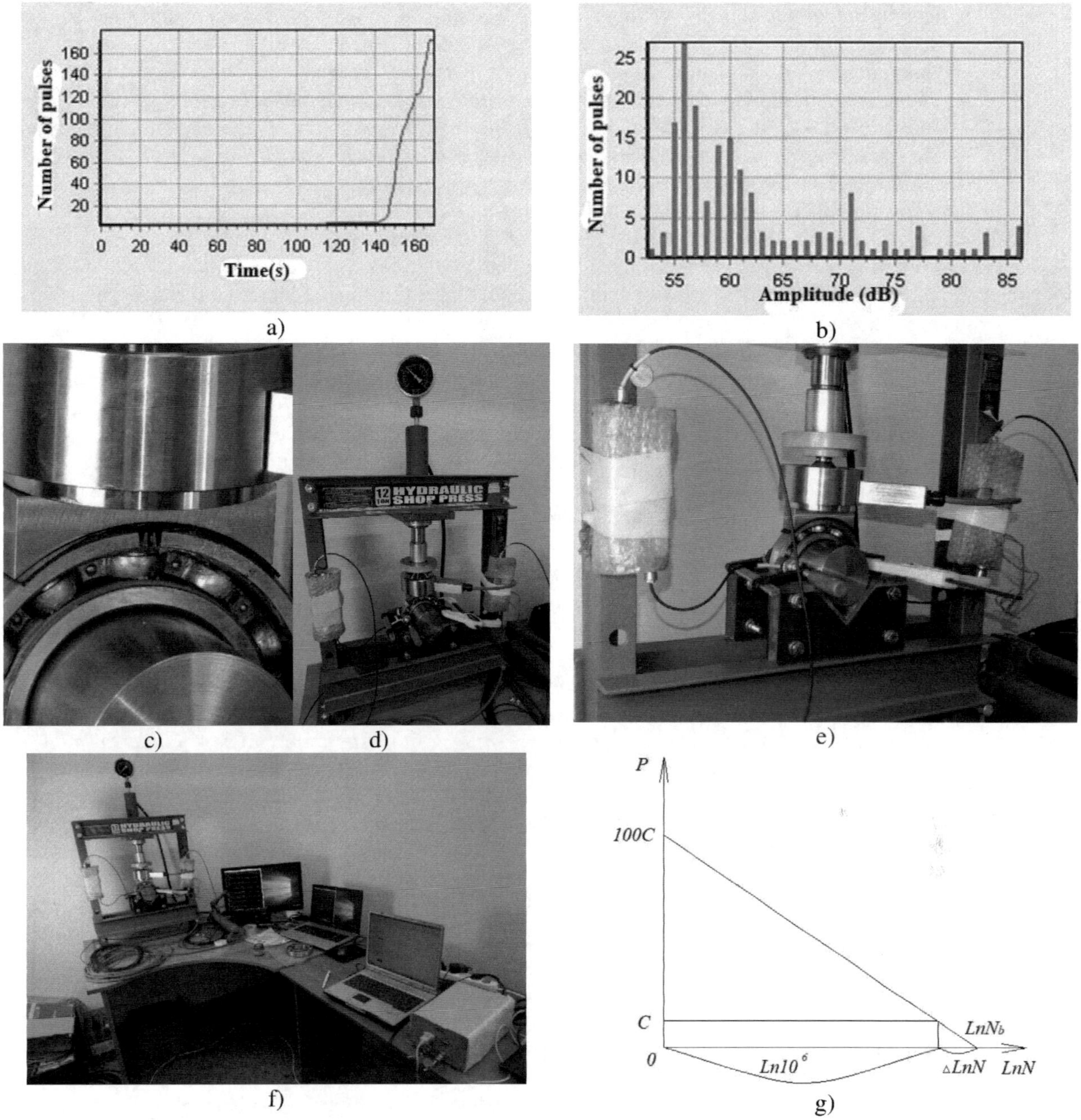

Figure 6. Results of registration of AE rolling bearing. a is the time dependence of the number of AE pulses; b is the amplitude distribution of AE signals; c, d, e, f is appearance and scheme of installation of piezo transducers, g is graphical illustration of the definition of the N_B parameter.

References

[1] Makhutov N A, Vorobyev A Z and Gadenin M M 1983 *Structural strength under low-cycle loading* (Moscow: Science) p 271

[2] Makhutov N A, Zatsarinny V V and Bazaras J L 1989 *Statistical laws of small-cycle destruction* (Moscow: Science) p 252

[3] Makhutov N A and MatvienkoY G 2016 Technical diagnostics of residual life and security of technosphere objects *Proceedings of the Russian Academy of Sciences and the business program of the "NDT 2016 Territory" forum* pp 22-31
[4] Lukyantsev M A, Vildanov R G and Iskhakov R R 2013 Estimation of the residual resource of reservoirs under low-cycle loading *Modern problems of science and education* **6**
[5] RD 26.260.004-91 1992 *Methodical instructions. Prediction of the residual life of the equipment to change the parameters of its technical condition during operation* (Moscow: Niihimmash) p 50
[6] STO GAZPROM 2-3.5-252-2008 *Organization standard. Methods of extending the safe operation of gas pipelines of JSC "GAZPROM"* (Moscow: LLC "VNIIGAZ") p 52
[7] Penkin A N, Galstyan L K, Milovanova L N and Penkin N S 2018 *Low-cycle fatigue of samples with stress concentrators* (Stavropol: Stavropol North Caucasus Federal University) p18
[8] Plotnikov D G, Sokolov S A, Borovkov A I and Mikhailov A A 2015 *St. Petersburg state polytechnic university journal of engineering science and technology* **1(214)** pp 186-193
[9] Badalyan V G, Bazulin E G and Vopilkin A K 2008 *Ultrasonic defectometry of metals using holographic methods* (Moscow: Mashinostroenie) p 368
[10] Makhutov N A and Razuvayev I V 2017 Integrated remote monitoring of the actual technical condition of FP I and II classes in oil refining, petrochemistry and chemistry using the method and systems of acoustic emission. Problems. Solutions. Results. *XXI All-Russian conference on non-destructive testing and technical diagnostics* (Moscow: Spectr) p 364
[11] Ivanov V I 2010 The prime problems of acoustic emission diagnostics of technical devices and constructions *10th European conference of non-destructive testing* (Moscow)
[12] Ivanov V I 2016 Basic problems and tasks of acoustic emission diagnostics *Proceedings of the Russian Academy of Sciences and the business program of the "NDT 2016 Territory" forum* (Moscow: Spectr) pp 132-140
[13] Nosov V V 2016 *Russian journal of nondestructive testing* **52(7)** pp 386–399
[14] Nosov V V 2016 Acoustic-emission diagnostics as the basis for improving the strength reliability of machines and equipment *Proceedings of the Russian Academy of Sciences and the business program of the "NDT 2016 Territory" forum* (Moscow: Spectr) pp 230-243
[15] Petrov V A, Petrov G V and Ryzhov O Y 2005 Comparative analysis of resource definition techniques *Proceedings of the St. Petersburg State Polytechnical University* **494** 8-14
[16] Leksovsky A M, Baskin B L and Yakushev P N 2016 Correlated non-linear interaction of micro / meso defects - a new systemically forming mechanism for the destruction of a heterogeneous system when it is deformed under conditions of constrained deformation *The fourth tectonophysical conference at the Institute of Physics of Physics, RAS. Tectonophysics and current issues of Earth sciences* (vol 1) ed Rebetsky Y L (Moscow: Institute of Physics of the Earth, Russian Academy of Sciences) pp 282-289
[17] Nosov V V, Makhmudov H F and Hotchenkov A S 2018 Acoustic emission diagnostics of loss of stability of loaded reinforced concrete beams *XXIII St. Petersburg readings on the problems of strength, dedicated to the 100th anniversary of the Physicotechnical Institute named after A.F. Joffe and the 110th birthday of the corresponding member USSR Academy of Sciences A.V. Stepanova* (Saint-Petersburg: VVM) p 325
[18] Nosov V V 2017 *Russian journal of nondestructive testing* **53(5)** 368-377
[19] Nosov V V 2014 *Russian journal of nondestructive testing* **50(12)** 719-729
[20] Nosov V V, Nominas S V and Zelensky N A 2015 *Peter the great St. Petersburg polytechnic university* **2(219)** pp 182-190
[21] Nosov V V and Potapov A I 2015 *Russian journal of nondestructive testing* **51(1)** pp 50-58
[22] Nosov V V and Zelensky N A 2017 *Russian journal of nondestructive testing* **53(2)** pp 89-95
[23] Nosov V V, Samigullin G Kh, Yamilova A R and Zelensky N A 2016 Micromechanical model of acoustic emission as a methodological basis for the prediction of welded joints fracture *Oil and gas business* **14(1)** pp 244-253

[24] Bobrov A L, Stepanova L N 2013 Study of the parameters of acoustic emission signals in the control of cast parts of a freight car truck *Defektoskopiya* **12** pp 61–68

[25] Plotnikov V A and Makarov S V 2010 Activation volume and acoustic emission during high-temperature deformation of aluminum *Tomsk state university journal* **15(3)** pp 1068-71

[26] Ainabekov A I, Raimberdiev T P, Dzhumabaev A A and Eshimbetov Sh T 2006 Engineering calculations of structural elements for durability with low-cycle fatigue, taking into account the influence of temperature *Scientific articles Kazakhstan* **5** pp 56-61

[27] Buylo S I 2017 *Physico-mechanical, statistical and chemical aspects of acoustic emission diagnostics: monograph* (Taganrog: Publishing House of the Southern Federal University) p 184

[28] Leontev A, Aleshin M, Klyavin O and Borovkov A 2018 MATEC web of conferences **148** pp 1–4

Application of polymeric materials for abating the environmental impact of mine wastes

M A Pashkevich[1], **A V Alekseenko**[1], **T A Petrova**[1]
[1] Saint Petersburg Mining University, 2, 21 Line, St Petersburg, 199106, Russia

E-mail: Alekseenko_AV@pers.spmi.ru

Abstract. Insulation methods, helping to prevent the negative environmental impact of mine dumps, tailings and other waste storages, are considered and their drawbacks are revealed. The proposed two ways of environmentally sound and cost-effective preserving are described. The first approach is to form polymeric screens over the waste surface, as well as under a storage facility. The conducted experiments included the investigation of changes in physical and mechanical properties of polymers depending on their processing temperature. To study the suitability of low-density polyethylene, high-density polyethylene and polypropylene, the damage to the coating structure integrity as a result of normal load application is assessed. The second approach is to cover slopes of the dumps with a three-dimensional erosion control mat. The subsequent hydroseeding allows ensuring complete coverage and high strength of the slope surface in a short time and at the lowest cost. In order to prevent mine dump erosion and minimise associated costs, the analysis of key properties was performed. It was revealed that polypropylene and polyester used as a geonet are characterised by the lowest volume of fertile soil needed to cover.

1. Introduction

The necessity to investigate possible ways to cover the dumps was revealed according to the notion of landscape disturbance under the impact of mining and processing industry [1, 2, 16, 17]. The assessed methods of abating the environmental impact of mine wastes were determined by the local factors of chemical element migration and accumulation: geomorphological structures of urban area and types of land use [26-36]. Monitoring research in Novorossiysk city revealed that the change of these factors had caused a significant redistribution of chemical elements and their maximal accumulation in soils of residential areas [8]. Pollutants are brought into the soil directly from the source of contamination, as well as due to the migration from landscapes located hypsometrically above [11, 18, 20, 21]. Due to the listed conditions, the need to indicate a long-term, commercial, and environmentally-friendly dump reclamation and insulation technique emerged. Secondary granulated polyethylene of low density (LDPE) and high density (HDPE), as well as polypropylene (PP), were used as the studied materials. Thus, two methods of applying polymeric materials were tried: insulation by LDPE, HDPE and polypropylene plastic wastes and soil stabilisation by a three-dimensional erosion control mat [5, 9, 13, 15, 22, 24].

The choice of materials is determined by both economic (cost of polymers, market saturation and availability) and environmental factors (safety for the natural environment) [3, 6, 10]. The cost of one kilogram of granular polymers is in the range of 30-40 rubles.

Low-density polyethylene (LDPE) is a colourless wax-like material, which is industrially obtained by polymerisation of gaseous polyethylene. LDPE is a thermoplastic polymer with a density of 910-930 kg/m^3. According to the published data, such polyethylene has a relatively high tensile strength, strength at multiple bending, impact strength and resistance to low temperatures.

High-density polyethylene (HDPE) is a less wax-like material characterised by high resistance to fats and oils, low resistance to shock load and strength at multiple bending. Resistance to compression and stretching is comparable to the levels of LDPE; the density is 940–960 kg/m^3.

Polypropylene (PP) is a linear polymer obtained by polymerisation of propylene in the presence of catalysts. The density of polypropylene is 900–920 kg/m^3. It is characterised by high resistance to shock load, low strength at multiple bending; PP becomes brittle at negative temperatures.

2. Methods

Physical and mechanical properties of polymer materials and their products are a set of properties that characterise their behaviour under the influence of external mechanical loads. Such properties of polymers are determined in the course of tests, the result of which is the revealed dependence of stresses on deformation (e.g. a tensile diagram). Analysing these dependencies, the main characteristics of elasticity, strength and plasticity (modulus of elasticity, fracture resistance, ultimate strength) are found.

Compared with non-plastic materials, the physical and mechanical properties of polymers have a number of features [19]:

– the ability to develop large reversible deformations, reaching hundreds or even thousands of percent;
– relaxation properties, i.e. the dependence of strains and stresses on the duration of an external action;
– the dependence of the physical and mechanical properties of the polymer on the conditions of its production, processing and pre-treatment, which is determined by the existence of various supra-molecular forms with the long adjustment time in polymers;
– the ability of polymers to acquire a sharp anisotropy of properties under mechanical action due to the orientation of macromolecules;
– the ability to undergo chemical transformations under the mechanical impact, e.g. the development of destruction processes.

In all cases, it is necessary to know what requirements the finished product must meet when producing insulation coatings from polymeric materials. The insulation coating for mine waste must have a number of very specific properties:

– the created coating should have high strength characteristics since the mass of the ore stack is over 300 thousand tonnes, which determines the high static load on the polymer material laid in the base;
– the polymer coating must be resistant to aggressive media, as it is exposed to an alkaline solution under the process of leaching;
– the strength of the coating, under the influence of a significant temperature difference (from –30 to +30 °C), should remain unchanged.

After analysing a number of research outcomes [2, 7, 14, 23], it can be concluded that the structure of polymeric materials is determined by two main levels: molecular and supra-molecular.

At the stage of polymer synthesis, its molecular structure is formed. By varying the polymerisation conditions (pressure, temperature, etc.), it is possible to regulate the molecular weight of the polymer. The supra-molecular structure is formed at the stage of semi-finished product processing into the finished one. Types and sizes of supra-molecular formations are determined by the method of polymer processing, temperature, pressure, cooling mode, etc.

As already mentioned, two types of polyethylene (low and high density) and polypropylene were chosen as the study objects. Samples of the material were produced by extrusion of granular polymers in the shape of sheets of 200×600 mm, followed by cooling at speeds that simulate real field conditions.

During the sample preparation for each polymer, three sheets with varying thickness (1.0, 3.0 and 6.0 mm) were produced to determine the dependence of the strength characteristics of the formed coating on the thickness of the formed layer. Based on the literature data [2, 14, 25], the melting temperatures of polymers are different and lie in the range from 105 to 175 °C.

It was experimentally found that for the studied polymers the temperature of transition from the highly elastic state to the viscous one is 120 °C for LDPE, 135 °C for HDPE and 170 °C for PP. Based on the established melting temperatures, the range from 120 to 220 °C in increments of 10 °C was studied to determine the dependence of the physical and mechanical characteristics of the polymer samples on the coating formation temperature. Thus, the total number of specimens to determine the tensile strength of the studied materials was 60.

In the course of the research, the following indicators characterising the strength of the polymer product were determined:

– breaking stress σ_b;
– tensile yield stress σ_y,
– breaking elongation ε_b,
– tensile yield elongation ε_y.

The determination of indicators characterising the strength of polymeric materials was carried out on a universal testing machine H75K-S for static testing of materials for tension, compression and bending. The tensile strength of the samples was tested in accordance with the GOST 11262-80 Plastics. Tensile test method.

For the tests, samples with a double-edged blade (dumbbell-shaped) with the following parameters were used:

– the total length L1 = 250 mm;
– the distance between the marks determining the position of the clips on the sample L2 = 170 mm;
– the length of the working part L3 = 80 mm;
– the sample head width b1 = 25 mm;
– the working part width b2 = 10 mm;
– The thickness of the sample (d) varies from 1 to 6 mm.

The rupture rate was also determined on the basis of the GOST and was 200 mm per minute. The specimens were prepared by mechanical cutting of the sheets produced by extruding. The transformation of the polymer material takes place phase-by-phase.

At the first stage, the raw material is loaded. Feedstock supplied to the hopper can be presented in the form of tapes or powders or granules. Granular polymers are the best raw material for obtaining a marketable product. Processing of tapes and powders is complicated by the fact that these raw materials are prone to caking, and, consequently, they form "plugs" in the hopper. Therefore, when processing powder and tape materials, stirrers are often placed in bins. In addition, the flowability of materials depends on the humidity: the higher the humidity, the less the flowability. Therefore, materials must be dried before feeding into the hopper [19].

The loading of the inter-turn space under the hopper funnel takes place on the screw segment of (1–1.5)×D, where D is the screw diameter.

At the second stage, the material is compacted in the feeding zone. The granules move due to the difference in the friction force created by the friction of the polymer on the inner surface of the cylindrical body and friction on the surface of the working body, i.e. the screw. Since the friction force of the granules on the surface of the working body is higher than the surface of the cylinder, the granulate ceases to move along the axis of the screw and rotates with it. The solutions to this problem are (i) the increase of the cylinder wall temperature and (ii) the parallel screw cooling in the first zone. Under optimal temperature conditions in the feeding zone, the processed polymer is pressed and compacted, forming a dense "plug" in the interturn space. Typically, the length of the load zone is (2–10)×D.

In the third zone, the granular compacted polymer is melted and plasticised. This process occurs both due to the heating of the cylinder surface and due to the heat released by the viscous friction of the material in the thin layer of the melt. When the polymer is rubbing against the cylinder wall, intense shear deformation occurs, resulting in the subsequent plastification. This process is graphically shown in Figure 1.

Journal of Physics: Conference Series **1384** (2019) 012039 doi:10.1088/1742-6596/1384/1/012039

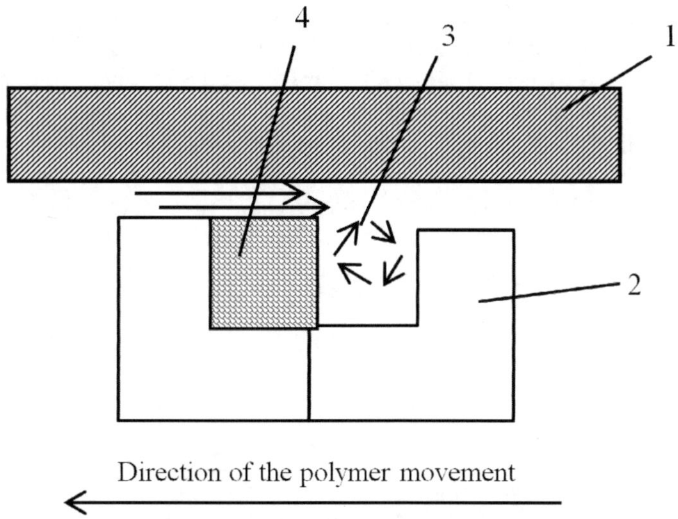

Figure 1 – Diagram of plug melting in the plastification area in the inter-track cross-section of the screw: *1 – the cylinder wall; 2 – comb auger; 3 – the flow of polymer melt; 4 – pressed solid polymer (plug) in the extruder*

3. Results

Damage to the formed coating can occur during its production, as well as during subsequent installation, but the largest number of fractures is most often observed during the insulation of a mine dump. Damage consists of punctures, disruptions and deformations (dents). Damage is caused by the presence of inhomogeneities on the insulated surface or non-compliance with the construction technology of a mine dump. Waterproofing base in the process of filling the stack and subsequent operation is subjected to loads leading to the violated integrity of the polymer material.

The determination of the strength of the material and its resistance to damage is performed in two ways:

– semi-production tests, by forming the coating and applying appropriate loads to it;
– laboratory experiments, by determining the strength parameters of the materials.

During the formation of a mine dump, the waterproofing coating undergoes tension, compression and shear stresses, caused by the displacements that occur between a polymer coating and a drainage layer. Compressive and tensile loads depend on the mass of the waste to be stored, the particle size distribution of a drainage layer and the term of dumping.

The stability of polymeric materials to load application vary significantly depending on several internal and external factors [4, 12, 19]. LDPE films with a thickness of not more than 0.2 mm were normally considered. Studies of polymer coatings based on LDPE and PP have been carried out as well. The method of determining the degree of damage by the number of defects per 1 m^2 of the test sheet sample is quite widespread in the study of polymer waterproofing materials. The experiment was conducted to determine the strength of the studied polymeric materials.

When conducting the research, the studied material was placed in a container with dimensions appropriate to the subject sample, simulating the conditions of laying the polymer coating, and brought to the destruction by the load generated by a metal stamp. The experiment tank was made of stainless steel with a wall thickness of 3 mm, thus, the tests exclude the possibility of expanding the model and the movement of sand on the surface of the polymer sheet. The proposed loading scheme characterises the stage of construction and operation of a waterproofing layer. Sands with an average particle size of 0.58 mm were used for testing.

The studies were carried out using a hydraulic press D2430B; Figure 2 schematically shows the installation for testing the sheet polymer material. A steel container (3) is filled with a 60 mm-thick layer of sand (2), which presence simulates a protective layer under the test sample. Sand is compacted, and the prepared polymer sheet (4) is laid on the surface. Sand covers the polymer sheet with a layer of the same thickness, simulating drainage. Dimensions of the tank are 400×200×150 mm. After that, the stamp (390×190 mm) is placed on the model to generate pressure on the sample. The container is installed between press plates (5) and (1), and the model is subjected to pressure at a constant speed of 3 mm/min.

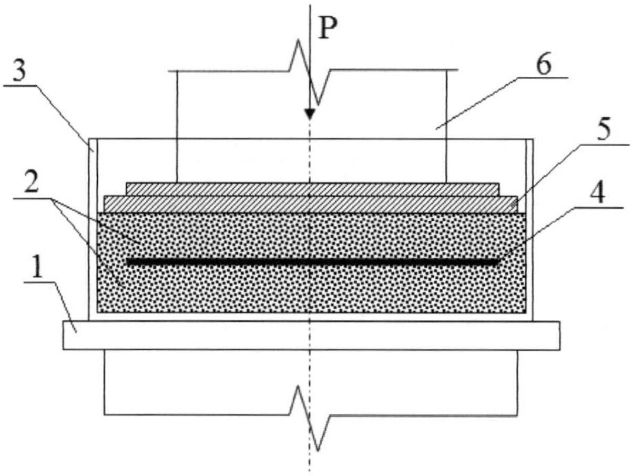

Figure 2. Design diagram of the installation for punching sample testing of polymer sheets

According to the state requirements, the thickness of the waterproofing material, based on the conditions of ensuring the continuity, will be determined by the formula (1):

$$\sigma = 0.1 \cdot d_{grain} \frac{q}{E_f} = 0.1 \cdot 2 \cdot 0.71 = 0.15 \text{ mm} \quad (1)$$

Where σ is the calculated film thickness; d_{grain} is the minimum diameter of the largest fraction of the soil scattered with the use of standard sieves, mm; E_f is the efficiency factor of additional protective fillers. q is the load taken for the screen as the greater of the two values calculated for the construction period and the operating period.

For the construction period, the load on the waterproofing layer depends on the pressure of the machines transmitted by the protective soil layer; for the operating period, it depends on the pressure of a dump. In our case, the pressure of the operating period will be determined by the mass of a dump and the area of its base.

When calculating the film thickness, the E_f is taken to be equal to one, since the dominating particle diameter is less than 10 mm and the use of an additional geotextile filler is not required. Based on the calculation, the studied insulation coating thickness (1.0, 3.0 and 6.0 mm) will meet the conditions of continuity of the formed layer in accordance with regulatory instructions. The value of the load created by loading machines depends on air pressure in tires. In the calculation, we take the value q = 0.71 MPa, as the largest at a tire pressure of 0.6 MPa. The estimated load generated by a dump per unit of base area is 0.2 MPa. The load was calculated in accordance with the state requirements. The pressure varies from 0.2 to 2.5 MPa; such value exceeds the range of real loads created by construction mechanisms, the application time of the created load is 4 min.

The other considered option is to cover the dumping sites – especially the slope surfaces – with geosynthetics subsequently topped with soil layer. In order to prevent soil erosion and minimise associated costs, the analysis of key properties was performed (Table 1). The data show the market prices for the year 2018.

Table 1. Comparative characteristics of geosynthetics

Cover type / **Properties**	Polypropylene and polyester	Polypropylene	Polyester
	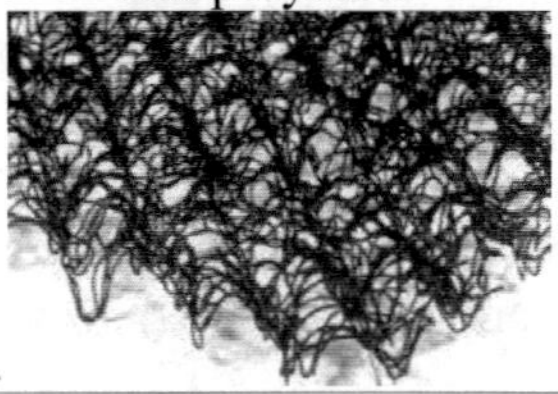		
Tensile strength, kN/m (avg., min.-max.)	50 (20–200)	20 (18–28)	20 (30–150)
Elongation,%	13	30–50	15
Thickness, mm	13–15	50–100	12
Permissible operating temperature, °C	-30…+100	-60…+55	-15…+250
Surface density, g/m^2	Not less than 550	Not less than 570	Not less than 300
Necessity of backfilling with soil, mm	25–50	100–150	80–100
Weight, g/m^2	400–800	320–1520	250–950
Price, rubles/m^2	80–155	70–358	50–124

As it can be seen from the data presented, polypropylene and polyester used as a geonet are characterised by the lowest volume of fertile soil needed to cover it; the other properties allow using this type of material for covering unstable slope surfaces to halt the fugitive dust emission from the studied dumps.

4. Conclusion

According to the results of the research on mine dump insulation and stabilisation, the following conclusions can be drawn:

- polymer coating manufacturing is accompanied by a non-linear strength change at different temperatures;
- the values of the tensile strength of polymer samples obtained at temperatures close to the melting point are not maximum, due to incomplete homogenisation of the melt and the presence of excess moisture and volatile impurities at 96–98%;
- the optimum temperature values for the processing of polymer granulate lie in the found range of 180–200 °C;
- the decrease in the strength characteristics of the studied samples at temperatures above 200 °C is determined by thermal and thermo-oxidative destruction;
- the considered approach to stabilising the slope surfaces of dumping sites with geosynthetics subsequently topped with soil layer can prevent soil erosion and minimise associated costs;
- polypropylene and polyester used as a geonet for mine dump stabilisation are characterised by the lowest mass of fertile soil required and acceptable expenses.

Acknowledgments
The analysis was performed using the equipment of the Common Use Centre of the Saint Petersburg Mining University.

References

[1] Aleksandrova T N, Talovina I V and Duryagina A M 2019 Gold–sulphide deposits of the Russian Arctic zone: Mineralogical features and prospects of ore benefication *Chemie der Erde* DOI: 10.1016/j.chemer.2019.04.006

[2] Alekseenko V A, Maximovich N G and Alekseenko A V 2017 Geochemical Barriers for Soil Protection in Mining Areas *Assessment, Restoration and Reclamation of Mining Influenced Soils* 255–274

[3] Aman A, Yaacob M M, Alsaedi M A and Ibrahim K A 2013 Polymeric composite based on waste material for high voltage outdoor application *International Journal of Electrical Power and Energy Systems* **45(1)** 346-352

[4] Azimi E A, Al Bakri Abdullah M M, Ming L Y, Hussin K and Aziz I H 2015 Review of geopolymer materials for thermal insulating applications *Key Engineering Materials* **660** 17-22

[5] Benkreira H, Khan A and Horoshenkov K V 2011 Sustainable acoustic and thermal insulation materials from elastomeric waste residues *Chemical Engineering Science* **66(18)** 4157-4171

[6] Chen L, Han W, Li Z, Wei T and Xiao C 2011 Preparation and properties of alkali stimulated geopolymer and its application in thermal insulation coating *Advanced Materials Research* **233-235** 2443-2446

[7] Chen L, Wang Z, Wang Y and Feng J 2016 Preparation and properties of alkali activated metakaolin-based geopolymer *Materials* **9(9)** 767

[8] Datta J and Włoch M 2017 Recycling of Polyurethanes *Polyurethane Polymers: Blends and Interpenetrating Polymer Networks* 323-358

[9] Goldfein M D and Kozhevnikov N V 2014 A commentary on the concepts of the physical chemistry of polymers in technologies and environment protection *Chemical and Biochemical Technology: Materials, Processing, and Reliability* 175-186

[10] Jittabut P 2017 The study on physical and thermal properties of geopolymer pastes *Key Engineering Materials* **751 KEM** 538-543

[11] Jordán M M, Bech J, García-Sánchez E and García-Orenes F 2016 Bulk density and aggregate stability assays in percolation columns *Journal of Mining Institute* **222** 877-881

[12] Kalužová A, Pěnčík J and Matějka L 2013 Analysis of environmental impacts on thermal conductivity of insulation block from secondary raw materials *Advanced Materials Research* **649** 223-226

[13] Kamseu E, Mohamed H, Sofack J C, Chaysuwan D, Tchakoute H K, Djobo J N Y, Rossignol S and Leonelli C 2018 Moisture Control Capacity of Geopolymer Composites: Correlation of the Bulk Composition–Pore Network with the Absorption–Desorption Behavior *Transport in Porous Media* **122(1)** 77-95

[14] Łach M, Korniejenko K and Mikuła J 2016 Thermal Insulation and Thermally Resistant Materials Made of Geopolymer Foams *Procedia Engineering* **151** 410-416

[15] Lv X, Guo P, Liu H, Cui L and Cui X 2018 Preparation of paraffin-based phase-change microcapsules and application in geopolymer coating *Journal of Coatings Technology and Research* **15(4)** 867-874

[16] Nagornov D O and Kremcheev E A 2018 Framework to evaluate efficiency of peat processing using roller-disc grinding machines *Journal of Physics: Conference Series* **1118(1)** 012028

[17] Nagornov D O and Kremcheeva D A 2017 Assessment of gravity component of moisture flow in raw peat in conditions of convective and radiative-convective heat input *Journal of Industrial Pollution Control* **33(1)** 749-752

[18] Nagornov D O, Kremcheev E A and Kremcheeva D A 2019 Research of the condition of regional parts of massif at longwall mining of prone to spontaneous ignition coal seams *International Journal of Civil Engineering and Technology* **10(1)** 876-883

[19] Pashkevich M A, Petrova T A 2017 Reclamation by Containment: Polyethylene-Based Solidification *Assessment, Restoration and Reclamation of Mining Influenced Soils* 235–253

[20] Perminova T, Sirina N, Laratte B, Baranovskaya N and Rikhvanov L 2016 Methods for land use impact assessment: A review *Environmental Impact Assessment Review* **60** 64-74

[21] Robertus Y V, Udachin V N, Rikhvanov L P, Kivatskaya A V, Lyubimov R V and Yusupov D V 2016 Indication by environmental components the pollutant transboundary transfer to Gorny Altai *Bulletin of the Tomsk Polytechnic University, Geo Assets Engineering* **327(9)** 39-48

[22] Su Z, Guo L, Zhang, Z and Duan P 2019 Influence of different fibers on properties of thermal insulation composites based on geopolymer blended with glazed hollow bead *Construction and Building Materials* **203** 525-540

[23] Wongsa A, Sata V, Nematollahi B, Sanjayan J and Chindaprasirt P 2018 Mechanical and thermal properties of lightweight geopolymer mortar incorporating crumb rubber *Journal of Cleaner Production* **195** 1069-1080

[24] Zhang Z, Provis J L, Reid A and Wang H 2015 Mechanical, thermal insulation, thermal resistance and acoustic absorption properties of geopolymer foam concrete *Cement and Concrete Composites* **62** 97-105

[25] Yusupov D V, Bolshunova T S, Mezhibor A M, Rikhvanov L P and Baranovskaya N V 2017 The use of Betula Pendula R. Leaves for the assessment of environmental pollution by metals around tailings from a gold deposit (Western Siberia, Russia) *International Multidisciplinary Scientific GeoConference Surveying Geology and Mining Ecology Management, SGEM* **17(41)** 665-672

[26] Bushuev Y Y, Leontev V I and Machevariani M M 2018 Geochemical features of Au-Te epithermal ores of the Samolazovskoye deposit (Central Aldan ore District, Yakutia) *Key Engineering Materials* **769 KEM** 207-212

[27] Krasotkina A O, Machevariani M M, Korolev N M, Makeyev A B and Skublov S G 2017 Typomorphic features of niobium rutile from the polymineral occurrence Ichetju (the Middle Timan) *Zapiski Rossiiskogo Mineralogicheskogo Obshchestva* **146(2)** 88-100

[28] Beloglazov I I and Kuskova Y V 2017 Simulation of Roasting Metallurgical Concentrates in Fluidized Bed Using CFD-DEM *IOP Conference Series: Materials Science and Engineering* **221** 012017

[29] Raupov I R and Korobov G Y 2018 Research of polymer compositions rheological properties for oil production *Acta Technica CSAV (Ceskoslovensk Akademie Ved)* **63(3)** 493-500

[30] Rogachev M and Kondrashev A 2015 Experiments of fluid diversion ability of a new waterproofing polymer solution *Petroleum Exploration and Development* **42(4)** 554-559

[31] Raupov I R and Shagiakhmetov A M 2019 The results of the complex rheological studies of the cross-linked polymer composition and the grounding of its injection *International Journal of Civil Engineering and Technology* **10(2)** 493-509

[32] Alabyev V R and Korshunov G I 2017 Improving methods for calculating air temperature in mine workings equipped with cold water pipelines *Ecology, Environment and Conservation* **23(1)** 444-450

[33] Ponomarenko T V, Nevskaya M A and Marinina O A 2017 Project-based approach to the implementation of low-waste technology at the mining companies *International Multidisciplinary Scientific GeoConference Surveying Geology and Mining Ecology Management, SGEM* **17(43)** 163-170

[34] Limanskiy A V and Vasilyeva M A 2016 Using of low-grade heat mine water as a renewable source of energy in coal-mining regions *Ecological Engineering* **91** 41-43

[35] Nevskaya M A and Marinina O A 2015 Regulatory aspects of mining waste management in the Russian Federation *Biosciences Biotechnology Research Asia* **12(3)** 2619-2628

[36] Moradi S S T, Nikolaev N I and Leusheva E L 2018 Improvement of cement properties using a single multi-functional polymer *International Journal of Engineering, Transactions A: Basics* **31(1)** 181-187

Development of an operational environmental monitoring system for hazardous industrial facilities of Gazprom Dobycha Urengoy

M A Pashkevich[1], **T A Petrova**[1]

[1] Saint Petersburg Mining University, 2, 21st Line, St Petersburg, 199106, Russia

E-mail: petrova9@yandex.ru

Abstract. The article presents an assessment of the technogenic impact on the atmospheric air inflicted by oil and gas extracting and processing facilities in Western Siberia. It is revealed that heterogeneity and areal dispersion of emission sources are the key features of the considered industry. Mathematical statistics techniques allowed grouping sources of air pollution and determining the major air pollutants at the considered production facilities. It is proposed to design an automated operational environmental monitoring system with a hierarchical two-level structure for operational and reliable control of the natural environment in the impact area of the studied facilities.

1. Introduction

Over the past 20 years, global energy consumption has tended to increase continuously [1]. Thus, in 2017 alone, global energy consumption increased by 2.9 %, with the share of hydrocarbons in the global energy balance of 80-85% [2] Due to this, extraction, processing and transportation of hydrocarbons is one of the main factors of global environmental pollution [3, 4, 5, 6].

Oil and gas companies inflict complex and intense environmental impact primarily through the atmosphere, lithosphere and hydrosphere. The effects of such exposure are found often at a considerable distance from the sources [7, 8, 9]. The negative impact of oil and gas facilities is due to the toxicity of natural hydrocarbons and related resources, and a variety of chemicals used in technological processes.

The main feature of oil and gas production and processing enterprises is the presence of flows of fire- and explosion-hazardous products and raw materials, leading to major accidents, environmental and economic damage from which amounts to hundreds of millions of dollars [10, 11, 12, 13].

The production oil and gas facilities of increased environmental and technogenic danger were studied at one of the fields of Western Siberia; the enterprise produces natural gas, gas condensate, oil, prepares hydrocarbon raw materials for transportation, builds the raw material base. The resource base of the enterprise is provided with reserves of hydrocarbons of Urengoyskoe, Yen-Yakhinskoe, Severo-Urengoyskoe and Pestsovoe fields located in the Yamalo-Nenets Autonomous Okrug of the Russian Federation.

Air protection is the top significant and multifaceted environmental aspect of the production activities of oil and gas companies [14, 15, 16]. The priority in this direction is the work aimed at reducing the emissions of major greenhouse gases (carbon dioxide and methane).

In this regard, the main objectives of the research were to study the landscape-geochemical situation [17, 18, 19] under the influence of intensive anthropogenic load on the territories of the production facilities; to identify the main causes and consequences of pollution, taking into account the specifics of the production process; and to improve the organisational structure of industrial environmental control based on the creation of GIS-technology for the safe functioning of the studied objects.

2. Materials and methods

The sources of influence on the atmospheric air during the development of the field and primary processing are numerous areal and linear objects of the main and auxiliary technological process, as well as the accompanying infrastructure [20, 21]: complex gas treatment plants (CGTP), condensate stabilisation units (CSU) central production facilities (CPF), booster pipeline pumping station (BPPS), well clusters, pipelines, etc.

During the five-year period, studies were carried out at the production facilities to monitor the main emission sources by using a mobile environmental laboratory.

The mobile laboratory is designed for sampling, measurements and data processing by the crew both in stand-alone mode and for delivery of samples and data to the place of further stationary processing. The car-based mobile environmental laboratory is equipped with:

- gas analysis stand;
- modern gas analysers;
- meteorological complex;
- set of air intake devices;
- and data collection and processing unit.

As a result of the conducted research, the data on emission sources, their intensity, location, and range of emitted harmful substances were collected, taking into account climatic conditions, which allows assessment of the environmental load on the study area.

Currently, the total number of pollutant sources at the enterprise is 6213, 6022 of which are organised. All the organised emission sources are low-height (up to 50 m high).

We propose to group the losses (costs) of gas directly or indirectly resulting in emissions into the atmosphere as follows:

1) direct gas losses caused by the production activities of all facilities. These include the consumption of fuel gas in the operation of fuel-using equipment (diesel plants, boilers, gas pumping units, etc.);
2) irretrievable losses of gas with emissions into the atmosphere, which are divided into two subgroups:
   2.1. technological losses of gas with emissions into the atmosphere, taking place in the performance of various technological operations at the facilities (for example, as a result of technological losses of extracted, processed and transported hydrocarbon raw materials);
   2.2. technical losses, the presence and volume of which depends on the level of construction, operation, reliability of equipment and other factors. This subgroup includes gas losses due to gas leaks from process equipment and leakages of shut-off valves, as well as losses in case of accidents, breaks and equipment damage.

The interpretation of the received field materials showed that the main pollutants of atmospheric air from the considered enterprises are [23]:

- components of oil and natural gas, i.e. hydrocarbons and carbon dioxide released into the air pool through leaks during extraction, processing and transportation of oil and gas condensate;
- carbon oxides, nitrogen oxides, sulphur oxides, resulting from the operation of heat consuming equipment;
- soot coming into the atmosphere from plants for the production of carbon black;
- toxic chemicals used in the extraction and transportation and at different stages of preparation and

3. Results

We propose to create an automated operational environmental monitoring system (OEM) for continuous reliable control of the state of the natural environment in the impact area of the studied oil and gas field.

The OEM system, formed using GIS software, is an information and measuring system designed to control the natural environment in the zone of interaction of production facilities with the components of the natural environment. The system carries out operational collection of measurement data on the state of the observed natural objects, their processing and analysis, as well as the dissemination of monitoring results between different production units of an enterprise [24].

The operational environmental monitoring system is based on technical, software, information and organizational means, based on the following principles:

- the system provides comprehensive control over the entire set of observed components of the natural environment that are exposed to negative effects from either hazardous production facilities or those that can adversely affect these facilities;
- the control of the natural environment in the system is based on a combination of point and line ground observations and remote information, which enables the area coverage and extrapolation of observations;
- the monitoring data processing at all the stages from primary measurements, data collection and accumulation to decision support on environmental safety and security management of facilities is based on single information technology;
- the system monitors the current state of the components of the natural environment, evaluates the dynamics of its development and allows making environmentally sound and economically sound management decisions.

To design the optimal OEM structure, the issue of choosing the number and location of control points becomes relevant, for which it is proposed to divide the emission sources into homogeneous groups so that within the groups the quantitative characteristics of emissions should be subject to the normal distribution law [25].

The division of emission sources by independent groups is proposed to be carried out not by separate features, but by their cumulative effect, for which there is a need for multiple correlations. For performing such classifications, cluster analysis is the optimal solution.

If there are p objects and m characteristics are measured, then the data set forms a matrix of p X m order. After that, a similarity level or match criterion is calculated between each pair of objects. A standardised m-dimensional Euclidean distance d_{ij} is used as the similarity quotient, which is calculated as (1):

$$d_{ij} = \sqrt{\frac{\sum_{k=1}^{m}\left(X_{ik} - X_{jk}\right)^2}{m}} \quad (1)$$

Where: X_{ik} is the value of the k-th variable at the i-th object, X_{jk} is the value of the k-th variable at the j-th object.

A small value for this distance indicates that the objects are similar or "close" to each other, while a large value indicates no similarity.

The next task is to obtain a hierarchical grouping of objects in which the objects with the highest similarity quotient are placed together. The selected groups of objects are then joined to the new ones with which they are most closely related, and this continues until a complete classification of objects is obtained.

After being divided into clusters, the average values of harmful substances in emissions are calculated for each cluster and graphs of the dependence of these averaged contents on the type of harmful substances are built for each cluster. The comparison of the graphs allows understanding the parameters by which the clusters are radically different from each other, as well as determining the contribution of each cluster to environmental pollution [26].

Taking a set of objects belonging to one cluster as an area pollution source, it is possible to compute the location of observation posts (the point with the maximum surface concentration) according to the *Ecologist* software (Version 3.0).

To monitor industrial objects of oil and gas field we propose installing an observation station at each site, equipped with gas analysers for determining a full range of polluting components and operating in automatic mode. Stationary posts are proposed to be located at the point of the total maximum surface concentration of a certain substance generated by emission sources, taking into account the statistical grouping of sources. The choice of substances is determined by the priority list of pollutants [27].

The data obtained by an automatic gas analyser will be received every second and processed at the automatic control point, which will allow, in case of destabilisation of the environmental situation at the facility, conducting a survey from the route posts located in the areas allocated by means of correlation grouping.

Location of a sensor is selected based on the need to determine the surface concentration of harmful substances in the atmosphere at an altitude of 1.5–3.5 m. The following meteorological parameters are determined simultaneously with air sampling: wind direction and speed, air temperature, weather conditions.

The automated environmental monitoring system provides a complete observation programme and has a hierarchical two-level structure. The lower level consists of devices and sensors, as well as software and hardware transmitting equipment. A central computer of the system and receiving equipment (a central point of data collection and processing) form the upper level. Switched telephone wires carry out the communication between the system levels. Information transfer from stationary posts is carried out by a complex of software and hardware for collecting and issuing output information in a convenient form for the operator of the central point.

The creation of the GIS-based software system for a comprehensive evaluation of environmental conditions in areas of influence of the considered enterprises suggests the following algorithm (Fig. 1). processing of gas (methanol, diethylene glycol, etc.).

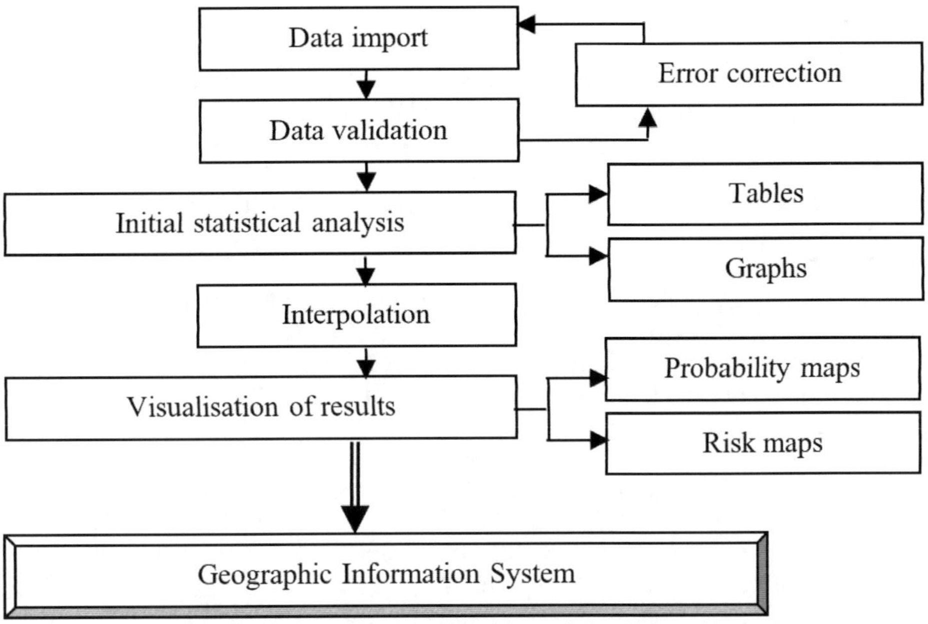

Figure 1. Functioning algorithm of the GIS technology

The analytical unit is a part of the structure of the information management system of the OEM that implements the ability to manage environmental safety [28] on the basis of GIS and includes the following operation sequence:
– data import (in particular, the results of monitoring studies) via GIS allows creating and modifying databases, as well as importing standard Excel and Access files and processing various graphic files;
– data verification, checking the test point coordinates for the data integrity [29-32];
– initial statistical analysis is carried out to identify representative samples, remove the so-called top-cut grade values from the calculations, establish patterns of distribution of concentrations of a pollutant, the effect of proportionality, the general trend, variability and other statistical parameters;
– data interpolation [33-37].

The analysis and processing of spatially distributed data should be done by geostatistics techniques, which will give an answer to a number of specific issues:
– to evaluate the value at the point where no measurements were taken [38-40];
– to draw a map, draw contours (define values on a dense grid);
– to estimate the interpolation error;
– to take into account measurement errors;
– to determine the probability of exceeding the specified level;
– to perform a joint spatial analysis of correlated variables;
– to describe spatial variability and uncertainty;
– to visualise results: score and error maps, contour and mosaic maps, and probability and risk maps.

4. Conclusion

The conducted research allowed the solution of a relevant R&D issue of improving the organisational structure of industrial ecological control and safe functioning of an oil and gas field:

1. the regularities of the formation of halos and pollution flows are revealed with the use of standard and original methods of numerical simulation;
2. an original method of optimising the location of environmental observation posts is developed based on pollution source categorisation by cluster analysis;
3. an optimal hierarchically organised structure of OEM based on GIS technology for the safe operation of oil and gas field facilities is proposed.

Acknowledgments

The reported study was carried out according to the state task № 5.12850.2018/8.9 of the Ministry of Science and Higher Education of the Russian Federation.

References

[1] Slepnev V N and Maksimenko A F 2019 Organizing the quality management system for the processes of prevention, localization and elimination of accidents at pipeline transport facilities *Neftyanoe Khozyaystvo - Oil Industry* **2** 106-111

[2] Neftegaz.RU Web Resource, https://neftegaz.ru/news/energy/218112-a-novak-dolya-uglevodorodov-v-mirovom-energobalanse-snizitsya-do-75-k-2035-g/

[3] Darbra R M, Palacios A and Casal J 2010 Domino effect in chemical accidents: Main features and accident sequences *Journal of Hazardous Materials* **183(1-3)** 565-573

[4] Hirschberg S, Burgherr P, Spiekerman G and Dones R 2004 Severe accidents in the energy sector: Comparative perspective *Journal of Hazardous Materials* **111(1-3)** 57-65

[5] Montiel H, Vílchez J A, Arnaldos J and Casal J 1996 Historical analysis of accidents in the transportation of natural gas *Journal of Hazardous Materials* **51(1-3)** 77-92

[6] Athar M, Mohd Shariff A, Buang A, Shuaib Shaikh M and Ishaq Khan M 2019 Review of Process Industry Accidents Analysis towards Safety System Improvement and Sustainable Process Design *Chemical Engineering and Technology* **42(3)** 524-538

[7] Leoni L, BahooToroody A, De Carlo F and Paltrinieri N 2019 Developing a risk-based maintenance model for a Natural Gas Regulating and Metering Station using Bayesian Network *Journal of Loss Prevention in the Process Industries* **57** 17-24

[8] Bianchini A, Guzzini A, Pellegrini M and Saccani C 2018 Natural gas distribution system: A statistical analysis of accidents data *International Journal of Pressure Vessels and Piping* **168** 24-38

[9] Ren C, Wu Q, Zhang C and Zhang S 2018 A normal distribution-based methodology for analysis of fatal accidents in land hazardous material transportation *International Journal of Environmental Research and Public Health* **15(7)** 1437

[10] An J and Peng S 2018 Prediction and verification of risk loss cost for improved natural gas network layout optimization *Energy* **148** 1181-1190

[11] Liu D, Zhou W and Pan X 2016 Risk evaluation for city gas transmission and distribution system based on information revision *Journal of Loss Prevention in the Process Industries* **41** 194-201

[12] Bariha N, Mishra I M and Srivastava V C 2016 Hazard analysis of failure of natural gas and petroleum gas pipelines *Journal of Loss Prevention in the Process Industries* **40** 217-226

[13] Gallagher M E, Down A, Ackley R C, Phillips N and Jackson R B 2015 Natural gas pipeline replacement programs reduce methane leaks and improve consumer safety *Environmental Science and Technology Letters* **2(10)** 286-291

[14] Chen Y, Bai Y and Lai X 2014 The applied research on quantitative risk assessment models and software for high pressure gas pipeline network in Urumqi, Xinjiang, China *2nd CCPS China Conference on Process Safety 2014* 274-279

[15] Siler-Evans K, Hanson A, Sunday C, Leonard N and Tumminello M 2014 Analysis of pipeline accidents in the United States from 1968 to 2009 *International Journal of Critical Infrastructure Protection* **7(4)** 257-269

[16] Vianello C and Maschio G 2014 Quantitative risk assessment of the Italian gas distribution network *Journal of Loss Prevention in the Process Industries* **32(1)** 5-17

[17] Alekseenko V A, Bech J, Alekseenko A V, Shvydkaya N V and Roca N 2018 Environmental impact of disposal of coal mining wastes on soils and plants in Rostov Oblast, Russia *Journal of Geochemical Exploration* **184** 261-270

[18] Nagornov D O, Kremcheev E A and Kremcheeva D A 2019 Research of the condition of regional parts of massif at longwall mining of prone to spontaneous ignition coal seams *International Journal of Civil Engineering and Technology* **10(1)** 876-883

[19] Alekseenko A V and Pashkevich M A 2016 Novorossiysk agglomeration landscapes and cement production: Geochemical impact assessment *IOP Conference Series: Earth and Environmental Science*, **43(1)** 012050

[20] Ma L, Li Y, Liang L, Li M and Cheng L 2013 A novel method of quantitative risk assessment based on grid difference of pipeline sections *Safety Science* **59** 219-226

[21] Cheng C W, Yao H Q and Wu T C 2013 Applying data mining techniques to analyze the causes of major occupational accidents in the petrochemical industry *Journal of Loss Prevention in the Process Industries* **26(6)** 1269-1278

[22] Zhao Y, Jianbo L and Lu L 2011 Numerical modeling of fires on gas pipelines *Applied Thermal Engineering* **31(6-7)** 1347-1351

[23] Batzias F A, Siontorou C G and Spanidis P M P 2011 Designing a reliable leak bio-detection system for natural gas pipelines *Journal of Hazardous Materials* **186(1)** 35-58

[24] Kremcheev E A, Gromyka D S and Nagornov D O 2018 Techniques to determine spontaneous ignition of brown coal *Journal of Physics: Conference Series* **1118(1)** 012021

[25] Vianello C and Maschio G 2011 Risk analysis of natural gas pipeline: Case study of a generic pipeline *Chemical Engineering Transactions* **24** 1309-1314

[26] Huang X, Li B, Peng S and Zhang J 2010 Assessment methods of failure probability on gas pipelines *Shiyou Xuebao/Acta Petrolei Sinica* **31(4)** 664-667

[27] Brito A J and de Almeida A T 2009 Multi-attribute risk assessment for risk ranking of natural gas pipelines *Reliability Engineering and System Safety* **94(2)** 187-198
[28] Caputo A C and Pelagagge P M 2003 Using neural networks to monitor piping systems *Process Safety Progress* **22(2)** 119-128
[29] Danilov A S, Smirnov Y D and Korelskiy D S 2016 Promising method of remote environmental monitoring of Russian oil and gas industry facilities *Neftyanoe Khozyaystvo - Oil Industry* **2(1)** 121-122
[30] Argimbaev K R 2016 Monitoring of the industrial process impact on the environment in an open pit *International Journal of Ecology and Development* **31(3)** 23-28
[31] Safina E B and Lebedeva O Yu 2016 Improvement of economic tools of promoting and inspiring environmental activities in the mining industry *Gornyi Zhurnal* **6** 67-72
[32] Bazhin V Y and Titov O V 2017 Intelligent monitoring system of temperature condition of the coke furnace *Ugol'* **2** 50-53
[33] Kozmenko S, Teslya A and Fedoseev S 2018 Maritime economics of the Arctic: Legal regulation of environmental monitoring *IOP Conference Series: Earth and Environmental Science* **180(1)** 012009
[34] Vvedenskij R V, Gendler S G and Titova T S 2018 Environmental impact of the tunnel construction *Magazine of Civil Engineering* **79(3)** 140-149
[35] Sarapulova G I 2018 Environmental Geochemical Assessment of Technogenic Soils *Journal of Mining Institute* **234** 658-662
[36] Bolshunova T S, Rikhvanov L P, Mezhibor A M, Baranovskaya N V and Yusupov D V 2017 Biogeochemical features of epiphytyc lichens from the area of the tailing of a gold-polymetallic deposit (Kemerovo region, Russia) comparative to a reference area *International Multidisciplinary Scientific GeoConference Surveying Geology and Mining Ecology Management, SGEM* 17(51) 165-172
[37] Yusupov D V and Karpenko Yu A 2016 REE, Uranium (U) and Thorium (Th) contents in Betula pendula leaf growing around Komsomolsk gold concentration plant tailing (Kemerovo region, Western Siberia, Russia) *IOP Conference Series: Earth and Environmental Science* **43(1)** 012053
[38] Alekseev V I, Polyakova E V, Machevariani M M and Marin Y B 2014 Evolution of zircons from postorogenic intrusive series with Li-F granites, Russian Far East *Geology of Ore Deposits* **56(7)** 513-530
[39] Bushuev Y Y, Leontev V I and Machevariani M M 2018 Geochemical features of Au-Te epithermal ores of the Samolazovskoye deposit (Central Aldan ore District, Yakutia) *Key Engineering Materials* **769 KEM** 207-212
[40] Beloglazov I 2018 Automation experimental studies of grinding process in jaw crusher using DEM simulation *Journal of Physics: Conference Series* **1118** 012007

Study of electronic and optical-mechanical levels

Yu I Pimshin[1], Yu V Zayarov[1] and D M Arsenyev[2]

[1]National Research Nuclear University "MEPhI"
[2]Don State Technical University DSTU

E-mail: YIPpimshin@mephi.ru, YVZayarov@mephi.ru, dima81-07@mail.ru.

Abstract. The article presents the study of electronic and optical-mechanical levels. It describes the vertical comparator and the method of its use in the study of levels. In addition, a variant of the universal comparator which improves stationary capabilities is proposed. In the study of levels the errors in measuring exceedances at specified working distances are determined, including a description of the procedure for determining errors of undercompensation of the level slope and the total error of the course correctness of the telescope focusing lens and the errors of applying divisions of the range poles for optical levels, and scales with Rab-code divisions for electronic ones. The article is written as part of the implementation of the NRNU MEPhI Competitiveness Enhancement Program.

1. Introduction

Describing the current state of the measuring equipment, it is necessary to note its rapid development. There is an update of measuring instruments, using new physical methods, principles of measurements, means of microelectronics and software every five, maximum ten years. However, a common feature of a new technology is its high cost compared to previous generations. Therefore, the update technology is somewhat behind the pace of its development. Production tasks are tightened in the requirements for accuracy of measuring, in its turn, without focusing on the difficulties, the complexity of the of new measuring equipment introduction. Under these conditions, there is a previously developed approach used for a long time to improve measurement accuracy. It is a study of existing equipment, the definition of achievable, accuracy-limiting characteristics and the use of the accuracy of devices at the limit when solving production problems. It almost always gives a positive result to solve high-precision measurement tasks and to provide some time to purchase new high-precision equipment when its cost is more affordable.

The levels are instruments designed to measure elevations. Today, this type of devices is widely used in geodesy, construction, installation, heavy mechanical engineering, etc. The problem of providing high and precise accuracy remains relevant for these devices [1–4].

The research carried out to increase the accuracy of measurements when leveling remains an urgent issue today.

2. Theoretical basis of the proposed method

The proposal of the technique is based on the use of a vertical comparator.

Figure 1 shows an example of a vertical comparator circuit, where S_1, S_2 is the distance from the instrument station to the first and last fixed point respectively, and $S_1 = S_2$ in this comparator. There is horizontal distance between the points of the comparator B_1, B_2, B_3, B_4, B_5, B_6, B_7, B_8 where $B_1= B_2= B_3= B_4= B_5= B_6= B_7= B_8$. Accuracy of performance of all distances of the comparator is ±10 mm. Readings on the rail by level, installed at the lower station are $A_{H1}...A_{H9}$ and readings on the rail by level, installed at the top station are $A_{B1}...A_{B9}$. Moreover

$$\begin{cases} A_{H1}= a_{H1}+d_{H1},\ \dots,\ A_{H9}= a_{H9}+d_{H9}; \\ A_{B9}= a_{B9}+d_{B9},\ \dots,\ A_{B1}= a_{B1}+d_{B1}; \end{cases} \tag{1}$$

where $a_{H1},\dots,a_{H9}$; $a_{B1},\dots,a_{B9}$ are values of theoretical readings on rails corresponding to the horizontal position of the sighting axis;

$d_{H1},\dots,d_{H9}$; $d_{B1},\dots,d_{B9}$ are errors caused by undercompensation of the angle of level inclination.

Whereby

$$\begin{cases} \Delta_{dH} = d_{H2} - d_{H1}, \\ \Delta_{dB} = d_{B8} - d_{B9}. \end{cases} \tag{2}$$

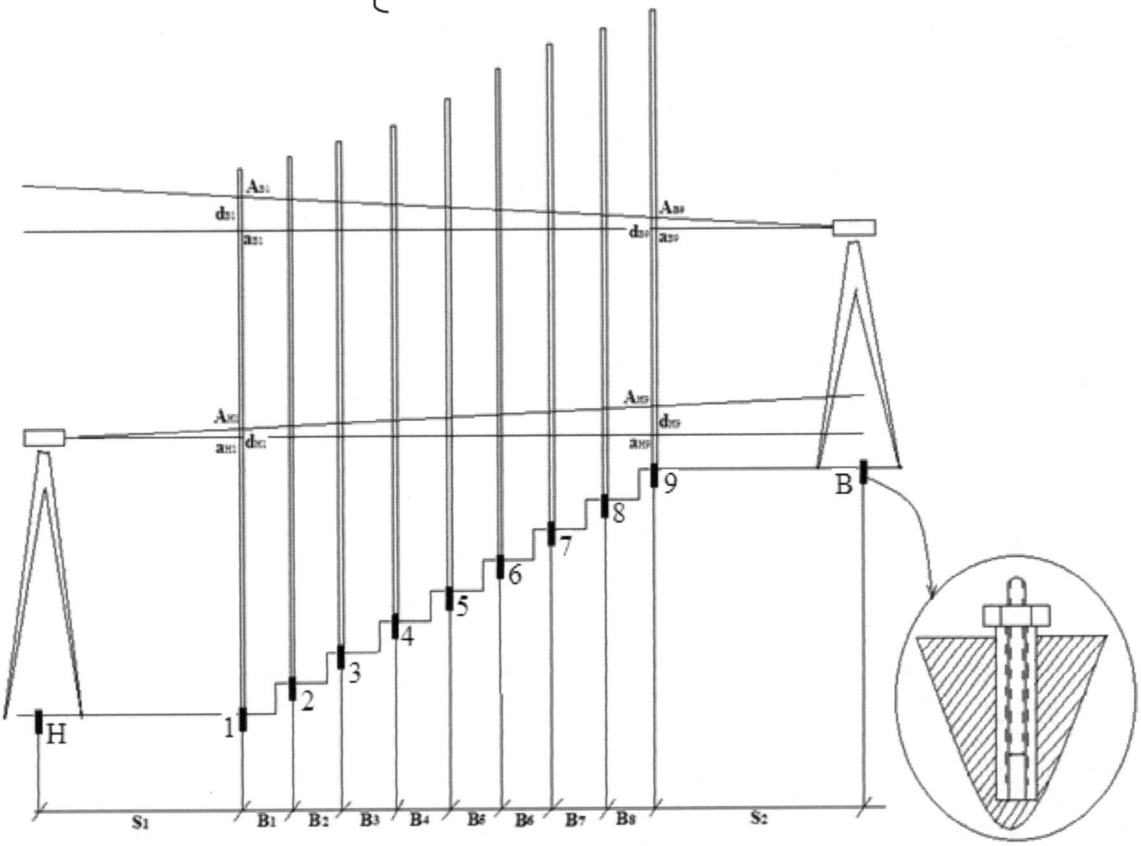

Figure 1. Step vertical comparator diagram

Based on the constructive condition of the comparator $B_1= B_2= B_3= B_4= B_5= B_6= B_7= B_8$ и $S_1= S_2$ we have $d_{H1}= d_{B9}$, $\Delta_{dH}= \Delta_{dB}$, then

$$\begin{cases} d_{H2}= d_{H1}+ \Delta_{dH},\ d_{H3}= d_{H1}+2\cdot\Delta_{dH},\ d_{H4}= d_{H1}+ 3\cdot\Delta_{dH},\ \dots,\ d_{H9}= d_{H1}+ 8\cdot\Delta_{dH}; \\ d_{B8}= d_{B9}+ \Delta_{dB},\ d_{B7}= d_{B9}+ 2\cdot\Delta_{dB},\ d_{B6}= d_{B9}+ 3\cdot\Delta_{dB},\dots,\ d_{B1}= d_{B9}+ 8\cdot\Delta_{dB}. \end{cases} \tag{3}$$

In this case, the readings by level $A_{H1}...A_{H9}$ and $A_{B1}...A_{B9}$ will be defined as:

$$\begin{cases} A_{H1}= a_{H1}+d_{H1},\ A_{H2}= a_{H1}+d_{H1}+ \Delta_{dH},\ A_{H2}= a_{H1}+d_{H1}+2\cdot\Delta_{dH},\ \dots,\ A_{H9}= a_{H9}+ d_{H1}+ 8\cdot\Delta_{dH}; \\ \end{cases} \tag{4}$$

$A_{B1}= a_{B9}+ d_{B9}+ 8\cdot\Delta_{dB}, \ldots, A_{B7}= a_{B9}+ d_{B9}+ 2\cdot\Delta_{dB}, A_{B8}= a_{B9}+ d_{B9}+ \Delta_{dB}, A_{B9}= a_{B9}+d_{B9}.$

Then the total count at each point of the vertical comparator will be equal to:

$$\begin{cases} A_1= (a_{H1}+ a_{B9})+(d_{H1}+d_{B9}+ 8\cdot\Delta_{dB}); \\ A_2 = (a_{H1}+ a_{B9})+(d_{H1}+ d_{B9}+\Delta_{dH}+7\cdot\Delta_{dB}); \\ A_3 = (a_{H1}+ a_{B9})+(d_{H1}+ d_{B9}+2\cdot\Delta_{dH}+6\cdot\Delta_{dB}); \\ A_4 = (a_{H1}+ a_{B9})+(d_{H1}+ d_{B9}+3\cdot\Delta_{dH}+5\cdot\Delta_{dB}); \\ A_5 = (a_{H1}+ a_{B9})+(d_{H1}+ d_{B9}+4\cdot\Delta_{dH}+4\cdot\Delta_{dB}); \\ A_6 = (a_{H1}+ a_{B9})+(d_{H1}+ d_{B9}+5\cdot\Delta_{dH}+3\cdot\Delta_{dB}); \\ A_7 = (a_{H1}+ a_{B9})+(d_{H1}+ d_{B9}+6\cdot\Delta_{dH}+2\cdot\Delta_{dB}): \\ A_8 = (a_{H1}+ a_{B9})+(d_{H1}+ d_{B9}+7\cdot\Delta_{dH}+\Delta_{dB}); \\ A_9 = (a_{H1}+ a_{B9})+(d_{H1}+ d_{B9}+8\cdot\Delta_{dH}). \end{cases} \quad (5)$$

The excess at each stage of the vertical comparator is determined from the expressions:

$$h_{1\text{-}2} = (A_1\text{-}A_2)/2,\ h_{2\text{-}3} = (A_2\text{-}A_3)/2,\ldots,\ h_{7\text{-}8} = (A_7\text{-}A_8)/2,\ \ h_{8\text{-}9} = (A_8\text{-}A_9)/2. \quad (6)$$

The component of undercompensation of the angle of level inclination installed at the lower station is determined by the formulas:

$$\delta_{H1\text{-}2} = (A_{H2}+ h_{1\text{-}2})\text{-}A_{H1},\ \delta_{H1\text{-}3} = (A_{H3}+ h_{1\text{-}2}+ h_{2\text{-}3})\text{-}A_{H1},\ \delta_{H1\text{-}4} = (A_{H4}+ h_{1\text{-}2}+ h_{2\text{-}3}+ h_{3\text{-}4})\text{-}A_{H1},\ldots,$$

$$\delta_{H1\text{-}9} = (A_{H9}+ h_{1\text{-}2}+ h_{2\text{-}3}+ h_{3\text{-}4}+ h_{4\text{-}5}+ h_{5\text{-}6}+ h_{6\text{-}7}+ h_{7\text{-}8}+ h_{8\text{-}9})\text{-}A_{H1},$$

The component of undercompensation of the angle of level inclination installed at the top station is determined by the formulas:

$$\delta_{B9\text{-}8} = (A_{B9}+ h_{8\text{-}9})\text{-}A_{B8},\ \delta_{B9\text{-}7} = (A_{B9}+ h_{7\text{-}8}+ h_{8\text{-}9})\text{-}A_{B7},\ \delta_{B9\text{-}6} = (A_{B9}+ h_{6\text{-}7}+ h_{7\text{-}8}+ h_{8\text{-}9})\text{-}A_{B6},\ldots,$$

$$\delta_{B9\text{-}1} = (A_{B9}+ h_{1\text{-}2}+ h_{2\text{-}3}+ h_{3\text{-}4}+ h_{4\text{-}5}+ h_{5\text{-}6}+ h_{6\text{-}7}+ h_{7\text{-}8}+ h_{8\text{-}9})\text{-}A_{B1}.$$

The analysis of the total error of the course correctness of the focusing lens and the errors of applying divisions of the range poles for optical levels, and scales with Rab-code divisions for electronic ones, are carried out by examining for equality of values $\Delta_{H2\text{-}3},\ldots,\Delta_{H8\text{-}9}$ and $\Delta_{B7\text{-}8},\ldots, \Delta_{B8\text{-}9}$.

$$\Delta_{H2\text{-}3} = \delta_{H1\text{-}3}\text{-}\ \delta_{H1\text{-}2}, \Delta_{H3\text{-}4} = \delta_{H1\text{-}4}\text{-}\ \delta_{H1\text{-}3},\ \Delta_{H4\text{-}5} =\delta_{H1\text{-}5}\text{-}\ \delta_{H1\text{-}4},\ \Delta_{H5\text{-}6} = \delta_{H1\text{-}6}\text{-}\ \delta_{H1\text{-}5},\ldots,\ \Delta_{H8\text{-}9} = \delta_{H1\text{-}9}\text{-}\ \delta_{H1\text{-}8}.$$

$$\Delta_{B7\text{-}8} = \delta_{B9\text{-}7}\text{-}\ \delta_{B9\text{-}8},\ \Delta_{B6\text{-}7} = \delta_{B9\text{-}6}\text{-}\ \delta_{B9\text{-}7},\ \Delta_{B5\text{-}6} = \delta_{B9\text{-}5}\text{-}\ \delta_{B9\text{-}6},\ \Delta_{B4\text{-}5} = \delta_{B9\text{-}4}\text{-}\ \delta_{B9\text{-}5},\ldots,\ \Delta_{B8\text{-}9} = \delta_{B9\text{-}1}\text{-}\ \delta_{B9\text{-}2}.$$

Thus, the levels are studied with the definition of corrections to the values of readings at given distances. The introduction of amendments allows increasing the accuracy of measurements. At the same time, it is possible to examine levels at different working distances by changing S_1 and S_2.

The vertical comparator can be made in a universal version, see Fig.2. The design of the comparator can be represented as the main frame of the comparator (1) on which the frame (2) is installed, with the possibility of lifting it. The steps (3) are permanently fixed on the frame (2), with a uniform pitch ($B_1= B_2= B_3= B_4= B_5= B_6= B_7= B_8= B_9$). Using the stop (4) and the universal ram pin (6), the length of which is determined by the sum of the lengths of the parts of its components, the frame (2) can be rotated relative to the frame of the comparator (1) at a given angle. The main frame of the comparator (1) is adjusted to a horizontal position with the help of lifting screws (5). The total height *h* of the triangle frame of the comparator is equal to

$$h^2 = H_b^2 + \Delta^2, \quad (7)$$

where

$$\Delta = \frac{H_b^2 - B^2 + L_b^2}{2 \cdot L_b^2}$$

H_b is total length of universal ram pin;

Lb is the length of the triangle base of the comparator frame;

$$B = \sum_{i=1}^{9} B_i.$$

Then the *h* excess will be equal to $h_{i\text{-}(i+1)} = (B/9)$ at each step of the comparator.

All other stages of the implementation of research levels correspond to the above.

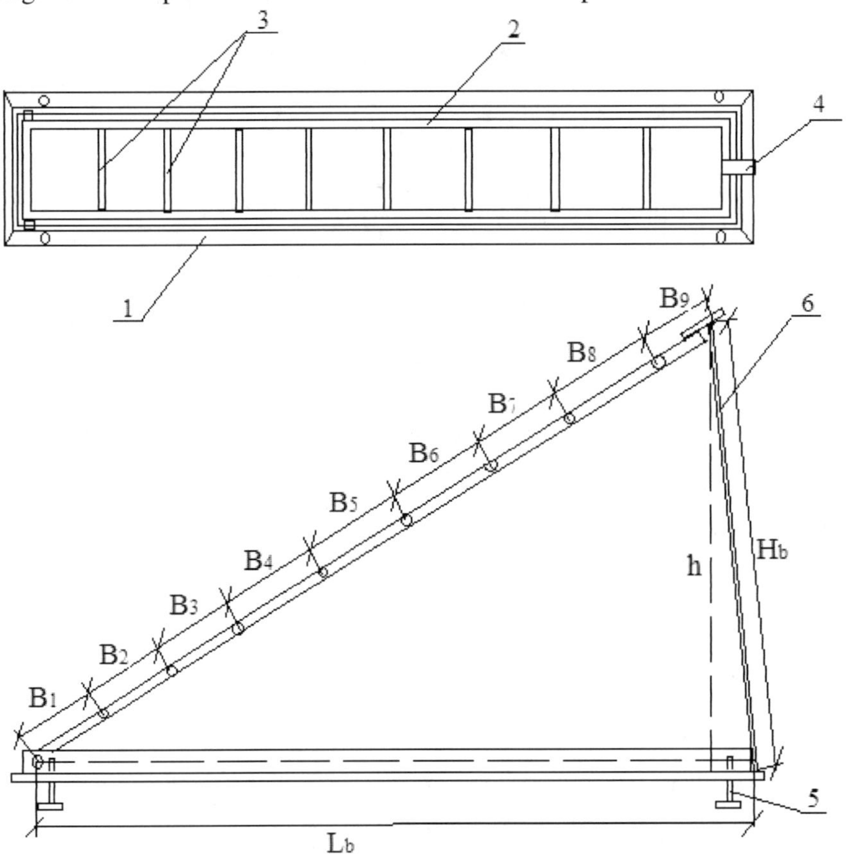

Figure 2. Universal step vertical comparator: 1 – Main Comparator frame; 2 – Frame steps; 3 – Steps of the comparator; 4 – Stop; 5 – Lifting screws; 6 – Stacked stock

3. Conclusion

The use of vertical comparators in the study of optomechanical and electronic levels provides a definition of the limiting instrumental accuracy at given working distances. A reliable knowledge of instrumental accuracy allows providing a reliable solution to production issues for determining exceedances or for controlling the horizontal position of elements of controlled objects. At the same time the implementation of the described methodology and the means of ensuring it do not require significant capital investments of the organization using them.

References

[1] SCNP (GNTA) 17-195-99 *Instructions for carrying out technological calibration of geodetic instruments* (Moscow: Scientific and Technical Center of Geodesy, Cartography)

[2] Pimshin Yu I 1993 *The method of checking the level* Patent 2003042, G 01 C 5/00, published 11/15/93, bull. no 41–42

[3] Staroverov S V 2018 *Development of compact means of geodesic metrology for operational calibration and research of levels and tacheometers* (Moscow: Geodesy) 24 p

[4] Yambaev Kh K, Golygin N Kh, Travkin S V and Stepochkin A A 2008 *The stand for checking and calibration of levels and reek* Patent 2341772, G01C 5/00 (2006.01), publ. 20.02.2008, Bull. no 35

Radio interferometric location finding of VLF signal transmitters

A S Poletaev[1], **D A Chensky**[1], **A G Chensky**[1]

[1]Irkutsk National Research Technical University, 83, Lermontov Street, Irkutsk, 664074, Russia.

E-mail: sardaukar9@yandex.ru

Abstract. The article presents some experimental results of using a very low frequency (VLF) interferometer designed in INRTU for radio direction finding of VLF transmitters. It is evaluated that amplitude angular-position measurements provides azimuth accuracy of several degrees. Distant VLF stations (3000 – 5500 km) can be located in a 2-6 km area if the signal azimuth is known with 0.001° accuracy. In order to minimize errors associated with antenna's angular offset the authors developed a correcting algorithm. Using both radio direction finding methods and additional geophysical data (year and daily variations of solar terminator moving, signal disturbances during solar flares) the authors ascertained that 24.1 kHz and 25.0 kHz signals are emitted from the transmitter in Mokpo (South Korea). The phase measurements give about two times lower error of 1.5–2° for the NWC station (19.8 kHz, Australia) which is located more than 8200 km away from the VLF interferometer.

1. Introduction

Very low frequency (VLF, 3-30 kHz) radio waves are commonly used for radio navigation and communications with ships and immersed submarines, for transmitting standard time and frequency signals. These signals propagate over long distances via reflecting from the lower ionospere and the Earth surface which makes the use VLF sounding as an effective metod for remote geophysical diagnostics of pocesses in D and E layers in the ionophere [1]. VLF radio interferometry method can be also used for electrical exploration of different geological structures [2]. Many geophysical researches require polarization separation of radio wave components and estimation of signal reception azimuth.

2. Interferometric measurements

For long term monitoring of the ionospere we designed in INRTU a VLF receiver [3, 4] with special algoritm for amplitude and phase measuaremes of MSK, GMSK and CPFSK signals [5]. Currently a three element VLF interferometer is deployed near Lake Baikal as it is shown in the Figure 1. The bases of the interferometer form a rectangular triangle with 16.891 km, 30.946 km, and 35.818 km sides. The red lines indicate expected VLF signals propagation paths along the great circle. The transmitters are located 3000 – 8600 km away from the receivers, a list of operational VLF stations is availibale for instance in [6]. If the signal reception azimuth is known we can apply the triangulation method [7] to find the transmitter location.

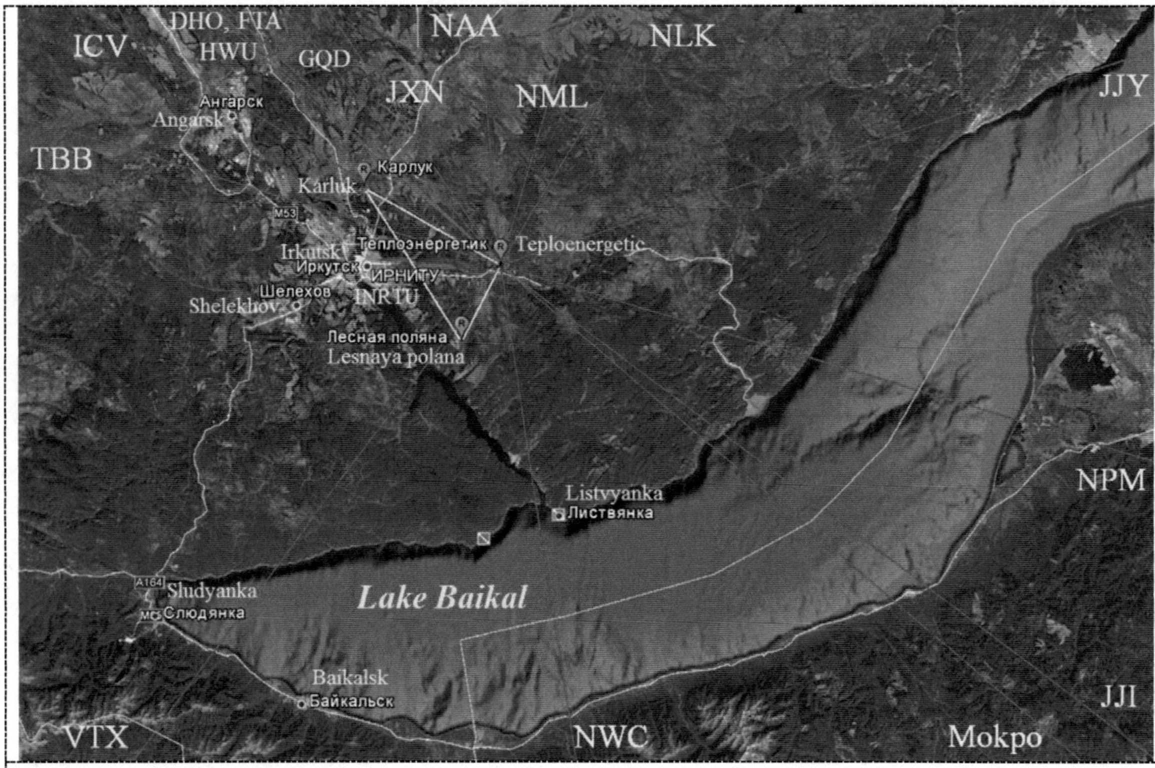

Figure 1. Configuration of location of the VLF interferometer and expected VLF signals arrival paths.

In the Table 1 the calculated sides of sperical triangles are shown for JJI and VTX transmitters. As it can be seen long distances 3000 – 5500 km require azimuth acurancy of 0.001° to get 1-3 km erros. There are two common techniques for receiving signal azimuth estimation: amplitude and phase radio location finding methods.

3. Amplitude method of VLF transmitter location finding

This method is based on using signal amplitude measurements with directional antennas. Each of our VLF receivers have two orthogonal magnetic 80 cm round loops with 36 turns of copper 0.5 mm^2 wire. One of loop antennas is oriented in the north-south (NS) direction, the othe one is for west-east (WE). The azimuth can be found as $\operatorname{tg} A = B_{WE} / B_{NS}$. There is a indeterminacy of the angle quarter as loop antenna directional diagram has two maxima and amplitudes are always positive. However it can be evaluated if an approximate direction is known.

The Figure 2 shows some detection results for a 24.1 kHz signal. At night and especially in the evening (dashed lines) the signal amplitude has deep interferometric minima which lead to significant azimuth errors. In the daytime this polarization effect is less prominent as the radio signal propagates

Table 1. Results of estimation of VLF transmitter location via triangulation method.

Station	Teploenergetic (T)		Lesnaya polana (L)		Karluk (K)	
	map	calculated L_1 (km)	map	calculated L_2 (km)	map	calculated L_3 (km)
JJI	126.572° 3077.289 km	**3074.376 (L)** 3103.782 (K)	126.198° 3074.451 km	**3071.537 (T)** 3069.691 (K)	126.186° 3108.964 km	3134.457 (T) 3103.205 (K)
VTX	216.373° 5455.164 km	5469.392 (L) 5463.184 (K)	216.307° 5438.511 km	5452.739 (T) **5439.467 (K)**	215.817° 5451.228 km	5459.248 (T) **5452.183 (K)**

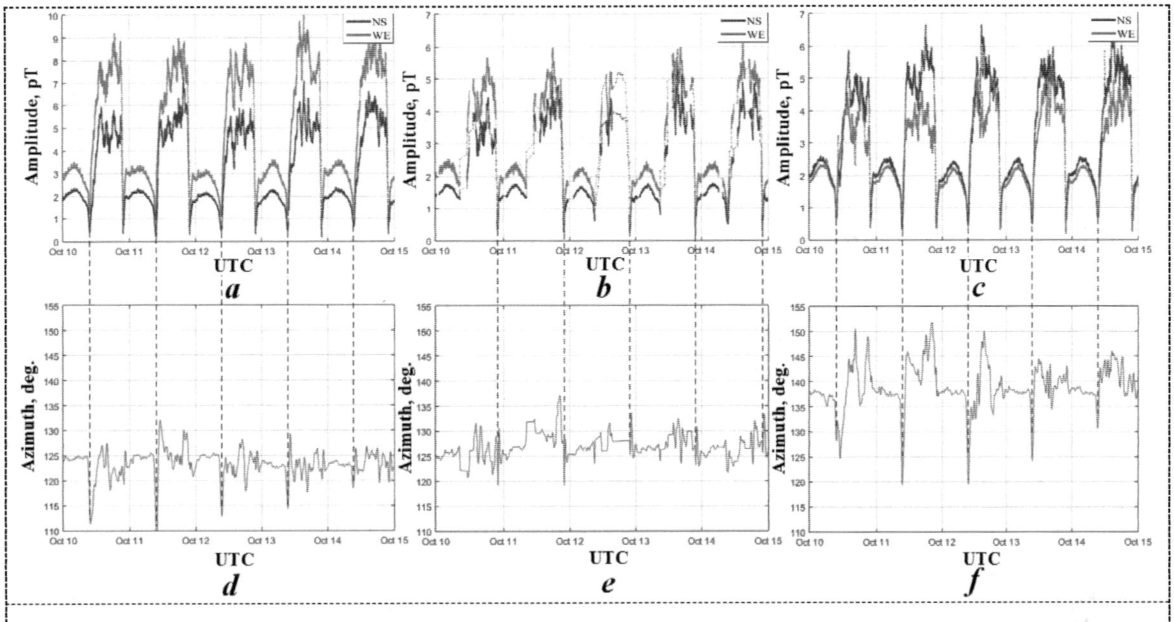

Figure 2. Amplitude and azimuth of 24.1 kHz radio signal received in Teploenergetic (a, d), Lesnaya polana (b, e) and Karluk (c, f)

primarily in the one-mode conditions [8]. The round circles in the Figure 4 show azimuths measured with day amplitudes for several VLF transmitters and their evaluation errors. One of reasons why the absolute errors reach 10° is antennas angular offset. The actual position of loops' planes is not strictly orthogonal and their orientation is slightly different from NS-WE directions.

4. Correction of antennas' angular offset

Let ε be the angle between the ideal and real NS antenna positions and ξ represents angular offset of the WE antenna. These variables are positive for counter clockwise shifts and negative when it is clockwise. The Poynting vector and the OX axis (see the Figure 3) form signal reception angle $\varphi = \varphi' - \xi$. The amplitudes ratio is given then as

$$K' = B'_{NS} / B'_{WE} = \sin(\varphi - \varepsilon) / \cos(\varphi - \xi) . \tag{1}$$

From the (1) let's express ε and write the equation for two signals with φ_1 and φ_2 arrival angles

$$\varphi_2 - \varphi_1 = \arcsin\left[K'_1 \cos(\varphi_1 - \xi) \right] - \arcsin\left[K'_2 \cos(\varphi_2 - \xi) \right] . \tag{2}$$

The equation (2) can be solved graphically or using numerical methods. When ξ is known then

$$\varepsilon = \varphi + (-1)^n \arcsin\left[K' \cos(\varphi - \xi) \right] + \pi n,\ n \in \mathbb{Z} \tag{3}$$

forms a number of solutions for ε. The angles ε and ξ can be found more precisely for different pairs of signals φ_1 and φ_2. Finally the signal reception angle with correction is expressed as

$$\operatorname{tg} \varphi = \frac{\sin \varepsilon + (\pm K') \cos \xi}{\cos \varepsilon - (\pm K') \sin \xi} , \tag{4}$$

where (+K′) is used for the signals in the I and III coordinate plane quarters, and (–K′) is for the II and IV quarters. The antennas in Teploenergetic have proved to have $\xi_T = -12.1°$ and $\varepsilon_T = -6.3°$ angular offsets. The corrected azimuths in the figure 4 have residual errors of 2–3°. The location finding has shown that 18.3 kHz and 21.75 kHz signals are belong to the same transmitter (HWU). The 24.1 kHz signal source is not known, some studies [9, 10] refer to the east seaboard of China. But its azimuth is close to the direction at the 25.0 kHz transmitter in Mokpo, South Korea (128.235° and 128.883° respectively in Teploenergetic, 127.830° and 128.817° in Lesnaya polana). The amplitude variations of these signals during 2017 year are compared in the Figure 5.

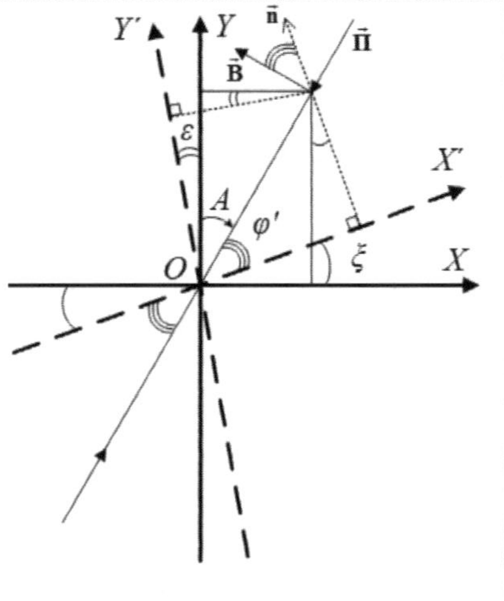

Figure 3. Estimation of angular offset of reception antennas for signals in the I coordinate quarter.

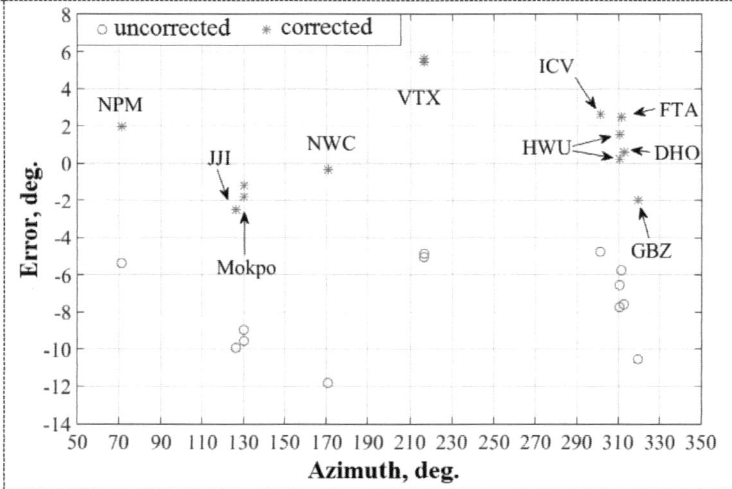

Figure 4. Residual errors of location finding in Teploenergetic before (circles) and after (asterisks) correction of antennas angular offsets $\xi_T = -12.1°$ and $\varepsilon_T = -6.3°$.

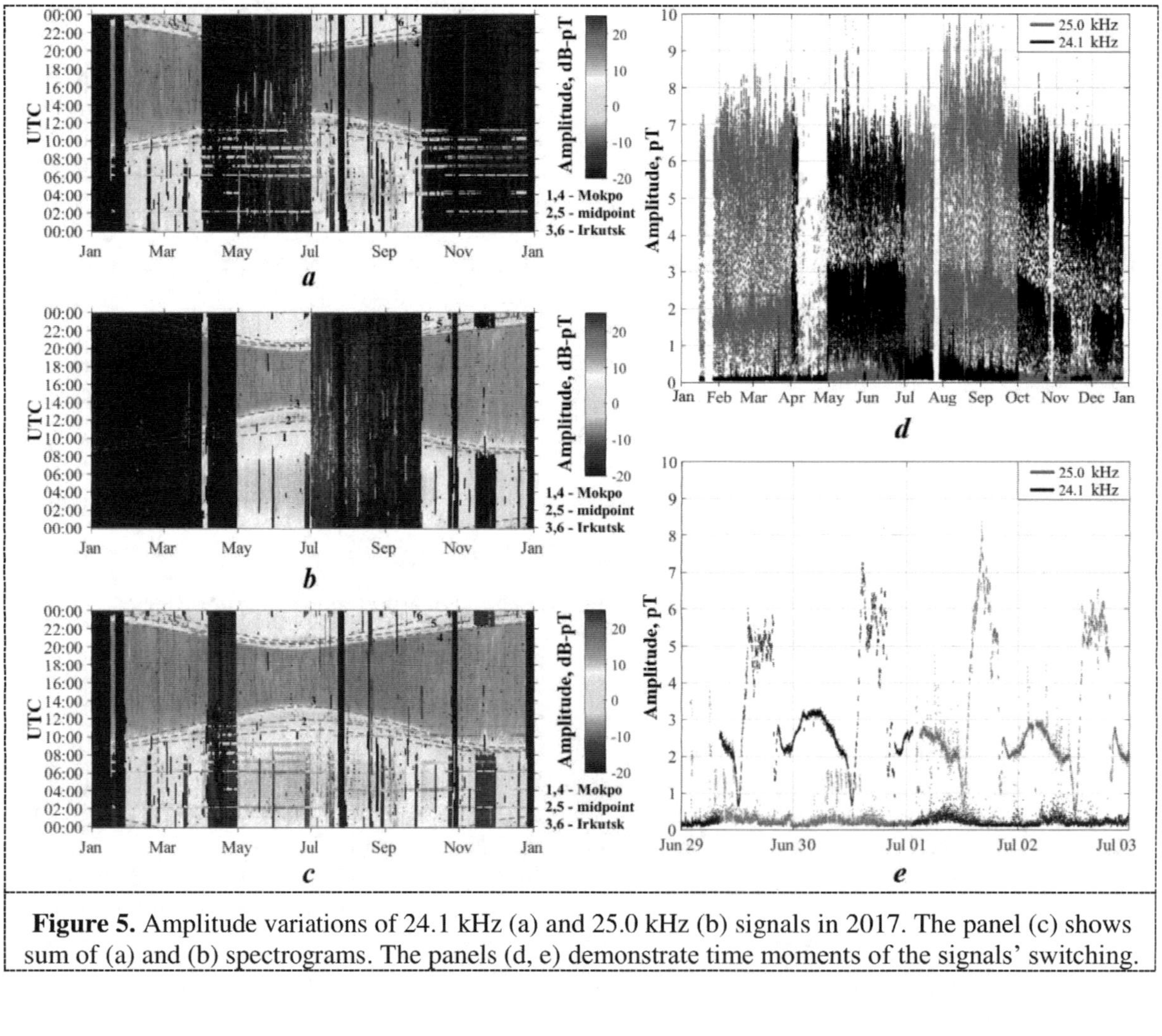

Figure 5. Amplitude variations of 24.1 kHz (a) and 25.0 kHz (b) signals in 2017. The panel (c) shows sum of (a) and (b) spectrograms. The panels (d, e) demonstrate time moments of the signals' switching.

As it can be seen from the figure 5 the 24.1 kHz and 25.0 kHz signals have the same power of radiation, their annual amplitude variations are described with identical cycles of solar terminator mooving. The analysis of amplitude distrurbances during solar flares confirm that the transmitter is located in the east from the VLF interferometer.

5. Phase angular-position measurements

The phase method of location finding requires identical heliogeophysical conditions of VLF signal paths. By this reason interferometer should have short baselines [11]. The angle between an interferometer baseline ℓ and normal to the signal path are found as

$$\alpha = \arcsin\left(\frac{\lambda \cdot \Delta\varphi}{2\pi\ell}\right) = \arcsin\left(\frac{kc \cdot \Delta\varphi}{2\pi f \ell}\right), \qquad (5)$$

where $\Delta\varphi$ is phase difference for two receivers (rad), f is signal frequency (Hz); $\lambda = \upsilon / f$ is wavelength (m); $\upsilon = kc$ – mean phase velocity (m/s); c – speed of light (m/c); k – wavelength shorting coefficient, experimental reviews in [12] adduce for VLF range $k \approx 0.9978... 0.9985$ in the daytime and $k \approx 0.9951... 0.9974$ at night.

This technique was applied to find the direction of wave surface propagating for NWC signal (Australia, 19.8 kHz). The scheme is presented in the Figure 6 while the results of phase measurements are shown in the Figure 7. The phase method has demonstrated errors less then 2° which is a good result for such a distant transmitter location (8260 km).

6. Conclusion

The location finding of VLF signal sources is useful for geophysical researches and in some practical applications. The triangulation method requires 10^{-3} degree accuracy for short baseline interferometry and distant VLF stations (3-5 thousand km). The amplitude technique of azimuth evaluation has shown residual errors of 2–3° when the correction algorithm is applied. The phase angular-position measurements are more accurate, the errors are less than 2°. Using both radio location finding method and geophysical data about signal propagation conditions have helped to identify that 25.0 kHz and 24.1 kHz signals are radiated by the same transmitter located in Mokpo, South Korea.

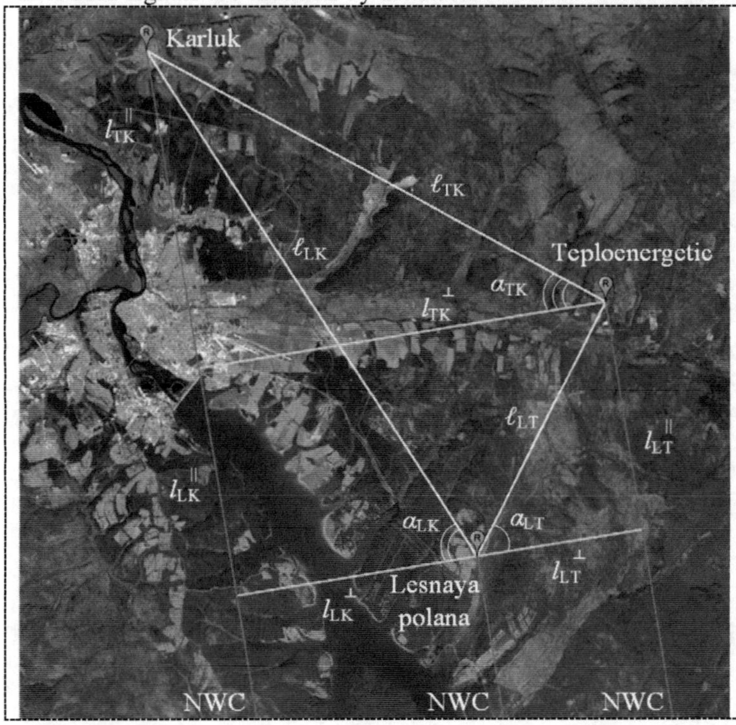

Figure 6. A scheme of wave front arrival for NWC signal (Australia, 19.8 kHz). The side of triangles are $l_{TK}^{\parallel} = 19.145$ km, $l_{LK}^{\parallel} = 32.849$ km, $l_{LT}^{\parallel} = 13.704$ km, $l_{TK}^{\perp} = 24.347$ km, $l_{LK}^{\perp} = 14.398$ km, $l_{LT}^{\perp} = 9.949$ km. The angles are $\alpha_{TK} = 38.179°$, $\alpha_{LK} = 66.332°$, $\alpha_{LT} = 54.021°$. The expected signal azimuths are $A_T = 170.879°$, $A_L = 170.766°$, $A_K = 170.502°$. The measured values are $A_T = 169.167°$, $A_L = 166.201°$, $A_K = 171.323°$ which mean errors $\varepsilon_T = -1.71°$, $\varepsilon_L = -4.57°$, $\varepsilon_K = 0.82°$ respectively. Measured wave front delays between receivers are $\tau_{TK} = 64.0$ μs, $\tau_{LT} = 45.7$ μs, $\tau_{LK} = 109.8$ μs. As the sampling interval is 10 μs these delays mean correlation method is not effective because signals are shifted in time domain only to several samples.

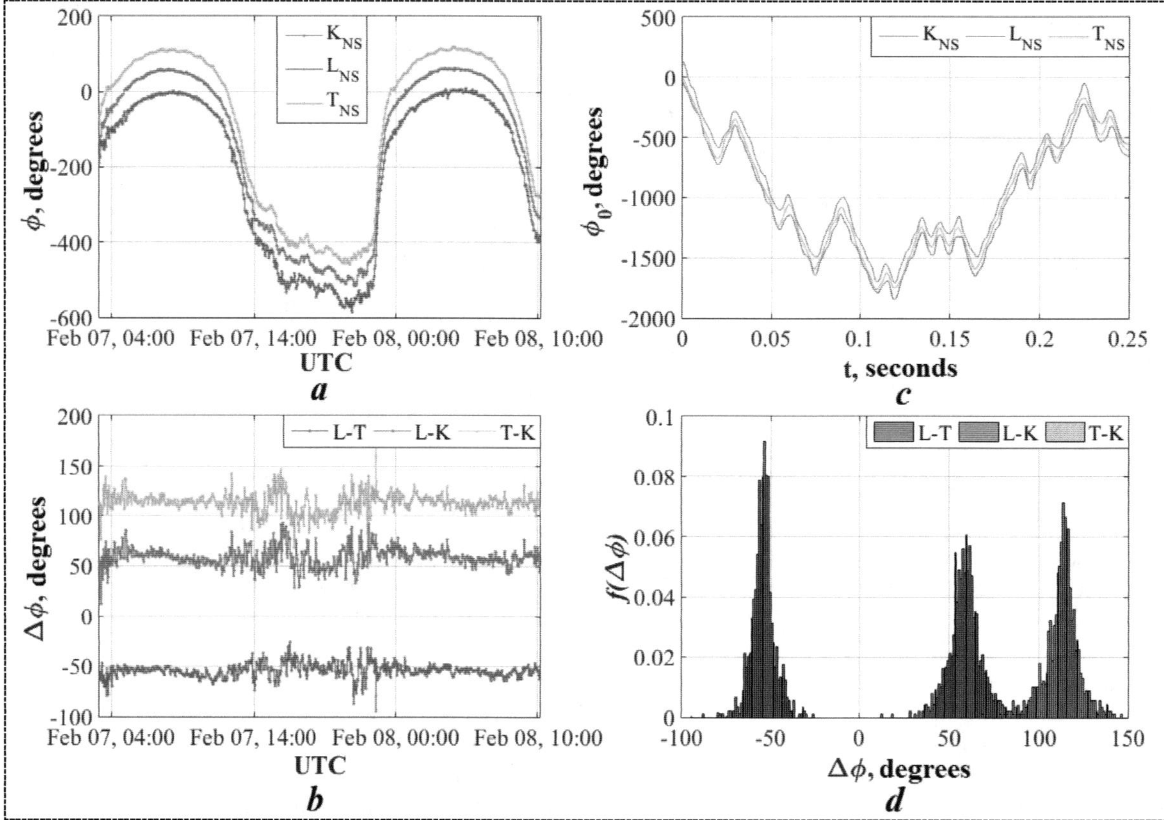

Figure 7. Results of phase measurements for 19.8 kHz signal (a), phase difference time diagram (b) and histogram (d). A 250 ms fetch (c) of the carrier phase simultaneously detected by all receivers.

7. Acknowledgments
The reported study was funded with the expense of grant No. 01-FPK-19 for the financial support of research and teaching staff of the INRTU.

References
[1] Shafer D C 1994 *Spread Spectrum VLF Remote Sensing of the Ionosphere* (California: Stanford University) p 115
[2] Davydov V A 2014 *Engineering Survey* **2** 6570
[3] Poletaev A S, Chensky A G, Tokmachev D A 2018 *Sensors and Systems Journal* **6 (226)** 5160
[4] Tokmachev D A, Poletaev A S, Bezrukin A G, Chensky A G, Zasenko V E, Gubin N A 2014 *Instruments and Experimental Techniques* **5** 587-93
[5] Poletaev A S, Chensky A G, Tokmachev D A 2018 *14th Int.Scientific-Technical Conf. APEIE* vol 1, part 3 (Novosibirsk: NSTU) pp 101-4
[6] SID Monitoring Station 2019 URL: https://sidstation.loudet.org/stations-list-en.xhtml.
[7] Kozlov V I, Karimov R R, Mullayarov V A 2008 *Sovr. Probl. DZZ Kosm.* **1** 132-7
[8] Kumar S 2009 *Research Letters in Physics* **2009** 13
[9] SID Manual 2017 URL: http://solar-center.stanford.edu/SID/docs/VLF-stations.pdf.
[10] Reeve Observatory 2017 URL: http://www.reeve.com/Documents/Articles%20Papers/Reeve_VLF-LFStationList.pdf.
[11] Ryakhovskiy I A, Gavrilov B G, Ermak V M, Poklad Yu V 2016 *Trigger responces in geospheres* **8** 161-7
[12] Alpert Ya L 1972 Electromagnetic Waves Propagating and Inosphere (Moscow: Nauka) p 563

Journal of Physics: Conference Series **1384** (2019) 012043 doi:10.1088/1742-6596/1384/1/012043

Substantiated and interrelated technical tolerances needed in regulatory documents on the construction of automobile roads and their geodetic support

L A Pronina, Yu V Stolbov, A I Uvarov and N A Parkhomenko

FSBEO HE OmSAU

E-mail: la.pronina@omgau.org, yuv.stolbov@omgau.org, ai.uvarov@omgau.org, na.parkhomenko@omgau.org

Abstract. The necessity of a regulation of substantiated and interrelated technical tolerances through regulatory documents regarding the construction of automobile roads and their geodetic support is proven. The discrepancies in the requirements for the implementation of construction and geodetic works in the arrangement of the structural layers of road pavement foundations and coatings are pointed out. The regulation of accuracy standards in the construction of automobile roads, which does not correspond to the basic principles of probability theory, mathematical statistics and the theory of mathematical processing of measurements, is given as an example. In the case of exploitation of road-building machinery with automatic systems for the maintenance of leveling points, the execution of level runs with a class III accuracy is recommended, while in the case of exploitation of road-building machinery without automatic systems, the execution of level runs with a class IV accuracy is recommended.

1. Introduction

Nowadays in construction industry, the accuracy of the geometric parameters in the construction of buildings and structure is one of the main indexes of their quality.

In accordance with the requirements of GOST 21780-81 [1], during the design planning of buildings, structures and their individual elements, as well as during the development of technologies, the production of elements, and the construction of buildings and structures, the necessary means and rules of technological accuracy should be observed and applied in the production. The tolerances of the geometric parameters of the elements of buildings and structures are regulated in state technical standards (in Russian 'GOST'), construction norms and regulations ('SNiP'), sets of rules ('SP'), and construction and planning documentation. The main regulatory document for the construction and operational acceptance of automobile road since January 1, 1985 has been *SniP 3.06.03-85* [2], and since from July 1, 2015 its updated version, i.e. the set of rules *CP 78.13330.2012* [3].

At the present time, in case of new construction, the main regulatory document for operational acceptance of roads is *CP 78.13330.2012*, while in case of reconstruction of previously built roads, the current one remains *SniP 3.06.03-85* [2]. With the introduction of *CP 78.13330.2012*, accuracy regulations in the construction of automobile roads are significantly increased if compared with *SniP*

3.06.03-85. This particularly concerns the regulation of the accuracy standards of the altitude position (heights) of the structural layers of the road pavement bases and coatings.

If the assurance of the accuracy of the horizontal position did not cause particular difficulties in *SniP 3.06.03-85* and does not as well in *CP 78.13330.2012*, then to ensure the required accuracy of the altitude position it is necessary to regulate through regulatory documents substantiated tolerances for the deviation of heights from the design project values at all stages of the execution of geodetic works and construction of automobile roads. This is primarily related to the assurance of the evenness of the structural layers of the bases and coatings of automobile roads as one of the main indexes of the quality of their construction.

In the normative documents *SniP 3.06.03-85* and *CP 78.13330.2012*, the acceptable values of deviation from the project design ones for the construction-finished structural layers of bases and coatings are given without differentiation between them when performing geodetic and construction works. In order to ensure the regulated altitude position of the structural layers of the road pavement bases and coating, substantiated technological tolerances are required at the different stages of the construction of automobile roads.

2. Substantiation of the necessity of substantiated and interrelated technical tolerances in regulatory documents

According to the works [4–6], it is necessary to substantiate the norms on the accuracy of geodetic works: for the execution of level runs along the routes of automobile roads with the establishment (outside the area of excavation during their construction) of working benchmarks every 500 meters; the recovery of the leveling points (heights) of pickets on the road routes from the working benchmarks; the detailed staking of the leveling points (heights) of the surfaces of the structural layers of the road pavement bases and coatings; the geodetic control of the altitude position of the surfaces of the structural layers of bases and coatings for the operational acceptance and the quality assessment of their arrangement.

In the regulatory documents on road construction *SniP 3.06.03-85* and *CP 78.13330.2012*, the accuracy of the structural layers of the road pavement bases and coatings is regulated depending on the road category, the application of road-building machinery with and without an automatic system for the maintenance of leveling points (heights), leveling steps, while in *SniP 3.01.03-84*, Geodetic works in construction industry [7], and its updated edition *CP 126.13330.2012* [8], the accuracy of geodetic measurements is regulated taking into account undeveloped or built-up areas. In these regulatory documents a discrepancy between the requirements for the implementation of construction and geodetic works in the arrangement of the bases and coatings of automobile roads can be observed. In *CP 78.13330.2012*, the acceptable discrepancies of the altitude position of bases and coatings from design project values are regulated with a confidence probability of P=0.90, while in *CP 126.13330.2012*, their geodesic support is provided for with a control of P=0.95; 0,99 and 0,9973, where during the transition from the marginal errors to the RMS (divisor) and back from the RMS to the marginal (multiplier) respectively t=2,0; 2,5 and 3,0. Therefore, reasonable as well as interrelated tolerances for construction and geodetic works in the arrangement of bases and coatings of automobile roads are needed.

In the regulatory documents, a regulation of the norms of accuracy in the construction of automobile roads which does not correspond to the main provisions of probability theory, mathematical statistics, and the theory of mathematical processing of measurements can be observed [9, 10]. It is a well-known fact that during the erection of buildings and structures, the discrepancies (errors) of geodetic and construction works are basically subject to the law of normal distribution. The examples of incosistency in the regulation in *SniP 3.06.03-85* of the acceptable deviation of the altitude position (heights) of the surfaces of the bases and coatings in their arrangement with the exploitation of road-building machinery with and without systems for the maintenance of leveling points are reported in the work [11].

The same situation can be observed in *CP 78.13330.2012*, too, as regards the regulation of the acceptable deviation of the altitude position (heights) of the surfaces of the bases and coatings of automobile roads.

For example, in the arrangement of the bases and coating of automobile roads with the exploitation of road-building machinery without an automatic system for the maintenance of leveling points (heights), in *CP 78.13330.2012* the accuracy norms are doubled if compared to *SniP 3.06.03-85*, and it contemplated that 90% of the deviations of heights can be as high as δ_n=± 25mm, while the remaining 10% can be as high as δ_n=± 50mm. Thus, the RMS errors of the altitude position of the surfaces of the structural layers ($m_n=\delta_n/t$, where t=1.645 with P=0.90) will have the values m_n=25/1.645=15.2mm, and the maximum acceptable deviations (limit error), according to the "three sigma rule", will be δ_{pd}=15.2·3=45.6mm, but not δ_n=±50mm (regulated in *CP 78.13330.2012*).

The substantiation of tolerances in the construction of automobile roads depends also on the methods applied in their calculation, that is it takes into account those factors that affect the accuracy of the technological processes for the arrangement of the structural layers of the road pavement bases and coatings. Technological processes in the arrangement of road pavements are characterized by the coefficients of their accuracy in the execution of construction and geodetic works, and hence by the confidence probabilities of the assurance their regulated tolerances from design project values as well.

In literature published until 2016 [12, 13, 14], it is indicated that during the construction of automobile roads the execution of level runs should be carried out by means of technical leveling. At the present time, electronic total stations and digital levels are widely used in the construction of automobile roads, and trials about their application and research are presented in [15-18].

The calculation methods for construction and geodetic works are presented in the works [4] taking into account the accuracy of the technological processes for the arrangement of the bases and coating of automobile roads, while in the works [5, 6] the substantiation and assurance of the accuracy of the detailed staking of the level points of the structural layers of the road pavement bases and coatings, the recovery of leveling points (heights) of pickets on the routes of automobile roads from working benchmarks, and the execution of level runs along the route or through the route are carried out.

According to the works [5, 6], in accordance with the regulation of acceptable deviation in the surface heights from design project values in *CP 78.13330.2012*, in the construction of roads with the application of road-building machinery with automatic systems for the maintenance of leveling points, the execution of level runs with a class III accuracy is necessary, while with the exploitation of road-building machinery without automatic systems the execution of level runs with a class IV accuracy is needed.

3. Conclusion

The necessity of the regulation of substantiated and interrelated technical tolerances in regulatory documents has been proven. The execution of level runs with a class III accuracy in the construction of roads in case of exploitation of road-building machinery with automatic systems for the maintenance of leveling points, and of level runs with a class IV accuracy in case of exploitation of road-building machinery without automatic systems has been proposed.

Reference

[1] GOST 21778-81 (ST SEV 2045-79). *System for ensuring the accuracy of geometric parameters in construction. The main provisions*(Moscow:Standards Publishing House)

[2] SNiP 1985 *SniP 3.06.03-85.Avtomobil'nye dorogi. Gosstroy SSSR*(Moscow : TsITP Gosstroya SSSR)

[3] SP 2012 *CP 78.13330.2012. Aktualizirovannaya redaktsiya SNiP 3.06.03-85. Avtomobil'nye dorogi*(Moscow: Minregion Rossii)

[4] Stolbov YuV, Stolbova SYu, Nagaev DO and Kokulenko KS2010The rationale for the tolerances on the construction and surveying work to ensure a high-altitude position of motor roads*Newsofuniversities "Construction"***9**75–80

[5] Stolbov YuV, Stolbova SYu, Pronina LA and Starovoitov IE2016Analysis of methods for calculating the tolerances for geodetic work in the survey and construction of roads to ensure their high-altitude position*Vestnik SibADI***5(51)**124–30

[6] Stolbov YuV, Stolbova SYu, Pronina LA and Starovoitov IE2016Ensuring the accuracy of the laying of leveling moves in the search and removal of pickets marks in the construction of roads*Vestnik SibADI***2(48)** 120–5

[7] SNiP 1985*SniP 3.01.03-84.Geodetic works in construction* (Moscow: Stroiizdat)

[8] SP 2012 *CP 78.13330.2012. Aktualizirovannaya redaktsiya SNiP 3.01.03-84. Geodetic works in construction*(Moscow: Minregion Rossii)

[9] Gmurman E V 1977*Teoriya veroyatnostey i matematicheskaya statistika*(Moscow: Vysshaya shkola)

[10] Bolshakov VDand Gaydaev PA1977*The theory of mathematical processing of geodetic measurements* (Moscow: Vysshaya shkola)

[11] Stolbov Yu V and Stolbova S Yu 2015O reglamentatsii tochnosti geometricheskikh parametrov vysotnogo polozheniya osnovaniy i pokrytiy avtomobil'nykh dorog v normativnykh dokumentakh *Vestnik SibADI***3(37)** 59–63

[12] Klyushin EB, Kiselev M.I, Mikhelev DSh and Feldman VD2014*Geodesy*edD ShMihaleva(Moscow: Akademiya)

[13] Maslov AV, Gordeev AV and Batrakov Yu2013 *Geodesy*(Moscow: Spica)

[14] Budenkov NA, Nekhoroshkov PA and Shchekova OG 2014*Engineering Geodesy Course*ed NA Budenkova(Moscow: Forum: infra)

[15] ZouJ, ZhuY, XuY, Li Q, MengL and LiH 2017 Mobile precise trigonometric levelling system based on land vehicle: an alternative method for precise leveling *Survey Review***49(355)** 249–58

[16] Nestorovic Z and Delcev S2014 Comparison of height differences obtained by trigonometric and spirit leveling method*Geonauka***2(4)** 30–7

[17] Shults R and Roshchyn-Kyiv O 2016Preliminary determination of spatial geodetic monitoring accuracy for free station method*Geodetski List***70(4)** 355–70

[18] Takalo M andRouhiainen P On *Digital Levelling Technique Applied in Water Crossing*Retrieved from: http://www.fig.net/resources/proceedings/fig_proceedings/fig2006/papers/ts53/ts53_03_takalo_rouhiainen_0507.pdf

Development of metrological methods for determining the balance of water and heat of generating companies

A S Shamaitis[1], E A Kremcheev, D O Nagornov, DA Kremcheeva

[1] St. Petersburg Mining University, 2, 21-line VO, St. Petersburg, 199106, Russia

E-mail: alexsham2@gmail.com

Abstract. The paper considers the causes of unbalance at the heat source during the normative consumption of resources, and also presents the results of the analysis of the metrological method of reducing the balance with statistical data and calculating the economic effect. Additionally, there are three options for solving the problem of unbalance, indicating the estimated economic effect for the station. The paper presents studies on the assessment of the imbalance in the supply and consumption of heat of an energy generating enterprise with a mixed consumption structure (instrument accounting, regulatory accounting). The possible solutions proposed in this work are based on the existing approach to estimating the uncertainty of instrumentation and are tested on arrays of statistical data from thermal power plants. For the proposed solutions to the problem of unbalance, an assessment of economic efficiency is made.

1. Introduction

Regardless of the billing period, at any source of heat, the problem of inconsistency of the results of accounting for consumed cold water and make-up water supplied to heating networks arises. This phenomenon is called the "occurrence of unbalance." The more powerful the heating main, the higher the level of divergence in the results of accounting for cold and make-up water. Unfortunately, in the current regulatory documentation, methods for solving this problem are not specified.

At the 6th All-Russian scientific and technical conference "Metrological support of energy resources accounting", the ideas about introducing metrological methods of reducing the balance of water and heat in the legal field were presented[9]. Unfortunately, over the past 14 years, methodological foundations and regulatory documentation have not been developed that would regulate the sequence of actions when taking measurements and distributing unbalances in energy resources. If we assume that at the station all the measurements were made correctly, but the imbalance is present, then with a high probability the cause of this problem is the systematic errors of the measuring instruments.

The study of the causes of the inconsistency of the accounting results will reveal the stage at which the error occurs and analyze possible solutions. Taking into account the fact that the majority of household consumers are calculated according to the consumption standard, the occurrence of unbalance is inevitable, since the average volume of water consumption by standard consumers is less than paid, respectively, the station receives a positive unbalance from them. Depending on the size of the settlement, the consumption standard may vary. Often, regional water tariffs exceed tariffs in Moscow and St. Petersburg, which is not a positive factor, since the standard of living in the regions and average wages are lower than those in the capital.

The factors that cause unbalance when accounting for the released and consumed water at the CHPP include the following:

1. Systematic errors of the used measuring instruments (provided that at the CHPP all necessary measurements are organized and performed correctly). Justified, in this case, will be the reduction of water balances taking into account the normalized metrological characteristics of all measuring instruments, the readings of which to one degree or another can affect the amount of unbalance. In this case, the correction of the obtained results in order to eliminate the unbalance should be carried out taking into account the following rules: only those measurement results that can cause the unbalance should be adjusted; the sign of the amendment to be introduced should be such that its application helps to reduce the discrepancies; the relative magnitude of the corrections should be directly proportional to the limits of permissible relative measurement errors of the corresponding quantities; the proportionality coefficient should be the same for all metering devices and determined by the ratio of the actual unbalance to its maximum permissible value.

2. Consumers using water according to the standard. Despite the need to install meters in each apartment, a large part of the population pays for water in accordance with the standards assigned to a particular region (city). Often, the consumer does not consume the amount of water that was allocated to him according to the documentation, which in turn provokes a positive unbalance, since the unspent water in fact is returned back to the station, at the time when it is officially registered as waste by consumer. Hence the use of "excess water", returning to the station in the status of spent.

2. Materials and methods

The main objective of the study is to eliminate imbalances in the heat source in the presence of contracts with household subscribers who consume resources according to standards.

The most common are two situations. The most common, in which the station does not conclude agreements on the provision of heating and water supply services directly with the subscriber, but uses an intermediary in this chain represented by the Heating Network organization, which in turn is already providing services in this area (Figure 1)

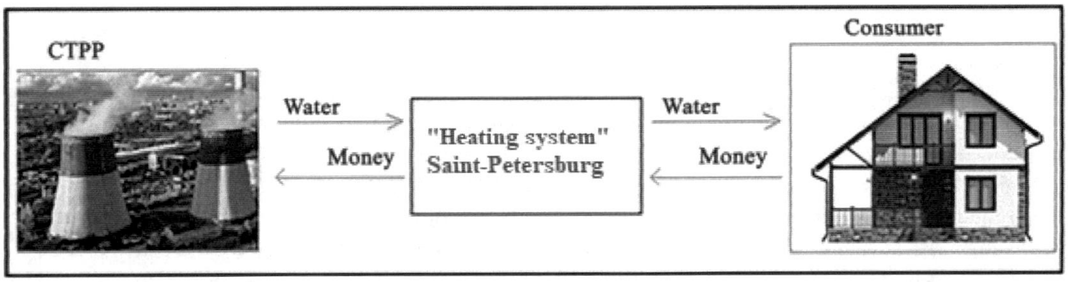

Figure 1. Intermediate Services Scheme

The second option is the conditions under which there is no intermediate link, as a result of which the CHPP directly concludes a contract with the consumer [14].

Depending on the accepted initial situations, options for solving the problem of unbalance will be different. One of the solutions to this problem is the transition from standard consumption to the use of instrumentation for all consumers. In this case, it is possible to reduce the unbalance, based on the permissible relative errors of the measuring instruments.

The second solution to the problem of unbalance is to increase the uncertainty component in the presence of regulatory consumers by 50%, and the excess water is supposed to be used, as indicated below, in the third solution [7].

The third solution option is to redistribute the resulting water into waste tanks [17] with further processing and use for the needs of the CHPP.

Consider the ways to solve the problem of inconsistency in the results of accounting for released and consumed water in highways. In particular, the approach to reducing unbalance according to the first scheme, that is, in the case when the CHPP does not conclude a direct contract for the provision of services with a substitute, but uses an intermediate link - the company «Teplosety».

In this case, the company "Teplosety" buys water from the CHPP and provides it to consumers. Regardless of whether they use water according to the standard or keep records of consumption by meters, consumers make payments for water to the Heat Network company. Surplus water arrives at the station marked "not used/not paid". A station cannot use such a resource for its own needs; excess water is not a source of positive unbalance in this case. Thus, the only possible way to reduce the imbalance is to take into account the values of the permissible relative errors of the MI. It will be necessary to evaluate the errors of all MI and the unbalance coefficient for further amendments to the accounting.

In a situation where the CHPP concludes an agreement on the provision of services directly with consumers, a positive unbalance at the station is provoked [19].

In such a situation, one of the solutions to the problem will be the rejection of normative consumption with a complete transition to instrument metering (using meters). This method will be able to facilitate the process of reducing imbalance by metrological methods using the values of the maximum permissible relative errors of measuring instruments (MI). Given that household consumers are reluctant to independently install water meters, this task can be assigned to the parent company, which includes the CHPP. An assessment was made of the number of normative consumers of one of the stations in St. Petersburg operating in two districts of the city. The total number of consumers reaches 750 thousand people, of which about 30% are regulatory consumers. The costs of the parent company for the provision of instrument metering for the studied group of consumers will amount to more than 200 million rubles. during one inter-verification interval, or, on average, more than 40 million rubles / year.

When implementing this set of measures, it is possible to simplify the process of reducing the unbalance at the CHPP, but unfortunately the station will not receive an additional economic effect, except for reducing the cost of compensating for the unbalance.

The essence of the second unbalance reduction scheme is to increase uncertainty by 50% when taking into account the water consumed by regulatory consumers. By supplying water, we can not reliably determine the amount of water used by a particular consumer who does not have the means of metering. This means that there is an additional component of the total uncertainty of water accounting. The general formula for calculating the total uncertainty of the mass flow of water can be written as:

$$u`q = \{u`c^2 + u`\kappa_r{}^2 + u`\kappa_c{}^2 + \left(\frac{2*\beta^4}{1-\beta}\right)^2 * u`D^2 + \left(\frac{2}{1-\beta^4}\right)^2 * u`d^2 + 0{,}25 * \left(u`^2_{\Delta p} + u`^2_{p}\right)^2\}^{0,5} \quad (1)$$

where, $u`c^2$ – relative standard uncertainty of the expiration coefficient; $u`\kappa_r{}^2$ - the uncertainty of the correction factor, taking into account the roughness of the inner surface of the measuring pipeline; $u`\kappa_c{}^2$ - the uncertainty of the correction factor, taking into account the blunting of the input edge of the diaphragm; β - is the relative diameter of the opening of the constricting device [1]; $u`D^2$ - the uncertainty of the dimensions of the outer diameter; $u`d^2$ - the uncertainty of the dimensions of the inner diameter; $u`^2_{\Delta p}$ - uncertainty of the measurement result; $u`^2_{p}$ - the uncertainty of the measurement result of absolute pressure [4].

By adding to this formula the coefficient u_p, which is responsible for the uncertainty of consumption at standard consumption and equal to 50%.[20]

As another alternative solution to the problem of unbalance, redistribution of the received excess water from the subscriber to waste tanks can be made with a view to its further processing and use for the station's needs [19]. With normative consumption, in most cases, the consumer does not use the amount of water that is allocated to him according to regulatory documentation. This in turn provokes a positive unbalance at the station. The solution to this problem will be the transfer of the resulting water to the category of waste for the purpose of its further processing. After the station purifies such water, it can use it for its needs. The average volume of water used at the station for domestic consumption is about 100 thousand cubic meters. m., which is less unbalance from regulatory consumers. Thus, the returned water is sufficient to meet the needs of the station and it is possible not to buy water from the water utility, which has an additional economic effect.

3. Discussion

In accordance with GOST 8.586.5-2005, the relative expanded uncertainty for the measured value at P = 95 will be calculated by the formula

$$U`q = 2 * u`_q \qquad (2)$$

Intermediate variables for formula 1 are determined by the following formulas:

$$u`c = 0{,}5 * (U`c_0 + U`_L + U`_{l+} + U`_{\text{вх}} + U`_h) \qquad (3)$$

where, U`c0 - is determined in accordance with table 1.

Table 1. Formulas for determining the uncertainty of the expiration coefficient depending on β [2]

1	$U`c_0 = 0{,}7 - \beta$	at $0{,}1 \le \beta < 0{,}2$
2	$U`c_0 = 0{,}5$	at $0{,}2 \le \beta \le 0{,}6$
3	$U`c_0 = 1{,}667*\beta - 0{,}5$	at $0{,}6 < \beta \le 0{,}75$

Since the flow meters of variable differential pressure (FVDP) [8] is a high-precision metering device, the remaining parameters do not introduce additional uncertainty into the expiration coefficient. If a different flow meter is installed in the metering unit (electromagnetic [16], ultrasonic, but it is necessary to rely on the metrological characteristics of a particular measuring instruments) [15].

In accordance with GOST 8.586.5-2005, clause 10.3.2 of the value u`d = 0.02%, and u`D = 0.1%, the uncertainty of the measurement result is determined by the formula:

$$u`_{\Delta p} = \left(\sum_{i=1}^{n} \left(\vartheta_i * u`_{yi}\right)^2\right)^{0{,}5} \qquad (4)$$

where, n - is the number of measuring instruments used to measure the differential pressure; ϑ_i - is the sensitivity coefficient of the i-th measuring device; u`yi - uncertainty introduced by the i-th measuring device.

The next step will be to determine the absolute error of mass measurement.

To determine the magnitude of our amendments, we must calculate the unbalance coefficient by the formula 5.

$$К_{ub} = \frac{К_f}{К_{max}} \qquad (5)$$

where, K_f is the value of the actual unbalance present at the station; Kmax is the maximum possible unbalance at the station.

Knowing the unbalance coefficient, we make the necessary corrections to the relative errors, multiplying the unbalance coefficient by the permissible relative error of the measuring instrument. After making adjustments, the unbalance value should tend to zero. To achieve a positive result of adjustments, it is necessary to take into account the amendments made, in the heat released on the highways [6].

For the above methods of solving the problem of inconsistency in the results of accounting for discharged water by highways and used water, economic indicators were calculated [13].

The event that has the greatest economic effect should be considered the use of water that was not used by regulatory consumers for the needs of the CHPP plant by transferring the resource to waste, with the aim of cleaning it and further using it. The estimated annual economic effect in prices of 2018 will be about 30.0 million rubles, which is a significant amount in the balance of the studied enterprise.

4. Conclusion

In order to eliminate the imbalance caused by systematic errors of the measuring instruments used (by which the incoming and outgoing water is recorded), the maximum permissible errors of the flow meters are calculated according to the method specified in the current regulatory documentation, namely GOST 8.585.5-2005, and the necessary amendments are made to initial measurement results.

The methods for reducing imbalance described in this work can have their practical application both independently and in combination, which will depend only on the available array of source data. After analyzing all the methods of reducing imbalance, it should be noted that the most economically promising is the one in which the CHPP directly concludes a contract with the subscriber and uses its excess water for its needs, which saves money by stopping the need to buy water for its needs from Vodokanal and the need to compensate for unbalance at their own expense.

References

[1] Standardinform 2007 State system for ensuring the uniformity of measurements. Measurements of liquids and gases flow rate and quantity by means of orifice instruments. Part 1. Principle of the method of measurements and general requirements (GOST 8.586.1-2005).

[2] Standardinform 2007 State system for ensuring the uniformity of measurements. Measurement of liquids and gases flow rate and quantity by means of orifice instruments. Part 2. Orifice plates. Technical requirements (GOST 8.586.2-2005).

[3] Standardinform 2007 State system for ensuring the uniformity of measurements. Measurements of liquids and gases flow rate and quantity by means of orifice instruments. Part 5. Measurement procedure (GOST 8.586.5-2005)

[4] Standardinform 1978 Water. Density at atmospheric pressure and temperatures from 0 to 100 degrees Celsius (SSRDS 2-77)

[5] Standardinform 1989 Tables of standard reference data. Water. Dynamic viscosity coefficient at temperatures of 0 ... 800 ° C and pressures from the corresponding rarefied gas up to 300 MPa (SSRDS 6-89)

[6] Standardinform 1999 Water. Specific volume and enthalpy at temperatures of 0 ... 1000 ° C and pressures of 0.001 ... 1000 MPa (SSRDS 187-99)

[7] Standardinform 1997 Recommended Practice - National Measurement Standards - Piped Hot Water Heat Supply Systems - Equations for Measurement of Thermal Energy and Quantity of Heat Transfer Agent (MI 2412-97)

[8] Lyapyavko A P 2004 *Academy of Standardization Metrology and Certification. Variable differential pressure flow meters* (Moscow: Academy)

[9] Lupey A G 2004 On the problem of water imbalances in heat sources and how to solve it (St Petersburg)

[10] Lupey A G 2004 On the error of measurements of the difference in mass of coolants in open heat supply systems (St Petersburg)

[11] Kremlevskiy P P 2004 *Flow meters and substance counters* (Moscow: Politechnika)

[12] Vodokana 2016 Tariffs for drinking water, industrial water and wastewater disposal of the state unitary enterprise "Vodokanal of St. Petersburg" in the territory of St. Petersburg, for 2016-2020. URL: http://www.vodokanal.spb.ru/dlya_abonentov/tarify1/.

[13] Rogalev N, Sukhareva E, Mentel G, Brojina Y 2018 Economic approaches to improving the structure of the electricity market *Terra Economicus* **16** 140-149

[14] Salustiano Martim A L S 2019 Electromagnetic flowmeter evaluation in real facilities: Velocity profiles and error analysis *Flow Measurement and Instrumentation* **66** 44-49

[15] Wang B 2019 Simulation and experimental verification on dynamic calibration of fuel gear flowmeters *Measurement: Journal of the International Measurement Confederation* **138**

Journal of Physics: Conference Series **1384** (2019) 012044 doi:10.1088/1742-6596/1384/1/012044

570-577.
[16] Huang J 2019 Development of a flow control valve with digital flow compensator *Flow Measurement and Instrumentation* **66** 157-169.
[17] Hvasta M G 2019 Calibrationless rotating Lorentz-force flowmeters for low flow rate applications *Measurement Science and Technology* **29**
[18] Zavyalov G V 2018 Mathematical model of heat exchange processes for heat protective cooling suit of a rescuer *Journal of Mining Institute* **231** 326-332
[19] Gasumov R A 2016 The specifics of operating minor deposits *Journal of Mining Institute* **220** 622-626

Diagnosis of mining and oilfield equipment by excited oscillations analysis technique

D I Shishlyannikov[1], A E Pushkarev[2]

[1] Perm National Research Polytechnic University, 29 Komsomolsky Avenue, Perm, 614990, Russia
[2] Saint-Petersburg State University Of Architecture And Civil Engineering, 4 Vtoraya Krasnoarmeiskaya St., Saint Petersburg, 190005, Russia

E-mail: 4varjag@mail.ru

Abstract. The paper presents relevance of application of non-destructive testing methods in the assessment of technical condition of the mining and oilfield machine components is justified. The prospects of diagnostics of process equipment by the excited oscillations analysis technique are proved. Design and operation algorithm of the multichannel synchronous analyzer Kamerton by LLC NPP ROS (Perm) are described. Methodological bases of defect finding using the multichannel Kamerton analyzer are described. The basic quality criteria characterizing the diagnostics-based state of technical elements are listed. It is proved that the mining and oilfield equipment technical condition assessment method by excited oscillations analysis allows carrying out operative diagnostics of technical facilities without shutdown. This method has no analogues in terms of efficiency compared to the traditional equipment nondestructive testing methods: magnetic particle, metal magnetic memory, ultrasonic flaw detection, ultrasonic thickness measurement, radiographic, metallographic, etc. In some cases, for example, for structures made of heterogeneous materials such as reinforced plastic or reinforced concrete, it is the only possible diagnosis method.

1. Introduction

Reliability improvement of process machines, reducing the costs of equipment operation, and increasing the efficiency of repair and rehabilitation work remain relevant challenges for mining and oil producing enterprises. Insufficient accuracy of estimation of the service life of parts and units of process machines, complexity and high labor intensity of diagnostics leads to an increase in the number of downtimes due to the occurrence of sudden failures and growth of material costs to maintain the efficiency of equipment. These problems can be solved on the basis of development and wide use of nondestructive testing [1, 2].

The prediction of individual service life of process equipment parts includes a set of challenges: assessment of the current state of the object, forecasting the development of this state for the near future and determination of the residual life [3, 4]. To date, one of the most promising ways to diagnose the components of mining and oilfield machines is the assessment of the technical state by the excited oscillations analysis method [1, 5, 6].

2. Research methodology

The process machine components diagnostics by excited oscillations analysis is based on oscillations propagation analysis inside the structure and is related to the internal structure of the object. Pulse excitation (produced by a hammer or impulse source of oscillation) leads to the development of oscillations in the target object, which differ in harmonic composition and damp out at different rates [7]. Mono-oscillations prevail in a homogeneous environment after impulse exposure, damping out for a long time. Presence of cracks (cavities, other discontinuity flaws), as well as the weakening of the internal structure of the environment due to the material ageing, increases the number of harmonics in the signal spectrum, reduces oscillation time due to energy absorption in the flawed areas [10]. Measurement of the wave movement timing parameters inside the object allows analyzing the environment parameters and to identify the flawed area. The method is used to detect internal defects, cracks, cavities in various equipment and structures, including metal, ceramic, plastic and multi-component parts, welded joints, bedding points, internal joints of equipment components and structures, allows revealing signs of material fatigue, localize probable flaw locations.

Figure 1 show that for the practical implementation of diagnostics by excited oscillations analysis the multichannel diagnostic tool Kamerton by LLC NPP ROS is used. The diagnostic complex includes 2 to 32 piezoelectric accelerometer and (or) acoustic sensors for measuring excited oscillations, synchronous analog-to-digital converter for normalization and primary signal conversion, analyzer on the basis of a laptop or a special controller, and oscillation exciter.

The oscillation is excited by a hammer, less often by a special pulse oscillation source. Pulse excitation is the most rational method: lack of impact on the target object of the variable load created by the weight of the pulse vibration source is a significant advantage in lightweight structure testing, as the point-to-point load change can cause the shift of frequencies in different measurement cycles. Besides, lack of necessity to use fixtures for installation of pulse oscillation source provides convenience of carrying out nondestructive testing directly at the place of process equipment operation [7].

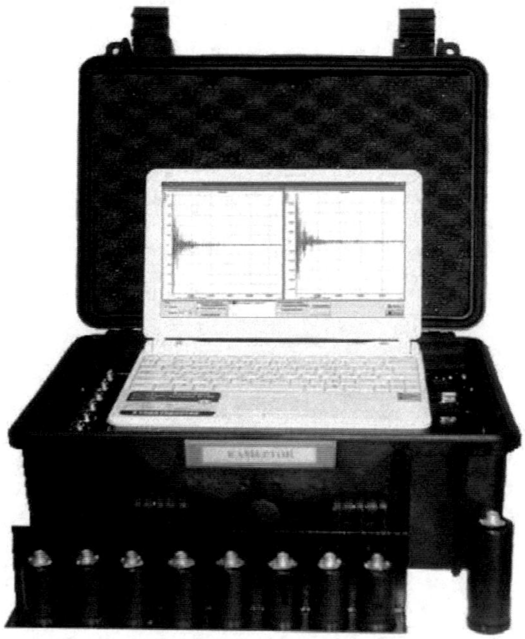

Figure 1. Multichannel analyzer Kamerton by LLC NPP ROS

The distance between the sensors varies depending on the size of the elements and the density of the structure material. E.g., in the assemblies made of hard metals and alloys, the sensors can be installed at considerable distances from each other, because in solid media the excitation signal can be transmitted at considerable distances, and the softer the material, the closer the sensors should be to each other.

3. Research procedure and analysis

Let's consider the algorithm of diagnostics of the oilfield equipment by the resonant oscillations excitation, as in the case of SK8 horsehead attachment unit. Figure 2 show that technical condition control method provided for installation of three sensors directly on the target element of the object. No cracks or other defects have been detected during the visual inspection of the horsehead attachment unit.

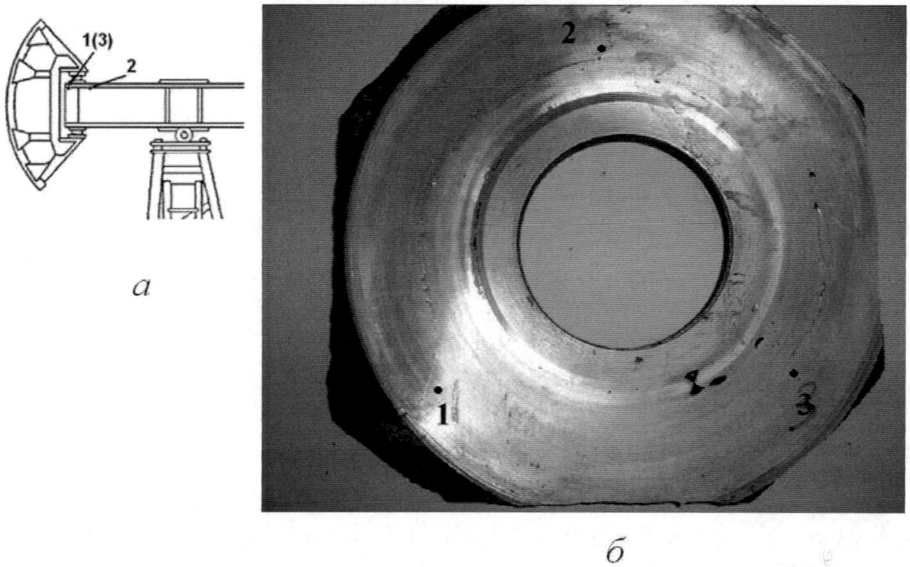

Figure 2. Diagnosis of SK8 horsehead attachment unit: *a* – sensor mounting layout of Kamerton diagnostic complex; *b* – 1.5 mm deep cut over the cross section of the horsehead attachment unit

Following installation of the sensors oscillations were excited on the element by a hammer with a head weight of about 0.1 kg. At least four to five similar measurement cycles in a row were carried out to increase the reliability of diagnosis and eliminate accidental interference. Figure 3 show that normal registrations are smoothly attenuating harmonic signals with no sharp local emissions, breaks, overswings.

Pulsed excitation causes the resonant wave processes in the product, which have characteristic features of space- and time development. In a uniform density environment, the time, speed and frequency parameters of the response signal are identical to those of the excitation signal. In the presence of cracks (cavities, discontinuities), and upon ageing of the material (when there is a rarefaction or weakening of its structure), the conditions for the passage of pulse-induced oscillations in different parts of the object are significantly different from each other, which allows identifying homogeneous and heterogeneous areas in the object. Analysis of the energy components of each recorded signal allows not only a qualitative but also a quantitative assessment of the object condition.

Signals registered by the device was processed in the expert diagnostic system Kamerton, which carries out the analysis of the spectral composition of oscillations of structural elements and the calculation of generalized parameters of the element quality.

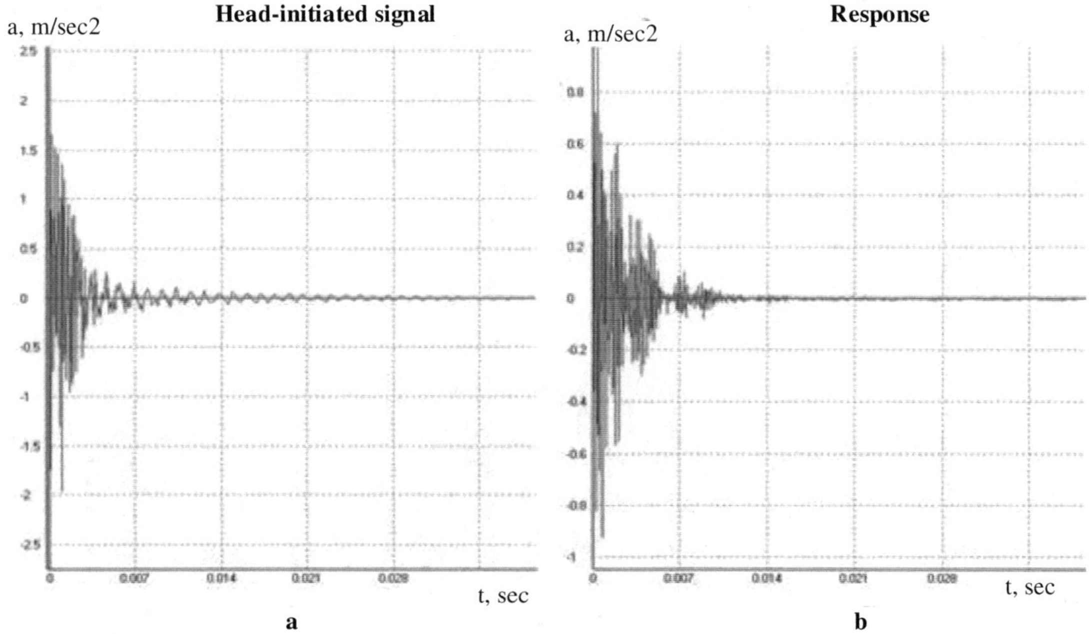

Figure 3. Head-initiated oscillation signal (*a*), and response (*b*) registered by piezoelectric accelerometer sensor

Figure 4 show that when analyzing the condition of a machine part or a structural element, each signal determines the damping rate, shock (carrier) frequency, deviations of the shock frequency from the normal range of values for a given material, the spectral pattern of the response signal is analyzed. The spectral pattern of defects in individual elements of technical objects is the same, which allows using this method to determine the state of materials of various hardness and composition.

The analysis of the technical condition of SK8 horsehead attachment unit was carried out by estimation of signal attenuation degree in the connection, and comparison of the spectral parameters of the excitation signal and the response signal. Each measurement was analyzed twice in different signal directions, when the excitation was carried out on both sides of the connection alternately.

After the full cycle of measurements all registration signals were processed by mathematical statistics methods. The expert software uses the following parameters to define the element and connection quality criteria:

– mean resonance carrier frequency for all available signals for the element or connection;

– average damping rate of an element or connection;

– mutual frequency deviation in the connection and the element;

– the probability of wall thinning or density reduction, weakening the structure of the element material;

– the probability of cracks, cavities, discontinuities in the element or connection [7].

According to the results of diagnostics of SK8 horsehead attachment unit it is established that the areas of welded joints between the points of installation of sensors *1–2* and *2–3* (see Fig. 2) have signs of cracks. Conclusion by the expert system: the object condition is unacceptable; no further operation of the horse-head pump is possible. Specialists of the company's service department decided to shutdown the horse-head pump, and to open the target unit in order to visually detect defects. A 1.5 mm deep cut over the section of the horsehead attachment unit exposed a U-shaped crack with kinkings (see points *1, 2, 3*, Fig. 2, b), and an incomplete fusion of the weld joint between the bushing and the horsehead horizontal metal sheet [8, 9].

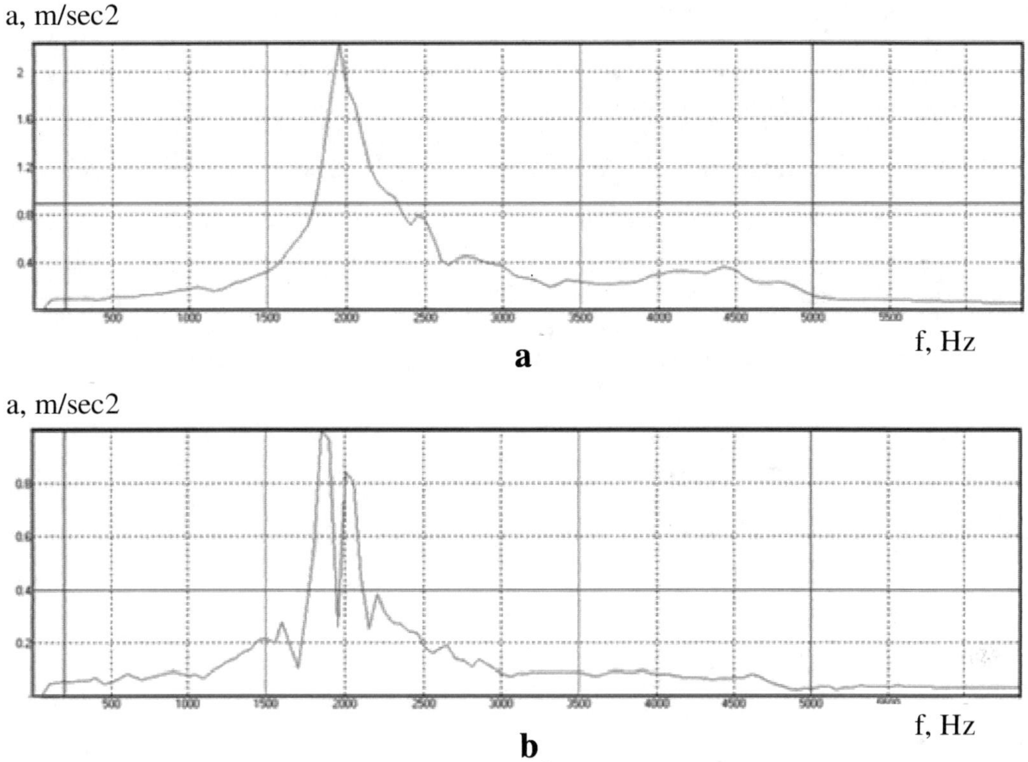

Figure 4. Examples of response signal spectra obtained during examination by the resonance oscillations excitation method on the SK8 horsehead attachment unit: *a* – no defect found; *b* – metal cracking

It should be noted in particular that the earlier performed diagnosis of the same unit by ultrasonic control method did not allow detecting the defects determined using the described method.

4. Conclusion

Diagnostics of mining and oilfield equipment units by means of the excited oscillations analysis allows an estimation of the technical condition of machines without their shutdown. This method has no analogues in terms of efficiency compared to the traditional equipment nondestructive testing methods: magnetic particle, metal magnetic memory, ultrasonic flaw detection, ultrasonic thickness measurement, radiographic, metallographic, etc. In some cases, for example, for structures made of heterogeneous materials such as reinforced plastic or reinforced concrete, it is the only possible diagnosis method.

References

[1] Belyaev A, Pushkarev A and Kuzmichev A 2018 *Proc. Int. Conf. "Organization and Traffic Safety Management in large cities", (SPbOTSIC edir)* vol **36** (Amsterdam: Elsevier B.V.) pp 27-29

[2] Pushkarev A, Podoprigora N and Dobromirov V 2018 *Proc. Int. Conf. "Organization and Traffic Safety Management in large cities", (SPbOTSIC edir)* vol **36** (Amsterdam: Elsevier B.V.) pp 634-639

[3] Shishlyannikov D and Vasilyeva, M 2017 *IOP Conference Series: Materials Science and Engineering* **10** *Cep "International Conference on Mechanical Engineering, Automation and Control Systems 2016"* 012-045

[4] Kuvshinkin, S, Zvonarev I and Ivanova P 2018 *Journal of Physics: Conference Series* **1118(1)** 012-054
[5] Shishlyannikov D and Lavrenko S 2017 *Proc. Int. Conf. on Industrial Engineering, Applications and Manufacturing (ICIEAM 2017)* p 8076122
[6] Timofeev I.P., Vasilyeva M.A. 2017 *Procedia Engineering* **206** pp 1023-1028
[7] Kremcheeva D., Kremcheev E. 2005 *Journal of Mechanical Engineering Research and Developments* **41(1)** 151-155
[8] Nagornov, D.O., Kremcheev, E.A. 2018 *Journal of Physics: Conference Series* **1118(1)** p 012028
[9] Bolobov, V.I., Bochkov, V.S. 2019 *Gornyi Zhurnal* **1** p 57-60
[10] Vasilyeva M.A., 2019 *Eurasian Mining* **1** 34-36
[11] Vasilyeva M. 2019 *Vibroengineering Procedia* **25** 111-115

Journal of Physics: Conference Series **1384** (2019) 012046 doi:10.1088/1742-6596/1384/1/012046

Use of heading-and-winning machine parameter loggers to predict gas-dynamic phenomena at potash mines

D I Shishlyannikov[1], **A E Pushkarev**[2]

[1] Perm National Research Polytechnic University, 29 Komsomolsky Avenue, Perm, 614990, Russia
[2] Saint-Petersburg State University Of Architecture And Civil Engineering, 4 Vtoraya Krasnoarmeiskaya St., Saint Petersburg, 190005, Russia

E-mail: 4varjag@mail.ru

Abstract. The paper presents analysis of the main instrumental monitoring methods of potassium beds outburst hazard. The main disadvantages of the existing means of predicting gas-dynamic phenomena in potassium beds are the low automation level and the cyclicality of measurements at a certain pitch or time interval, which leads to a decrease in the productivity of mechanized combine systems and adversely affects the safety of processes in the shrink stopping. The main requirements are laid down, which shall regulate the development of methods and techniques for predicting gas-dynamic phenomena in potassium beds. Design and operating principle of VATUR software logging system developed by members of the Mining Electrical Engineering Department of the Perm National Research Polytechnic University is described. The results of experimental researches on estimation of size and character of change of energy parameters of a potassium bed fragmentation by operating cutters of Ural-20R heading-and-tunneling machines are presented. The researches have shown that the measurement technique and the equipment reveal deviations of specific power inputs of a potassium bed fragmentation by the machine's operating members with high accuracy and speed. The results of the experimental studies made it possible to substantiate the prospects of using the energy parameter loggers of cutter-loaders at predicting gas-dynamic phenomena at potassium mines. The suggested concept of the software measuring complex can be used in the development of on-board parameter loggers of the heading-and-winning machines in potassium mines.

1. Introduction

For enterprises mining potassium ore underground using mechanized combine systems, the safety of mining operations at increasing both the winning machine performance and the face output. The successful solution of this challenge is facilitated by the development and active introduction of onboard performance monitoring systems for mining machines, providing for registration and recording of information about the load of the cuter-loader drives, signaling the unacceptable development of events and violation of the potassium ore mining process.

The changes in the stress-strain state (SSS) of the potassium salt bed during intensive cleaning operations cause an increase in the probability of gas-dynamic phenomena (GDP), accompanied by collapses and releases of significant amounts of rock mass into excavations. Dynamic phenomena mainly occur in the face area, threatening the health and life of miners.

2. Research methodology

Nowadays, the main way to prevent GDP at potassium mines in Russia and the CIS countries is the current predication, which provides for the identification of outburst-hazardous areas of the potassium bed by instrumental techniques during preparatory and stopping works. The following current GDP predication methods were tested and applied at the Verkhnekamskoe potassium-magnesium salt deposit.

The barometric GDP predication method in the potassium bed is based on the measurement of the pressure buildup rate and the amount of gas in the boreholes drilled and sealed in the roof rocks. Due to the simplicity of the equipment and measurement techniques, the method has become widespread in potassium mines. However, the insufficient reliability of the barometric method shall be noted, due to the fact that it is based on the GDP manifestation concept due to the action of non-associated gas alone, regardless of SSS and physical and mechanical properties of rocks. Barometric control devices impose strict requirements on the hole mouth diameter, which affects their sealing reliability. However, it is disregarded that if gas is actively drained from the potassium bed, the stress inside the mined formation is relieved, and the outburst hazard is reduced. For the same reason, the potassium bed outburst hazard control method based on the measurement of combustible and toxic gases concentration in the underground environment by installation of onboard stationary gas control equipment, e.g., SD-1 sensors or IKG-9 gas concentration meters on the machine, shall be considered as insufficiently reliable.

The GDP acoustic predication method is based on the dependence between the potassium bed outburst hazard level and the volume of micro-occluded gas localized in salt crystals. The method is implemented by recording the acoustic signal when dissolving salt crystals and releasing gas into the brine. The disadvantages of the described GDP predication method are low reliability caused by the temperature effect and the influence of insoluble residue on the acoustical activity of salts. This method would be reasonable to use in the relative estimation of the micro-occluded gas content in rocks.

GDP predication methods based on the assessment of the mechanical state of saliferous rocks are characterized by high reliability. Outbursts are known to be more likely to occur the closer the rocks around the mine are to critical SSS. Assessment of the strength of saliferous rocks, taking into account their potential energy accumulated from the rock and gas pressure forces shall be carried out by contact method and well drilling with core sampling. The contact method involves pressing a spherical indenter into the walls of wells drilled along the perimeter of the workings, obtaining and analyzing of saliferous rock fracture diagrams. The core sampling method evaluates the nature of fracture and the size of core fragments collected while drilling wells [1].

The main disadvantages of the above methods (except for the method providing for installation of gas control equipment on the machine) are the low automation level and cyclicality of measurements at a certain interval or period of time during which the GDP may occur. In order to obtain reliable information at mines with multiple movable faces significant staff is required, to carry out measurement campaign, frequent transfer and installation of equipment. The use of existing GDP predication tools in close proximity to the face is hampered by the need to conduct the basic potassium ore mining and transportation processes, and may negatively impact the overall performance of the mechanized complex.

Taking into account the above mentioned aspects, the development of the current GDP predication methodology in the potassium bed shall be carried out in accordance with the following requirements:

– comprehensive accounting of all factors influencing the GDP formation (strength and strain parameters of rocks, presence of occluded gas and changes in the rock SSS during mining operations);

– continuous measurements for the purpose of GDP predication;

– measurement campaign should not impede the potassium ore mining and transportation processes.

Potassium bed outburst hazard predication technique suggested by B.V. Laptev and M.M. Bey meets these requirements to the fullest extent (Perm, All-Russia Mineral Salt Production Research and

Development Establishment) [2, 3]. The method is based on the provision that the saliferous rock in the outburst hazard are of the bed is characterized by reduced strength and increased tension due to the presence of rock and gas pressure. The results of numerous researches carried out by the members of the All-Russia Mineral Salt Production Research and Development Establishment allowed establishing the unique dependence of compressive strength, cutting resistance, mineralogical composition and the saliferous rocks outburst hazard. It follows from the above that GDP zones can be predicted through continuous monitoring of specific energy consumption for the potassium bed fragmentation by operating cutters of heading-and-tunneling machines.

3. Research procedure and analysis

The research aimed at to determine the energy parameters of the potassium bed fragmentation process by Ural-20R machines operated at the mines of PJSC Uralkali were carried out by members of the Mining Electrical Engineering Department of the Perm National Research Polytechnic University, in cooperation with experts from LLC RCC (Perm). Figure 1 show that the experimental research methodology provided for the use of VATUR portable logging system, which measures and logs instantaneous currents and voltages of electric motors onboard of the operational heading-and-winning machine [4, 5].

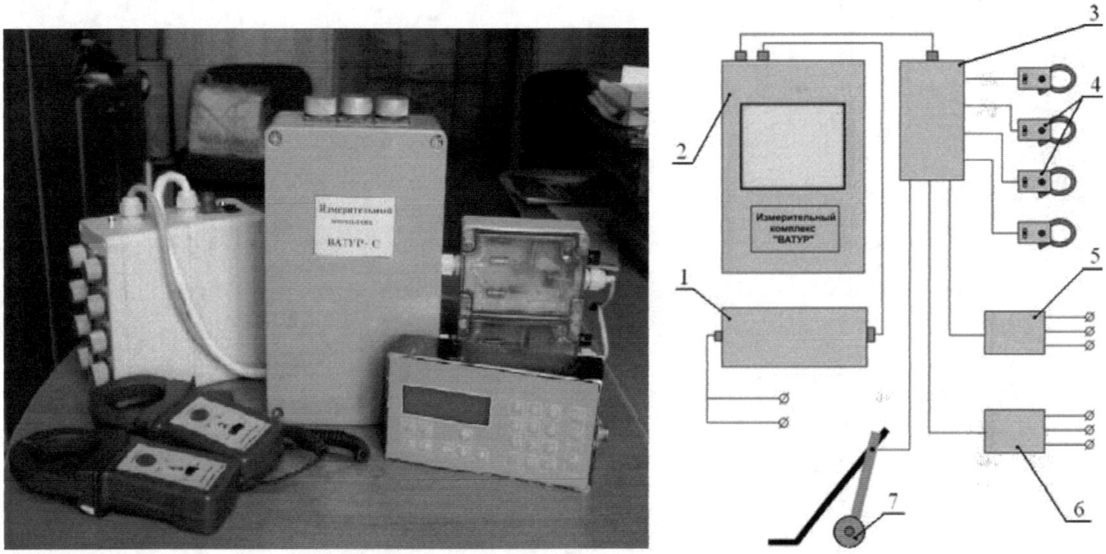

а) б)

Figure 1. VATUR software logging system: a) general view; б) functional chart; 1 – power supply unit; 2 – processor unit; 3 – switching unit; 4 – current sensors (current clamps); 5, 6 – voltage sensors; 7 – distometer

VATUR complex consists of a processor unit, power supply and switching units, current clamps, voltage sensors and a distometer. The processor unit consists of an industrial computer for portable systems, and an analog-to-digital converter (ADC). The current clamps ensure that the input current is converted into an output voltage of 1A : 1mV. Voltage sensors convert input voltages up to 1000 V into output voltages at 1 V : 3mV. Signals from current clamps and voltage sensors are fed to the ADC input connected to the computer. The distometer is based on an incremental PUF-6 rotary encoder with 360 strokes per revolution. The sensor is mounted on the side skid and ensures that the machine’s movement is recorded.

Within a 20 ms power supply period, 100 measurements take place. Every 0.5 second the effective current, voltage, active and total power values are calculated and recorded to the computer’s hard disk.

Pre-measurements on the machines have shown that the supply voltage and load are symmetrical, so a single wattmeter method has been adopted to measure the active power in a three-phase circuit. Figure 2 show that visualization and processing of the received data is carried out by means of specially developed VATUR-off software. The following graphs show the changes in the active power consumption of the electric motors of the machine's cutter wheels relative motion; the starting (*A-B*), idling (*B-C*), cutting (*C-D*) and steady state (*D-E*) areas can be identified.

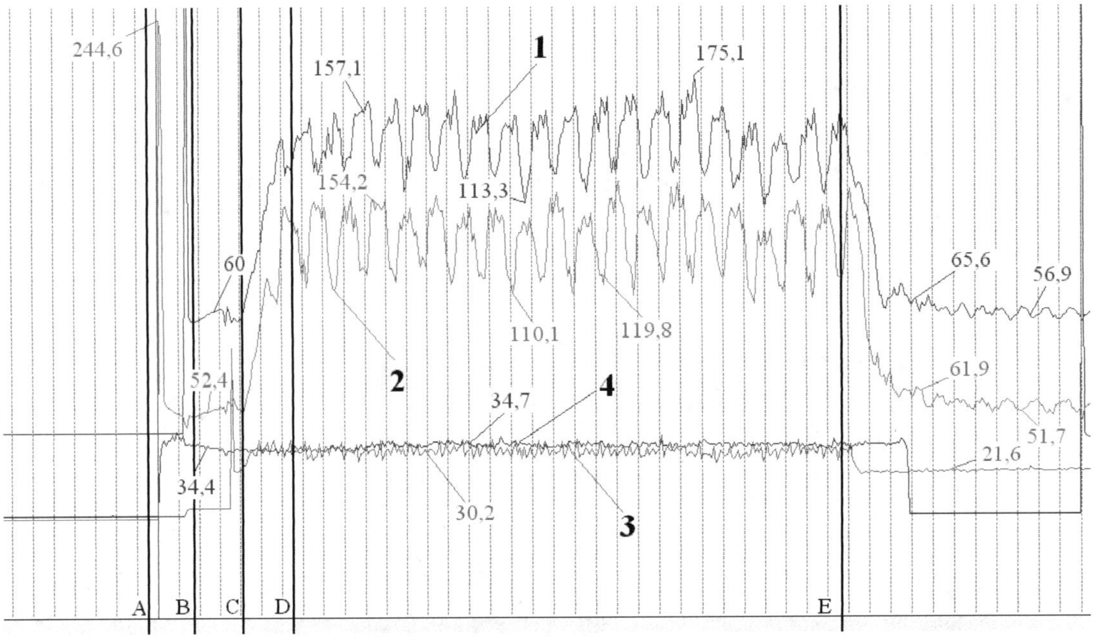

Figure. 2. Graphs of active three-phase electric motor capacities at the average speed of Ural 20R machine – 10.5 meters per hour: 1, 2 – cutter wheel relative movement drives; 3 – breasting device drive; 4 – conveyor drive. Grid line interval – 5 s. Footnotes refer to instantaneous active power values, kW.

The energy parameters of Ural-20R machines were registered by VATUR complex at SKRU-3 mine, Krasnyi-II formation of PJSC Uralkali. The researches were conducted in accordance with the manufacturer-approved technique. The machine was running a continuous face with a conservative feed rate. Single measurement included the machine's operation at breaking of three to seven ore cars with a given speed of 4 to 12 m/h.

Figure 3 show that the machine's specific energy consumption depends significantly on the feed rate. Due to insufficient capacity of self-propelled cars, machines operate at reduced feed rate $V_п$=6–7 m/h. Machine's operation with nominal engine load and performance close to the nameplate load at $V_п \approx 12$ m/h leads to a 1.5-times reduced energy consumption for mineral extraction.

The researches have shown that the technique and the equipment reveal deviations of specific power inputs of a potassium bed fragmentation by the machine's operating members with high accuracy and speed. Bed fragmentation-related energy consumption values correlated with the machine's feed rate can be saved as reference in the memory of the machine's software logger. Further, during stopping works the reference values are compared with the actual readings and at reduction of specific energy consumption of bed defragmentation by more than 25% the system signals about possible GDP occurrence [6-8].

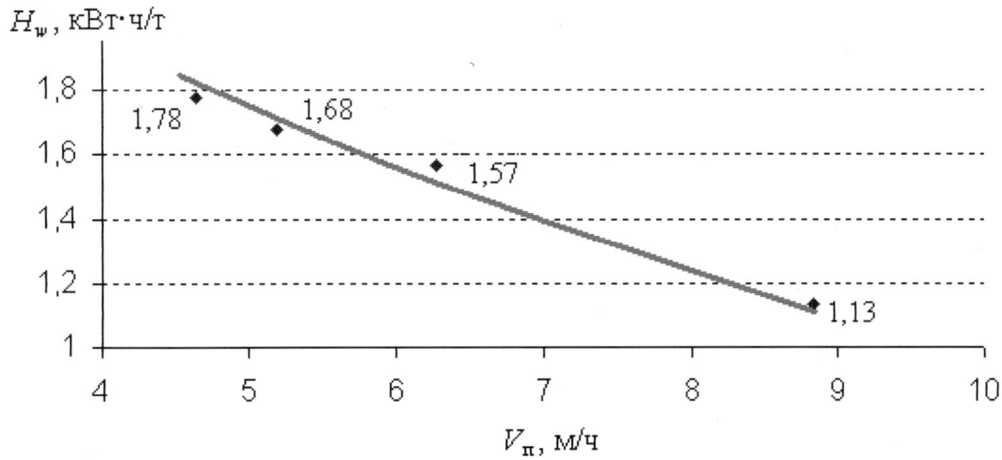

Figure 3. Dependence specific energy H_w consumption at potassium bed fragmentation with an epicyclical disk on Ural-20R machine’s feed rate $V_п$

VATUR software logger ensures creation and storage of data sets containing the information on quite long operation periods of the heading-and-winning machine, enables control the operating parameters, determine the technical performance and operating time of the machine. The analysis of wattmetergrams characterize the rhythmicity of mineral extraction processes and its transportation into the shrink stopping enables to estimate the work organization level and the equipment loading level, to reveal negative tendencies and factors influencing the performance of the mechanized complex. In case of emergency, the wattmetergram is a basis to review and identify the causes of the accident and the actions of personnel [9].

The spectral analysis of the current signals received via VATUR software logger, enables to reveal the frequency components from 0 to 100 Hz, characterizing the vibrations of operating device – gearbox – drive motor kinematic chain. Defects in work units and mechanical transmissions cause variable loads, which causes new spectral components. Periodic measurement of values in the current spectrum characterizing specific defects in the drive motor and mechanical transmission, allows assessing the technical condition of the winning machine and, if necessary, carry out repairs aimed at preventing emergency failures [10].

4. Conclusion

Thus, the use of the winning machine performance loggers providing for continuous control of the energy parameters of the potassium bed fragmentation process allows the easiest way to carry out the current GDP predication and identify the outburst hazardous zones in the bottom-hole area, which leads to an increase in the safety of drifting and stoping operations at potassium mines. The suggested concept of the software measuring complex can be used in the development of on-board parameter loggers of the heading-and-winning machines in potassium mines.

References

[1] Laptev B V 1994 *Prevention of gas-dynamic phenomena in potassium mines* (*Moscow: Nedra*) p. 138
[2] Laptev B V and Bey M M 1989 Rock bed outburst hazardous zones current predication method *USSR Patent 1458571 Newsletter No. 6*
[3] Chekmasov N V, Shishlyannikov D I and Trifanov M G 2013 *News of Higher Educational Institutions; Mining magazine* **6** pp. 103–107.
[4] Limanskiy, A.V., Vasilyeva, M.A. 2016 *Ecological Engineering* **91** p 41-43

[5] Shishlyannikov, D.I., Vasilyeva, M.A. 2017 *IOP Conference Series: Materials Science and Engineering* **177 (1)** pp. 535-538
[6] Ivanov S L 1999 *Extending the service life of the transmissions of mining machines based on the assessment of the energy load of their elements* (St. Petersburg: St. Petersburg Mining University) p. 92
[7] Trifanov G D, Knyazev A A, Chekmasov N V and Shishlyannikov D I 2013 *Mining equipment and electromechanics* **2**, pp. 41-44.
[8] Barkov A V, Barkova N A and Borisov A A 2012 *Methodology for diagnosing electrically operated mechanisms by current consumption* (St. Petersburg: North-West Education Center) p. 68
[9] Kremcheev, E.A., Gromyka, D.S., Nagornov, D.O. 2018 *Journal of Physics: Conference Series* **1118 (1**) p 012021
[10] Nagornov, D.O., Kremcheev, E.A., Kremcheeva, D.A. 2019 *International Journal of Civil Engineering and Technology* **10 (1)** 876-883

Forecasting indicators of thermal oxidative stability of lubricating oils

B I Kovalsky, O N Petrov, V G Shram*, Yu N Bezborodov, A N Sokolnikov, N N Lysyannikova

Siberian Federal University, 82 Svobodny Avenue, Building 6, Krasnoyarsk, 660041, Russia

E-mail: Shram18rus@mail.ru

Abstract. The results of testing partially synthetic Total Quartz 10W-40 SL / CF motor oil for thermo-oxidative stability in terms of optical density, volatility and coefficient of thermo-oxidative stability are presented taking into account the thermal energy absorbed by the products of oxidation and evaporation. The linear dependences of the decimal logarithm of thermal energy absorbed by the products of oxidation and evaporation and the total thermal energy absorbed by these products from the decimal logarithm of the test time in the temperature range from 160 to 190 ° C are established. A graphical and analytical model is proposed for predicting the indicators of thermo-oxidative stability, determined by the dependence of the decimal logarithm of the thermal energy absorbed by the products of oxidation and evaporation on the temperature of thermostating.

1. Introduction

As a result of the study of the mechanism of oxidation of lubricating oils [1-3], it was found that the formation of water, resins, ester acids increases the acidity of the oil, which many authors have taken as a criterion for assessing the intensity of oxidation processes [4, 5]. In [6], a formula was proposed for calculating the time to reach the set value of the acid number at temperature T_x based on data obtained at temperatures T_1 and T_2:

$$\lg\frac{t_x}{t_1 - t_c} = \frac{T_2}{T_x}\cdot\frac{T_1 - T_x}{T_2 - T_1}\cdot\lg\frac{t_1 - t_c}{t_2 - t_c}, \qquad (1)$$

where t_x – the required time to reach the accepted values of the acid number at the test temperature T_x, ч; t_1 – the time to reach the accepted values of the acid number at the test temperature T_1, h; t_2 – the time to reach the accepted values of the acid number at the test temperature T_2, h; t_c – oxidation resistance time, h.

In the research work [7], an alternative method for predicting the indices of thermo-oxidative stability using photometry and an analytical model is proposed

$$\frac{\lg t_1 - \lg t_2}{T_1 - T_2} = \frac{\lg t_1 - \lg t_x}{T_1 - T_x}, \qquad (2)$$

where t_1, t_2, t_x – the time to reach the set values of thermal oxidative stability at temperature T_1, T_2, T_x, h.

Using formula (2), it is possible to determine the established values of the optical density, volatility and coefficient of thermo-oxidative stability, taking into account the processes of oxidation and evaporation for the selected temperature T_x based on these temperatures T_1 and T_2. It should be noted that lubricating oil cannot absorb thermal energy indefinitely. Therefore, its excess is discharged in the form of oxidation and evaporation products. The amount of thermal energy Q absorbed by the lubricating oil is determined by the product

$$Q = T \cdot t, \tag{3}$$

where T – test temperature, °C; t – test time at which excess thermal energy was released, h.

The amount of thermal energy absorbed by the oxidation products is determined by the product

$$Q_D = T \cdot t \cdot D, \tag{4}$$

where D – optical density of lubricating oil during the test t.

Amount of thermal energy absorbed by evaporation products G

$$Q_{K_G} = T \cdot t \cdot K_G, \tag{5}$$

where K_G – evaporation coefficient;

The total amount of thermal energy absorbed by the products of oxidation and evaporation is determined by the formula

$$Q_{K_{tos}} = T \cdot t \cdot K_{tos}, \tag{6}$$

where K_{tos} – value of the coefficient of thermal oxidative stability during the test t.

The purpose of this research is to substantiate a graph-analytical model for predicting thermal oxidative stability indicators based on data obtained at three test temperatures or a reduced temperature control time.

2. Materials and methods

The universal, all-weather partially synthetic Total Quartz 10W-40 SL/CF engine oil was chosen for the study. As a means of control and testing were used: thermostat, photometric device and electronic scales. A sample of constant weight oil (100 ± 0.1 g) was thermostated, sequentially at temperatures of 160 °C, 170 °C, 180 °C and 190 °C with stirring with a mechanical stirrer with a speed of 300 rpm. At regular intervals (10 hours), a sample of oxidized oil was weighed, the mass of evaporated oil was determined; a part of the sample (2 g) was taken for direct photometry and determination of optical density D

$$D = \lg \frac{300}{R}, \tag{7}$$

where 300 – photometer readings in the absence of oil in the cuvette, μA; R – photometer readings with an oil-filled cuvette, μA.

The coefficient of thermo-oxidative stability was calculated from the data of optical density and evaporation K_{tos}

$$K_{tos} = D + K_G, \tag{8}$$

where K_G – evaporation coefficient

$$K_G = \frac{m}{M}, \tag{9}$$

where m – mass of evaporated oil during the study t, g; M – mass of oil sample before testing, g.

3. Study of the optical density of lubricating oils

The investigated engine oil was tested at temperatures of 190 and 180 °C to a value of optical density equal to 0.5...0.6, and at temperatures of 170 and 160 °C the tests were carried out for 30 hours and after every 10 hours of testing, samples of oxidized oil were taken for determination of optical density, volatility and coefficient of thermo-oxidative stability, then, according to the data obtained, the indicators of thermo-oxidative stability were predicted after 10 hours to 90 hours of testing for a temperature of 170 °C and 140 hours for a temperature of 160 °C. At the same time, engine oil tests were continued to verify the convergence of experimental and calculation methods for determining the indicators of thermo-oxidative stability.

Figure 1a shows the dependences of the decimal logarithm of the thermal energy absorbed by the oxidation products on the decimal logarithm of the time and temperature of the engine oil test. At a temperature of 190 °C (direct 1), the oil was tested for 30 hours, and at a temperature of 180 °C - 60 hours (direct 2). These dependences are described by linear equations for temperatures:

190 °C $$\lg Q_D = 2.44(\lg t - 0.07), \quad (10)$$

180 °C $$\lg Q_D = 2.37(\lg t - 0.21). \quad (11)$$

For temperatures of 170 °C (direct 3) and 160 °C (direct 4), the test oil was tested at the beginning of 30 hours, while the analysis was carried out after 10 hours. Based on the results of three values of the decimal logarithm of the thermal energy absorbed by the oxidation products, graphical dependences on the decimal logarithm of time are constructed, which are described by linear equations for temperatures:

170 °C $$\lg Q_D = 2.24(\lg t - 0.32), \quad (12)$$

160 °C $$\lg Q_D = 2.20(\lg t - 0.49), \quad (13)$$

where 2.44, 2.37, 2.24 and 2.2 – coefficients characterizing the rate of change of the decimal logarithm of the thermal energy absorbed by the oxidation products from the decimal logarithm of the time and temperature of the test, h^{-1}; 0.07, 0.21, 0.32 and 0.49 – coefficients characterizing the start time of the change in the decimal logarithm of the thermal energy absorbed by the oxidation products, h.

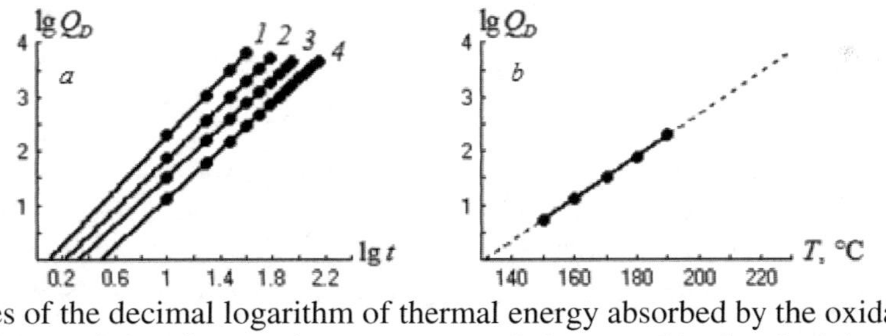

Figure 1. Dependences of the decimal logarithm of thermal energy absorbed by the oxidation products on the decimal logarithm of the time and temperature of the test (b) absorbed by the oxidation products after 10 hours of testing the partially synthetic engine oil of the partially synthetic engine oil Total Quartz 10W-40 SL/CF: 1 - 190 °C; 2 - 180 °C; 3 - 170 °C; 4 - 160 °C

According to the formulas 12 and 13, the values for $\lg Q_D$ the test oil test time were calculated at a temperature of 170 °C up to 90 hours of testing ($\lg t = 1.95$) and a temperature of 160 °C up to 140 hours ($\lg t = 2.14$) with an increase of time by 10 hours and sampling of oils for analysis.

For example, at a temperature of 160 °C, it is necessary to determine the optical density during the test time of 50 hours ($\lg t$ = 1.7) using the formula (13) or Figure 1. First, the decimal logarithm of the thermal energy absorbed by the oxidation products is determined

$$\lg Q_D = 2.21(1.7 - 0.49) = 2.67, \quad (14)$$

As $Q_D = T \cdot t \cdot D$ and antilogarithm 2.67 is 470.98, but $470.98 = T \cdot t \cdot D = 160 \cdot 50 \cdot D$, then $D = 470.98 / 8000 = 0.06$.

Comparing with the experimental data (table 1), at a test time of 50 hours, the optical density was 0.06. The relative error was 1.66%.

Further tests of the test oil at temperatures of 170 and 160 °C showed that the experimental data fell on lines 3 and 4.

In order to predict the values of thermal energy absorbed by the oxidation products and determine the optical density at other test temperatures, a graph-analytical model is presented, represented by the dependence of the decimal logarithm of the thermal energy absorbed by the oxidation products on the test temperature (Figure 1b).

This dependence is described by a linear equation

$$\lg Q_D = 0.0388(T - 131) \tag{15}$$

where 0.0388 – coefficient characterizing the rate of change of the decimal logarithm of thermal energy from the test temperature, h^{-1}; 131 - the value of the temperature of the onset of oxidation processes, °C.

Using formula (15), it is possible to determine the value of thermal energy absorbed by the oxidation products Q_D and the value of optical density at other temperatures. For example, at a test temperature of 150 °C, the decimal logarithm of thermal energy was 0.7372, and the amount of thermal energy absorbed by the oxidation products (antilogarithm of the number 0.7372) was 5.46, and since $5.46 = 150 \cdot 10 \cdot D$, then $D = 0.0364$.

For a temperature of 200 °C, the decimal logarithm of thermal energy was 2.6772, and the amount of thermal energy absorbed by the oxidation products was $Q = T \cdot t \cdot D = 475.6$, $D = 0.238$.

The dependences of the decimal logarithm of the thermal energy absorbed by the products of evaporation (Figure 2a) on the decimal logarithm of the test time are constructed in a similar way. These dependences are described by linear equations for temperatures:

190 °C $$\lg Q_{K_G} = 1.8062 \lg + 0.1007, \tag{16}$$

180 °C $$\lg Q_{K_G} = 1.8743 \lg + 0.2661, \tag{17}$$

170 °C $$\lg Q_{K_G} = 1.9340 \lg + 0.6677, \tag{18}$$

160 °C $$\lg Q_{K_G} = 1.9675 \lg + 1.0834. \tag{19}$$

where 1.8062, 1.8743, 1.9340 and 1.9675 – coefficients characterizing the rate of change of thermal energy absorbed by the evaporation products from the decimal logarithm of the time and temperature of the test, h^{-1}; 0.1007, 0.2661, 0.6677 and 1.0834 – coefficients characterizing the value of thermal energy at $\lg t = 0$, or the start time of the change in thermal energy absorbed by the evaporation products, h.

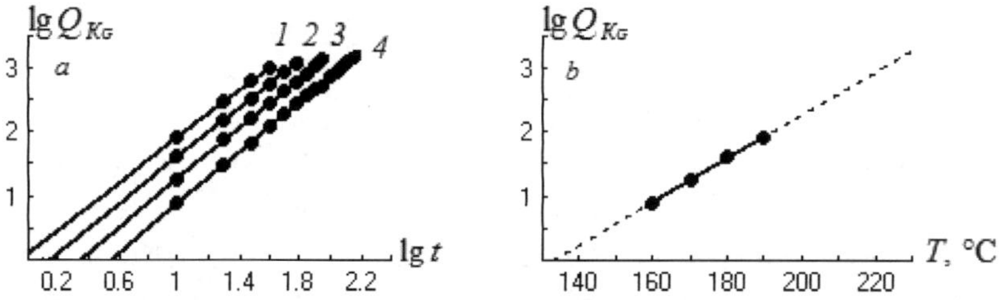

Figure 2. Dependences of the decimal logarithm of thermal energy, (a) absorbed by the evaporation products on the decimal logarithm of the time and temperature of the test (b) absorbed by the evaporation products after 10 hours of testing the partially synthetic engine oil Total Quartz 10W-40 SL/CF: 1 - 190 °C; 2 - 180 °C; 3 - 170 °C; 4 - 160 °C

Formulas 18 and 19 calculated the values $\lg Q_{K_G}$ for the test time of the test oil at temperatures of 170 and 160 °C. For example, at a temperature of 160 °C it is necessary to determine the value of volatility during a test of 50 hours ($\lg t$ = 1.7) using formula (19) or Figure 2a. In this case, the decimal logarithm of the thermal energy absorbed by the products of evaporation is determined

$$\lg Q_{K_G} = 1.9675 \cdot 1.7 + 1.0834 = 2.2614, \quad (20)$$

As $Q_{K_G} = T \cdot t \cdot K_G = 2.2614$, and antilogarithm 2.2614 is 182.6, but $182.6 = T \cdot t \cdot K_G = 160 \cdot 50 \cdot K_G$, then $K_G = 182.6/8000 = 0.0228$.

Comparing with the experimental data (table 1), at a test time of 50 hours, the volatility was 0.0227 g. The relative error was 0.44%.

To predict the values of thermal energy absorbed by the products of evaporation and to determine the volatility at other temperatures, we use a graph-analytical model represented by the dependence of the decimal logarithm of the thermal energy absorbed by the products of evaporation on the test temperature of the test engine oil (Figure 2b).

This dependence is described by a linear equation

$$\lg Q_{K_G} = 0.0337\,(T - 133), \quad (21)$$

where 0.0337 – coefficient characterizing the rate of change of the decimal logarithm of thermal energy absorbed by the products of evaporation, g·°C; 133 - value of the temperature of the beginning of evaporation of the investigated engine oil, °C.

Using formula (21), it is possible to determine the decimal logarithm of the thermal energy absorbed by the products of evaporation and the value of volatility at other temperatures for 10 hours of testing. So, for example, at a test temperature of 150 °C, the decimal logarithm of thermal energy was $\lg Q_{K_G} = 0.0337\,(150 - 133) = 0.573$, but $Q_{K_G} = 3.7$, as $3.7 = 150 \cdot 10 \cdot K_G$, then $K_G = 0.0025$.

At test temperature 200 °C $\lg Q_{K_G} = 0.0337\,(200 - 133) = 2.2579$, then $Q_{K_G} = 181.1$, as $181.1 = 200 \cdot 10 \cdot K_G$, then $K_G = 0.0906$.

The dependences of the decimal logarithm of the total thermal energy absorbed by the products of oxidation and evaporation on the decimal logarithm of the test time are constructed in the same way (Figure 3a). These dependences are described by linear equations for temperatures:

190 °C $$\lg Q_{K_{tos}} = 2.5373\,(\lg t - 0.06) \quad (22)$$

180 °C $$\lg Q_{K_{tos}} = 2.2990\,(\lg t - 0.13) \quad (23)$$

170 °C $$\lg Q_{K_{tos}} = 2.1739\,(\lg t - 0.22) \quad (24)$$

160 °C $$\lg Q_{K_{tos}} = 2.1570\,(\lg t - 0.38) \quad (25)$$

where 2.5373, 2.2990, 2.1739 and 2.1570 – coefficients characterizing the rate of change of the total thermal energy absorbed by the products of oxidation and evaporation from the decimal logarithm of the time and temperature of the test, h^{-1}; 0.06, 0.13, 0.22 and 0.38 – coefficients characterizing the decimal logarithm of the onset of oxidation and evaporation from the test temperature, h.

Using formulas 24 and 25, we calculate the values $\lg Q_{K_{tos}}$ for the test time of the test oil at temperatures of 170 and 160 °C. For example, at a temperature of 160 °C, it is necessary to determine the value of the coefficient of thermo-oxidative stability K_{tos} during a test of 50 hours ($\lg t$= 1.7) using formula (25) or Figure 3a. The decimal logarithm of the total thermal energy absorbed by the products of oxidation and evaporation is determined

$$\lg Q_{K_{tos}} = 2.1570\,(1.7 - 0.38) = 2.85, \quad (26)$$

As $\lg Q_{K_{tos}} = T \cdot t \cdot K_{tos} = 2.85$, and antilogarithm 2.85 equals 707.95, but $707.95 = T \cdot t \cdot K_{tos} = 160 \cdot 50 \cdot K_{tos}$, then $K_{tos} = 707.95 / 8000 = 0.089$.

Comparing the calculated data of the coefficient of thermal oxidative stability K_{tos} = 0.089 with experimental (table 1) K_{tos}= 0.09, found that the relative error was 4.5 %.

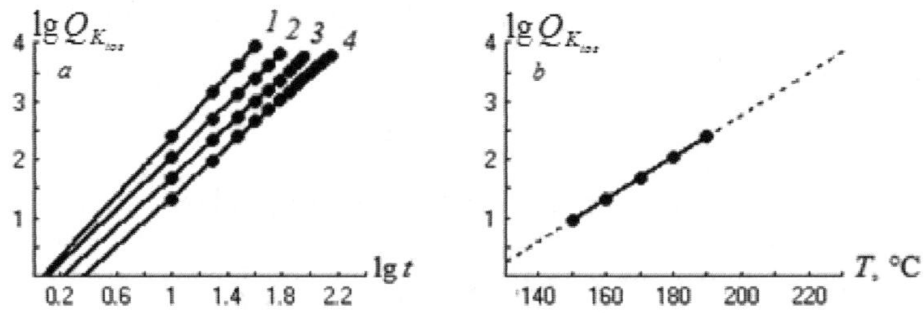

Figure 3. Dependences of the decimal logarithm of thermal energy absorbed by the products of oxidation and evaporation on the decimal logarithm of the time and temperature of the test (b) absorbed by the products of oxidation and evaporation after 10 hours of testing the partially synthetic engine oil Total Quartz 10W-40 SL/CF: 1 - 190 °C; 2 - 180 °C; 3 - 170 °C; 4 - 160 °C

Using equations 10-25 thermo-oxidative stability values can be predicted *D, G* и K_{tos} for any test time, using the dependencies obtained over three test time ranges.

To predict the values of the total thermal energy absorbed by the products of oxidation and evaporation, and to determine the coefficient of thermal oxidative stability at other temperatures, we use a graph-analytical model represented by the dependence of the decimal logarithm of the total thermal energy absorbed by the products of oxidation and evaporation on the test temperature of the test engine oil (Figure 3b).

This dependence is described by a linear equation

$$\lg Q_{K_{tos}} = 0.033(T - 123.5), \quad (27)$$

where 0.033 – coefficient characterizing the rate of change of the decimal logarithm of the total heat energy from the temperature of the test engine oil, °C; 123.5 – temperature of the start of conversion processes in lubricating oil, °C.

Using formula (27), it is possible to determine the values of the decimal logarithm of the total thermal energy absorbed by the products of oxidation and evaporation, and the value of the coefficient of thermo-oxidative stability at other temperatures for 10 hours of testing. So, for example, at a test temperature of 150 °C, the decimal logarithm of thermal energy was $\lg Q_{K_{tos}} = 0.033(150 - 123.5) = 0.956$, but $Q_{K_{tos}} = 9.045$, as $9.045 = 150 \cdot 10 \cdot Q_{K_{tos}}$, then $Q_{K_{tos}} = 0.26$. At test temperature 200 °C $\lg Q_{K_{tos}} = 2.7609$, then $Q_{K_{tos}} = 576.63$, as $576.63 = 200 \cdot 10 \cdot K_{tos}$, then $K_{tos} = 0.288$

4. Conclusion

The provided method for predicting the performance of a thermo-oxidative system takes into account thermal energy, which allows obtaining a graphical and analytical model for predicting the performance of thermo-oxidative stability for any temperatures obtained in three time values using the decimal logarithm of thermal energy and liquid temperature thermostating test oil tests.

The proposed forecasting method allows temperature studies being carried out without testing and determining the decimal logarithm of thermal energy, thermostating and the values of thermo-oxidative density studied by lubricating oil, as well as determining the temperature at which oxidation and evaporation processes begin.

Table 1. Test results and design data

Time, h	lg*t*	*D*	$Q_D=T{\cdot}t{\cdot}D$	$lgT{\cdot}t{\cdot}D$	K_G	$Q_{KG}=T{\cdot}t{\cdot}G$	$lgT{\cdot}t{\cdot}K_G$	K_{tos}	$Q_{\Pi Toc}=T{\cdot}t{\cdot}K_{tos}$	$lgT{\cdot}t{\cdot}K_{tos}$
Total Quartz 10W-40 SL/CF T_{test} = 190 °C										
10	1,00	0,102	194,6	2,29	0,0414	78,66	1,90	0,1337	254,00	2,40
20	1,30	0,278	1056,4	3,02	0,0765	290,57	2,46	0,3288	1420,44	3,15
30	1,48	0,525	2992,5	3,48	0,1046	596,02	2,78	0,7178	7091,46	3,61
40	1,60	0,843	6406,8	3,81	0,1245	945,90	2,98	1,1659	8860,84	3,95
Total Quartz 10W-40 SL/CF T_{test} = 180 °C										
10	1,00	0,041	74,1	1,87	0,0219	39,42	1,60	0,0619	111,42	2,05
20	1,30	0,103	370,8	2,57	0,0416	149,80	2,18	0,1432	515,52	2,70
30	1,48	0,180	972,0	2,99	0,0600	324,24	2,51	0,2382	1286,28	3,11
40	1,60	0,272	1958,4	3,29	0,0771	555,43	2,74	0,3466	2495,52	3,40
50	1,70	0,379	3411,0	3,53	0,0927	834,68	2,92	0,4687	4218,30	3,63
60	1,78	0,500	5400,0	3,73	0,1067	1152,08	3,06	0,6043	6526,44	3,81

table continuation

Time, h	lg*t*	*D*	$Q_D=T{\cdot}t{\cdot}D$	$lgT{\cdot}t{\cdot}D$	K_G	$Q_{KG}=T{\cdot}t{\cdot}G$	$lgT{\cdot}t{\cdot}K_G$	K_{tos}	$Q_{\Pi Toc}=T{\cdot}t{\cdot}K_{tos}$	$lgT{\cdot}t{\cdot}K_{tos}$
Total Quartz 10W-40 SL/CF T_{test} = 170 °C										
10	1,00	0,019	31,6	1,50	0,0108	18,36	1,26	0,0281	47,77	1,68
20	1,30	0,048	162,5	2,21	0,0214	72,87	1,86	0,0660	224,40	2,35
30	1,48	0,079	400,9	2,60	0,0303	154,75	2,19	0,1053	537,03	2,73
40	1,60	0,111	754,1	2,88	0,0399	271,22	2,43	0,1459	992,12	3,00
50	1,70	0,145	1230,0	3,09	0,0492	418,24	2,62	0,1878	1596,30	3,20
60	1,78	0,180	1836,0	3,26	0,0584	595,28	2,77	0,2310	2356,20	3,37
70	1,85	0,217	2579,9	3,41	0,0673	801,40	2,90	0,2755	3278,45	3,51
80	1,90	0,255	3469,4	3,54	0,0760	1034,09	3,01	0,3299	4486,64	3,65
90	1,95	0,295	4512,0	3,65	0,0846	1294,77	3,11	0,3685	5638,05	3,75
Total Quartz 10W-40 SL/CF T_{test} = 160 °C										
10	1,00	0,008	13,5	1,13	0,0049	7,84	0,89	0,0099	20,89	1,32
20	1,30	0,018	58,9	1,77	0,0093	29,91	1,48	0,0308	98,56	1,99
30	1,48	0,031	148,3	2,17	0,0138	66,38	1,82	0,0515	247,20	2,39
40	1,60	0,045	290,6	2,46	0,0183	116,80	2,07	0,0722	462,08	2,66
50	1,70	0,060	480,8	2,68	0,0227	181,67	2,26	0,0927	741,60	2,87
60	1,78	0,075	720,0	2,85	0,0272	261,18	2,42	0,1132	1086,72	3,04
70	1,85	0,090	1008,0	3,00	0,0316	354,39	2,55	0,1335	1495,20	3,17
80	1,90	0,105	1344,0	3,13	0,0361	462,24	2,66	0,1537	1967,36	3,29
90	1,95	0,121	1735,2	3,24	0,0341	490,94	2,69	0,1739	2504,16	3,40
100	2,00	0,136	2174,4	3,34	0,0449	718,02	2,86	0,1939	3102,40	3,49
110	2,04	0,152	2668,2	3,43	0,0497	873,93	2,94	0,2138	3762,88	3,58
120	2,08	0,167	3212,2	3,51	0,0546	1048,19	3,02	0,2336	4485,12	3,65
130	2,11	0,183	3812,6	3,58	0,0586	1219,92	3,09	0,2534	5270,72	3,72
140	2,15	0,199	4466,6	3,65	0,0631	1413,64	3,15	0,2730	6115,2	3,79

References

[1] Chernozhukov N I, Kerin S E, Losikov B V 1995 *Chemistry of mineral oils ML L* (Mosow: Gotoptekhizdat).
[2] Nepogodyov A In 1997 *Chemistry and technology of fuels and oils* **4** 34-38
[3] Shram V G, Lysyannikov A V, Kovaleva M A 2016 *Procedia Engineering* **150** 458-463.
[4] Bakunin VN, Popova ZV, Oganesova E Yu, Kuzmina G N 2001 *Petrochemicals* **1** 41-46.

Journal of Physics: Conference Series **1384** (2019) 012047 doi:10.1088/1742-6596/1384/1/012047

[5] Kondakov LA 1982 *M Mechanical Engineering* **216**
[6] Kowalsky B I, Shram V G, Lysyannikov A V 2018 *Lecture Notes in Mechanical Engineering* 4th International Conference on Industrial Engineering 873-880.
[7] Kovalsky BI, Petrov ON, Nikolenko GA, Agrovichenko D 2012 *A method for predicting the thermal oxidative stability of lubricants* (Russian Patent No. 2649660).

Effect of coal-containing additives on the rheological properties of limestone-nepheline charge in the technology of alumina production

E V Siziakova, P V Ivanov, T Yu Nikitina

Saint-Petersburg Mining University, Vasilyevsky Island 21st line, 2, Saint-Petersburg, 199106, Russia

E-mail: Ivanov_pv@pers.spmi.ru

Abstract. The article is devoted to the influence of brown coal additives on the sintering of the lime-nepheline charge in the technology of alumina production from nepheline. The process of extraction of humic compounds from brown coal is considered, the effect of humic additives on the rheological properties of pulp depending on the size and moisture content of carbon-containing additives is studied.

1. Introduction

Due to the sufficient contain of bauxite-containing ores, the actual production of alumina from nepheline-containing raw materials is a topical issue for the Russian aluminum industry [1].

One of the most important revamping of this technology is the process of nepheline sintering with limestone in a tubular rotating furnace [2-4]. This operation is associated with significant fuel and energy consumption, the reduction of which is one of the urgent tasks for the intensification of production. As a solution to this problem, it may be suggested to reduce the moisture content of the alumina-containing pulp entering sintering, while maintaining its mobility at the level necessary for transportation by pipelines, by adding lignite charcoal additives containing surfactants-humic acids.

From the literature data [5,6] it is known that humic compounds obtained from natural coal containing materials by alkaline treatment are effective liquefier of nepheline-lime pulps. Conducted studies proved that introduction of coal-alkaline reagent (CAR) into the nepheline-limestone pulp allows reducing moisture content of the pulp from 30% (current industrial value) to 24.5%, while retaining mobility allowing transportation by pipelines without increasing energy consumption[7-11].

For this purpose it is conceivable to use the additions of humic-containing lignite charcoal without preliminary alkaline treatment due to the interaction of humic acid with alkalis of the pulp liquid phase. To this end, it is necessary to determine conditions under which the humates extraction proceeds to the maximum extent, determine the optimal composition of the pulp salt phase, the effect of charcoal additives on rheological properties [12-14], depending on the content of humic acids in it, and also determine the amount of additives that have maximum diluting effect on the pulp and minimum critical humidity of the pulp with additives, at which its mobility for transportation is maintained.

2. The influence of particle size of coal on the rate of extraction of humic compounds

Limestone-nepheline pulps of industrial composition were used in studies. The research purpose was to study the effect of humic compounds on the rheological properties of limestone-nepheline pulps. It is known from the literature that humic compounds, being surface-active substances, have a liquefying effect based on their adsorption on the surface of dispersed phase particles, preventing the aggregation of these particles and reducing strength of the coagulation structures, adsorption blocking bonding sites of the particles.

To study the effect of surface-active additives on pulp rheology in laboratory conditions, the humates were used, previously obtained from charcoal by treating it with alkali. The action of NaOH is first manifested in decomposition of the complex formed from residual charcoals and humates, and then in ion exchange between insoluble cations of calcium and aluminum and sodium ions. As a result of the interaction of alkali with lignite charcoal, a coal-alkali reagent is formed, consisting of three parts: liquid; coarse-grained, easily precipitated; slowly precipitating gelatinous suspension. The amount of alkali necessary to saturate the exchange complex of humic acids with sodium cations is 0.12 g per 1 gram of humic substances. Dissolution of humic acids in alkaline solutions is preceded by peptization. At a certain ratio between humic acids and an alkaline solution, a polydisperse system is formed, containing the humate of the alkaline solution and salt of the humic acids. With acidification of alkaline solutions, as well as with introduction of calcium and magnesium ions, humic acids coagulate. At low pH values, humic acids are coupling into large micelles, so the viscosity of aqueous solutions of humates decreases with pH decreasing, and the surface tension is also decreases.

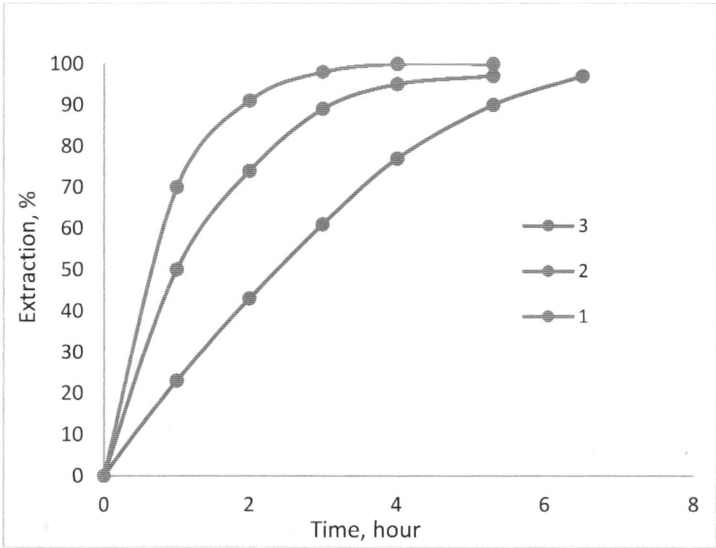

Figure1. Extraction rate of humic compounds from brown coal of various sizes:
1 - particles less than 50 μm
2 - particles less than 1 mm
3 - particles less than 5 mm

Figure 1 shows the dependence of charcoal solubility via the particle size. These data prove that the charcoal particles size significantly affect transition of humates to the solution. When the coal is crushed to a particle size of 50 μm, about 90% of the humates passes into solution within 2 hours. With a high content of humates in coal (60 - 90%), short time leaching of coal is sufficient to ensure the necessary mobility of low humidity pulps. This fact is confirmed by the experiments of the determination of the diameter of the circle spreading of coals of different sizes. In this case, an alkaline solution of humates was obtained by treating the lignite charcoal with a liquid pulp phase at temperature of 60°C for 2 hours. For comparison, a solution of humates obtained by leaching charcoal of various concentrations separated from the solid charcoal phase were added to the dry charge, as well as solution after leaching, containing particles of residual charcoal (such as an coal-alkali reagent) of various sizes.

Coal solution containing solid phase significantly reduces the diluting effect of humates. This may be due to the pulp concentration increase caused by the increase of the solid phase content, the adsorption properties of solid charcoal particles, or increased viscosity of the colloidal component of the reaction product of charcoal with alkali. In the case of finer grinding of charcoal in a solution (to a particle size of less than 50 μm), its negative effect on the pulp mobility was reduced.

When pulp is stored for 1-3 days, its rheological characteristics improves significantly due to the ongoing process of humates isolating, further prolongation of the aging time does not significantly affect it. However, the complete release of humates is accomplished in 6 to 7 days, and the pulps stored during this time have a somewhat better fluidity than those stored for 3 to 5 days. This is especially evident for pulps that initially did not have the ability to flow, in which the structure is destroyed the next day by the process of humates releasing reducing its viscosity.

Dispersion composition of recommended size charcoal (not larger) after humic compounds leaching is as follows: less than 3 μm - 85% of the particles; of them - less than 1 μm - 30%, from 1 to 2 - 40%, from 2 to 3 μm - 15%. Density of such charcoal is 1.6 g/cm^3.

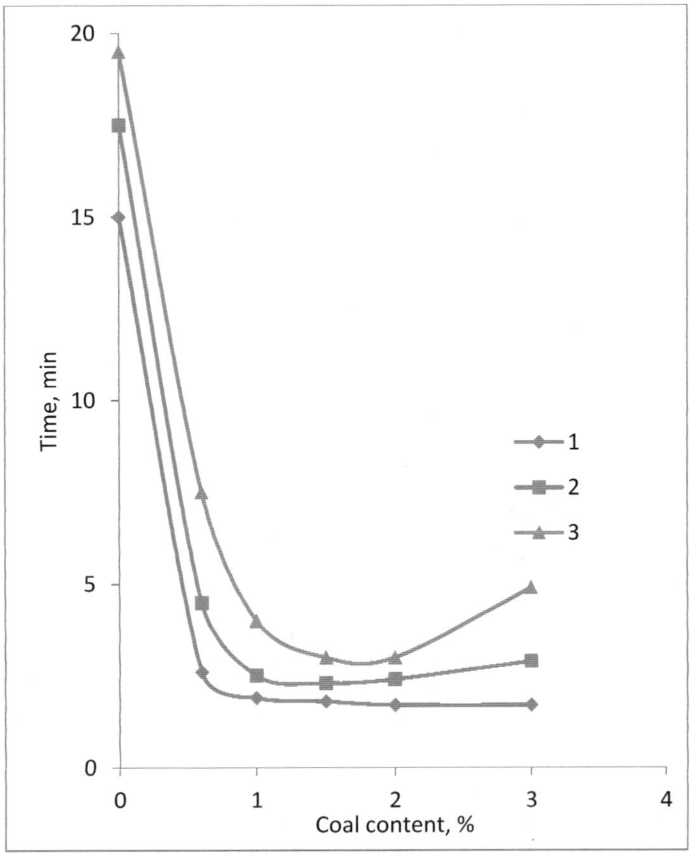

Figure 2. Dependence of pulp flow time at 25% humidity via amount of humic lignite charcoal additive: 1 - nepheline ore; 2 - limestone; 3 - nepheline-limestone charge.

3. Study of the effect of brown coal additives of different moisture on the fluidity of limestone-nepheline pulps

In order to determine the rheological properties of the charge, into which the charcoal was introduced during grinding, a capillary viscometer with a hole of 3 mm in diameter was used. The obtained dependences of flow time via the coal additive amount at humidity of 25% are shown in Figure 2, the dependence on humidity for various coal additives is shown in Figure. 3.

In order to establish on which component of a charge the charcoal exerts greater effect, the experiments were carried out separately with nepheline and limestone pulps, and then with nepheline-limestone pulp. According to the experimental data, it may be argued that the limestone slurry has a structure more prone to fracture and reorientation of the particles. For limestone, small amounts of charcoal (about 0.25% in terms of coal content) is sufficient, and further increase of coal content does not result in any changes of material consistency. The lamellar structure of limestone not only promotes formation of stable structures, but also their destruction. Probably, limestone particles move apart along the cleavage plane in the course of humic salts adsorption and reorient, causing the structure to rearrange.

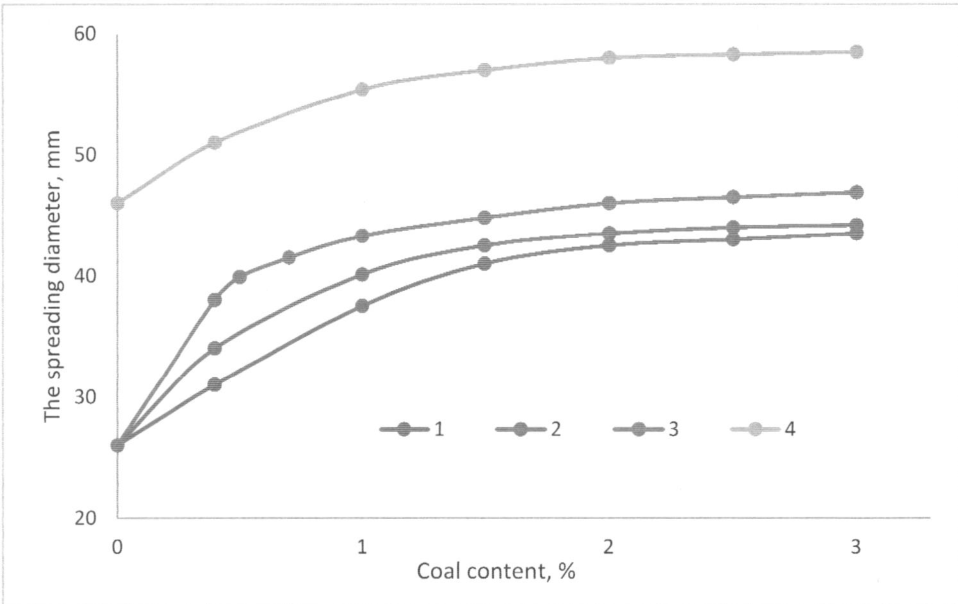

Figure 3. Effect of the humic lignite charcoal on the fluidity of limestone-nepheline pulps at humidity of: 1 – 24.5%; 2 – 25%; 3 – 26%; 4 – 30%.

Nepheline has a hexagonal framework structure, so nepheline suspension, in order to achieve the maximum degree of dilution, requires a larger amount of surfactant — 1.5-2% (in terms of coal content) to reorient the particles of this form. With charcoal content increasing from 3% or higher, fluidity deteriorates again, the pulp loses mobility.

When considering the pulps mobility in presence of humic lignite charcoal addition, a range of moisture values below critical value is of interest to illustrate possibility of its reducing while maintaining liquid-flow properties.

Figure 4 shows dependence of lime-nepheline charge flow time when charcoal is added in the amount of 0.25% via moisture at size particle of less than 50 μm and on the particle size at 30% moisture.

Since the limestone component predominates in the charge (ratio of the limestone to nepheline amounts is approximately 2:1), the effect exerted by coal on its rheological properties is determining. Presumably, such an effective of small additive is explained by the fact that this amount of humates, extracted from coal, forms a monomolecular layer on the surface of reoriented particles sufficient to destroy the structure of the suspension and change its properties.

Adsorbed on solid particles of the suspension, humic salts interact through hydrogen bonds and electrostatic forces with adsorbed cations and anions that are part of limestone and nepheline lattices. That is why, it is sufficient to form a monomolecular layer on the surface of the particles of the solid phase.

High content of alkalis in the liquid phase has a retarding effect on the humic compounds activity in limestone-nepheline pulp because presence of electrolytes in the solution promotes increasing of humic acids formation, and as with solution pH increasing, the viscosity of aqueous humates solutions increase. Dependence of pulp fluidity in presence of charcoal humic on moisture has the form of a hyperbola. Dependence of pulp flow at constant value on the amount of coal additive up to the limit value, which has a diluting effect, can be described by equation:

$$\tau = A \cdot e^{BX+CX2}$$

where τ – time of pulp flow, X – content of coal additive, A, B, C – empirical coefficients.

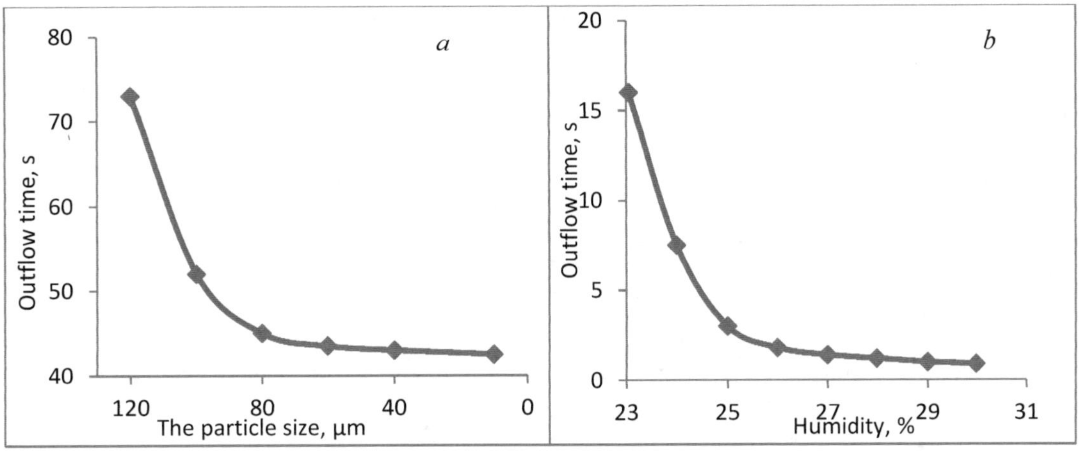

Figure 4. Effect of humic lignite charcoal on the fluidity of limestone-nepheline pulp with 0.5% coal addition: a - depending on the particle size of charcoal at 30% humidity and b- depending on the moisture content of the pulp when particle size charcoal is less than 50 µm.

The essential factor for pulp liquefaction is the amount of humic acids contained in charcoal, because their number fluctuates depending on the formation. Therefore, for a guaranteed positive effect, the required amount of charcoal must be somewhat greater than the minimum.

The behavior of humates is determined by their content in the lignite charcoal. Charcoal with high humic acids content should be ground to sufficient finely, enough to increase the degree of humates extraction, which contributes to the higher effectiveness of surfactants on the rheological properties of the pulp. The extraction of sufficient amount of humates from the coals with reduced humic acids content in the solution is possible with the increase in the coal addition, but in this case the positive effect of additive is levelled due to the increase of the solid phase mass that does not contain humates and increase in the pulp water demand. Even fine grinding of charcoal does not improve the pulp fluidity, on the contrary, it contributes to its delamination during storage due to emergence of small coal particles due to their lower density compared to the pulp solid phase. In case of high humic acids content in the coal, fine grinding does not lead to the delamination of the pulp because most of the coal is dissolved in alkali and the solid residue consists mainly of mineral part of the coal, the graphitized residue being not more than 5% of the graphitized coal part.

From the foregoing, it can be concluded that the introduction of lignite charcoal into a limestone-nepheline pulp with a high alkali content in the liquid phase has a diluting effect associated with the humates formation and is based on the fact that they are adsorbed on the surface of the particles of the dispersed phase and reduce the strength of the coagulation structures. The addition of lignite charcoal contributes to the increase in pulp flow, reduction of the flow time and decrease in shear stress. In this case, pronounced effect is achieved by the addition of charcoal with high content of humic acids (60-80%). The introduction of humates exerts greater effect on the limestone component of the pulp due to the specific features of the limestone particles structure. The range in which the charcoal addition is effective on rheology is from 0.5 to 2% in terms of coal content.

The production tests proved that also the energy consumption for transportation by pipeline is maintained, the addition of 1% of lignite charcoal in terms of coal content makes it possible to reduce moisture content of the pulp from 30 to 26%, which reduces the fuel consumption for moisture evaporation in the sintering furnaces.

4. Conclusions

1. The introduction of lignite charcoal into a limestone-nepheline pulp with a high alkali content in the liquid phase has a diluting effect associated with the humates formation and is based on the fact that they are adsorbed on the surface of the particles of the dispersed phase and reduce the strength of the coagulation structures.
2. The addition of lignite charcoal contributes to the increase in pulp flow, reduction of the flow time and decrease in shear stress. In this case, pronounced effect is achieved by the addition of charcoal with high content of humic acids (60-80%). The introduction of humates exerts greater effect on the limestone component of the pulp due to the specific features of the limestone particles structure.
3. The range in which the charcoal addition is effective on rheology is from 0.5 to 2% in terms of coal content.
4. The production tests proved that also the energy consumption for transportation by pipeline is maintained, addition of 1% of lignite charcoal in terms of coal content makes it possible to reduce moisture content of the pulp from 30 to 26%, which reduces the fuel consumption for moisture evaporation in the sintering furnaces.

References

[1] Arluk B I 1978 *Improvement of processing aluminate-alkaline nepheline* sintering (Moscow: Institute Nonferinformation) 52
[2] Makarov G N 1986 *Chemical technology of solid fossil fuels* (Moscow.: Chemistry) 496
[3] Zenkova N A 1988 Reducing energy costs for transportation glinozemservice slurries *Nonferrous metals* **9** 60-61
[4] Zenkova N A, Siziakov V M 1988 Effect of addition of brown coal on the rheology and grinding of limestone-nepheline slurry *Nonferrous metallurgy* **1** 17-19
[5] Rainer M 1988 *Rheology* (Moscow: Sience) p 223
[6] Cheremisina O, Lutsky D and Fedorov A. 2017 Comparison of extraction methods for extraction of iron, aluminum, manganese and titanium using carboxylic acids and natural vegetable oils from water-salt systems *International Multidisciplinary Scientific GeoConference Surveying Geology and Mining Ecology Management SGEM* (Albena; Bulgaria) **17(11)** 803-810
[7] Neroslavskaya L L, Gusarov A L, Barbakadze L G 1988 Sintering of limestone-nepheline charge of low humidity *Nonferrous metallurgy* **23** 24-25
[8] Siziakov V M 2016 Chemical and technological regularities of the processes of sintering of alkaline aluminosilicates and hydrochemical processing specs, *J. of Mining Inst.* **217** 102-112
[9] Sizyakov V M, Korneev V I 2000 Scientific basis and technology of new materials with the addition of hydrocarboaluminate calcium *Proc. of the Int.-Conf.* (Moscow: MGU) 515-521

Journal of Physics: Conference Series **1384** (2019) 012048 doi:10.1088/1742-6596/1384/1/012048

[10] Siziakova E V, Ivanov P V 2017 Multifunctional coagulants based on hydrocarboaluminates calcium, *J. of Ecolog. Eng.* **18(2)** 16–20

[11] Siziakova E V, Ivanov P V and Boikov A V 2019 Application of calcium hydrocarboaluminate for the production of coarse-graded alumina *Journal of Chemical Technology and Metallurgy* **1(54)** 200 – 203

[12] Petrov G V, Fokina S B, Boduen A Ya, Fidarov B F, Zotova I E 2019 Aspects of electrochemical reaction mechanis of magnetite reductive leaching in sulfuric acid medium *International Journal of Mechanical Engineering and Technology* **10(1)** 1595-1601

[13] Melnichuk M S, Fokina S B, Boduen A Ya, Petrov G V 2018 Co-recovery of platinum-group metals and chrome in processing of low-grade dunite ore material *Obogashchenie Rud* **1** 50-55

[14] Petrov G V, Fokina S B, Boduen A Y, Zotova I E, Fidarov B F 2018 Arsenic behavior in the autoclave-hydrometallurgical processing of refractory sulfide gold-platinum-bearing products *International Journal of Engineering and Technology(UAE)* **7(2)** 35-39

Development of the beneficial utilisation of urban sewage sludge using modern analysis methods

Yu D Smirnov, M V Suchkova

Saint Petersburg Mining University, Vasilievsky Island, 21 line 2, Saint Petersburg, 199106, Russia

E-mail: cjgreykot@gmail.com

Abstract. The article presents Russian and foreign experience in the recycling of sludge of urban sewage wastewater with the prospect of the beneficial use of both the direct waste and the products of its processing. Special attention is also paid to the problem of recycling of ashes of sewage sludge incineration. The article also considers the possibility of applying modern methods of analysis (such as X-ray fluorescence analysis, spectroscopy, biotesting) to develop a composition of the ash-based soil mixture. The study is realized on the laboratory and experimental base of the accredited Center for the collective use of high-tech equipment of Saint-Petersburg Mining University (as a part of state task for 2018-2019 year).

1. Introduction

One of the most pressing problems in the metropolis is the issue of disposal and recycling of the ever-increasing amount of waste in a limited capacity of the objects of their placement. In particular, the activities of urban sewage treatment plants are not limited only to cleaning of drains; an integral part is also the management of sewage sludge.

In large cities, the volume of urban wastewater is calculated in the millions and the amount of sewage sludge - thousands of cubic meters per day. The reduction of sewage sludge formation and its further transformation into a harmless product are the main tasks assigned to modern technologies.

A feature of sewage sludge as a waste is the presence of organic and mineral phases. The fourth hazard class is usually due to the high content of heavy metals in the sludge [1]. Therefore, most often this waste is deposited in landfills and can not be reused. Available data demonstrate the relevance of this problem and the need to search for new approaches and technologies to involve sewage sludge in economic circulation.

In 2005, the main pathways for the disposal of sewage sludge in European Union was reused in agriculture (52%), incineration (38%) and landfills (10%) (Sewage Sludge Directive 86/278/EEC) [2].

Landfill disposal of sewage sludge is carried out when the waste can not be reused. In the countries where land resources allow disposal sites being formed, this is the main method of waste treatment [3]. After the operation of the landfill is completed, the owner is obliged to reclaim disturbed land in the manner prescribed by the legislation of the Russian Federation. Composts and soils based on sewage sludge can be successfully used for biological remediation as well as to creating green buffer strips [4, 5].

Composting is an aerobic process of bacterial decomposition for the stabilization of organic waste and the production of humus. It is a reliable method of disinfecting sewage sludge and producing useful products such as compost and fertilizer.

In particular, composting technologies available in the Northwestern region of the Russian Federation range from simple open windrow systems with a small effort in terms of process structures to fully enclosed composting plants with accelerated treatment processes, complete enclosure and a high quality treatment of exhaust air. The most feasible for sludge handling are windrow composting and tunnel composting technologies. The operation and maintenance of both methods are relatively simple and only require a basic understanding of the biology and biochemistry of composting [6].

Pyrolysis is one of the most promising areas in the disposal of sewage sludge. It is the thermal decomposition of carbon-containing materials at high temperatures in an inert atmosphere. Pyrolysis products are used in a wide variety of industries and economy. In example, biochar can be used as a fuel or source of phosphorus and nitrogen. Pyrolysis oil after processing may be a source for producing carboxylic acids, paraffin, coke dust, etc. [7].

Due to the presence of high concentrations of phosphorus and nitrogen sewage sludge is a good fertilizer. In agriculture, it is possible to widely use sewage sludge as a valuable fertilizer, turn poor soils into fertile areas and use activated sludge as a protein feed for livestock and poultry.

Netherlands, Spain and Sweden use more than 60% of sludge in agriculture, and Denmark, England and Switzerland use more than 45% for similar purposes. Their experience shows that the negative environmental effects of using sludge in agriculture can be minimized, subject to rules and strict standards [8]. There are also examples of effective fertilization and irrigation using both treated wastewater and sewage sludge [9].

However, sludge used as a fertilizer can be a source of pollution. Until the mid-1980s, collective and state farms actively used the treated sludge as a fertilizer, but in the early 1990s, permissible limit of heavy metals in soil were established and information on the increased content of heavy metals in the urban sewage sludge was made public. All this has led to difficulties in the registration of the waste as fertilizer and this is one of the reasons why in recent years the incineration has become increasingly widespread. Sewage sludge is incinerated if its recovery is impossible or economically impractical, also if the territory for its storage is limited or absent, and in those cases when it is required for sanitary and hygienic reasons [6].

Russia's attempt to adopt foreign combustion experience provided the following results: the volume of the solid phase decreased by only 20% while simultaneously emitting large amounts of gaseous toxic substances and products of combustion into the atmosphere. In this regard, in Russia, as in all other CIS countries, the main way to manage sewage sludge remains its disposal at landfills.

The negative impact that waste has on the environment consists of the alienation of land areas for landfills and the attendant pollution of the atmosphere and water bodies. Thus, it should be noted the need for a more thorough approach to the search for new technologies to involve sewage sludge in the natural economic.

Nowadays, in order to reduce waste generation, State Unitary Enterprise (SUE) «Vodokanal of St. Petersburg» uses incineration of sewage sludge in fluidized bed furnaces. Despite the fact that this technology allows reducing the waste volume by a factor of ten, the annual volumes of the incinerated sewage sludge ash (SSA) are still quite large and amount to about 50 thousand tons.

The study focused on the topical issue of urban sewage sludge utilization with the prospect of using waste in the national economy. The purpose of the study is to assess the possibility of beneficial use of sewage sludge.

The scientific novelty of the study is the establishment of a positive effect, which has the introduction of incinerated sewage sludge ash (SSA) within 20% by weight as a component of the waste-based soil mixture (at the rate of 28.53 t/ha per year), on plant growth (by example *Trifolium praténse*).

The practical significance of the study is the development of the SSA-based organic-mineral mixture taking into account the possibility of its further use for the improvement of urban areas and land reclamation.

The study was carried out using a set of research methods, including a system analysis of the problem based on studies by Russian and foreign scientists as well as laboratory methods for studying the structure and properties of sewage sludge. The study is realized on the laboratory and experimental base of the accredited Center for the collective use of high-tech equipment of the Saint-Petersburg Mining University.

2. Generation of the data

The South-West wastewater treatment plant is under the jurisdiction of the branch "Water disposal of St. Petersburg" of the State Unitary Enterprise Vodokanal St. Petersburg and put into operation in 2005. The daily volume of urban wastewater is about 330 thousand m^3. In 2007, the sludge incineration plant was commissioned; its design capacity is 310 m^3 of dry sludge (cake) per day. A cake is a mixture of sediment of primary sedimentation tanks and compacted sludge of aerotanks. Incineration is carried out in fluidized bed furnaces [10]. SSA externally is a brown fine powder. The basic composition of SSA represented by SiO_2, phosphates of Fe and Ca, silicates of certain metals; the heavy metals are concentrated mainly in the fine fraction. Particle size is 1-50 microns; bulk density is 0.68 g/cm^3 (at transport humidity).

The most promising ways of beneficial using SSA in agriculture is land reclamation, or as a component in building materials. Nowadays, the growing demand for building material requires an alternative way to obtain it from various sources, including SSA. There are researches on such areas of using SSA as the production of bricks and tiles, raw materials for cement, aggregates for concrete and a component for the synthesis of light materials etc. [9].

However, if the acidic environment affects the waste, the transition of heavy metals into a soluble form and their migration into the environment becomes possible. This is explained by the fact that with an increase in acidity in the soil solution, the amount of free fulvic acids of the active fraction increases. They bind many heavy metals, forming stable complex compounds. Therefore, metals pass into a pseudo-soluble state and become available for plants [11]. For this reason, due to the high content of heavy metals in the SSA of the Vodokanal of St. Petersburg, the supervisory authorities do not recommend the beneficial use of ash, despite the established IV hazard class. The continuous growth of waste generation, which is accompanied by a shortage of free space for its storage, determines both the relevance of the study and the interest of the Vodokanal St. Petersburg in actively addressing the issue of ash disposal.

Nowadays in St. Petersburg, according to the Decree of the Government of St. Petersburg N 989 (dated December 11, 2013), about 50,000 t/year of SSA are generated as a result of the activities of three incineration plants and then completely removed to landfills. Vodokanal of St. Petersburg is in charge of the «Volkhonka-2» landfill (currently closed for disposal of waste) and the «Northern» landfill. Placing landfills in close proximity to residential areas is a significant environmental issue. At the same time, the filling volume of the «Northern» landfill amounts to more than 91 percent of the design capacity, which has a negative impact on the environment. The termination of the ash storing in the landfill and possible disposal of already stored waste will not only free up some areas, but also reduce the negative impact of the object on the environment in the future.

3. Results

In order to assess the degree of contamination of ash with heavy metals, it is necessary first of all to establish their content in the sample and compliance with quality standards. In order to assess the degree of contamination of ash with heavy metals, it is necessary first of all to establish their content in the sample and compliance with quality standards [12, 13]. Sampling is performed according to the requirements [14]. On the basis of the obtained data on the ash composition (using the method of x-ray fluorescence analysis with a portable metal analyzer Niton XLt 898), a list of elements was selected for quantitative analysis in accordance with the atomic absorption spectroscopy method: Pb, Zn, Mn, Fe, Cu, Mo [15]. AAS-7000 spectrometer (Shimadzu, Japan) was used for measurements.

The assessment of the degree of contamination of the sample performed by calculating the concentration ratio K according to the formula (1):

$$К = \frac{С}{Ск} \quad (1)$$

where C – the element content in the sample, mg/kg;

C_k – normatively permissible value (MAC or APC) in the medium, mg/kg.

The following exceedances of the relevant standards were recorded: Zn (7.9 APC), Pb (6.3 MPC), Cu (3.7 APC). The results are presented in table 1.

The total pollution indicator Z was calculated by the formula (2):

$$Z = \sum К - (n - 1), \quad (2)$$

where n – number of elements taken in the calculation.

Table 1. Content of heavy metals in the SSA

№	Me	C_i	MPC_i	APC_i	Concentration ratio
			mg/kg		
1	**Zn**	1740	-	220	**7.9**
2	**Pb**	200	32	-	**6.3**
3	**Cu**	492	-	132	**3.7**
4	Mn	965	1500	-	0.6
5	Mo	16,5	Informative		
6	Fe	270000			

The indicator is equal to 15.9 (low degree of pollution) [16].

According to the results of the research, the assignment of waste to the IV class of hazard was confirmed using the software "Calculation of the hazard class of waste 2.0" (c) INTEGRAL 2001-2003 [17]. The IV hazard class of waste is also confirmed according to the results of the biotesting of the aqueous extract of waste (using the chlorella alga culture) [18, 17].

In accordance with the research work [19], the calculation of allowable doses of heavy metals in the soil of the Leningrad Region was carried out, taking into account their content in the waste.

Permissible dose D_p of admission to the soil of a heavy metal according to the formula (3), g/ha:

$$D_p = (0{,}8 \cdot MAC\ (APC) - B) \cdot M \quad (3)$$

where 0.8 – a correction factor reducing the normative intake of heavy metals into the soil by 20%;

B – the content of heavy metal in the soil before the introduction of waste (background concentration), g/t;

MAC (APC) – the maximum permissible (approximately permissible) concentration of the metal in the soil according to [12] or [13], g/t;

M – the mass of the arable horizon of the soil in terms of dry matter, t/ha (M = 3000 t/ha).

The average annual dose of waste in the soil in terms of dry matter for 10 years, t/ha, was defined as:

$$D_a = \frac{D_P}{T \cdot C_{SSA}} \quad (4)$$

where T – the maximum total term for depositing waste for the same area, years;

C_{SSA} – the concentration of metal in the ash sample, g/t.

It should be noted that when a useful component is introduced into the soil in the arable horizon, the normative permissible values of heavy metals should not be exceeded:

$$B + A \leq MAC\ (APC) \tag{5}$$

where F – the content of heavy metal in the soil before the introduction of waste (background concentration), mg/kg;

A – the additional flow of heavy metal into the arable horizon, weighing 3,000 t/ha.

From the obtained calculated data (table 2), it follows that when SSA is introduced into the soil of the Leningrad Region at the rate of 28.53 t/ha per year, the content of heavy metals in the arable horizon will not exceed the normative permissible values (where 28.53 t/ha is the annual dose of Zn in soil).

In order to assess the effect of SSA as a useful component of the waste-based soil mixture on the plants growth, Klever meadow (*Trifolium praténse*) of the *Fabáceae* family was selected as a plant capable of accumulation of heavy metals (based on literature data). Also *Trifolium praténse* is a siderate, i.e. additionally contributes to the enrichment of the soil with nitrogenous compounds and strengthening.

Table 2. Average annual dose of the test component in the soil

Me	Regulatory allowable value	Soil	SSA	D_p	D_a	B+A
	mg/kg			g/ha	t/ha*year	mg/kg
Mn	1 500	30	965	3 510 000	363.73	52.29
Pb	32	3	200	678 00	33.90	7.62
Cu	132	3	492	307 800	62.56	14.37
Zn	**220**	**10.5**	**1 740**	**496 500**	**28.53**	**50.69**

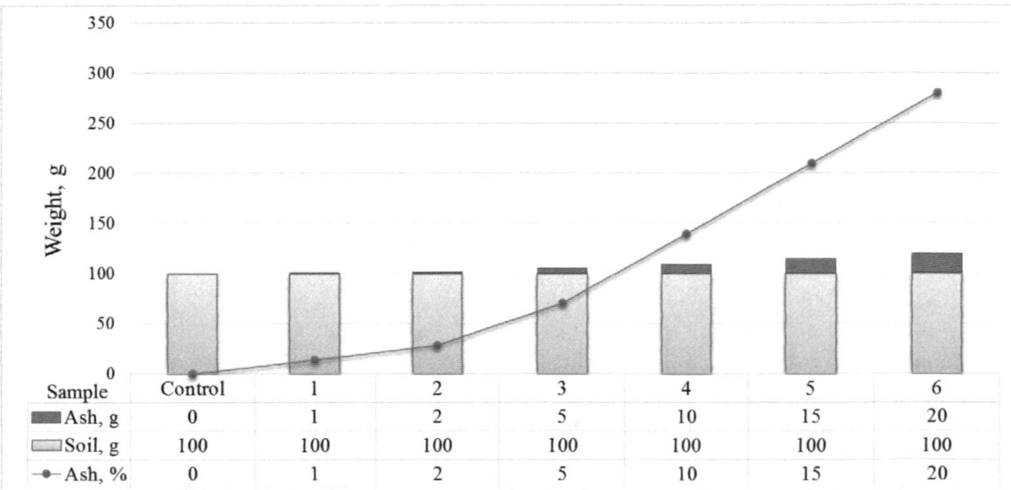

Figure 1. Composition of soil mixture model samples

Based on the established allowable doses of waste into the soil, the composition of model samples was calculated (Fig. 1). *Trifolium praténse* seeds were sown in a 80x90x90 mm tank with a mass of initial soil in each sample of 100 g. The seed seeding rate of *Trifolium praténse* (as a lawn grass) is 10 g/m^2; 25 seeds were planted on each model sample. Regular watering was done three times a week (volume 50 ml) to maintain humidity.

The first seedlings are marked on the third day. The germination in the control sample reached 88%. In the remaining samples (except for the sample with a dose of ash "20 g"), the germination rate exceeded the control one. The best dynamics of germination was observed in samples "5 g" and "10 g", while it should be noted that the sample "10 g" reached 100% germination on the fifth day, and the sample "5 g" on the sixth day. Other model samples did not reach 100% germination until the end of the experiment (fig. 2).

In the course of the experiment, soil acidity control was also performed in accordance with the method [20]. While the initial pH value of the salt extract of the control sample (soil of the Leningrad region) was 5.03 pH. When SSA was introduced into the soil, the pH of the salt extract increased to 2 pH compared with the control; at the same time, the growth of pH depends on the dose of the introduced SSA.

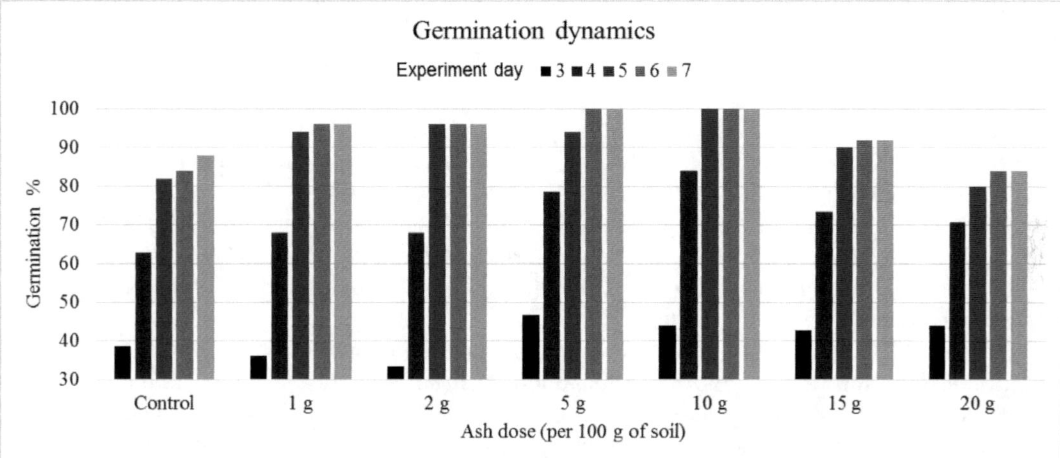

Figure 2. Dynamics of germination of Trifolium praténse

4. Discussion and conclusion

SSA-based soil mixture can be used as fertilizer for planting trees along the roads, for recultivation of disturbed lands, slopes of roads, landfills for municipal solid waste. All of these ways are extremely relevant in the metropolis. In the long term the use of product has a significant resource saving effect by recycling waste and reducing the amount of original soil required for reclamation processes, as well as the economic benefit, because the cost of waste-based soil mixture is significantly lower than the market value of soil.

The data obtained in the course of the experiment demonstrate the positive effect that SSA from the Vodokanal of St. Petersburg has on the *Trifolium praténse* growth in the soil of the Leningrad Region.

The economic effect from the implementation of the results of theoretical and scientific-practical developments at the facilities of the Vodokanal of St. Petersburg and in large metropolitan areas is determined taking into account the following aspects.

The SSA-based soil mixture is proposed as a development result. The product is most promising for use as a fertilizer of the second group - for the recultivation of disturbed land, road slopes, and solid municipal waste landfills. This makes the product suitable for use by housing and communal services, as well as by private enterprises, which need a useful component for the land reclamation. The use of the product has a significant resource-saving effect due to the disposal of waste and reducing the amount of initial soil required for remediation, as well as the economic effect, since the cost of soil mixture is much lower than the market value of the initial soil.

As an example, the authors consider the data of the Address Program of repair of roads for 2018 at the expense of the budget of St. Petersburg for 2018-2020 (for the target item "The cost of repairing roads"). It follows from the document that during this period it is planned to carry out road construction works that require landscaping improvement on a total area of 237.75 hectares. Taking into account the fact that the average annual allowable amount of SSA applied to the landscaped area will be 28.53 t/ha, the total amount of waste required for the improvement of the specified area will be 6,783 tons. At the same time, about 20 m^3 of SSA per day or about 4964 tons/year is generated only at The South-West wastewater treatment plant.

The removal of heavy metals from SSA and the search for ways of its beneficial use have been announced as one of the priority scientific topics for the maintenance of the Committee on Energy and Engineering of St. Petersburg. Further research is supposed to be directed to the development of a methodological approach to the choice of a rational composition of the SSA-based organic-mineral mixture as well as assessment of the effectiveness of its use in the conduct of recultivation in the field with filing application for the invention.

Sewage sludge management is an integral part of the work of any modern wastewater treatment plants. It is important to recover the nutrients contained in the sludge, to use it as a material or energy source and to dispose of it effectively.

References

[1] Barkan M Sh, Kuznetsov V S, Fedoseev I V 2007 Research of physical and chemical parameters of deposits of city sewage *Journal of Mining Institute* **V 172** pp 214–216

[2] Healy M G, Clarke R, Peyton D, Cummins E, Moynihan E L, Martins A, Béraud P, Fenton O 2015 *Resource recovery from sewage sludge. Sewage Treatment Plants: Economic Evaluation of Innovative Technologies for Energy Efficiency* (eds Katerina Stamatelatou, Konstantinos P Tsagarakis) (London, UK: IWA Publishing) 376 p

[3] Antonova I A, Gryaznov O N, Guman O M, Makarov A B, Kolosnitsina O V 2014 Geological conditions for allocation of solid municipal and industrial waste disposal sites in the Middle Urals *Water Resources* **№ 41 (7)** pp 896–903

[4] Federal Agency on Technical Regulation and Metrology 2011 State standard R 54535-2011 *Resources saving. Sewage sludge. Requirements for waste dispose and use at landfills* (Moscow: Standards Publishing House) 12 p

[5] Smola M, Kulczyckab J, Henclika A, Gorazdac K, Wzorek Z 2015 The possible use of sewage sludge ash (SSA) in the construction industry as a way towards a circular economy *Journal of Cleaner Production* **Vol 95** pp 45–54

[6] Lyuft Y, Tuuli O [etc] 2012 *Sewage Sludge Treatment: Good Practices and Practical Tips* (eds L Naumenko) (Turku, Finlyandiya: Project on Urban Reduction of Eutrophication) 125 p

[7] Vodakanazer 2016 *Recycling of sewage sludge and its disposal* [online] Available at: http://vodakanazer.ru/kanalizaciya/osadok-stochnyx-vod-eto.html [Accessed on 07.10.2019]

[8] Vigneswaran S , Sundaravadivel M 2009 *Recycle and Reuse of Domestic Wastewater Wastewater Recycle, Reuse, and Reclamation* (eds Saravanamuthu Vigneswaran) (Oxford, UK: EOLSS Publishers/UNESCO) 468 p

[9] Sousa G , Fangueiro D , Duarte E , Vasconcelos E 2011 Reuse of treated wastewater and sewage sludge for fertilization and irrigation *Water Science & Technology* **№64 (4)** pp 891–894

[10] Karmazinov F V, Izmajlova V N, Krasnov I A 2008 Incineration of sewage sludge is the solution to their disposal *Water Supply and Sanitary Technique* **№ 9** pp 19–24

[11] Korelskiy D S, Chukaeva M A 2013 Estimation of the condition of the soil-vegetative complexes having stress at atmospheric impact *Journal of Mining Institute* **Vol 203** pp 174–177

[12] Chief State Sanitary Physician of the Russian Federation 2006 *GN 2.1.7.2041-06 Maximum permissible concentration (MPC) of chemical substances in soil* (Moscow: Federal Center of Hygiene and Epidemiology of the Rospotrebnadzor) 16 p

[13] Chief State Sanitary Physician of the Russian Federation 2009 *GN 2.1.7.2511-09 Approximate permissible concentration (APC) of chemical substances in soil* (Moscow: Federal Center of Hygiene and Epidemiology of the Rospotrebnadzor) 12 p

[14] FSI Environmental Control and Analysis Center 2003 *ERD F 12.1:2:2.2:2.3:3.2-03 Sampling of soils, soils, sediments of biological wastewater treatment plants, sludges of industrial wastewater, bottom sediments of artificially created water bodies, storage ponds and hydraulic structures* (Moscow: FSBI «FCAO») 15 p
[15] IS «MEGANORM» 2008 *M-MVI-80-2008 Methods for measuring the mass fraction of elements in soil samples, soils and bottom sediments by atomic emission and atomic absorption spectrometry* [online] Available at: http://meganorm.ru/Index2/1/4293824/4293824289.htm [Accessed on 07.10.2019]
[16] Chief State Sanitary Physician of the Russian Federation 1999 *MU 2.1.7.730-99 Hygienic evaluation of soil in residential areas* (Moscow: Department of sanitary and epidemiological monitoring of Ministry of Health of the Russian Federation)
[17] Ministry of Natural Resources and Environment of the Russian Federation 2014 Order of December 4, 2014 N 536 «On approval of the Criteria for the classification of waste to I–V hazard classes according to the degree of negative impact on the environment»
[18] FSI Environmental Control and Analysis Center 2004 *ERD F T 14.1:2:4.10-04 Toxicological analysis methods. The method of determining the toxicity of drinking water and waste water, water extracts from soils, sewage sludge and waste of production and consumption on the change in optical density of the culture of algae (Chlorella vulgaris Beijer)»* (Moscow: FCBI «FCAO») 36 p
[19] Federal Agency on Technical Regulation and Metrology 2012 *State standard R 54651-2011 Organic fertilizers on the basis of sewage sludge. Specifications* (Moscow: Standards Publishing House) 20 p
[20] The Soviet Ministry of Agriculture and Food 1985 State standard 26483-85 *Soils. Preparation of salt extract and determination of its pH by CINAO method* (Moscow: Standards Publishing House) 6 p

Metrological support management quality assessment

E E Smirnova, T A Strizhova

Saint-Petersburg Mining University, 2, 21st Line, St Petersburg 199106, Russia;

E-mail: smiree@mail.ru

Abstract. The article presents data on assessing metrological support management quality based on the probabilistic approach. Assessing event probability can be additional important information in planning and analyzing complex processes such as metrological management. The analysis was carried out by comparing the requirements for metrological support with information about real situations. Requirements for metrological support are developed on the basis of regulatory legal acts, company management instructions, optimization of resource consumption, as well as taking into account company development prospects. To a large extent, the achievement of metrological support goals depends on management quality. The final step in assessing the quality of metrological support management is development of the results of assessment and formulation of conclusions (recommendations).

1. Introduction

Modern requirements for the quality of production technologies and high-tech industrial products and services determine a role of metrological support. To a large extent, the achievement of metrological support goals depends on management quality [1-6].

There are a lot of works on metrological support and quality management. But they do not deal with the assessment of metrological support management quality. The "assessment of metrological support management quality" is identification of the degree of metrological support management and shortcomings of the process. Assessment can improve management, its quality and quality of metrological support in general.

This article studies the impact of quality assessment of metrological management in companies.

2. Materials and methods

The method of assessing the mathematical probability of an event and statistical methods for quantifying quality of the company management process were used. The research material is the "metrological potential" of industrial enterprises.

There is an urgent problem of development of the method for assessing quality of metrological support management. This method should be universal, easy to use and take into account requirements of legal acts and specifics of metrological support at a particular enterprise. Management is considered as a complex continuous process performed by the metrological support subsystem.

3. The study of the structure of the modified lead-tin-base bronze

Currently, product quality assurance is based on the use of quality management systems that meet the requirements of international standards ISO 9000 [7, 8]. The most important element of the quality management system (QMS) is its metrological support system, which is an informational and analytical basis for managing processes and facilities. It determines the effectiveness of the QMS of an enterprise. Metrological support has a lot of requirements within the quality management systems, including the ubiquitous presence of measurements at the stages of incoming inspection of raw materials, when monitoring and analyzing the accuracy control of technological processes and products, conducting an internal audit and determining adequate control actions [9]. Effective enterprise management should be implemented on the basis of reliable objective information, which can be obtained only on the basis of timely measurements that have required indicators of accuracy and sufficient completeness.

To ensure reliable quality assessment, it is necessary to assess quality of the main interrelated actions that make up the management process:

1. Collection of information on the state of metrological support.
2. Analysis of information. Identification of a problem.
3. Making management decisions.
4. Informing performers about the management decisions.
5. Implementation of management decisions.
6. Monitoring.
7. Analysis of control results and, elimination of deficiencies.
8. Expenditure of time and resources on the management process.
9. Assessment of the degree of goal achievement.

The choice of indicators to assess quality of the above actions is a complex task. Obviously, the indicators should be determined on the basis of company requirements for management and meet the goals and objectives of metrological support.

There are three main options for assessing quality of management of metrological support:

The first option is the simplest and most common. The assessment is based on the opinions of officials. Moreover, these officials can be heads of the enterprise or specially appointed inspectors. The assessment depends on the objectivity of the officials, their experience, competency, knowledge of the situation, requirements and development prospects, ability to draw conclusions and make decisions. That is, the probability of a subjective assessment is high.

The second option is implemented using the method of group expert assessments. It is important to meet basic requirements for experts: a high level of competence in metrological support, independence and objectivity of assessment. That allows for a reliably assessment of the quality.

But within the one enterprise (organization), this is a rather complicated and laborious method that requires a lot of time and a trained working group for its implementation. It should be noted that for a reliable assessment of the quality of management, taking into account all the features of metrological support, it is necessary to develop separate methods for conducting group examinations of the quality of management for each area of activity.

The third option uses statistics. The statistical approach is based on a multitude of quantitative and qualitative indicators of quality assessment. In addition, statistics is used to assess probability of an event, i.e., the probability that the system (process) will meet requirements for the quality level of the functions performed in a given period.

Assessment of the quality of a process by assessing the event probability is of interest and may be additional important information in planning and analyzing such complex processes as metrological management. For example: the probability that the management subsystem will make the best management decision. However, the developed methods for quality assessment through the event probability depend on the reliability of statistical data for the past period do not take into account changes in external conditions; results obtained still require the use of additional indicators and quality assessment methods [10, 11].

It is still rational to use quantitative indicators. Since they are easy to obtain using statistical information; in contrast to qualitative indicators, they are more reliable and less dependent on the influence of various subjective factors.

When using quantitative indicators for evaluation, it is necessary to determine their required minimum or maximum values and (or) compare them with indicators for the last estimated period (year, quarter, month).

It is also rational to assess the fulfillment of company management requirements (the higher system) and requirements of regulatory legal acts for metrological support based on the principle: "the requirement has met" or "the requirement has not met".

Quantitative indicators and management requirements must be presented to the inspecting official as a formalized assessment form which can simplify the assessment and reduce time costs.

At the same time, the assessment indicators should express the degree of achievement of the management goal. In turn, the goal of management is to eliminate problems in achieving (ensuring) the required quality of metrological support. Moreover, the problem is the discrepancy between the existing and the required state of metrological support.

The analysis is carried out by comparing the requirements for metrological support with the collected information about the real situation. Requirements for metrological support are formed on the basis of regulatory legal acts, instructions of the enterprise management, resource consumption, development prospects of the enterprise (organization).

Based on the requirements, selected assessment indicators and minimum or maximum values, formalized forms of management quality assessment are compiled. Further, the inspecting official determines real values of the indicators and implementation of the requirements [12].

Assessment of management decisions is based on the objective function [13].

Any decision is aimed at achieving a specific goal. It provides some useful effect associated with costs. Comparison of the beneficial effect with the cost allows us to identify the quality of the decision.

Therefore, the most common indicator is an integral one, the ratio of the beneficial effect (E) to the total cost (C). It is the objective function:

$$G=\frac{E}{C}. \tag{1}$$

The solution maximizing the objective function is the best according to the optimality criterion (2).

$$G=\frac{E}{C}=\max \tag{2}$$

The synthesis of the optimal solution by the maximum of the objective function (2) suggests two possibilities.

The first one is elucidating the conditions that maximize the beneficial effect at fixed costs. The second one is to minimize costs that provide the same beneficial effect. Thus, criterion (2) splits into two particular criteria, which are satisfied by two different solutions (3) and (4):

$$G_1=\frac{E=\max}{C=const} \tag{3}$$

$$G_2=\frac{E=const}{C=\min} \tag{4}$$

If there is a choice, the optimal solution is one that provides a greater value of the objective function.

A variety of practical problems and approaches to their solution do not allow for further consideration of this issue. In each case, a particular method for the synthesis of an optimal solution is developed according to the criterion of the maximum of the objective function.

Thus, the assessment of the quality of management of metrological support can be carried out on the basis of a comprehensive assessment of the most significant indicators of the quality of management decisions.

The key indicators are as follows:

1) The degree of goal achievement due to the fulfillment of the estimated management decision in

Journal of Physics: Conference Series **1384** (2019) 012050 doi:10.1088/1742-6596/1384/1/012050

managing metrological support.

2) resources spent on the adoption and implementation of the management decision.

3) time spent on the adoption and implementation of the management decision.

The degree of goal achievement can be expressed quantitatively in the form of a comprehensive indicator that takes into account all important private indicators (5):

$$Q_j = \sum_{l=1}^{n} q_j g_l \tag{5}$$

where Q_j - comprehensive assessment of the degree of goal achievement; q_j - the private indicator for assessment; g_j - the normalized weight coefficient of this indicator.

The quantitative value of the normalized weight coefficient (g_j) can be determined by

1) setting priorities for the tasks performed by the system required to achieve the desired goal;

2) based on the views of competent officials (experts).

In the first case, it can be determined on the basis of statistical data on the functions of the system using which you can draw conclusions about tasks which are most important to achieve the desired goal.

In the second case, you can use formula (6) provided that the competence of officials (experts) is the same, that is, at equal values of the weighting coefficients of officials (experts):

$$g_i = \sum_{j=1}^{m} r_{ij} \Big/ \sum_{i=1}^{n} R_i , \tag{6}$$

In formula (6), the numerator is the sum of scores of the i-th indicator given by all m - officials (experts), and the denominator - the total sum of scores for all n-indicators given by m experts (r_{ij} - score by j-th expert for the i-th indicator).

If it is determined that for any evaluated solution, the goal is achieved, for the subsequent assessment of the quality of these solutions only the resources and time are used [13]

4. Conclusion

The final step in assessing the quality of metrological support management is the development of assessment results and conclusions (recommendations). This information can serve as the basis for making a management decision to eliminate deficiencies and improve quality of metrological management [14 - 16].

It is necessary to consider the metrological support of the enterprise as a set of measures aimed at achieving the required quality of measurements. When developing, manufacturing and operating products, one has to deal with a broad understanding of the content of metrological support, since the quality of measurements is determined by their accuracy and duration; feasibility of certain methods and means of measurement depends on the cost of measurements and complexity of their technical implementation .

References

[1] Kremcheeva D A, Kremcheev E A, Grudinin N N 2017 Use of a quality management system at a metallurgical enterprise *Modern high technology* **8** 25-29

[2] Altushkin, I.A., Korol, Y.A., Cherepovitsyn, A.E. 2013. Methodical approaches to economic assessment of sustainable development of mining and metallurgical holding. *Non-ferrous Metals* 2 3-9.

[3] Bilyatdinov K Z, Krivchun E A 2014 Assessment of the quality of management of organizational and technical systems *Notes of the Mining Institute* **209** 152-155

[4] Kremcheeva D A, Kremcheev E A 2018 Use of a Quality Management System at the Iron and Steel Enterprise *Journal of Mechanical Engineering Research and Developments* **1, 41** 151 - 155

[5] Vinogradova A A, Leontyuk S M 2018 Evaluation of current status in field of electrical voltage measurements *Journal of Physics: Conference Series* **1118** 1-5

[6] Zhambalova S Ts, Vinogradova A A 2018 Accuracy increase of self-compensator *IOP* Conf. *Series: Materials Science and Engineering* **327** 20-24

[7] ISO 9001. *Development, implementation, certification, improvement of the quality management system. A practical guide for quality professionals.* (M.: Publishing. Media Forum)

[8] Novokupova I N, Mekhdiev Sh Z 2016 Evaluation of the quality and effectiveness of enterprise and personnel management *Internet Journal of Science* **8-1** http://naukovedenie.ru/PDF/35EVN116.pdf

[9] Usik N I, Belorukov A E, Vasilenok A V 2016 The importance of the quality management system at enterprises Scientific journal *NRU ITMO. Series "Economics and Environmental Management"* **4** 70-77

[10] Magkiev Z I 2016 Development and implementation of a quality management system in accordance with ISO 9001-2015 *Scientific and methodological electronic journal "Concept"* **2** 376-380

[11] Shklyarskiy, A.Y., Zamyatin, E.O., Rastvorova, J.V. 2018. Developing of electric power quality indicators evaluation and monitoring intellectual system. Conference of Russian Young Researchers in Electrical and Electronic Engineering, *ElConRus* Volume 2018-January, 14 March 2018, 761-762.

[12] Bilyatdinov K Z, Fortinsky A G 2013 Improving the quality of management decisions. Communication technology. *Scientific and technical collection* **2 (141)** 186-189

[13] Sokerina M R, Charushina E V, Shestakova N V 2009 Evaluation of the effectiveness of the quality management process *Omsk Scientific Bulletin* **3** 9-11

[14] Fedyunina A A, Averyanova Y V 2018 Empirical analysis of competitiveness factors of russian exporters in manufacturing industries *Ekonomicheskaya Politika* **13 (6)** 102-121

[15] Fedoseeva E A, Charykova A S 2017 Implementation of yje quality management system at small enterprises. *Young scientist* **18** 187-189

[16] Afanasieva, N.V., Rodionov, D.G., Vasilev, Y.N. 2018. System of indicators of coal enterprise competitiveness assessment. *Espacios* Volume 39, Issue 36, 10p

Journal of Physics: Conference Series **1384** (2019) 012051 doi:10.1088/1742-6596/1384/1/012051

Development of a model for predicting the dynamic effect on the stability of rock excavation

M A Karasev, R O Sotnikov, V Yu Sinegubov, N A Egorova, K V Makarov and A I Thorikov

Saint-Petersburg Mining University, 2, 21 line, Vasilievsky island, St. Petersburg, 199106, Russia

E-mail: karasevma@gmail.com, ross61@mail.ru, vs.geotechburo@gmail.com, s175043@stud.spmi.ru, xmarcix@yandex.ru, dron15_117@mail.ru

Abstract. The present work is devoted to the prediction of the excavation stability and analysis of the geo-mechanical processes near the excavation under dynamic impact. The development of a numerical model of the forecast of the development of a stress-strain state with allowance for the dynamic effect taking into account the geometric parameters of the mining development, the speed and nature of the application of the load, and the formation of zones of intense fracturing in its vicinity are performed. The influence of the intensity of the dynamic effect on the nature of the manifestation of geo-mechanical and geodynamic processes in the vicinity, the movement of the contour of the rock excavation, and the probability of ejection of the rock mass are determined.

1. Introduction

The development of ore deposits involves the occurrence of geo-mechanical and geodynamic activity manifested in various forms of excavations stability. The loss of stability in overstressed rock masses composed of rocks predisposed to brittle fracture can be manifested in the form of significant local deformations of the rock contour related to the zones of active rocks fracture [1]. Rock fracture can be accompanied by the release of rock mass from the vicinity zone of the excavation, namely, rock bump takes place. Currently, the fastening system including dynamic (flexible) anchors and support elements that ensure the transfer of load from the rock mass to the anchor from the perimeter of excavations is preferable to be used in order to ensure the stability of excavations located in such conditions. Various variants of the anchor system were developed and tested both in laboratories and in mines, where they proved their worth. At the same time, the problem of choice of the type of dynamic support and its parameters has not been solved yet and is mainly based on empirical dependencies.

The solution of the task requires the study of the following problems: the first – to study the development of geo-mechanical and geodynamic processes in the area of excavations under dynamic impact; the second – to develop the model of dynamic anchor fastening system into account its actual mechanical characteristics; the third – to analyze the results of the application of dynamic anchorage and its parameters on the stability of excavations. The paper focuses on the first problem, that is, the forecast of the development of geodynamic processes in the area of excavations.

As the result of rock bump, the following forms of the loss of excavations stability can be observed: the formation of new cracks in the rock mass in the vicinity; stratification of rocks in the vicinity, rock crushing; extrusion of rocks in the sides and arch; rock uplift in excavation bases; drop-out of rock mass along the cracks; release of rocks, etc.

The degree of excavations damage as a result of rock bump can be divided into three types [2, 3]. The first type is characterized by minor damage when the rocks in the vicinity excavation can be broken or exfoliated by an amount of not more than 0.25 m. The second type is defined by moderate damage when the rocks in the vicinity zone are considerably broken. The significant displacement of the rock contour is observed. In this connection, the zone of rock exfoliation is in the range from 0.25 to 0.75 m. The third type is characterized by the extensive damage when the size of the zone of rock exfoliation is more than 0.75 m. The support of the excavation can also be considerably damaged. The third type of rock damage in the vicinity excavations caused by local extrusion of rocks from the zone of intensive stratification and cracking involves rapid development of the rock contour displacement which can reach 300 mm or more. The degree of excavation damage determines the type and parameters of the support to ensure the stability.

The process of energy transfer from the seismic event to the rock in the vicinity zone resulting in deformation, damage and fracture of the rock is described by the numerical model of dynamic impact of the remote seismic event on the excavation contour. The numerical modeling was made in a plane deformation (fig. 1). The model included the set of solid elements responsible for the parameters of rock mass in the area of the excavation and the set of damping elements providing the condition of seismic waves moving outside the model without their reflection from the boundaries.

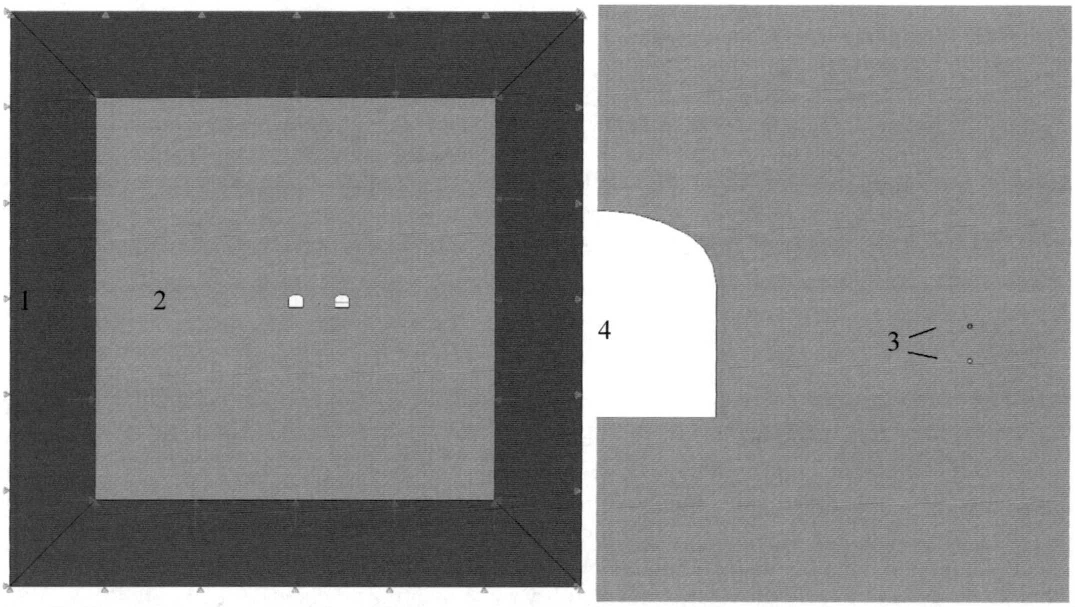

Figure 1. Geometric representation of the numerical model of the forecast of stress-strain state in the excavation area under seismic impact: 1 – zone of seismic waves damping; 2 – zone of the main rocks; 3 – sources of seismic impact; 4 – excavation

The task was solved in the conditions of the initial stress state field. The vertical stress was taken to be equal to 15 MPa and horizontal stresses of 30 MPa. In the model, the rock mass was considered to be monolithic undisturbed mass having the average strength of 100 MPa in the conditions of uniaxial compression.

The Johnson-Holmquist Model was used to describe the mechanical behavior of rocks under the dynamic impact. It allows taking into account the influence of the loading rate of the material on the strength and the degree of rock damage.

In general, the equation of the surface of the plastic flow of the Johnson-Holmquist model [4, 5] can be written as:

$$\sigma^* = \sigma_i^* - D\left(\sigma_i^* - \sigma_f^*\right),$$

where

$$\sigma_i^* = A\left(p^* + T^*\right)^N \left(1 + C\ln\varepsilon^*\right) \le \sigma_i^{\max};$$
$$\sigma_i^* = B\left(p^*\right)^M \left(1 + C\ln\varepsilon^*\right) \le \sigma_f^{\max}; \quad (1)$$

where σ^* - normalized equivalent stresses, $\sigma^*=\sigma/\sigma_{HEL}$; σ_{HEL} – the *Hugnoit* elastic limit in terms of equivalent stresses; D – index of the degree of the material damage, 0..1; p^* - normalized pressure, $p^*=p/p_{HEL}$; p – average stresses; p_{HEL} – average stresses at *HEL*; T^* - normalized strength at hydrostatic tension, $T^*=T/p_{HEL}$; T - strength at hydrostatic tension; σ_i^* - normalized strength of the not damaged material in terms of equivalent stresses, $\sigma_i^*=\sigma_i/\sigma_{HEL}$; σ_f^* - normalized strength of the damaged material in terms of equivalent stresses, $\sigma_f^*=\sigma_f/\sigma_{HEL}$; $\sigma_i^{\max*}$ – normalized maximum strength of the not damaged material in terms of equivalent stresses, $\sigma_i^{\max*} = \sigma_i^{\max}/\sigma_{HEL}$; $\sigma_f^{\max*}$ - normalized maximum strength of the damaged material in terms of equivalent stresses , $\sigma_f^{\max*} = \sigma_f^{\max}/\sigma_{HEL}$; A, B, C, M, N – parameters of strength condition.

Let us transform the above presented equations using the indicator of strength in uniaxial compression for the convenience of numerical computations. The following equations determining the mechanical behavior of rocks under dynamic impact are obtained:

$$q^* = \left(A(1-D) + B\left(p^*\right)^N\right)\left(1 + C\ln\dot{\varepsilon}\right);$$

$$q^* = \frac{q}{\sigma_{c.i}};\ p^* = \frac{p}{\sigma_{c.i}};\ T^* = \frac{T}{\sigma_{c.i}};\ \varepsilon^* = \frac{\dot{\varepsilon}}{\dot{\varepsilon}_0};\ E = 0.043\rho^{3/2}\sqrt{\sigma_{c.i}};\ G = \frac{E}{2(1+\nu)}; \quad (2)$$

$$K_1 = \frac{E}{3(1-2\nu)};\ T = 0.62\sqrt{\sigma_{c.i}};\ \mu = \frac{\rho}{\rho_0} - 1;\ p_{crush} = \frac{\sigma_{c.i}}{3};\ \mu_{crush} = \frac{p_{crush}}{K_1},$$

When performing numerical calculations, the following parameters of the Johnson-Holmquist model were taken into account to determine the resulting indicators of dynamic impact (table 1).

Table 1. Parameters of the Johnson-Holmquist model

Line 1	ρ_0, kg/м^3 3000	G, Pa 30.5e9	A 0.30	B 1.73	N 0.79	$\sigma_{c.i}$, Pa 100e6	C 0.005	ε_0 1
Line 2	T, Pa 5e6	σ_{max} 7	P_{crush}, Pa 50e6	μ_{crush} 0.002	P_{lock}, Pa 3.47e9	μ_{lock} 0.11		
Line 3	D_1 0.04	D_2 1.0	$\varepsilon^{pl}{}_{f.max}$ 999	$\varepsilon^{pl}{}_{f.min}$ 0.01	FS -			
Line 4	K_1, Pa 33.3e9	K_2, Pa 0	K_3, Pa 0					

The simulation of the formation of seismic wave by the detonation of explosives in boreholes was carried out using the JWL state equation (the John-Lee-Wilkerson model) [6, 7]. The state equation in the units of internal energy stored in mass unit E_m can be written as

$$p = A\left(1 - \frac{\omega\rho}{R_1\rho_0}\right)\exp\left(-R_1\frac{\rho_0}{\rho}\right) + B\left(1 - \frac{\omega\rho}{R_2\rho_0}\right)\exp\left(-R_2\frac{\rho_0}{\rho}\right) + \omega\rho E_m, \quad (3)$$

where A, B, R_1, R_2, ω - are the model constants, selected for each type of explosive; ρ_0 - is the density of the explosive; ρ is the density of the explosion products. The JWL model constants for emulsion explosives are summarized in table 2.

Table 2. Constants of state equation of explosives

The explosive	The constants of state equation of explosives									
	ρ_0, kg/m^3	A, Pa	B, Pa	R_1	R_2	ω	c_d, m/s	E_0, J/m^3	p_b, Pa	K_{pd}, Pa
Emulsion explosive	1140	3.85e11	5.045e9	5.487	1.171	0.24	5573	3.26e6	2.1e+010	3.3e9

When performing the numerical simulation, the area of blasting operations was distanced from the excavation contour at the distance sufficient for the transition of the wave from spherical one to flat, which was equal to 10 m for the conditions under consideration [8]. The intensity of seismic impact was changed due to the change of E_0 index. The resulting indicator was the movement speed of pgv rock particles. The determination of the movement speed of the rock particles pgv, the displacement of the contour of the rock exposure u and the size of the zone of brittle rock fracture h_{fr} were calculated.

The results of the calculation of rock mass response in the area of the excavation are presented in the form of diagrams of displacements development, movement speed of rock particles and the rock mass damage. The displacements of the excavation contour (figure 2a) and the movement speed of the rock particles (figure 2b) are presented as the dependences of the change of these parameters on time from the moment of the occurrence of the seismic event. The results are presented for different intensity of seismic impact.

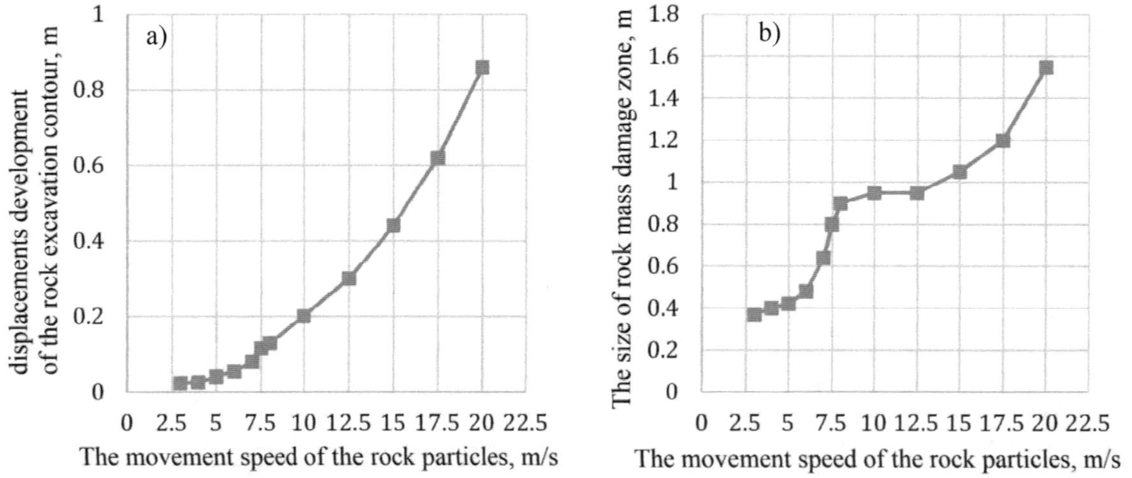

Figure 2. Displacements development of the rock excavation contour (a) and the formation of the damaged zone (b) as the functions depending on the movement speed of the rock particles

The pictures of the formation of longitudinal and shear cracks in the contour zone in the dependence on the intensity of seismic impact are shown in fig. 3.

Considering the development of displacements in time, continuous development of the displacements of the rock contour up to the intensity index from 0.7 to 1.0 (pgv_d = 0.7-1.0) is noticed. This shows that contouring of the rock mass section with its further release into the excavation space has taken place under the impact of seismic event.

Special attention should be paid to the kinetic energy extruded from the rock mass justifying the choice of dynamic support parameters under such influence. The decrease of the intensity of seismic impact below 0.7 still has a serious impact on the development of geodynamic processes and damage of rock mass in the vicinity of the mining excavation. However the release of rock mass is not observed.

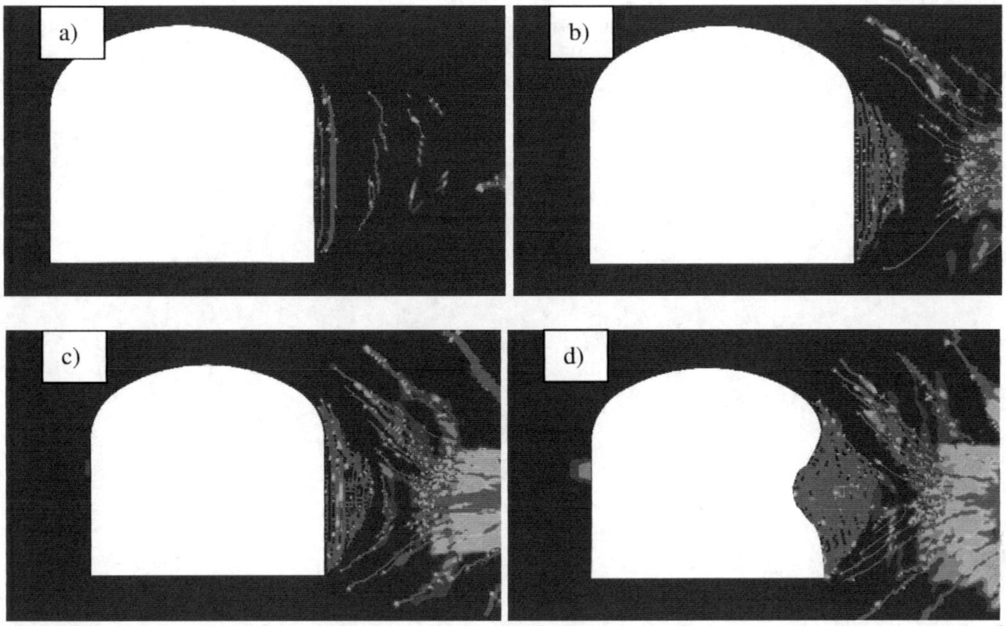

Figure 3. The distribution of damages in the area of the excavation: a – movement speed of rock particles 5 m/s; b – movement speed of rock particles, 10 m/s; c – movement speed of rock particles, 15 m/s; d – movement speed of rock particles 20 m/s

It should be noted in the conclusion that the paper presents the selected results of the research of the evaluation of seismic event impact on the stability of rock excavation. Thus, seismic event itself is considered as dot blast occurring in the vicinity of mining excavation. Empirical experiments permitting to ascertain the veracity of the backgrounds introduced in the paper for the simulation of rock damage in the vicinity under seismic impact will be made at the next stage. Further studies will be devoted to the development of the mathematical model of the functionality of dynamic anchors of various structures and to the combination of the forecast model of geodynamic processes in rock mass in vicinity and the developed model of mechanical behavior of dynamic anchors.

References

[1] *CAMIRO: Canadian Rockburst Research Program 1990–1995* 1995 A comprehensive summary of five years of collaborative research on rockbursting in hard rock mines (CAMIRO Mining Division)

[2] Tannant D D, McDowell G M, Brummer R K and Kaiser P K 1993 Ejection velocities measured during a rockburst simulation experiment In: *3rd Int Symp on Rockbursts and Seismicity in Mines* pp 129–33

[3] Yi X and Kaiser P K 1993 Impact testing of rockbolt for design in rockburst conditions. International *J. of Rock Mechanics and Mining Sci. Geomech Abstr.* **31** 671–85

[4] Johnson G R and Holmquist T J 1994 An improved computational constitutive model for brittle materials *High-Pressure Sci. and Technol.* (American Institute of Physics)

[5] Holmquist T J, Johnson G R, Cook W H 1993 A Computational Constitutive Model For Concrete Subjected To Large Strains, High Strain Rates, and High Pressures. *Int. Symp. on Ballistics* **14** 591–600

[6] Choi S, Wang J, Munfakh G and Dwyre E 2006 3D nonlinear blast model analysis for underground structures *GeoCongress 2006 ASCE* pp 1–6

[7] Liu H 2009 Dynamic analysis of subway structures under blast loading *Geotechnical and Geological Engineer.* **27** 699–711

[8] Gospodarikov A P, Vykhodtsev Y N and Zatsepin M A 2017 Mathematical Modeling of Seismic Explosion Waves Impact on Rock Mass with a Working *Zapiski Gornogo instituta* **226** 405–11 DOI: 10.25515/PMI.2017.4.405

Improvement of the system of industrial environmental monitoring of atmospheric air in the area of anthropogenic arrays impact

A V Strizhenok[1], **D S Korelskiy**[1]

[1] Saint Petersburg Mining University, 2, 21st Line, St Petersburg, 199106, Russia

E-mail: alexeystrizhenok@mail.ru

Abstract. Nowadays the waste management is one of the most pressing problems in the industry of mineral resources extraction and processing. The main method of waste disposal is storage on the surface of the Earth in the form of anthropogenic arrays. Anthropogenic arrays have an integrated and long-term impact on the atmospheric air. In this regard, the need for continuous monitoring of the state of atmospheric air in order to identify the main sources of dust emission and their quick elimination acquires particular relevance. This task can be performed by automating the existing system of industrial environmental monitoring. The article touches upon the issues of the implementation of the task of continuous monitoring of the dust content of atmospheric air in the areas of the anthropogenic arrays impact. The structure of the automated system of industrial environmental monitoring for the rapid detection of sources of intense dust emission, their ranking by the degree of potential danger and the measures for their timely elimination are proposed.

1. Introduction

Currently, one of the most significant sectors of Russian economy is the mineral-raw material complex. While the enterprises of the mineral resources extraction and processing are sources of an integrated and concentrated impact on the components of the natural environment and on the health of the inhabitants. Today the area of land which is disturbed by the activities of enterprises of the mineral-raw materials complex reaches more than 15 million hectares only in Russia. At the same time, about 10% of disturbed areas are occupied by storage sites of the wastes of mineral resources extraction and processing - anthropogenic arrays [1].

Anthropogenic arrays are sources of continuous and prolonged environmental pollution through the atmosphere, lithosphere and hydrosphere. There are disturbance and alteration of natural landscapes, a decrease in the species diversity of animals and plants, deterioration of population health indicators are observed (life expectancy, morbidity and mortality, congenital abnormalities) in a territory of their impact [2].

The production facilities of JSC Apatit, and first of all the alluvial technogenic array of the second apatite-nepheline processing plant (the ANOF-2 tailing dump) were the objects of research.

The ANOF-2 tailing dump was commissioned in 1963 and today it is the largest alluvial technogenic array in Russia according to the height of the enclosing dam and the volume of waste stored in it. The tailing dump is located 8 kilometers north-west of the city of Apatity (Murmansk region), its total area is about 8 square kilometers and the height of the enclosing dam at the beginning of 2018 reached 76 meters. About 8 million cubic meters of apatite-nepheline ore processing waste flow from the concentrating plant to the tailing dump every year, but more than 680 million cubic meters of processing waste had been accumulated in tailing dump from the beginning of its operation [3].

The main source of negative impact on the atmospheric air, lithosphere and hydrosphere is wind dusting of open beaches and sides of the dam of tailing dump, as a result of which more than 250 tons of inorganic dust with SiO_2 content from 70 to 20% gets into the air annually. This makes up more than 30% of the total mass of suspended substances emitted by the enterprise. In addition, there is an excess of the average maximum permissible concentration by 1.5-2 times in the territory of the city of Apatity provided the southeastern direction of the wind and its velocity of 6-8 meters per second. It should be noted that, the probability of southeastern winds for this region is 27% in accordance with the average annual wind rose, but the air flow rate in the summer months reaches 25 meters per second [4].

The company has successfully used a number of environmental protection measures that can significantly reduce the intensity of dust transfer from the territory of the alluvial anthropogenic array today [3,4].

In this regard it is necessary to introduce an automated production environmental monitoring system for the improving the effectiveness of existing environmental protection measures during periods of adverse meteorological conditions. Automated monitoring system is able to quickly and efficiently identify areas of intense dust formation and rank them by the degree of potential danger in order to eliminate them rapidly.

2. Materials and methods

The structure of the automated system of industrial environmental monitoring was developed in accordance with the regulatory document SP 11-102-97 "Engineering and environmental surveys for construction". Based on this document, industrial environmental monitoring should identify the main regular occurrences of quantitative and qualitative changes in the state of the environment in space and in time. In this regard, the automated production environmental monitoring system should perform the following tasks:

- systematic registration and control of indicators of the state of the environment in the locations of potential sources of pollution and areas of its possible distribution;
- the development of environmental regular occurrences of natural and man-made nature and the forecast of possible changes in the state of the components of the environment based on the identified regular occurrences;
- development of recommendations and proposals to reduce and prevent the negative impact of industrial facilities on the environment;
- control of the use and effectiveness of the adopted recommendations to reduce the anthropogenic load on the territory at issue [5,6].

The main research methods are: system analysis of fundamental scientific works of Russian and foreign scientists on this subject; environmental hardware monitoring of the state of atmospheric air in order to determine the layout of stationary monitoring posts; methods of mathematical statistics, cartographic modeling and ranking of sources of environmental pollution according to the degree of potential danger.

Environmental monitoring of atmospheric air was conducted in line with GOST 17.2.3.01–86 "Nature protection. Atmosphere. Rules of air quality control for populated areas" [7].

The method for determining the concentration of inorganic dust in atmospheric air included laser nephelometry, which is used in dust analyzers manufactured by TSI Company [8]. In addition, a particle size analysis of enrichment wastes was performed in the framework of scientific research for the revelation of the dynamics of dust transfer. It was carried out in accordance with GOST R 8.777-2011 "State System for Ensuring the Uniformity of Measurement. The disperse composition of aerosols and suspensions. The determination of particle size by laser diffraction" [9].

Multifactor analysis and methods of cartographic modeling were used to rank the sources of air pollution by degree of danger, as well as to determine the location of posts in the system of industrial environmental monitoring [4].

The scientific instrumental base for conducting field research and for results processing was provided by the Environmental Monitoring Laboratory of St. Petersburg Mining University [10].

3. Results

The system of industrial environmental monitoring is an analytical complex, which consists of an information-measuring unit, a source data processing unit, a settlement unit and a central management unit. The automated production environmental monitoring system carries out continuous quality control of the surface atmosphere in the territory of the production facility at issue. It continuously records concentrations of inorganic dust, air humidity, wind speed and direction at stationary monitoring stations and transmits the received data to a united information resource. The united information resource collects, stores and processes these data [11,12].

The number of monitoring posts and their positioning directly depends on the dusting intensity, the distance of the dust transfer and the waste parameters stored in the anthropogenic array [13]. The dependence of the specific blowing of dust on the moisture content of the stored material and air velocity was revealed based on experimental studies. The graph of the dependency is presented in Figure 1.

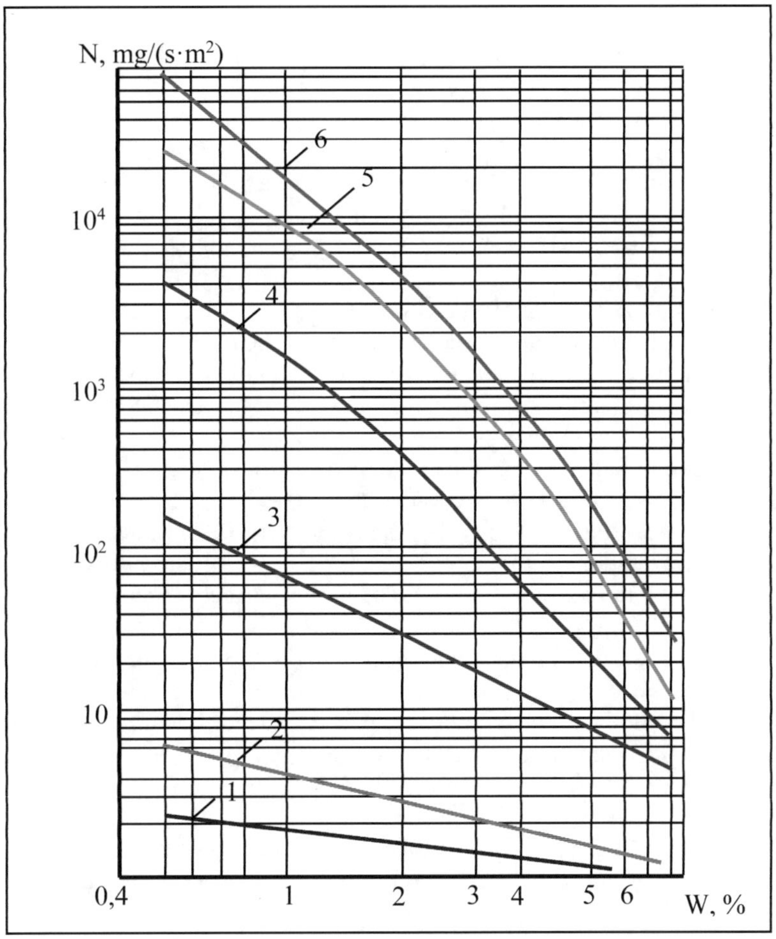

Figure 1. Graph of the dependency of the specific blowing of dust (N) on the moisture content of the stored material (W) and air velocity (V): 1 - 2 m/s; 2 – 3.5 m/s; 3 - 5 m/s; 4 – 6.5 m/s; 5 - 8 m/s; 6 – 9.5 m/s

The graph shows that when the wind speed is up to 5 meters per second, the specific blowing of dust is insignificant and its change is linear, however, with an increase in the wind speed, the specific deflection of dust increases along a parabola, but the dusting intensity increases manifold.

A network of atmospheric air monitoring posts in the production environmental monitoring system was developed on the example of the ANOF-2 tailing dump based on the established experimental dependency [14]. The network includes five stationary posts located on the territory of the alluvial beaches of the tailing dump. The monitoring posts are located taking into account the average annual wind rose (along the most frequently observed wind directions) so that, with any wind direction, the current concentration of dust is recorded in at least one monitoring station. The network of monitoring posts is designed in such a way as to control the dusting of the surface atmosphere throughout the tailing dump with minimal costs for monitoring stations equipment. Thus, five monitoring stations are the minimum number of control points required to cover the entire ANOF-2 tailing site and to create the possibility of rapid response in the event that sources of intense dust emission are formed on its territory under adverse meteorological conditions [12].

Each monitoring post is equipped with an automatic dust analyzer (for example, the DustTrak TSI-8533 dust analyzer), a digital wind speed measurement sensor (for example, WS 100) and a digital wind direction measurement sensor (for example, WD 100). Any other sensors similar in characteristics and determined parameters may be used for the equipment of the monitoring post. All devices are powered with 220 V.

The concentration of inorganic dust in the atmospheric air, wind speed and direction are updated every 20 seconds and transmitted to the central management unit as a digital signal, which eliminates the possibility of interference.

As a communication channel, it is supposed to use a channel for transmitting digital data of GSM 1800/900. This type of communication is one of the most reliable types of digital communication, ensuring high quality of the signal on the territory covered by the GSM operator. The GSM communication channel does not require significant capital expenditures, since the data channel requires only the installation of GSM signal transmitters at each monitoring station and a GSM signal-receiving sensor on the central management unit, and has relatively low operating costs. In addition, during the operation of the data channel via GSM, highly qualified specialists are not required to maintain equipment that ensures uninterrupted operation of the GSM communication channel, since the signal transmissions are carried out by repeaters of mobile operators, whose employees independently install and service their equipment. Thus, the use of a GSM communication channel eliminates the need to deploy your network of repeaters [15,16].

Any other type of communication can be used to data transfer, such as Wi-Fi, GPS, radio and others if it is impossible to use a GSM connection.

The central management unit controls the operation of the automated system of industrial environmental monitoring, including:

- a survey of sensors monitoring the quality of the surface atmosphere and meteorological parameters;
- primary processing of digital data and accumulation of measurement results on a single information resource;
- comparison of the obtained concentrations of inorganic dust with the maximum permissible concentration and the calculation of contrast ratios;
- calculation of the total dispersion of inorganic dust tailings with a particle size of less than 10 microns for various meteorological parameters;
- identification of regular occurrences of migration of inorganic dust and the development of a forecast of the state of the surface layer of the atmosphere in the residential area of the city of Apatity;
- identification of areas on the territory of the ANOF-2 tailing dump with the greatest potential danger of forming sources of intense dust emission under various meteorological conditions, taking into account the natural and technical security of the production facility;

- ensuring the dialogue between the operator and the industrial environmental monitoring system;
- change the mode of operation of the automated industrial environmental monitoring system;
- generation of messages and displaying them on the screen in a tabular and graphical (in the form of a map) form, as well as transmission of these messages, via radio channel, if it is necessary [12].

The identification of zones with the greatest potential danger of forming sources of intense dust emission is carried out using a computer program, which in its essence implements the solution of the logical function "IF ... THEN ...". That is, if the concentration of inorganic dust exceeds the value of the maximum permissible concentration at one or several monitoring stations, the tailing areas from which emission of inorganic dust could occur under certain meteorological conditions are automatically highlighted on the screen of the central management unit. Graphic information displayed on the screen of the central management unit is also necessarily duplicated by an audible signal, and an information message that contains the approximate area of the dusting territory and recommendations for its elimination, as well as a forecast of the state of the surface atmosphere in the residential area of the city of Apatity [17]. The message can be transmitted over radio channels if it is necessary.

A GIS software product, for example MapInfo Professional, ESRI ArcGis or Golden Software Surfer, carries out the processing of digital information and its conversion into graphic form automatically. The construction of the algorithm implemented by the automated system of industrial environmental monitoring is developed by the author of research and made up as a computer program [18,19]. It is set to the computer of central management unit as a basic function that the computer performs periodically as new digital data is received from atmospheric air monitoring stations.

The structure of the automated system of industrial environmental monitoring is presented in Figure 2.

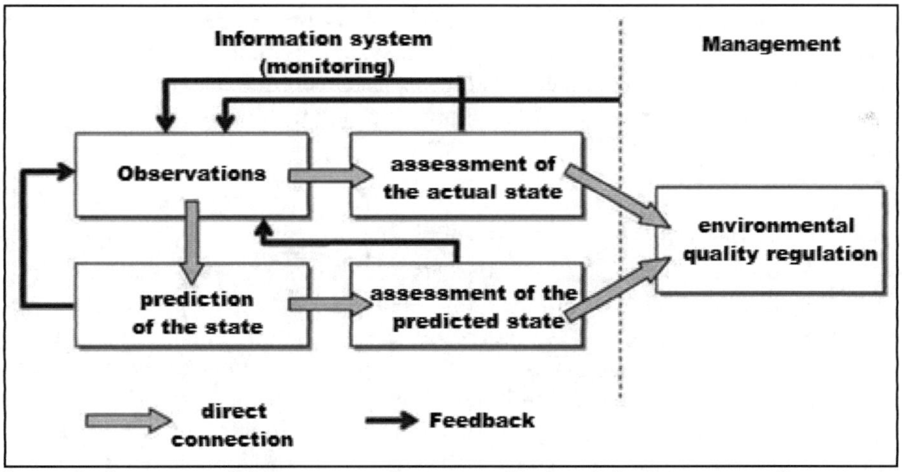

Figure 2. Structure of the automated system of industrial environmental monitoring

The information-measuring unit is a set of sources of information about the state of the surface layer of the atmosphere, the data from which are transferred to the source data processing unit via a GSM communication channel. The main sources of information are the network of automatic stationary stations for monitoring atmospheric air and stations for monitoring atmospheric air of the State Institution "Murmansk UGMS". In addition to the monitoring data, the information-measuring unit provides access to electronic databases of maximum permissible concentrations, archives of data of long-term observations of the quality of the surface atmosphere in the region and cartographic base of the ANOF-2 tailing dump. A mobile laboratory for monitoring the quality of the surface atmosphere can also be introduced into the automated system of industrial environmental monitoring if it is necessary or if there is insufficient data from stationary monitoring stations of atmospheric air [12].

The raw data processing unit includes a single information resource that receives, preprocesses and accumulates digital data of the industrial environmental monitoring system, data of atmospheric air monitoring stations "Murmansk UGMS" and standards for maximum permissible concentrations of the pollutants being analyzed. All data is archived and stored on the server for the required period.

The processing unit also includes an information resource that receives and stores cartographic bases. Periodic updating of the cartographic base is required because the height of the alluvial dam of an anthropogenic array increases proportionally with the accumulation of enrichment waste, which leads to a change in the area of the alluvial beaches of the tailing dump. These changes must be taken into account when organizing a system of industrial environmental monitoring in the area of anthropogenic arrays impact [20].

A block of mathematical, statistical and analytical calculations is a powerful personal computer, on the basis of which the parameters characterizing the quality of atmospheric air, the identification of environmental regular occurrences and the prediction of the state of the surface atmosphere based on the calculated parameters, as well as the visualization of the established regular occurrences are made. All operations are carried out automatically using the software package for mathematical calculations (MathLab, MathCad), statistical calculations (Statistica) and for visualizing production environmental monitoring data and established patterns (MapInfo, ArcGis, Surfer). Software products with similar functionality can be interchanged, for example, one of the listed applications in this category or any other software product with similar functionality can be used as a software application for data visualization [17, 19].

The central management unit is the head center of the entire system of industrial environmental monitoring. It is a separate building located on the territory of the land allotment of the ANOF-2 tailing dump. It contains all the blocks of the automated system of industrial environmental monitoring except for the information-measuring unit, which components are situated on the territory of the considered anthropogenic array. In addition, the operator's workplace is located in the central control center, ensuring uninterrupted operation of the entire system of industrial environmental monitoring. The operator of the central management unit can make certain management decisions and send commands via the radio link on the need to eliminate promptly the sources of intense dust emission based on the received information [12].

4. Conclusion

The automated system of industrial environmental monitoring is a modern and effective means of continuous monitoring of the quality of the surface atmosphere in the territory of intensive technogenesis. The introduction of an automated system of industrial environmental monitoring on the territory of existing anthropogenic arrays will allow obtaining timely and reliable information about the state of the surface layer of atmospheric air in the area of their impact. This information will allow a timely response to changes in the environmental situation and the rapid elimination of sources of intense dust emission. Thus, it becomes possible to manage the quality of atmospheric air in the area of impact of anthropogenic arrays and in the territory of nearby residential areas.

Acknowledgements

This scientific work is supported by the Scholarship of the President of the Russian Federation SP-3455.2019.3.

References

[1] State report 2019 About condition and protection of the environment of the Russian Federation in 2017. Retrieved from: https://nangs.org/analytics/minprirody-rossii-gosudarstvennyj-doklad-o-sostoyanii-i-ob-okhrane-okruzhayushchej-sredy-rossijskoj-federatsii-pdf (20.05.).

[2] Statistical collection 2018 *Environmental protection in Russia* **125**.

[3] Strizhenok A.V., Tcvetkov P.S. 2017. *Ecology-economical assessment of new reclamation method for currently working technogenic massifs.* Journal of Ecological Engineering. **18(1)** 58-64.

[4] Masloboev V.A. 2014. Numerical modeling of the processes of dusting of the tailings ANOF-2. *Vestnik MSTU* **17(2)** 376-384.

[5] Kuzhaeva A., Berlinskii I. 2018. *Effects of oil pollution on the environment.* International Multidisciplinary Scientific GeoConference Surveying Geology and Mining Ecology Management, SGEM. **18(5.1)** 313-320.

[6] Rosstandart 1998 Engineering and environmental surveys for construction SP 11-102-97.

[7] Gosstandart 1986 Nature protection. Atmosphere. Rules of air quality control for populated areas GOST 17.2.3.01-86.

[8] TSI-Russia 2019 Dust analyzer DustTrak TSI. Retrieved from: http://tsi-russia.ru/product/category/3/parent/22/products_id/57.

[9] Rosstandart 2011 State System for Ensuring the Uniformity of Measurement. Disperse composition of aerosols and suspensions. Determination of particle size by laser diffraction GOST R 8.777-2011.

[10] SpB Mining University 2019 Scientific instrument base of the laboratory of Environmental monitoring of the Saint-Petersburg Mining University. Retrieved from: http://old.spmi.ru/system/files/lib/sci/ckp/ekologicheskoe.pdf.

[11] Barkan M.S., Kornev A.V. 2017. *Prospects for the use of associated gas of oil development as energy product.* International Journal of Energy Economics and Policy. **7(2)** 374-383.

[12] Volkodaeva M.V., Kiselev A.V. 2017. *On the development of the system for environmental monitoring of atmospheric air quality*. Journal of Mining Institute. **227** 589–596.

[13] Dorokhov I.N., Smirnov V.N. 1998. Automated environmental monitoring system for an industrial area. *Software products and systems* 35-40.

[14] Nagornov D.O., Kremcheeva D.A. 2017. Assessment of gravity component of moisture flow in raw peat in conditions of convective and radiative-convective heat input *Journal of Industrial Pollution Control* **33(1)** 749-752

[15] Dzhevaga N.V., *Lobacheva O.L. 2018. Separation of the rare-earth elements (Ree) by flotation approaches.* International Multidisciplinary Scientific GeoConference Surveying Geology and Mining Ecology Management, SGEM. **18(5.1)** 801-808.

[16] Danilov A.S., Smirnov Y.D., Ivanov A.V. 2016. The new method of dust control on tailing dumps (Exemplified by Kovdor mine) *Proc. International Multidisciplinary Scientific GeoConference Surveying Geology and Mining Ecology Management (SGEM)* pp. 743-750.

[17] Chen P. 2019 Visualization of real-time monitoring datagraphic of urban environmental quality. Eurasip Journal on Image and Video Processing **2019(1)** 42.

[18] Leusheva E.L., Morenov V.A. 2015. *Combined oilfield power supplying system with petroleum gas utilization as an energy carrier.* Neftyanoe khozyaystvo - Oil Industry. **4** 96-100.

[19] Strizhenok A.V., Tcvetkov P.S., Kuznetsov V.S., Frolova A.Yu. 2018 *The program for calculating the total index of air pollution based on environmental monitoring data.* The patent for the computer program

[20] Danilov A.S., Pashkevich M.A., Petrova T.A. 2018. Environmental integrated monitoring system at reclamation of large open-cast coal mine. *Proc. of Innovation-Based Development of the Mineral Resources Sector: Challenges and Prospects - 11th conference of the Russian-German Raw Materials.* pp. 189-194.

Ecological and economic justification of the utilization of associated petroleum gas at small oil fields of Russia

A V Strizhenok[1], D S Korelskiy[1]

[1] Saint Petersburg Mining University, 2, 21st Line, St Petersburg, 199106, Russia

E-mail: alexeystrizhenok@mail.ru

Abstract. One of the main environmental problems in oil production in the world today is the utilization of associated petroleum gas (APG) by combustion it in flares. At the same time, this process has enormous economic damage, since APG is a potential raw material that can be used to produce marketable products, power or heat energy. More than 30 billion cubic meters of APG are combusted in flares annually only in Russia, which is about 20% of the total amount of combusted APG in the world. It should be noted that combustion of 1 billion cubic meters of APG is equivalent to the loss of marketable products worth more than 300 million dollars. In the present article, the authors carried out a selection of the optimal method for utilization of APG at small fields in Russia and its environmental and economic justification. The method of power generation directly at the field using gas turbine power station was adopted as the most cost-effective method. The implementation of such a project at the field reduces emissions of pollutants hundreds of times, and payback period does not exceed 2.5 years.

1. Introduction

The oil and gas industry is a dynamically developing sector of the world economy, but produced oil and natural gas is one of the main energy resources on our planet.

The oil and gas industry is one of the main industries ensuring budget formation in the Russian Federation. More than 500 million tons of oil and about 600 billion cubic meters of natural gas were produced in Russia in 2017, as the same time about 50% of crude oil and about 40% of natural gas were exported.

With such significant production volumes and not the largest oil reserves, the resource supply of Russia with oil, according to various estimates, ranges from 20 to 25 years with current production volumes. The situation with natural gas is a little better, as resource supply of Russia with natural gas, according to various estimates, ranges from 80 to 100 years [1].

However, one of the most pressing environmental problems of the oil and gas industrial complex in Russia is the utilization of associated petroleum gas (hereinafter referred to as APG). APG is one of the by-products of the process of oil production and its preliminary preparation for transportation. The main components of PNG are methane and other low molecular weight (volatile) alkanes [2].

Due to a number of technical and economic reasons, as well as due to the peculiarities of the legal regulation of the oil industry, the main method of utilization of APG in Russia and in some other countries today is it combustion in flares [3]. More than 30 billion cubic meters of APG are combusted in flares annually only in Russia, which is about 20% of the total amount of combusted APG in the

Journal of Physics: Conference Series **1384** (2019) 012053 doi:10.1088/1742-6596/1384/1/012053

world [1]. As a result, more than 17,000 flares are burning at oil fields and refineries around the world, annually emitting about 350 million tons of CO2, hundreds of thousands of tons of nitrous oxide, carbon monoxide, sulfur dioxide and products of incomplete combustion of hydrocarbons. Emissions of flares are more than 35% by volume all emissions to the atmosphere from the oil and gas industry. In addition, the combustion of APG in the subarctic regions leads to the formation of a large amount of soot deposited on the snow cover, which increases the absorption of solar energy and accelerates the melting of the Arctic ice [4,5].

Figure 1 shows a cartographic model of dispersion in the surface atmosphere of carbon monoxide, emitted by a flare at one of the small deposits of Siberia. Carbon monoxide is one of the major pollutants during the process of combustion of APG and has a toxic effect on living organisms [6]. The coefficients, calculated as the ratio of the current concentration of carbon monoxide to the normative indicator of its concentration in atmospheric air, are marked on contour lines. Thus, this ratio shows how many times the concentration is exceeded relative to the standard.

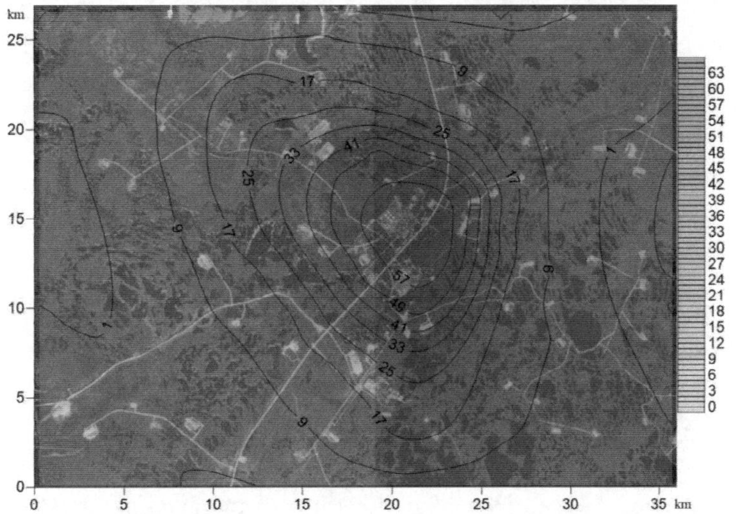

Figure 1. Cartographic model of carbon monoxide dispersion in the surface atmosphere

It can be seen in Figure 1 that within a radius of more than 10 kilometers the established standard for carbon monoxide is exceeded dozens of time. At the same time, several other flares of nearby fields fall within the radius of impact of the flare under consideration, the negative impact of which was not taken into account when creating the cartographic model. Consequently, the real environmental situation in the territory under consideration is several times worse.

However, the colossal negative impact on the natural environment is not the only reason for the observed trend to decreasing of the volume of APG combustion in flares and to the search for alternative ways to utilize APG with a positive economic effect. According to the calculations of the Institute of Energy Strategy of Russia, combustion of 1 billion cubic meters of APG is equivalent to the loss of marketable products worth more than 300 million dollars. This fact determines the relevance of the search and implementation of APG utilization methods with an economic benefit for oil producing enterprises and optimization of the ecological situation in regions with developed oil producing industry [7].

2. Materials and methods

There are several generally accepted methods of APG utilization today. The main of which are:

- processing of APG at gas processing plants (deep processing and shallow processing) with a production of marketable products;
- power generation;

- burning for own needs (as a rule, with receiving heat energy);
- injection back into the reservoir for enhanced oil recovery (reservoir pressure maintenance system);
- injection into production wells - using of "gas lift".

Analysis of publications and fund documents of oil producing enterprises allowed setting that the most effective from the point of view of environmental and economic justification is the processing of APG at the gas processing plant with a production of marketable products [8].

Table 1 presents a comparison of existing methods of utilization of APG based on the required capital costs and economic effect.

Table 1. Comparison of existing methods of utilization of APG

	Combustion	Injection back into the reservoir	Power generation	Shallow processing	Deep processing
Capital costs, USD per cubic meter APG	0.01	0.07	0.08	0.23	0.29
Economic effect, USD per cubic meter APG	0	0	0.07	0.17	0.31

From Table 1 it can be seen that the processing of APG at the gas processing plant with the production of marketable products (deep and shallow processing of APG) has an insignificant capital costs compared with the resulting final income. In addition, recycling APG on gas processing plant has a minimal negative impact on the components of the environment in comparison with other methods of utilization of APG [9,10]. However, Table 1 presents the ideal conditions under which gas-processing plant is provided with APG in an amount sufficient to operate it at design capacity and without outages. In addition, only the cost of the product was taken into account when calculating the economic effect, but transport costs, availability of sales markets and other secondary factors were not taken into account [11].

Unfortunately, such conditions on the territory of Russia are found only in very large oil fields with a high flow rate of APG (more than 1 billion cubic meters per year). There are no more than 10 such oil fields on the territory of Russia. Thus, this method is justified technically and economically only if these conditions are met [1].

The bulk of oil on the territory of Russia is produced in medium and small fields, which can be geographically located at a considerable distance from each other and from settlements, which makes it impractical to transport APG. In addition, such fields have a small production rate of APG (not more than 300 million cubic meters per year) and cannot ensure the permanency work of gas processing plant at the designed capacity [12].

The calculations made by the authors showed that for small fields with relatively small oil reserves and, accordingly, small volumes of extracted APG (up to 300 million cubic meters per year), this method is disadvantageous from an economic point of view, since it requires significant capital costs for transportation and storage APG, and also does not ensure the permanency work of gas processing plant at the designed capacity, which reduces the final profit. Thus, the payback period of technical event increases to several decades [13].

Thus, after analyzing fund materials of several dozens of small oil fields in Russia, their prospected reserves and transport accessibility, it was determined that the power generation is universal (in terms of field capacity of APG, taking into account the development of infrastructure, etc.) on the basis of generators with gas turbines [14]. Such installation can even operate on low-quality fuel, which is APG, thereby significantly reducing the operating costs of preliminary preparation of APG for combustion [15]. The generated power energy is used mainly for own needs. At the same time, exhaust gases can be used as a thermal agent in the central heating system of industrial facilities of the field, or it can be subjected to catalytic purification and released into the atmosphere with concentrations for the main pollutants not exceeding the established standards [16].

The evaluation of prospected reserves of oil, flow rate of oil and APG, as well as the chemical composition of produced oil and APG was carried out based on fund materials.

The calorific properties of APG and the possibility of its use for power generation were carried out by calculation based on the composition of APG.

All calculations were carried out in accordance with the methods certified in Russia.

3. Results

Being one of global largest suppliers of oil and gas, electricity tariffs in Russia are quite low and amount to about 0.05 USD per 1 kW. Nevertheless, the main advantage from an economic point of view is not saving on power energy charges, but the complete elimination of the costs of external power energy supply infrastructure, which, due to the considerable remoteness of deposits, requires considerable capital costs [17].

After analyzing the most popular methods of power generation, it was decided that the best in terms of ecological and economic justification would be to obtain power energy using a gas turbine installation (hereinafter referred to as GTI) [7].

Based on the fund data of gas producing enterprises in Russia, it was found that the average power consumption of small oil field is about 50 MW for every 45 days (period of 1 shift team). The calculations were carried out on the basis for an oil field production rate of APG about 100 million cubic meters per year, and also on the condition that all the produced power energy will be used on the needs of the oil field. Thus economic justification takes into account only the cost of electricity, which the manufacturing entities of field will no consume when the own source of power energy generation appears on the oil field [3,18].

The Russian market has a large selection of GTPP, both domestic and foreign production. Based on the energy intensity of the field, production rate and composition of associated gas, as well as the remoteness of the field from developed infrastructure networks, the choice of power plant was given to the domestic mobile GTPP with an output of power energy 500 MW per year [10].

A safety factor of 20% was also adopted when choosing a GTPP. It is necessary for the prospect of a potential increase in oil production, as well as for additional capacities of power energy during emergencies.

The hourly consumption of APG at full load of the GTPP 12 400 cubic meters, and the annual consumption is 109 million cubic meters. This power plant will fully meet the need of the field in power energy and partially in thermal energy. APG produced in the field is sufficient for the operation of GTPP with work loading about 92% of the design capacity, which is enough for normal operation of the field. If necessary, diesel fuel can be used as a backup fuel.

Exhaust gases from the GTPP enter the waste gas cleaning system, which efficiency reaches 95%. A cartographic model of carbon monoxide dispersion in the surface atmosphere after the project realization was created based on the actual emissions value obtained according to the calculation (Figure 2). The coefficients, calculated as the ratio of the current concentration of carbon monoxide to the normative indicator of its concentration in atmospheric air, are marked on contour lines. Thus, this ratio shows how many times the concentration is exceeded relative to the standard.

It can be seen in Figure 2 that the concentration of carbon monoxide slightly exceeds the established standard for carbon monoxide only within a radius of 1 km, which corresponds to the radius of the sanitary protection zone.

When implementing this project, the flare will completely decommissioned, which allows achieving such a significant reduction in emissions of pollutants into the atmosphere. However, the flare must remain in working condition in case of an emergency shutdown of a GTPP or a short-term increase in the APG flow rate at the field [19].

The cost of operations for the utilization of associated petroleum gas at a gas turbine power plant is determined, on the one hand, by capital costs and, on the other hand, by operating costs. Capital costs include the cost of the equipment itself and the cost of its delivery and installation costs (construction and installation works), while operating costs include wages and depreciation.

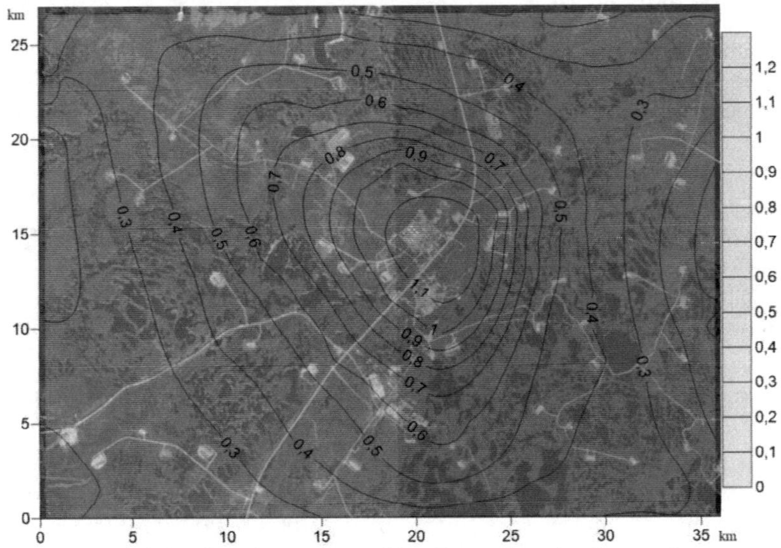

Figure 2. Cartographic model of carbon monoxide dispersion in the surface atmosphere after the project realization

The recoupment of the proposed project will consist of saving on power energy and from the reduction of environmental pays for atmospheric air pollution from APG combustion products in flare [1].

The calculation results of the economic efficiency of the proposed project are presented in Table 2.

Table 2. Calculation results of the economic efficiency of the proposed project

Capital costs, USD	480 000
Operating costs, USD per year	52 000
Reduction of environmental pays for atmospheric air pollution, USD per year	230 000
Saving on power energy, USD per year	15 500

Thus, based on the calculated data given in the Table 2, it can be concluded that the payback period of the proposed project at a small oil field will be about 2.5 years. At the same time, these calculations did not take into account the savings in the absence of the need to build an external power supply infrastructure, since all the considered fields are operating and the infrastructure on them has already been formed. At the same time, if this project is introduced at the field, where the operation is just beginning, then the absence of the need to build an external power supply infrastructure will save from 200 000 USD, which reduces the payback period by half [14].

4. Conclusion

The authors had conducted a search for the best way to utilize APG in small oil fields in Russia (flow rate of APG up to 300 million cubic meters per year) and had done its ecological and economic justification as part of the work. It can be drawn the following conclusions because of the study:

1. The most economically and environmentally justified method of utilization of APG is processing APG with the production of marketable products (Table 1), but due to many factors, such as poor infrastructure, nature and climate conditions, low volumes of produced APG, this

method is economically justified only for large oil fields (debit APG more than 1 billion cubic meters per year), which in Russia is not more than 10.

2. The most economically justified method of utilization of APG in small oil fields is power generation, while the produced power energy will be used on the needs of the oil field, without obtaining additional income from the sale of electricity. From the point of view of technical equipment, it is optimal to use GTPP as a power generation unit, since GTI are the most unpretentious to the quality of APG, have a block-modular (mobile) design, and are widely represented on the market.
3. Environmental calculations showed that when implementing the proposed method of utilization of oil and gas production, emissions at the field are reduced from thousands of tons per year to several tens of tons per year, which significantly improves the state of the environment and reduces the negative impact on living organisms near the field (Figure 2).
4. The economic justification for the project showed that with a capital expenditure of just under half a million USD and operating costs of about 50 000 USD per year, the payback period for this project will be about 2.5 years for the field being exploited (Table 2). The main economic indicators are savings on environmental pays for air pollution (230 000 USD per year) and savings on electricity (about 15 000 USD per year).
5. At the same time, if we consider the introduction of this method to the newly created field, then the absence of the need to build an external power supply infrastructure will save from 200 000 USD. In this case, the payback period of the project may be a little more than 1 year.

Acknowledgements
This scientific work is supported by the Scholarship of the President of the Russian Federation SP-3455.2019.3.

References

[1] Strizhenok A.V., Korelskiy D.S. 2019. *Estimation and Reduction of Methane Emissions at the Scheduled and Repair Outages of Gas-Compressor Units.* Journal of Ecological Engineering. **20(1)** 46-51.
[2] Pashkevich M.A., Petrova T.A. 2016. *Creation of a system for industrial environmental monitoring in hydrocarbon production and transporting companies of Western Siberia.* Journal of Mining Institute. **221** 737–741.
[3] Turbakov M.S., Kozhevnikov E.V. 2013. *Utilization of associated petroleum gas at small fields.* Neftyanoe khozyaystvo - Oil Industry. **8** 118-120.
[4] Litvinenko V.S., Kozlov A.V., Stepanov V.A. 2017. *Hydrocarbon potential of the Ural–African transcontinental oil and gas belt.* Journal of Petroleum Exploration and Production Technology. **7(1)** 1-9.
[5] Kuzhaeva A., Dzhevaga N. 2017. *Particulars of oil structural organization.* ARPN Journal of Engineering and Applied Sciences. **12(13)** 4157-4166.
[6] Semyachkov A.I., Slawikowskaja Y.O., Pochechun V.A. 2018. *Methodological features of the assessment of economic damage from adverse environmental consequences in conditions of territories with a developed mining complex*. Ecology and Industry of Russia. **22(4)** 46-51.
[7] Volkodaeva M.V., Kiselev A.V. 2017. *On the development of the system for environmental monitoring of atmospheric air quality*. Journal of Mining Institute. **227** 589–596.
[8] Danilov A.S., Smirnov Yu.D., Korelsky D.S. 2016. *A promising method of remote environmental monitoring of oil and gas facilities in Russia.* Neftyanoe khozyaystvo - Oil Industry. **2** 121-122.
[9] Kuzhaeva A., Berlinskii I. 2018. *Effects of oil pollution on the environment.* International Multidisciplinary Scientific GeoConference Surveying Geology and Mining Ecology Management, SGEM. **18(5.1)** 313-320.
[10] Bian H., Xu W., Li X., Qian Y. 2011. *A novel process for natural gas liquids recovery from oil*

field associated gas with liquefied natural gas cryogenic energy utilization. Chinese Journal of Chemical Engineering. **19(3)** 452-461.

[11] Drozdov A.N. 2014. *Utilization of associated petroleum gas with using of existing field infrastructure.* Neftyanoe khozyaystvo - Oil Industry. **4** 74-77.

[12] Barkan M.S., Kornev A.V. 2017. *Prospects for the use of associated gas of oil development as energy product.* International Journal of Energy Economics and Policy. **7(2)** 374-383.

[13] Turbakov M.S., Kozhevnikov E.V., Ryabokon E.P., Chernyshov S.E. 2012. *The main directions of the associated gas utilization.* Neftyanoe khozyaystvo - Oil Industry. **11** 130-132

[14] Vanadzina E., Gore O., Viljainen S., Tynkkynen, V.P. 2015. *Electricity production as an effective solution for associated petroleum gas utilization in the reformed Russian electricity market.* International Conference on the European Energy Market. **EEM 2015** article number 7216626.

[15] Nannarone A., Klein S.A. 2019. *Start-Up Optimization of a CCGT Power Station Using Model-Based Gas Turbine Control.* Journal of Engineering for Gas Turbines and Power. **141(4)** article number 041018.

[16] Welch M. 2014. *Selection criteria for power generation systems to minimise environmental impact.* Society of Petroleum Engineers - SPE Heavy and Extra Heavy Oil Conference. 845-856.

[17] Leusheva E.L., Morenov V.A. 2015. *Combined oilfield power supplying system with petroleum gas utilization as an energy carrier.* Neftyanoe khozyaystvo - Oil Industry. **4** 96-100.

[18] Rajović V., Kiss F., Maravić N., Bera O. 2016. *Environmental flows and life cycle assessment of associated petroleum gas utilization via combined heat and power plants and heat boilers at oil fields.* Energy Conversion and Management. **118** 96-104.

[19] Dmitriev V.T., Timukhin S.A., Simisinov D.I., Karyakin A.L. 2017. *Analysis of energy parameters of mine hoists.* Gornyi Zhurnal. **8** 70-72.

Accounting for the effects of third elements in the emission spectral analysis and construction of global analytical techniques

V S Sukhomlinov[1], A S Mustafaev[2], A N Popova[2] and H Koubaji[2]

[1]Saint-Petersburg State University, 7–9, Universitetskaya naberezhnaya, St. Petersburg, 199034, Russia

[2]Saint-Petersburg Mining University, 2, 21 line, Vasilievsky island, St. Petersburg, 199106, Russia

E-mail: popova_an@pers.spmi.ru

Abstract. The paper discusses the method of processing the metals and alloys spectra, which were obtained using emission spectrometers with spark excitation spectra. The proposed technique makes it possible to build a global method for analyzing any alloys, taking into account the so-called "third elements effects".

1. Introduction

During the determination of the elemental composition of substances by the method of emission spectral analysis, the so-called problem of "accounting for the influence of third elements" often arises [1–6]. Its essence is as follows. As is known, in order to implement this method using a specific instrument (emission spectrometer), it is necessary to calibrate it using working reference samples (WRS), i.e. samples of this type of substance with different (but known) content of elements to be determined. Calibration graphs are built on a set of WRS for each element separately and represent a smooth curve, which, as a rule, is approximated by a polynomial of some degree (usually no more than four). The calibration curve obtained in this way is the dependence of the concentration of the element being determined, on the intensity of its analytical spectral line. When examining a sample of unknown composition, the concentration of the impurity element is determined from the measured intensity of the analytical line using a calibration curve.

2. Materials and methods

However, when constructing calibration graphs in some types of samples, the dependence of the concentration of an element on the intensity of its analytical line is not monotonous. That is, WRS with a greater concentration of this element gives a lower intensity of the analytical line or vice versa. The reason is that the intensity of the analytical line of a certain element can be influenced by the fact that there is another impurity element in the WRS. Especially this effect is significant in the analysis of alloys using a low-voltage spark, which is the most common method for determining the elemental composition of metals [7–10]. It also manifests itself in the analysis of powder samples by the method of fractional evaporation from the crater of a graphite electrode. The simplest example is that if there are alkali metals in the powder sample as an impurity, the intensities of the analytical lines of all the elements being determined are significantly reduced (compared to the same sample without an alkali metal). This occurs, in particular, due to a decrease in the plasma temperature and, as a consequence, a decrease in the rate of excitation of the corresponding atomic (and ionic) levels. Thus, the results of the analysis for the content of a certain impurity in a sample by emission methods depend on the content of other impurities in this sample. This fact must be taken into account when analyzing samples with an unknown composition. This is the essence of the problem of the "third element influences".

The above factors are the reason that, for example, when analyzing metals and alloys, it is necessary to analyze different alloys on one basis under different conditions [11–12]: the parameters of the spectrum excitation source, the recording system, sample preparation, the set of analytical lines and reference lines. The examples include carbon and high speed steels, various types of bronzes, and the like. The above parameters in the analysis using the emission spectrometer are the so-called analytical methods, the choice of which in the software of a modern spectrometer is carried out by the operator manually. Each technique has individual calibration curves for the impurity elements to be determined, which allow to analyze only a specific type of alloy.

Thus, for accurate quantitative analysis, it is necessary to know the grade of alloy being analyzed, in order to relate it to a specific type, and select the appropriate analytical method. When the grade of the alloy is unknown, the quantitative analysis of the sample is substantially difficult if not impossible. The reason for this is the influence of the third elements in alloys of various types.

Thus, the methods of constructing so-called "global techniques" are relevant, in which, using uniform calibration graphs, it is possible to determine the content of impurity elements in any (or at least in large quantities) different types of substance on one basis, for example, in any alloys based on **Fe** by taking into account the influence of third elements. In addition, even within the same type of alloys, taking into account this phenomenon can significantly reduce the systematic errors in measuring the concentrations of impurity elements.

3. Discussion

In the software of modern Russian emission spectrometers, the influence of the third elements is taken into account according to a linear empirical theory, which does not allow building global methods covering a large number of types of substances with the same basis. The situation is somewhat better with imported devices, where even quadratic, but also empirical approximation formulas are used.

We are developing a general methodology for taking into account the influence of third elements from first principles, which is applicable to all types of substances. This technique was preliminarily tested by us on the Russian emission spectrometer SPAS-02, intended for the analysis of metals and alloys. SPAS-02 is a desktop instrument with average technical characteristics. Testing has shown that it is possible to build global methods for analyzing alloys, the systematic error in determining the composition of which does not exceed that in the method for analyzing one type of alloys. That is, for example, the determination of the concentration of an impurity element in the method of analysis of carbon low and medium alloyed steels is the same as in the global method of analysis of Fe-based alloys (high-speed steels, high-alloyed and high temperature steels, bleached cast iron, tool steels, etc.).

Some results of this testing are given below. Fig. 1 shows the dependence of calculated and passport concentrations of copper in WRS. The influence on the definition of Cu of the following elements was taken into account: Ni, W, Cr, V, since the high-speed steels contain high concentrations of W, V, and high-alloyed ones contain Ni, Cr. It can be seen that the general theory significantly reduces the error in determining the concentration of copper steels. The average standard deviation of the calculated concentration from the passport is reduced by 2.4 times – from 14.1 % to a value of 5.8 %. At the same time, the standard deviation for the set of low and moderately alloyed steels (where the influence of the third elements is not observed) is 5.6 %, that is, as noted, they are almost equal.

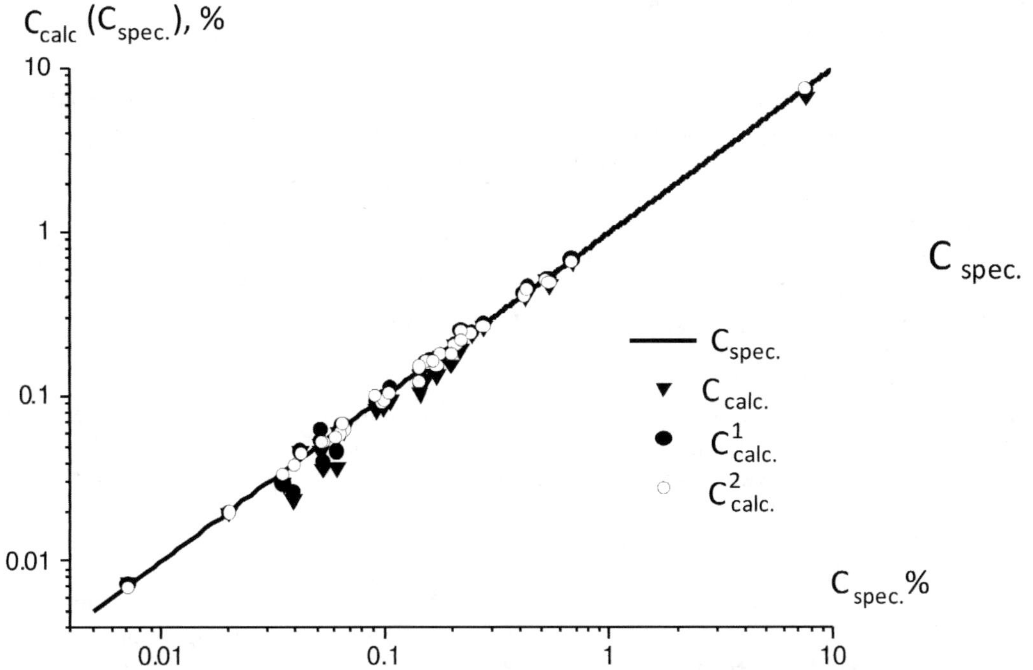

Figure 1. Results of the developed theory application for determining the copper content (high-speed steel, high-alloyed and high temperature steels, bleached cast iron, tool steels, etc.) in a single graph. The standard deviation in the global method (5.8 %) is almost equal to the same value but for the determination method in low- and medium-alloy steels (5.6 %). $C_{calc.}$ - calculation of the content without taking into account influences; $C^1_{calc.}$ - calculation using linear theory; $C^2_{calc.}$ - calculation using the general theory; $C_{spec.}$ – specified concentration.

4. Conclusion

The proposed theory in this particular case almost ideally takes into account the influence of the third elements. It should be noted that the average standard deviation of the calculated concentration from the nameplate, taking into account the influence of the same elements according to the empirical theory, is 11 %, that is, twice the corresponding value for the general theory.

It is interesting to note that the range of concentrations of Cu using the results of taking into account the effects of third elements on the definition of copper according to the general theory, as can be seen from Fig. 1 – 0.007–8 % – more than three orders of magnitude.

References

[1] Rosenberg E and Panne U 2014 Atomic Absorption Spectrometry (AAS) and Atomic Emission Spectrometry (AES) (Book Chapter) *Handbook of Spectroscopy: Second, Enlarged Edition* **2–4** 507–82

[2] Attri P, Arora B and Choi E H 2013 Utility of plasma: A new road from physics to chemistry *RSC Advances* **3**(31) 12540–67

[3] Weltmann K-D et al 2019 The future for plasma science and technology *Plasma Processes and Polymers* **16(1)** 1800118

[4] Mustafaev A.S, Sukhomlinov V S and Ainov M A 2017 Ion velocity distribution function in intrinsic gas plasma under conditions of resonance recharging as a main process *Experiment. High Temperature* **55(3)** 346–51

[5] Kondo H, Aimoto M, Wagatsuma K 2014 Optical emission spectrometry for optimization of steelmaking processes (Review) *Tetsu-To-Hagane/J. of the Iron and Steel Institute of Japan* **100(7)** 846–56

[6] Gibson B, Carter S, Fisher A S, Lancaster S, Marshall J and Whiteside I 2014 Atomic spectrometry update. Review of advances in the analysis of metals, chemicals and functional materials *J. of Analytical Atomic Spectrometry* **29(11)** 1969–2021

[7] Frydenvang J, Kinch K M, Husted, S, Madsen M B 2013 An optimized calibration procedure for determining elemental ratios using laser-induced breakdown spectroscopy *Analytical Chem.* **85(3)** 1492–500

[8] ElFaham M M, Okil M and Mostafa A M 2018 Limit of detection and hardness evaluation of some steel alloys utilizing optical emission spectroscopic techniques *Optics and Laser Technol.* **108** 634–41

[9] Usala J D, Maag A, Nelis T and Gamez G 2016 Compressed sensing spectral imaging for plasma optical emission spectroscopy *J. of Analytical Atomic Spectrometry* **31(11)** 2198–206

[10] He X, Dong B, Chen Y, Li R, Wang F, Li J and Cai Z 2018Analysis of magnesium and copper in aluminum alloys with high repetition rate laser-ablation spark-induced breakdown spectroscopy *Spectrochimica Acta – Part B Atomic Spectroscopy* **141** 34–43

[11] Deng J-H 2013 Glow discharge optical emission spectrometry and its applications in domestic iron and steel industry *Yejin Fenxi/Metallurgical Analysis* **33(10)** 24–33

[12] Bengtson A, Eklund A, Lundholm M and Saric A 1990 Further improvements in calibration techniques for depth profiling with glow discharge optical emission spectrometry *J. of Analytical Atomic Spectrometry* **5(6)** 563–7

Journal of Physics: Conference Series **1384** (2019) 012055 doi:10.1088/1742-6596/1384/1/012055

Instrumental virtual tools for measuring random signal characteristics

I I Sytko, E E Smirnova

Saint-Petersburg Mining University, 21 Line, St. Petersburg, Russia, 199106

E-mail: ivan-sytko@yandex.ru

Abstract. Based on modern computer technologies developed by National Instruments, a virtual device for adjusting, controlling and studying various noise characteristics has been developed. Signs helping identify the laws of random signal probability distribution are smoothness or sharpen-ing, asymmetry and uncertainty. Examples of identification of noise characteristics of elec-tronic devices are provided; instrument technologies are used to monitor the technical condi-tion of such devices.

1. Introduction

The relevance of the problem of measurement of characteristics of random signals is due to the complexity of choosing a random process model, technical implementation of the meter, sta-tionarity and non-stationarity, ergodicity and non-ergodicity.

The object is the existing methods and technical tools designed to measure probabilistic characteristics of random signals.

The methods for determining probabilistic characteristics of random signals are very diverse. They are based on the display of an indicator of the probability density distribution curve and determination of the distribution law, characteristics of the position and dispersion. Existing methods for measuring characteristics of random signals do not ensure operational control of smoothness, sharpening, asymmetry and uncertainty, as well as identification of the distribution law.

The scope of the meter is regulation, control and research of radio electronic devices accord-ing to noise characteristics and taking into account metrological requirements; these measuring instruments can be used in various industries. Therefore, the operational measurement of proba-bilistic characteristics of random signals taking into account metrological requirements is an ur-gent task.

Не используйте сноски и интерактивные ссылки.

2. Purpose of research

The purpose is to develop hardware and software for measuring probabilistic characteristics of random signals using computer technologies of virtual instruments developed by National In-struments (NI). The paper studies the procedure for identifying distribution laws using a priori and a posteriori information on the second, third and fourth order moments, characteristics of uncertainty, creates a working model and determines the scope of its application.

3. Materials and methods

To determine the type of the law of distribution of random signals, it is necessary to establish signs or criteria by which the laws of distribution of random signals are identified. Distributions are quite diverse: limited or unlimited; with a flat top or round, with or round [1, 2, 3, 4, 5, 6].

J. Kendal and A. Stuart classified distributions into five types: symmetric single-mode; symmetric two-mode; oblique; extremely short; other [4].

Mathematical models and hardware for measuring the characteristics of random signals are discussed in [1, 3, 7, 8, 9, 10, 11, 12, 13]. The problem of measuring probabilistic characteristics of random signals which make them inaccurate is a limited observation time [12, 13, 14, 15].

For classification and determining a form of the distribution of random signals, it is necessary to select identification signs that would show the mutual proximity or distance of the distributions and could be quantified [4, 7].

As a smoothness or aggravation of the distribution curve near its mode [2], an excess is used:

$$\varepsilon = \frac{\mu_4}{\mu_2^2}, \tag{1}$$

where $\mu_4 = \int_{-\infty}^{+\infty} x^4 \times p(x)dx,\quad \mu_2 = \int_{-\infty}^{+\infty} x^2 \times p(x)dx$ central moments of the fourth and second order;
x is the value of the random signal; *p(x)* – the random distribution law.

The excess for various distributions varies infinitely from 1 to $+\infty$), therefore the excess is uncomfortable. Therefore, we will perform its nonlinear transformation into a counter-excess value:

$$\chi = \frac{1}{\sqrt{\varepsilon}} \tag{2}$$

The counterexcess for any distributions is in the range from 0 (ε=∞) to 1 (ε=∞). Therefore, as the first sign of identification of distributions, we take the value of counterexcess χ.

An analysis of the results of calculations in the Mathcad showed that the counterexcess values for symmetric distributions range from 0.68 to 0.73 for trapezoidal distributions, from 0.67 to 0.882 for arcsinusoidal distributions, from 0.05 to 0.71 for exponential two-sided distributions and in significantly overlap in the area of flat-top and peaked distributions.

For asymmetric distributions, the counterexcess value ranges from 0.628 to 0.655 for the Beta distribution, from 0.327 to 0.493 for the Gamma distribution, from 0.532 to 0.621 for the Nakagami distribution, from 0.481 to 0.634 for the Rayleigh distribution.

The entropy coefficient [4] whose value is determined by the formula, is used as the second independent sign which characterizes the distribution form.

$$K = \frac{\Delta_э}{\sigma}, \tag{3}$$

where $\Delta_э = \frac{1}{2} \cdot \times \exp(H(x))$ – the random signal entropy value;

σ – the standard deviation of a random signal;

H(x) – random signal entropy.

The entropy of a random signal is a measure of its uncertainty, it depends on the type of distribution and is determined by formula:

$$H(x) = -\int_{-\infty}^{+\infty} p(x) \times \ln p(x)dx. \tag{4}$$

For the most common symmetric distributions, the value of the entropy coefficient varies from 1.83 to 2.00 for trapezoidal distributions, from 1.11 to 1.88 for arcsinusoidal distributions, from 0.09 to 1.87 for exponential two-sided distributions, from 0.35 to 1.12 for antimodal distributions. The values of the entropy coefficient for symmetric distributions overlap in the areas of flat-top and sharp-top distributions.

For asymmetric distributions, the entropy coefficient ranges from 3.268 to 3.458 for the Beta distribution, from 1.318 to 2.22 for the Gamma distribution, from 1.317 to 1.435 for the Nakagami distribution, and from 1.288 to 2.409 for the Rayleigh distribution.

As the third feature characteristic of the asymmetry of the distribution curve relative [2] is the asymmetry determined by formula:

$$S = \frac{\mu_3}{\mu_2^{3/2}}, \tag{5}$$

where $\mu_3 = \int_{-\infty}^{+\infty} x^3 \times p(x)dx$ – the third central moment.

Symmetric distributions have an asymmetry equal to zero. For asymmetric distributions, the asymmetry ranges from -0.51 to -0.071 for the Beta distribution, from 1.564 to 2.988 for the Gamma distribution, from 1.574 to 1.645 for the Nakagami distribution, from 0.518 to 2.015 for the Rayleigh distribution.

Thus, when using the counterexcess and the entropy coefficient as identification signs characterizing the form and type of distributions, the image point (or some area) with coordinates χ and K will always be within the rectangle limited by the values of the entropy coefficient and counterexcess.

Studies have shown that the form of the density curve for asymmetric distributions depends on the shape and scale parameters which have a functional dependence on the counterexcess, the entropy coefficient, and the asymmetry.

The form of the Beta distribution density curve is sensitive to the asymmetry. The mathematical model of the functional dependence of parameter α (β) with shape index $\beta = 3$ ($\alpha = 3$) is as follows:

$$\alpha = 3 \times exp(-1.39 \times S) \tag{6}$$

$$\beta = 3 \times exp(1.39 \times S) \tag{7}$$

For the Nakagami distribution in the range of asymmetry changes from 1.574 to 1.645, the shape of the distribution density curve is sensitive to the counterexcess. The mathematical model of the functional dependence of shape parameter α on the counterexcess value with scale parameter $\beta = 1$ is as follows:

$$\alpha = 30.394 - 127.987 \times \chi + 135.135 \times \chi^2 \tag{8}$$

Thus, after determining the type of asymmetric distribution according to three classification criteria, shape and scale parameters, it is possible to establish a specific type of asymmetric distribution.

4. Experimental results and discussion

The studies were carried out in the NI LabVIEW 2012 using real-noise recordings of electronic devices [16, 17, 18].

To conduct experimental studies, a virtual meter (VM) of probability characteristics was designed. Its structural diagram is shown in Fig. 1, and the appearance of the front panel (LP) is shown in Fig. 2.

The meter contains a normalizing amplifier (NA) - 1; a channel for measuring the mean square value (RMS MCH) - 2; an asymmetry measurement channel (A MCH) - 3; a channel for measuring the probability distribution density and the entropy coefficient (MCH PD and EC) - 4; a measuring channel for the counterexcess (MCH CA) - 5; a channel for measuring the parameters of the distribution form (MCH DCH) - 6; a decision making unit (DB) - 7; an indicator (I) - 8.

Experimental and metrological studies of measuring probabilistic characteristics of random signals and identifying the type of distribution were carried out and verified by semi-natural modeling using real-life recordings of radio electronic devices (Table).

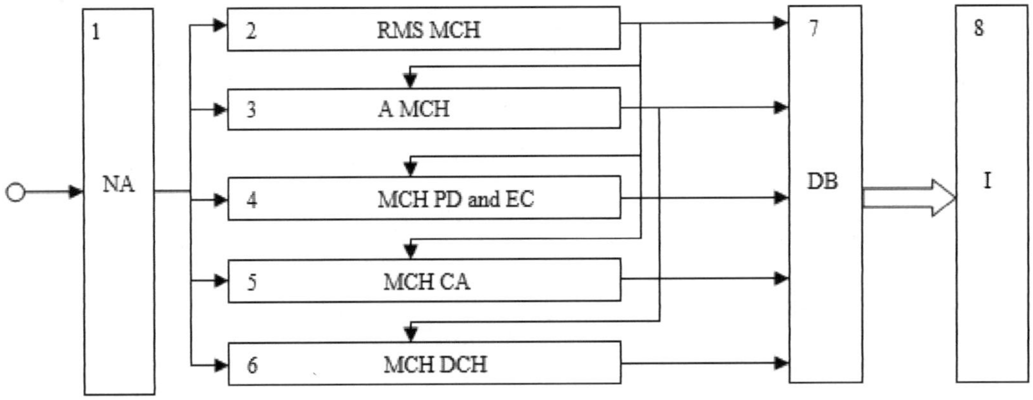

Figure 1. VM of probabilistic characteristics. Structural scheme.

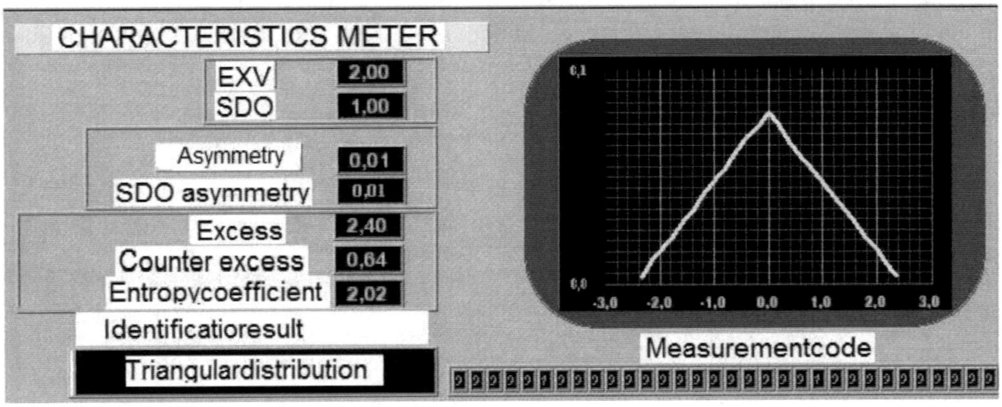

Figure 2. VM of probabilistic characteristics. The VM FP fragment.

The VM identifies the following types of symmetric distributions: uniform, triangular, arcsinusoidal-I, II, III, trapezoidal-I, II, III, antimodal-I, II, two-way exponential with a power factor of 1/4 to 7, Laplace and normal, asymmetric - Beta, Nakagami, Rayleigh, Gamma. The indicator displays a graph of the distribution curve, as well as information about the type and parameters of the distribution.

5. Conclusion

Probability characteristics meter allows for identification of 16 types of symmetric distributions, 5 types of asymmetric distributions, and 7 parameters. The probability of distribution identification is 0.95, the accuracy of parameters measurement is ± 1%. Accuracy of measurement and identification of the type of distribution increases with increasing time of observation of a random signal. The results establish a relationship between the results of measurements and identification in the data of the probabilistic characteristics meter and the real probabilistic characteristics of noise.

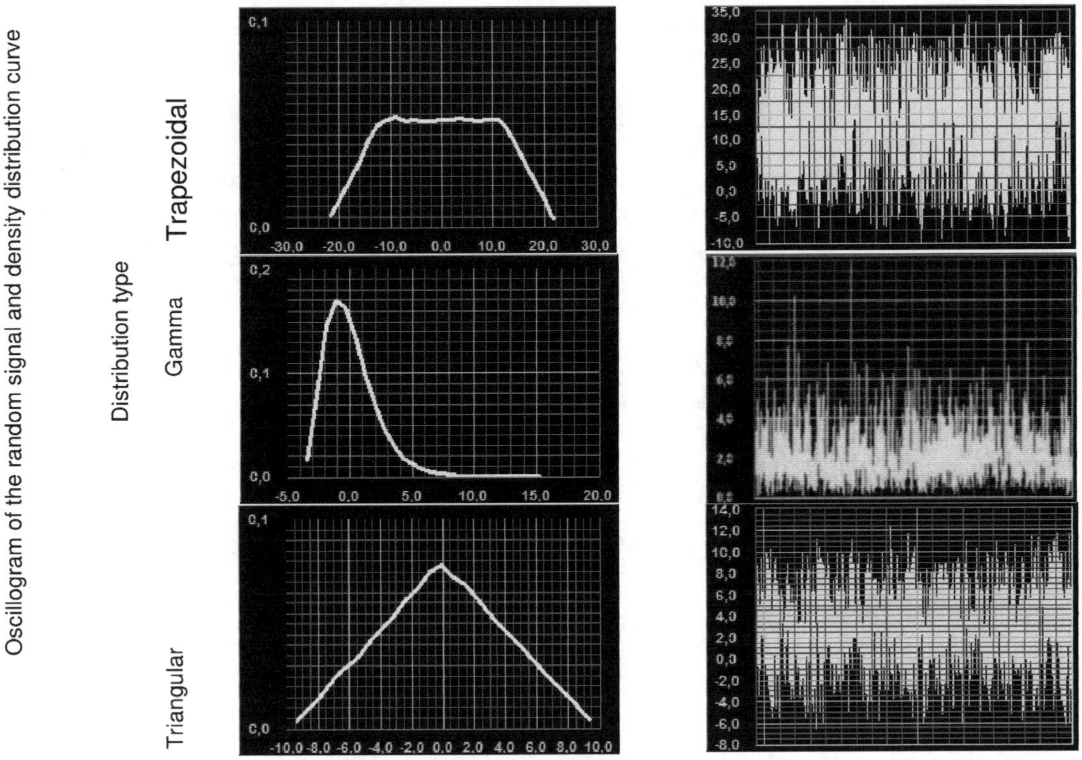

Figure 3. Experimental studies of random signals

References

[1] Bendat J, Piersol A 1974 *Measurement and analysis of random processes* (Moscow: Mir).

[2] Wentzel E S 1962 *Probability Theory* (Moscow: Physics and Mathematics).

[3] Zhovinsky V N 2009 *Express engineering analysis of random processes* (Moscow: Energy).

[4] Novitsky P V, Zograf I A 1991 *Error estimation of measurement results* (Leningrad: Energoatomizdat).

[5] Sytko, I. I., & Kremcheeva, D. A. (2017). Instrumentation for measuring the parameters and characteristics of four-poles. *International Journal of Mechanical Engineering and Technology, 8*(10), 844-854.

[6] Vystrchil, M., & Novozhenin, S. (2017). Estimation of autolock mode accuracy of robotic total stations with various configurations reflectors. Paper presented at the *International Multidisciplinary Scientific GeoConference Surveying Geology and Mining Ecology Management, SGEM, , 17*(22) 281-288.

[7] Antsev G V, Sytko I I, Sarychev V A, Sokolov G P 2006 A device for measuring characteristics of random processes, Patent No. 2280278 of the Russian Federation.

[8] Sytko I I, Shumakov P P, Naumenko N S, Latiy O V 2008 A device for measuring characteristics of random processes // Patent No. 2336562 of the Russian Federation.

[9] Sytko I I 2014 Characterization of random processes distributed according to the Nakagami law *Components and Technologies* **1** 172-173.

[10] Vinogradova A A, Leontyuk S M (2018) Evaluation of current status in field of electrical voltage measurements *Journal of Physics: Conference Series* **1118(1)** DOI: 10.1088/1742-6596/1118/1/012048.

[11] Kremcheev E A, Kremcheeva D A 2017 Methodological approaches to controlling the feed breakage process in drum and tube mills based on the results of vibroacoustic diagnostics

Journal of Industrial Pollution Control **33**(1) 787-791.

[12] G V Rudianov, E I Krapivsky, S M Danileiv 2018 Evalution of signal properties when searching for cavities in soil under concrete slabs by radio detection stations of subsurface investigation *Journal of mining institute* **231** 245-253.

[13] Manzhos V N, Shirman Ya D 2011 *Theory and technology of radiolocation information processing under noises* (Moscow: Radio and communication).

[14] Voytyuk, I. N., & Bolshunova, O. M. (2019). Ensuring energy efficiency of the belt-type conveyor electric motor drive operation due to noncontact control of mass discharge of cargo during its transport in mining condition. Paper presented at the *Proceedings of the 2019 IEEE Conference of Russian Young Researchers in Electrical and Electronic Engineering, ElConRus 2019,* 736-739.

[15] Sytko I I, Kremcheev E A, Makhov V E 2017 Device for measuring relative amplitude-frequency characteristics // Patent No. 2668951 of the Russian Federation.

[16] Makhov V E 2014 Nondestructive optical control diamonds on the basis of computer technology National instruments *Journal of mining institute* **209** 119-123.

[17] Vasil'ter Yu V, Zheltov S Yu, Knyaz V A, Khodarev A N, Morzhin A V 2007 *Processing and analysis of digital images with examples on LabVIEW IMAQ Visio*n (Moscow: DMK Press).

[18] Zhambalova S T, Vinogradova A A 2018 Accuracy increase of self-compensator *IOP Conference Series: Materials Science and Engineering* **327**(2) DOI: 10.1088/1757-899X/327/2/022109

Development of a local quasigeoid model for Vietnam land area using the global EGM2008 model

Tran Thanh Son[1], A A Kuzin[1], M G Mustafin[1]
[1] Saint Petersburg Mining University, 2, 21 Line, Saint Petersburg, 199106, Russian Federation

E-mail: kuzin_aa@pers.spmi.ru

Abstract. The article presents the results of research on the development of methods for obtaining normal heights in Vietnam using the global geoid model EGM2008 and software for processing GNSS measurements. The errors between the normal heights computed by the global model of the EGM2008 geoid and the heights found by GNSS measurements and geometric levelling on the territory of Vietnam are determined. The localisation method for the global geoid model EGM2008 for Vietnam's territory is suggested, such a geoid model is practically formed as the EGM2008_ TN file. Tests of the developed technique and the localised model of the EGM2008__TN geoid were carried out, showing the possibility of its application to obtain normal heights from GNSS measurements in Vietnam with the 4th levelling accuracy class.

1. Introduction

GNSS technologies allow determining geodetic heights for points on the earth's surface with high accuracy. However, the normal heights are often used in the industry, so the transition from geodetic to normal heights with the accuracy of geometric levelling should be performed by a series of transformations based on the data on anomalies of the height-difference surface of the quasigeoid and ellipsoid. The determination of normal heights from satellite levelling compares favourably with geometric levelling under complicated conditions, such as harsh weather, mountainous swampy areas, rugged terrain, etc. In addition, even if no results of ground geometric levelling are available, GNSS measurements, allow height determination for any point; in this case, a model of height anomalies of the required accuracy is necessary. In this regard, the actual task is to determine the normal heights using GNSS technology without a level run.

Such global geoid models as EGM2008, EIGEN-6C4, GECO and GAO-2012 can be used. The highest degree of geopotential decomposition (2190-degree) has the EGM2008 model with a grid size of 1x1 or 2.5x2.5 minutes of the National Geospatial-Intelligence Agency, NGA of the U.S. Department of Defense. Due to the high level of detail, EGM2008 is widely used in production. However, in mountainous areas, it is necessary to clarify this model because in some cases the result obtained from satellite levelling using the model of height anomalies does not coincide with the results of more reliable geometric levelling.

2. Materials and methods

In order to create a local quasigeoid model, more accurate than the global model (EGM2008), it is necessary to use a set of points with known normal heights at which satellite observations were made and geodetic heights were determined.[2, 5-7, 15]. In this case, the constructed model of the local quasigeoid (surface) is applicable to the local area only.

Heights are related by the expression:

$$H = H^{g} + N = H^{\gamma} + \zeta, \qquad (1)$$

where: H is for the geodetic height, H^{g} is for the orthometric (absolute) height (correlates with the geoid), H^{γ} is for the normal height (correlates with quasigeoid), ζ is for the height anomaly, N is for the geoid height.

According to the work [13], residual and wave methods can be applied for constructing a height anomaly model. The residual method based on the interpolation of the height anomaly from the EGM2008 model is the most common and simpler. It has higher accuracy [4, 8–11, 13, 14], especially at low density of GNSS-levelling points. The residual method gives a better model of elevation anomalies on a local scale (for a small area). This method adjusts the EGM2008 model to the actual values of elevation anomalies.

Knowing the exact model of the height anomaly ζ, it is possible to convert the geodetic height determined by GNSS technology to the normal height. To achieve this goal, a model of height anomalies ζ with a 1-cm accuracy is required to determine the height with a levelling accuracy of the 3rd or 4th classes [12,16–17].

The wave method is the construction of a local elevation anomaly model without improving the EGM2008 model. The local elevation anomaly model is based on interpolation to compute the ζ value at an arbitrary point from a set of GNSS points with the known normal heights and to equalise the heights of GNSS levelling points. The wave method is often used under a fairly large density of GNSS measurement points at which normal heights are known.

Model EGM2008 is sufficiently improved and can be used jointly with the data of the GNSS measurements to construct a local height anomaly model for Vietnam.

2.1. An adjustment algorithm that combines geodesic height, normal height and height anomaly

For the points of the network by the values of geodetic height, normal height and height anomalies, the deviation vector $\Delta\zeta_i$ can be computed by the formula:

$$\Delta\zeta_i = H_i - h_i - \zeta_i^{2008} = \zeta_i^{ITHCC/Hu6} - \zeta_i^{2008} \tag{2}$$

and the equations of adjustment amendments can be drawn:

$$\Delta\zeta_i = f(\phi_i, \lambda_i) + v_i = a_i^T x_i + v_i \tag{3}$$

The elements of the matrix A and the unknown, depending on the model parameters selected to describe the difference between the three elements of the height. In equations (2) and (3), H_i, h_i, ζ_i^{2008} are the geodesic height, normal height, and height anomaly, respectively (derived from the EGM2008 model) that correspond to point *i,* and $H_i - h_i = \zeta_i^{ITHCC/Hu6}$ is the GNSS-leveling height anomaly.

Accuracy estimation can be performed by 5 types of models: including 4 parameters, 5 parameters, a polynomial of 1st degree, a polynomial of 2nd degree and a polynomial of 3rd degree [3, 14].

In general case

$$a_i^T x_i = \sum_{m=0}^{M}\sum_{n=0}^{N} x_q(\varphi_i - \varphi_0)^n(\lambda_i - \varphi_0)^m \cos^m\varphi_i \tag{4}$$

The system of measurement equations and its solution can be drawn as a matrix:

$$L = Ax + V \tag{5}$$

$$x=-(A^TPA)^{-1}A^TPL=-(A^TA)^{-1}A^TL \tag{6}$$

where P is the weight matrix (the inverse variance matrix, i.e. covariance measurement matrix). We suppose that the measurements and their corresponding errors do not correlate between each other and heights in the same measurement system, i.e. they are independent. On the other hand, we also assume that the heights are measured with the same accuracy; in this case, the weight matrix turns into a unit matrix.

In addition to the above mentioned model, the interpolation method based on the spline function is also widely used in geodetic surveying [1, 3]. Spline interpolation is commonly used in the wave method to construct a local anomaly model.

Let us consider both cases with and without the use of the global EGM2008 model. The height anomaly of a point is interpolated by the points with the use of the spline function according to the following formula:

$$\Delta\zeta[P(x,y)] = \sum_{i=1}^{n} a_i r_{PP_i}^2 \ln(r_{PP_i}^2) + \tau_1 + \tau_2 x + \tau_3 y, \tag{7}$$

where: $r_{PP_i} = \sqrt{(x-x_i)^2 + (y-y_i)^2}$, $a_i (i = 1 \div n); \tau_1, \tau_2, \tau_3$ is the required parameters of the equation system:

$$\begin{bmatrix} 0 & g_{1,2} & \cdots & g_{1,n} & 1 & x_1 & y_1 \\ g_{2,1} & 0 & \cdots & g_{2,n} & 1 & x_2 & y_2 \\ \cdots & \cdots & \cdots & \cdots & \cdots & \cdots & \cdots \\ g_{n,1} & g_{n,2} & \cdots & 0 & 1 & x_n & y_n \\ 1 & 1 & \cdots & 1 & 0 & 0 & 0 \\ x_1 & x_2 & \cdots & x_n & 0 & 0 & 0 \\ y_1 & y_2 & \cdots & y_n & 0 & 0 & 0 \end{bmatrix} \times \begin{bmatrix} a_1 \\ a_2 \\ \cdots \\ a_n \\ \tau_1 \\ \tau_2 \\ \tau_3 \end{bmatrix} = \begin{bmatrix} \Delta\zeta_1 \\ \Delta\zeta_2 \\ \cdots \\ \Delta\zeta_n \\ 0 \\ 0 \\ 0 \end{bmatrix}, \quad (8)$$

where

$$g_{i,j} = g_{j,i} = \begin{cases} r^2_{P_iP_j} \ln(r^2_{P_iP_j}); (i \neq j). \\ 0; (i = j) \end{cases} \quad (9)$$

By using the above mathematical models, it is possible to compute the height anomalies $\Delta\zeta$ of the points and to move to normal heights by the results of GNSS-positioning:

$$h = H - (\zeta^{2008} + \Delta\zeta) \quad (10)$$

3. Results

The geodetic network based on the GNSS technology consists of 180 points (GNSS-levelling) used for the considered (test) territory of the Central Highlands in Vietnam. Measurements were carried out by two-frequency receivers during at least 6 hours for each reception. GNSS levelling points are distributed relatively uniformly across provinces from Quang Nam to Lam Dong, Binh Phuoc, and Ninh Thuan (Fig.1). The research is based on 163 points, the rest 17 control points were used to test the methodology.

3.1. Design of a local quasigeoid model using the EGM 2008 model

With regard to our data (163 points at the Central Highlands) we have the coordinates of the boundaries of the site: Southwest $B_1 = 11^0 41'$, $L_1 = 107^0 00'$; northeast – $B_2 = 15^0 21'$, $L_2 = 109^0 25'$. The region size $\Delta B = 3.67^0$, $\Delta L = 2.52^0$ with a cell size of 2.5' x 2.5', the number of grid points (N) will be 5251 (N = 89x59). At each point of the grid (i), 3 values can be found: latitude B_i, longitude L_i and the height anomaly (from the EGM2008 model) ζ_i (i = 1, 2 ..., 5251).

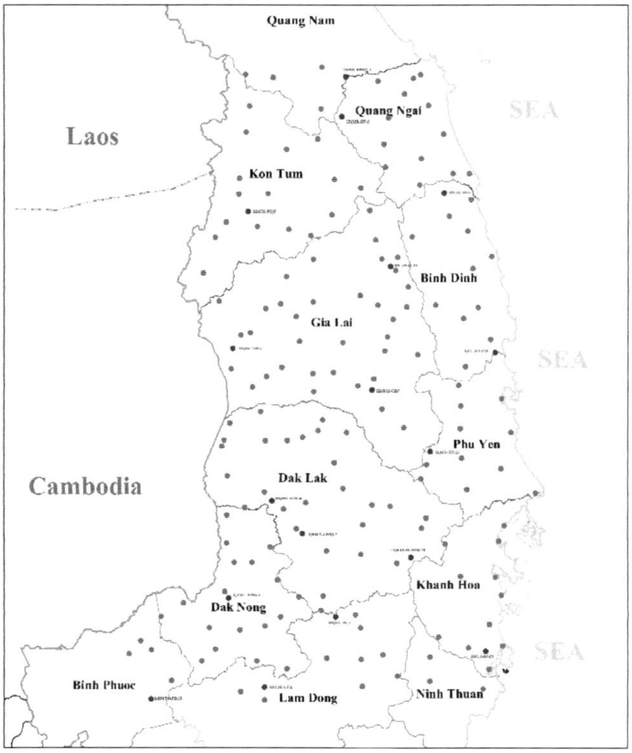

Figure 1. GNSS-levelling network (Central Highlands, Vietnam)

Correction computation for the grid points is done using the spline function (8, 10). Height anomalies in the global model EGM2008 are computed by the formula: $\zeta^{TN} = \zeta^{2008} + \delta\zeta$. (where ζ^{2008} is for the anomalous height, obtained from the EGM2008 model, $\delta\zeta$ is for the anomaly correction obtained by the wave method interpolation (on the basis of the spline function), ζ^{TN} is for the anomaly height of a continuous model). Height anomaly computation is shown in the flowchart (Fig. 2).

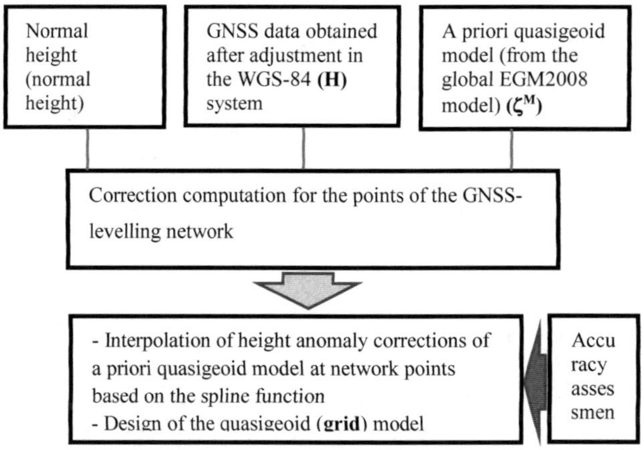

Figure 2. Design of a local quasigeoid model according to the global EGM 2008 model

After computation of the height anomaly for the network points (ζ^{TN}), a quasigeoid model of the Central Highlands (and the South Central Coast) has been revised.

The Central Highlands and the South Central Coast quasigeoid models are based on the EGM2008 model, latitude from $11^0 41'$ to $15^0 21'$, longitude from $107^0 00'$ to $109^0 25'$, area of ca. 80 000 km^2. A grid model (Table 1) has a cell size of 2.5' × 2.5' (including 5251 points).

Table 1. Height anomaly and correction for the grid points (cell model) (part)

No.	Name	B(^{0})	L(^{0})	ζ^{2008}(m)	$\delta\zeta$ (m)	ζ^{TN} (m)
1	1	15.3500	107.0000	-12.550	0.711	-11.839
2	2	15.3500	107.0417	-12.132	0.726	-11.406
3	3	15.3500	107.0833	-11.715	0.742	-10.973
…	…	…	…	…	…	…
5250	5250	11.6833	109.3750	5.370	1.314	6.684
5251	5251	11.6833	109.4167	5.519	1.364	6.883
				Min = -12.55 Max = 5.519	Min = 0.031 Max = 1.364	Min = -11.839 Max = 6.883

The cell size in the local EGM2008_TN quasigeoid model is 2.5′ × 2.5′, the circle radius is from 6 to 11 km. With this radius, the number of selected points will be between 4 and 13 cell points. Points within this radius will be used as reference points for interpolating GNSS measurement points.

Accuracy assessment of the interpolation results

The model was tested on 17 points of the satellite network that are not used for the modelling. First of all, we need to explore and choose the interpolation algorithm of the height anomalies of the EGM2008_TN.

According to the data obtained from 17 points, we calculate the difference in height between these points and then compare them with the tolerances for the 3rd and 4th levelling classes and technical levelling.

In practice, when measuring heights using GNSS technology, at least one point with normal height is needed in the network. For this point, it is necessary to determine the height anomaly $\Delta\zeta$ using the geoid model, then calculate the difference Δh using the formula:

$$\Delta h = \Delta H - \Delta\zeta, \tag{12}$$

where ΔH is the difference of geodetic heights at the considered point.

Let us denote $\Delta H_{i,k}$ as the difference of geodetic heights between two points; $\Delta h_{i,k}$ as the difference of normal heights, and $\Delta\zeta_{i,k}$ as the height anomaly determined by the global model (EGM2008) between two points.

The difference between the normal height of 2 points is computed as follows:

$$\delta_{i,k} = \Delta H_{i,k} - \Delta h_{i,k} - \Delta\zeta_{i,k} \tag{13}$$

Considering the weight measurement, the standard deviation for 1 km is computed using the following formula (17)

$$m_{км} = \pm\sqrt{\frac{[P\delta_i\delta_i]}{m}} \tag{14}$$

The weight is calculated by the formula:

$$P_{i,k} = \frac{1}{D_{i,k}}, \tag{15}$$

where $D_{i,k}$ is the distance between two points in kilometres, m is the number of point pairs to study. In this case, m should be not less than 20, and the distance between the points $D_{i,k}$ should be of different values. The results are shown in Table 2.

Table 2. Accuracy assessment for the determined normal heights using EGM2008_TN model

No.	Factors	EGM 2008	Weighted average	Polynomial of 1st degree	Polynomial of 2nd degree	Spline function
1	Number of points	136	136	136	136	136
2	Non-compliance with the technical levelling tolerance	9 (7%)	0 (0%)	0 (0%)	0 (0%)	0 (0%)
2	Compliance with the technical levelling tolerance	127 (93%)	136 (100%)	136 (100%)	136 (100%)	136 (100%)
3	Compliance with the tolerance level of the 4th class	108 (79%)	136 (100%)	136 (100%)	136 (100%)	136 (100%)
4	Compliance with the tolerance level of the 3rd class	54 (40%)	94(69%)	103(76%)	108(79%)	115 (85%)
5	Standard deviation (m) – m_{km}	0.0244	0.0119	0.0102	0.0096	0.0086

The interpolation results for 17 control points show that all four interpolation methods provide acceptable results. According to Table 2, ca. 80% of the routes have acceptable accuracy for the 3rd grade levelling under all four interpolation methods, even in the cases of mountain areas. The 100% accuracy is obviously provided for the technical levelling. In practice, we suggest applying the weight-average interpolation method.

4. Conclusion

The developed calibration procedure for height anomalies of the EGM2008 is intended for use on a large scale with the large number of points in a network. The computation method is relatively simple and improves the accuracy of height determination based on GNSS measurements.

The network points (GNSS points) should coincide with the level points of the 3rd class and above. The points should be distributed relatively uniformly and fairly densely.

The local EGM2008_TN quasigeoid model based on the local EGM2008 model is of higher accuracy and can be applied for various remote areas. According to the results of the accuracy assessment based on 17 test points, the EGM08_TN model showed an accuracy of 100% corresponding to the 4th levelling class, i.a. for mountainous areas; the 85% level corresponds to the 3rd class. The error in determining the normal height according to GNSS measurements is reduced from 0.0244 m/km to 0.0086 m/km.

References

[1]. Kuzin A A 2017 Accuracy evaluation of terrain digital models for landslide slopes based on aerial laser scanning results *Ecology, Environment and Conservation* **2**(23) 908 - 914

[2]. Gairabekov I G Kravchuk I M 2010 Evaluation of the accuracy of the geodetic height computation by satellite measurements *Geodesy and cartography* **6** 5-7.

[3]. Zhurkin, I G, Neiman Yu M 1988 *Computation methods in geodesy* (Moscow: Nedra).

[4]. Klyushin E B, Kravchuk I M 2009 Satellite levelling, In *Proceedings of the international scientific conference devoted to the 230th anniversary of the Foundation of MIIGAiK* vol. 2, pp. 40-42, (Moscow: MGIAK).
[5]. Kravchuk I M 2010 Normal height computation by the results of satellite *Unversities' Herald. Geodesy and aerial photography* **4** 35-40.
[6]. Kravchuk I M 2010 *Development of methods for normal height computation from satellite measurements in engineering and geodetic works* (Moscow)
[7]. Mustafin M G Than Shon Chan 2018 Method of determination of normal heights according to satellite definitions taking into account deviations of a steep line *Geodesy and cartography* **79(7)** 2-10.
[8]. Sugaipova L S 2011 Comparison of current models of the global gravity field of the Earth / L.S. Sugaipova *Unversities' Herald. Geodesy and aerial photography* **6** 14-20.
[9]. Mustafin M G et al 2018 Topographic-geodetic and cartographic support of the Arctic zone of the Russian Federation *Journal of Mining Institute* **232** 375-382.
[10]. Bogomolova E S et al 2013 Geodesic support of construction of cable-stayed bridges in the city Vladivostok *Journal of Mining Institute* **204** 33-36.
[11]. Khac Luyen Bui, Dinh Toan Vu 2014 Tinh toan di thuong do cao cho khu vuc Viet Nam tren co so su dung he so ham dieu hoa cau chuan hoa day du cua mo hinh the trong truong toan cau EGM *Tap chi KHKT Mo – Đia cha* **46** 77-84.
[12]. Duy Do Nguyen, Nam Chinh Dang 2012 Refinement anomalous elevation EGM2008 base on GPS-levelling data in local region Tay Nguyen and south central coastal areas *Journal Science Earth (Vietnam)* **34**(1) 85 - 91.
[13]. Tongcuc D 1998 Bao cao xay dung he quy chieu va he toa do quoc gia, (Hanoi: BTNMT).
[14]. Tziavos I N et al 2012 Adjustment of Collocated GPS, Geoid and Orthometric Height Observations in Greece. Geoid or Orthometric Height Improvement *Conference Paper. Greece* 481 – 487.
[15]. Khudiakov G I, Makarov G V 2012 Use of affinne coordinate conversion at the local geodetic surveys with applying of GPS-receivers. *Journal of Mining Institute* **204** 15-18
[16]. Manukhov V F, Razumov O S, Spiridonov A I 2011 *Satellite methods for determining the coordinates of points of geodetic networks* (Saransk: Publishing house of Mordovia University).
[17]. Potiuchliaev V G 2013 The calculation of accuracy of breaking the network using satellite navigation systems. *Journal of Mining Institute* **199** 325-328
[18]. Simenko E. V., Voronina M. V.2017 Constructive Methods of Forming Surfaces *International Journal of Applied Engineering Research* 6(12) 956-962.
[19]. Drebenstedt C., Alekseenko, A.V. 2019 Environmental impact of abandoned mine wastes on an urban area in nw Caucasus *Innovation-Based Development of the Mineral Resources Sector: Challenges and Prospects - 11th conference of the Russian-German Raw Materials*, **1** 223-229
[20]. Sidorenko A.A., Ivanov V.V., Sidorenko S.A. 2019 Numerical simulation of rock massif stress state at normal fault at underground longwall coal mining International *Journal of Civil Engineering and Technology* **10** (1) 844-851
[21]. Danilov A.S., Smirnov U.D., Pashkevich M.A. 2015 The system of the ecological monitoring of environment which is based on the usage of UAV *Russian Journal of Ecology* **46**(1), 14-19

Validation of the method of height transmission by laser tape measure in construction

Tran Thanh Son[1], **A A Kuzin**[1], **M G Mustafin**[1]

[1] Saint Petersburg Mining University, 2, 21 Line, Saint Petersburg, 199106, Russian Federation

E-mail: kuzin_aa@pers.spmi.ru

Abstract. The article deals with the features of the use of laser tape measure for the transfer of heights from the Earth surface to the upper levels in civil engineering and to a mine in surveying. The most suitable engineering solution for the transfer of height to the mine is justified. Certain experimental results are presented. The algorithm that increases the accuracy of the transmission marks to the mine in the construction of deep tunnels is developed.

1. Introduction

During the construction of high-rise buildings for civil and industrial purposes, tower-type structures (chimneys, silos, cooling towers, etc.), the construction of tunnels and mines, transferring of the high-rise base point coordinates to the installation horizons is needed. The current edition of the SNiP standard on geodetic works in construction [6] recommends solving the problem of transferring the coordinates upward by means of inclined, vertical projection or using GNSS technologies. In fact, it is possible to identify a greater number of ways to solve this problem. In order to ensure maximum efficiency in tunnel construction and within acceptable limits, it is necessary to solve the problem of improving the accuracy of the geodetic framework in the construction of tunnels.

In the construction of mines and tunnels, various methods of transferring height to the tunnel through a vertical well are used to create a geodetic high-altitude network [1–3]. According to the previous research data [4,5], the method of height transmission through a vertical well can be carried out by means of an electronic total station. However, this method has some disadvantages, such as the difficulty of installing a prism reflector at the shaft mouth, impossibility of its orientation and rotation at angles up to 45°, etc. In order to compensate for the disadvantages of this method, it is necessary to study and choose a height transmitting method through a vertical well with a distance measuring device Leica DISTO pro4a. This will improve the accuracy of the orientation of the tunnel and will help to comply with the terms and schedule of the construction of the tunnel in practice.

2. Materials and methods

The principle of height transfer from the ground to the tunnel is the task of two-level surfaces: one is on the ground, the other is in the tunnel. It is possible to transfer the height from the ground to the tunnel if the distance between the two surfaces is known.

The height transmission scheme by the distance measuring device DISTO pro4a and the level through a vertical well into the tunnel is shown in Figure 1.

The laser tape measure Disto (D) is suitable for the use at the shaft mouth, when hanging on a cord about 15 cm long so that the laser beam of the Disto tape measure coincides with a steep line. In the mine, at the projection point of the laser beam, it is necessary to set the prism or reflective paper (O).

Journal of Physics: Conference Series **1384** (2019) 012057 doi:10.1088/1742-6596/1384/1/012057

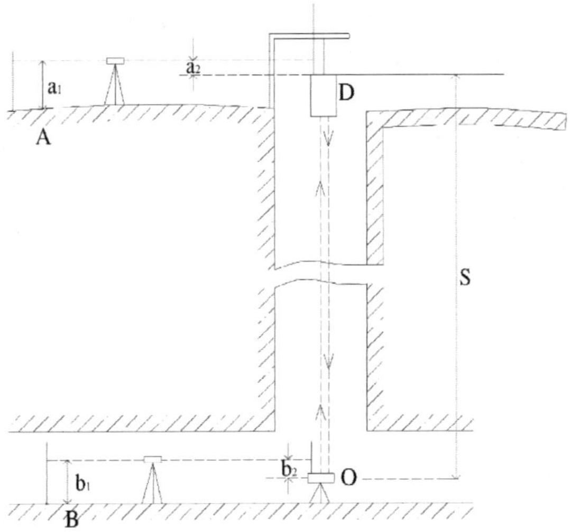

Figure 1. Scheme of height transmission to the tunnel through a vertical well

The order of measurements in the transmission of height from the surface of the earth to the mine is as follows. First, the excess between point A, located on the earth's surface at an altitude of H_A and the installation point of the Disto laser tape is measured. At the point A, the level rail is set and the count in mm (count a_1) is taken; at the point located at the base of the Disto, the count a_2 in mm is taken with a steel ruler with millimetre divisions. The distance S from the base of the Disto to the prism O is then measured. In the mine, it is necessary to measure the excess between the prism O to point B, the height of which must be determined when conducting centre work. At the same time, two counts are taken in the mine: count with the steel ruler located at point B (count b_1), and count b_2 in mm on the steel ruler located on the prism O (number value). Then the height of point B in the tunnel is calculated as follows:

$$H_B = H_A + (a_1 - a_2) - S - (b_1 - b_2), \tag{1}$$

where: H_A is the height of the point A on the surface; a_1, a_2 are the values of the marks on a steel ruler (mm) on the earth's surface; S is the distance from the base of the Disto the device to a prism; a_1, a_2 are the readings in mm on the steel ruler at the mine.

Formula (1) can be written as follows:

$$H_B = H_A + h_1 - h_2 - h_3, \tag{2}$$

where: $h_1 = a_1 - a_2$; $h_2 = S$; $h_3 = b_1 - b_2$.

2.1. *Method of accuracy assessment*

The formula (2) provides the following:

$$m_{H_B}^2 = m_{H_A}^2 + m_{h_1}^2 + m_{h_2}^2 + m_{h_3}^2, \tag{3}$$

where: is the average root-mean-square (RMS) error of the source point (if the RMS error is not considered in the initial point, $m_{H_A} = 0$); m_{h_1} are the RMS errors of measurements during the geometric levelling on the surface of the earth; are RMS errors of laser tape distance measure by Disto; are RMS errors of measurements during the geometric levelling in the mine.

By replacing: $m_{h_2}^2 = m_S^2$, on the principle of equal influence [8-9]:

$$m_h = m_0 = m_{a_1} = m_{a_2} = m_{b_1} = m_{b_2} \Rightarrow m_{h_1}^2 + m_{h_3}^2 = 4m_0^2 \tag{4}$$

By replacing in the same way in the formula (3):

$$m_{H_B}^2 = 4m_0^2 + m_S^2 \Rightarrow m_{H_B} = \sqrt{4m_h^2 + m_S^2}\,. \tag{5}$$

Thus, the accuracy of this method depends on the accuracy of the measurement of height difference on the surface of the earth, as well as in the mine (m_h) and the accuracy of the distance measuring device Disto (m_S). According to the manufacturer's specification, the Leica DISTO pro4a has a distance measurement accuracy of $m_S = \pm$ 1.5 mm. If the RMS error is the measurement of the excess equal to m_h = 1 mm, then m_{H_B} = 4.3 mm. However, experiments are needed to verify the RMS error value.

When the DISTO laser beam deviates from the vertical line, the difference between the measured distance (*D*) and the distance measured by the vertical line (*S*) occurs. Bringing measurements to a vertical line is calculated as:

$$S = \sqrt{D^2 - e^2}\ , \tag{6}$$

where: *D* is the measurement distance of Disto; $e = \frac{\beta'' D}{\rho}$.

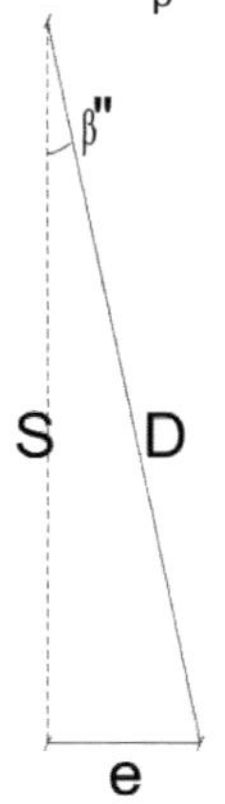

Figure 2. Difference between the distances

When using laser tape measure Leica DISTO pro4a with accuracy $m_S = \pm$ 1.5 mm + 0.03 mm/m, located on the measuring stand at the well to measure at a depth of 100 m, *e* = 40 cm, the distance difference (*D* – *S*) is 2.4 mm.

Obviously, the method of height transmission to the tunnel by Disto has many advantages over traditional methods (steel wire or level), but it is necessary to carry out experimental measurements to confirm the accuracy and the possibility of wide method application.

2.2. Experimental computation

Experimental research of the mark transfer technology using laser tape DISTO™ pro4a was carried out by the Vietnam Institute for Building Science and Technology, Ministry of Construction – IBST. The test distance – horizontal distance (measured three times) were simulated. The average value is used, the tension of the steel ruler when measuring is provided by a weight of 5 kg. When using a steel ruler for measurement, it is necessary to calculate the elongation factor of the steel ruler and calculate it according to the measurement results. The elongation factor of the steel ruler is calculated by the formula:

$$\Delta L = \frac{FL}{EA} \tag{7}$$

where: ΔL is the elongation factor of the steel ruler; *F* is the tension of a ruler by weights when measured (change of 0.1 kg = 9.81 N, i.e. 5 kg = 49.05 N); *L* is the ruler length (m); *E* is the modulus of elasticity of steel, E = 2.1 * 105 MPa; *A* is the width and thickness of a steel ruler (width of 10 mm; thickness of 0.5 mm).

For the length of 10 m it is: ΔL = (9.81 * 5 * 10000) / (2.1 * 105 * 10 * 0.5) = 0.467 mm; the factor is calculated for the lengths of 20 m; 30 m or more multiplied by the ratio. The results of the calculations are presented in Table 1.

Table 1. Calculation results

No.	Steel ruler distance with a tension of 5 kg at the test site (m)	ΔL	Steel ruler distance after correction (mm)	Distance with DISTO pro4a (m)	Difference (mm)
(1)	(2)	(3)	(4) = (2) + (3)	(5)	(6) = (5) - (4)
1	40.000	1.8	40.0018	40.001	- 0.8
2	50.000	2.3	50.0023	50.002	-0.3
3	60.000	2.8	60.0028	60.002	- 0.8
4	70.000	3.2	70.0032	70.002	- 1.2
5	80.000	3.7	80.0037	80.004	+ 0.3
6	90.000	4.2	90.0042	90.006	+ 1.8

Table 1 shows that it is possible to use the DISTO pro4a device to transmit the height to the tunnel through a vertical well at a depth of up to 90 m when the maximum error in the height transmission to the tunnel is ≤ 3 mm [6].

Experimental work was carried out using Leica DISTO™ Pro4a with an accuracy of $m_S = \pm 1.5$ mm+0.03 mm/m. The measurement is performed in accordance with the scheme shown in Fig. 1 (building block N01, 136 Ho Tung Mau, Hanoi, Vietnam). The experiment procedure is shown in Fig. 3.

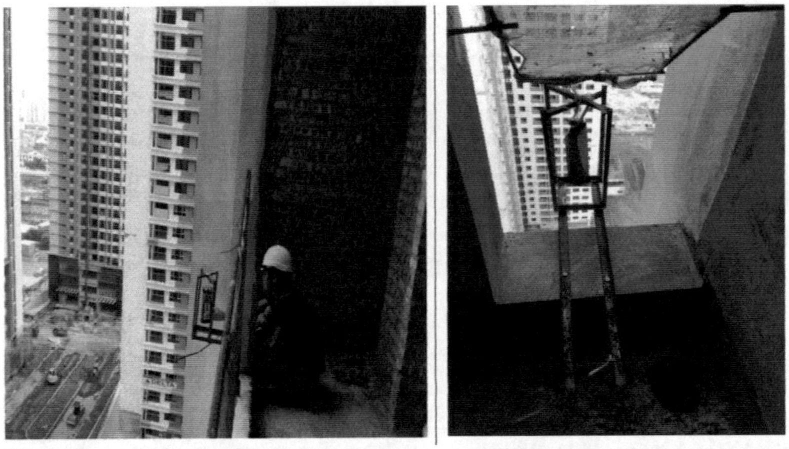

Figure 3. Experiment in blocks N01, 136 Ho Tung Mau, Hanoi, Vietnam

The Disto was placed in block N01 on the 16th, 25th and 32nd floors. After turning on the laser beam roulette, it is necessary to note the point (O) and to install there a prism or reflective film. The point (O) was located on the second floor. Accordingly, three distances were measured: from the 16th, 25th and 32nd floors to the 2nd floor. Then the height difference was measured by levelling: h_1 and h_3 (from the formula (2)). The results of the calculations are presented in Table 2.

The height transfer to the tunnel through a vertical well was also performed using a steel tape measure in the place where the experimental measurement was carried out by the Disto device. As a result, the height of point B on the second floor by measuring with a steel ruler (with tension by the weight of 5 kg) with a level is shown in Table 3.

Table 2. Experimental measurement results

	Nominal height (H_A) of the floor	Height difference between the reference point A and the device base (stand) ($a_1 - a_2$)	Distance from the Disto device to the film reflector (S)	Height difference between the film reflector and the benchmark B ($b_1 - b_2$)	Height of the benchmark B (H_B) on the 2nd floor
16th floor	0.000	0.328	45.139	– 0.126	– 44.685
25th floor	0.000	0.316.	73.953	– 0.126	– 73.511
32nd floor	0.000	0.322	96.340	– 0.126	– 95.892

Table 3. Comparison of the results of height transmission by two methods

	Height of the benchmark B by the steel ruler (m)	ΔL (mm)	Steel ruler distance after correction (mm)	Distance with DISTO pro4a (m)	Difference (mm)
(1)	(2)	(3)	(4) = (2) + (3)	(5)	(6) = (5) - (4)
16th floor	– 44.683	– 2.1	– 44.6851	– 44.685	– 0.1
25th floor	– 73.508	– 3.4	– 73.5114	– 73.511	– 0.4
32nd floor	– 95.886	– 4.5	– 95.8905	– 95.892	– 1.5

Comparison of Tables 2 and 3 shows that the height measured by the DISTO pro4a compared with the height measured by the traditional method has the results of deviations within the permissible limits (3 mm) at a depth of 95 m. This error meets the standard of height transmission in Russia [6] and Vietnam [7]. The experimental results show that it is possible to use the Disto device to transmit the height to the tunnel through a vertical well up to 95 m in depth to overcome the general technical difficulties when using other methods to transmit the height to the tunnel.

3. Conclusion

According to the results of theoretical and experimental analysis, the following conclusions can be drawn.

1) The use of the Disto device with the standard deviation up to m_S = 1.5 mm + 0.03 mm/m for height transmission through the vertical well of the tunnel is particularly effective when the tunnel depth is up to 95 m. Further research is needed when the depth of the vertical well exceeds 95 m.
2) For best results, place the Disto device on a vertical well so that the laser is aligned with the vertical line. Then the distance measurement between the Disto device and the reflector reaches the required accuracy.
3) Currently, laser roulette is widely available on the market. For example, Leica Disto D5 or Leica DISTO S910 with an accuracy of m_S = 1.0 mm + 0.025 mm/ m at a distance of 200 m and 300 m, respectively, will also provide the necessary accuracy of the transmission of marks to the depth of 95 m. The use of these roulettes for height transmission to the entire specified range of distance measurement should be further investigated.

References

[1]. Pisarenko V K, Shmelin N A Izmaylov R B 2014 Height transmission to the mounting horizons *Geodesy and cartography* **10** 2-4.

[2]. Klyushin E B, Gayrabekov I G, Markelova E Yu, Shlapak V V 2015 *Satellite measurement methods in geodesy* (Publishing house of the Moscow State University of Geodesy and Cartography).

[3]. Bogomolova E S et al 2013 Geodesic support of construction of cable-stayed bridges in the city Vladivostok *Journal of Mining Institute* **204** 33-36.

[4]. Tran Viet Tuan 2010 The improving research of method that carry elevation from the earth's surface into the tunel through vertical well by total stations *Journal of Mining and Earth Sciences* **44** 62 – 65.

[5]. Nguyen Quang Thang, Diem Cong Huy 2017 Some solutions to improve the efficiency of application of GPS technology in the construction of high-rise buildings and industrial works *Journal of Construction Science* **1**. 63–69.

[6]. SP 126.13330.2017. SNiP 3.01.03-84. Set of rules. Geodetic works in construction / The order of the Ministry of Construction of Russia, October 24, 2017 No. 1469. Moscow Ministry of Construction of Russia, 2017. 67 p.

[7]. Van Hien Phan, Viet Tuan Tran, Doan Mau Pham, Xuan Vinh Dinh 2013 Giao trinh Trac dia cong trinh ngam (Hanoi University of Natural Resources and Environment)

[8]. Mustafin M G, Tran Thanh Son 2018 Method of determination of normal heights according to satellite definitions taking into account deviations of a steep line *Geodesy and cartography.* 79(7) 2-10.

[9]. Kuzin A A, Kazantsev A I, Valkov V A 2018 Satellite-based techniques for monitoring of bridge deformations *Journal of Physics: Conference Series* **1118**(1) 1 - 6.

[10]. Kazantsev A I , Kochneva A.A. 2017 Justifcation of quality estimation method of creation of digital elevation models according to the data of airborne laser scanning when designing the motor ways *Journal of Industrial Pollution Control* **23**(2) 1000-1006

[11]. Kuzin A A, Kovshov S.V. 2017 Accuracy evaluation of terrain digital models for landslide slopes based on aerial laser scanning results *Ecology, Environment and Conservation* **23**(2) 908-914.

[12]. V. A. Belin, A.N. Kholodilov, A. P. Gospodarikov 2017 Methodical principles of prediction of seismic effect due to large-scale blasting *Gornyi Zhurnal* **2** 66-70

[13]. E.A. Pravdina, O.J. Lepikhina 2017 Laser scanner data capture time management ARPN Journal of Engineering and Applied Sciences **12** (5) 1649-1661

Automation of mass fraction determination of water in petroleum in the laboratory

M P Serzhantova[1], V S Tynchenko[1,2], V V Bukhtoyarov[1,2], E A Petrovsky[1], V V Kukartsev[1,2] and V V Tynchenko[1,2]

[1] Siberian Federal University, 79, Svobodny pr., Krasnoyarsk, 660041, Russian Federation
[2] Reshetnev Siberian State University of Science and Technology, 31, Krasnoyarsky Rabochy Av., Krasnoyarsk, 660037, Russian Federation

E-mail: vadimond@mail.ru

Abstract. Petroleum quality indicators should be monitored at all stages of petroleum production, preparation and transportation quickly and efficiently. Such a critically important indicator as the mass fraction of water in petroleum determines not only its quality, but also has a huge impact on the performance properties of the equipment and the pipeline. The article demonstrates the automation of laboratory research in determining the mass fraction of water in petroleum: the modernization of existing research using optical sensors and a data acquisition board. As a result, the complexity of the research is reduced, the work of laboratory research is optimized. The article presents the assembly diagram of the automated installation, compiled an algorithm for the software.

1. Introduction

Formation waters are companions of oil and gas fields [1, 2]. Water can be contained in petroleum and its products or in the form of a simple suspension, which is easily settled during storage, or in the form of an emulsion [3, 4].

The presence of water in petroleum products is undesirable, especially in products used at very low temperatures [5]. At these temperatures, the dissolved moisture begins to fall in the form of ice crystals, which can cause serious complications when using such oil products [6].

Oil delivered to refineries and intended for refining must meet the requirements of regulatory documents, which establish four indicators of oil quality: water mass fraction, chloride content, mechanical impurities, and vapor pressure at oil temperature at the point of delivery. [7, 8]

Modern petrochemical enterprises must have a laboratory that meets modern quality standards, performing its functions accurately, correctly and with minimal financial and time costs [9, 10]. Automation of technological processes of the laboratory will minimize their complexity, increase productivity and reduce the cost of living labour [11, 12].

The current method for determining the mass fraction of water in oil according to the Dean and Stark method is the most accurate, but remains very laborious, requiring the constant attention of a technician [13]. Laboratory research needs modernization: reducing man-hours spent by way of its automation [14].

2. Material and Methods

The quantitative content of water in the oil and in all petroleum products is determined by the method of Dean and Stark. [15]

The method consists in the fact that 100 g of the test oil is heated in a mixture with 100 cm3 of solvent in a Dean and Stark apparatus. The solvent, evaporating, carries with it the moisture contained in the oil product. Water and solvent vapors are condensed in a refrigerator, and distilled water settles to the bottom of the receiver - a graduated trap. By the amount of water in the trap, the percentage of its content in the oil product is calculated. [16]

The following apparatus, reagents and materials are used to determine the amount of water in petroleum and petroleum products:

- Apparatus for the quantitative determination of water content in petroleum, food and other products (Figure 1).
- Glassware.
- Electric heating device.
- Fridge.
- Solvents.
- Stopwatch.

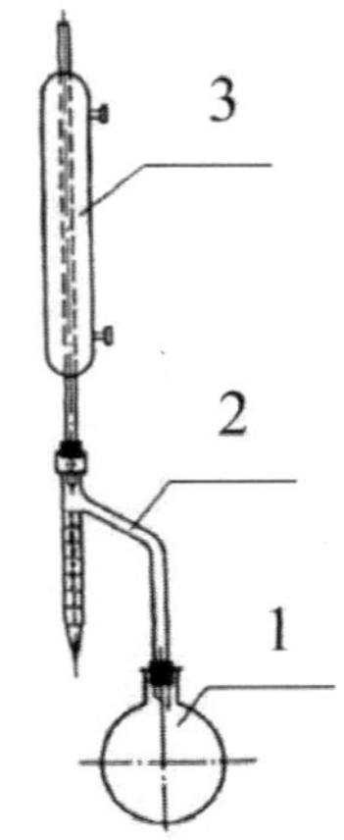

Figure 1. Apparatus for quantitative determination of water content:
1 – round bottom flask, 2 – receiver trap, 3 – refrigerator.

The result of the mass fraction in percent is calculated by the formula:

$$X = \frac{V_0}{m} \cdot 100 \quad (1)$$

where V_0 - is the volume of water in the receiver-trap, cm^3;
m - is the sample mass, g.

3. Results and Discussion

The laboratory research process in determining the mass fraction of water in oil is automated as follows: a sensor (4) for determining the liquid level is installed in the lower part of the refrigerator (3), and a sensor (6) that reacts to the formation of droplets is installed at the point where the beveled part ends refrigerator tubes, in the upper part of the receiver-trap (5).

The wire (7) connects the sensors to the data acquisition board and transmits the information to the processing computer, based on the computer’s counting results, the program decides whether the heating capacity of the heating mantle (9) is to be increased by means of the heating power regulator or signals the technician to participate in the analysis.

The task of automating the study is solved by installing it presented in the form of a diagram in Figure 2, where: (1) is a tripod, (2) are hoses for supplying and discharging coolant to a refrigerator.

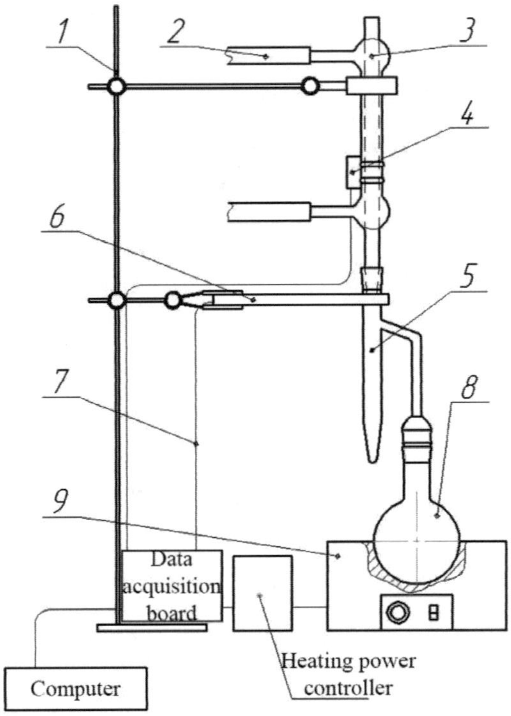

Figure 2. Apparatus for determining the proportion of water in oil with an automated process.

A fluid level sensor registers a critical boiling level. The dropping speed is determined by an optical sensor installed at the point where the beveled part of the refrigerator tube ends, at the top of the receiver-trap, and is calculated by the formula:

$$v\_ds = n/t \tag{2}$$

where n - is the number of drops registered by the sensor in t seconds, drops;

t - time, s.

The unit of power is set by calibrating the heating device and is determined empirically.

The automated laboratory research process for determining the mass fraction of water in oil is described by the algorithm in Figure 3.

Thus, because of the introduction of an algorithm for the automated process of determining the mass fraction of water in oil, the participation of a laboratory technician in the process is minimized. The laboratory technician remains to prepare the installation at the beginning of the test, check everything for leaks, turn everything on, and run the algorithm on the computer.

At the end of the analysis, the computer will independently notify the technician of the completion of the study and the need to measure the volume of accumulated water in the trap, and then disassemble the system.

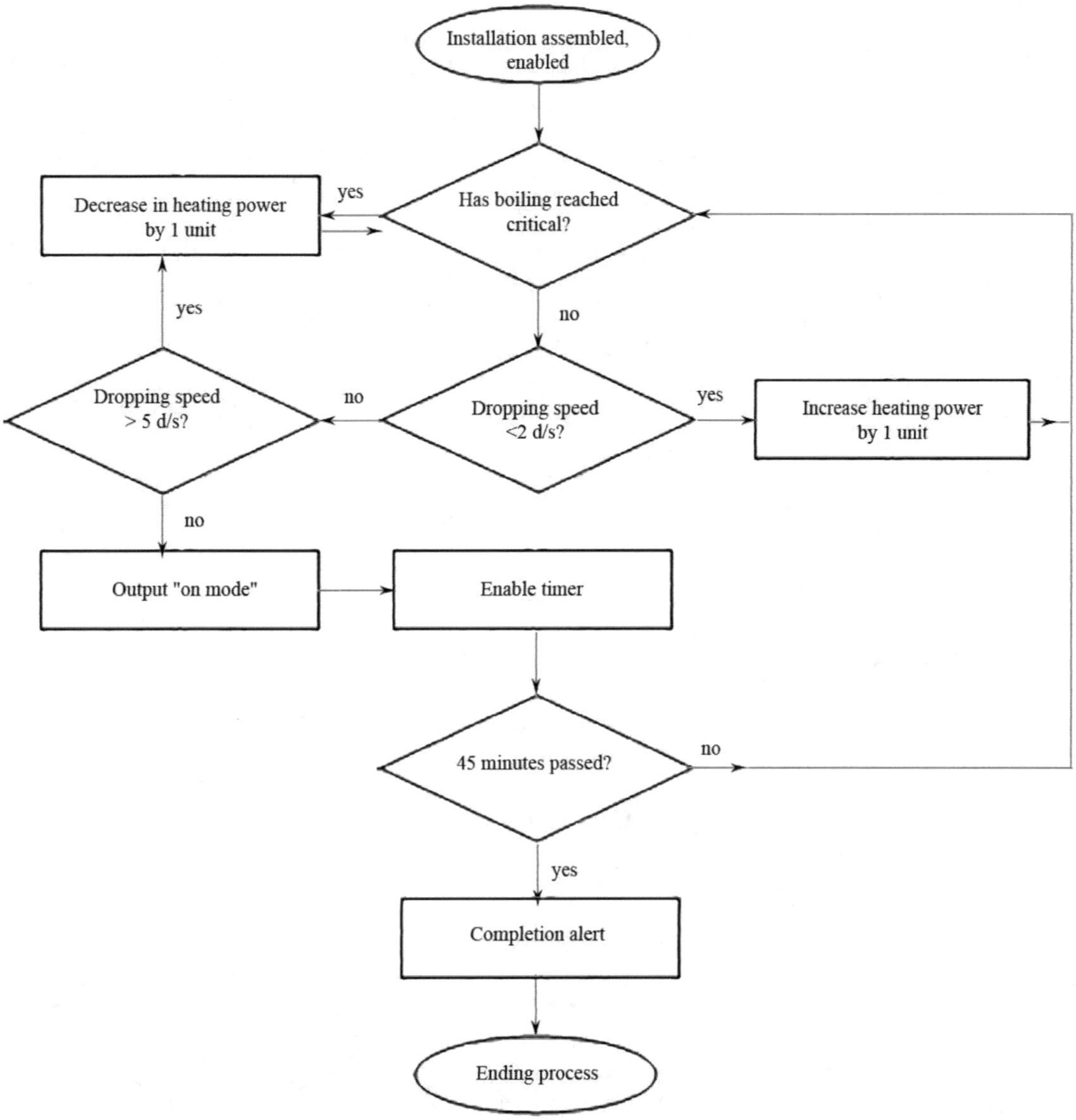

Figure 3. Algorithm for determining the share of water in oil in an automated process.

4. Conclusion

In the process, an automated system for determining the mass fraction of water in oil was developed. This system has the following main functions:

- Control of the automated process of heating the flask.
- Control of boiling level.
- Access to the working boiling rate and maintaining it in the desired range.
- Visualization of the processes.

The algorithm was developed for automated laboratory research software for determining the mass fraction of water in oil.

Thus, it improves the quality of the test and reduces the human factor.

Journal of Physics: Conference Series **1384** (2019) 012058 doi:10.1088/1742-6596/1384/1/012058

References

[1] Alboudwarej H, Muhammad M, Barrere C, Archer R, Hannan R and Jones T 2005 Viscosity of formation water: Measurement, prediction, and reservoir implications *Proceedings - SPE Annual Technical Conference and Exhibition* 2179-2186

[2] Nie C, Pu C, Zhang R, Zhou J and Ren L 2005 Study and application of techniques for gas recovery by liquid drainage in Banqiao condensate oil/gas field *Tianranqi Gongye/Natural Gas Industry* **25(6)** 83-86

[3] Ali M A J, Abeer A A, Ali A A and Suhaib A A 2018 The effect of suspended solid concentration and oil droplet size distribution on particle coalesces and emulsion stability *Society of Petroleum Engineers - SPE/DGS Saudi Arabia Section Technical Symposium and Exhibition 2011* 1-15

[4] Lord D, Luketa A, Wocken C, Schlasner S, Aulich T, Allen R and Rudeen D 2015 Literature survey of crude oil properties relevant to handling and fire safety in transport *Crude Oil Properties and the Hazards of Transport: Background, Data and Literature Summary* **January 01** 1-95

[5] Umar A A, Saaid I B M, Sulaimon A A and Pilus R B M 2018 A review of petroleum emulsions and recent progress on water-in-crude oil emulsions stabilized by natural surfactants and solids *Journal of Petroleum Science and Engineering* **165** 673-690

[6] Bui B 2012 Modeling the effect of pipe rotation on pressure loss through tool joint Society of Petroleum Engineers - *SPETT Energy Conference and Exhibition 2012* 158-172

[7] Kalashnikova Y V 2009 The control of parameters of oil quality *Interexpo Geo-Siberia* 18-26

[8] Bukhtoyarov V, Tynchenko V, Petrovskiy E, Bukhtoyarova N and Zhukov V 2018 Investigation of methods for modeling petroleum refining facilities to improve the reliability of predictive decision models *Journal of Applied Engineering Science* **16(2)** 246-253

[9] Schupak L V and Zhadan V V 2018 Modern problems of cost planning at the enterprise *Economy info* **15** 55-60

[10] Fomin G S and Fomina O N 2006 *Oil and petroleum products. Encyclopedia of international standards* (Moscow, Tread)

[11] Sverdlik G V 2006 The automation of technological process is a source of steady economic advantage *Tselliuloza, Bumaga, Karton/Pulp, Paper, Board* **8** 82-83

[12] Tynchenko V S, Kukartsev V V, Tynchenko V V, Chzhan E A and Korpacheva L N 2018 Automation of monitoring and management of conveyor shop oil-pumping station of coal industry enterprise *IOP Conference Series: Earth and Environmental Science* **194(2)** 022044

[13] Lu M, Zhang Z, Li J and Li K 2012 Study on the economic stripping ratio in oil shale open-pit mine used in refinery *Advanced Materials Research* **524-527** 1915-1920

[14] Bukhtoyarov V, Bashmur K, Nashivanov I., Petrovsky E and Tynchenko V 2018 Magnetic impact dampening of vibrations in technological equipment for oil and gas production *International Multidisciplinary Scientific GeoConference Surveying Geology and Mining Ecology Management, SGEM* **18(1.4)** 573-580

[15] Knunyants I L 1988 *Chemical encyclopedia in five volumes* (Moscow, Soviet Encyclopedia)

[16] Borovaya M S 1968 *Laboratory assistant at the oil and gas laborator* (Moscow, Nedra)

Automation of laboratory studies in the determination of the mass of mechanical impurities in petroleum

M P Serzhantova[1], V S Tynchenko[1,2], V V Bukhtoyarov[1,2], K A Bashmur[1], V V Tynchenko[1,2] and V V Kukartsev[1,2]

[1] Siberian Federal University, 79, Svobodny pr., Krasnoyarsk, 660041, Russian Federation
[2] Reshetnev Siberian State University of Science and Technology, 31, Krasnoyarsky Rabochy Av., Krasnoyarsk, 660037, Russian Federation

E-mail: vadimond@mail.ru

Abstract. The article solves the problem of automated control of laboratory research process in determining the mass of mechanical impurities in petroleum. The research analyzed the features of the test and identified the need for an automated control system of this test. The developed automated system controls the filtering process of the sample and pump power, allows simultaneous parallel studies on independent suction lines, while optimizing the pump motor power, depending on the number of used lines. The article provides a block diagram describing the basic system algorithm and scheme of the laboratory tests automated installation in determining the mass of mechanical impurities in petroleum. The result is a reduction in the complexity of the study, optimization of laboratory research.

1. Introduction

Natural combustible minerals, such as oil and gas, are the main energy carriers. They play a significant role in the economy of any state. [1]

Due to the rapid global development of the chemical and petrochemical industry, the need for oil is rising not only to increase the production of fuels and oils, but also as a source of valuable raw materials for the production of synthetic rubbers and fibers, plastics, surfactants, detergents, plasticizers, additives, dyes, etc. (more than 8% of the world production) [2, 3]. Oil is unique in its combination of qualities: high energy density (thirty percent higher than that of the highest quality coal), oil is easy to transport (compared to gas or coal, for example). Finally, it is easy to get a lot of the above-mentioned products from oil [4, 5].

The oil produced in the fields contains many impurities (water, sand, gas, pieces of rock, etc.) [6]. In order to submit this raw material to the system of trunk oil pipelines for further transportation to refineries or for export, first, it must be brought to the required level of quality [7, 8].

The presence in the produced oil of a large number of mechanical impurities complicates the operation of well bores, increases equipment wear, complicates well bore maintenance, and operating costs increase [9, 10].

The impurities contained in the pumped liquid are different in qualitative and quantitative composition: these can be products of reservoir destruction or cement ring or particles of different

Journal of Physics: Conference Series **1384** (2019) 012059 doi:10.1088/1742-6596/1384/1/012059

composition brought from the surface [11]. However, their effect on all equipment is identical: they clog up the filters of the pumps, first reducing, then completely stopping the flow of fluid into the pump, or acting as an abrasive, accelerating the process of wearing out items of equipment or seizing them [12]. In addition, mechanical impurities cause erosion of the inner surface of pipes during transportation, form deposits in the equipment, which leads to a decrease in heat transfer coefficient, increase the ash content of distillation residues (fuel oils, tars), promote the formation of stable emulsions [13]. In accordance with the requirements of existing regulations mechanical impurities should be absent in the commodity oil [14].

It is important to control such an indicator as the amount of mechanical impurities in oil at all stages of oil production, transportation, and preparation [15, 16]. Proceeding from the flow rate of the wells currently in operation and the volumes of oil on the market, there is a need to modernize existing methods, namely the automation of laboratory monitoring of the oil quality: the amount of mechanical impurities in the oil [17, 18].

2. Material and Methods

The existing method for determining the mechanical impurities in oil includes filtering the test samples with preliminary dissolution in a solvent and washing the filter cake with a solvent followed by drying and weighing. For the test, a porcelain funnel (1), a glass flask (2) and an air vent (3), which is connected to a vacuum pump (assembly scheme in Figure 1a), are used, then a prepared paper filter is installed (4, Figure 1b).

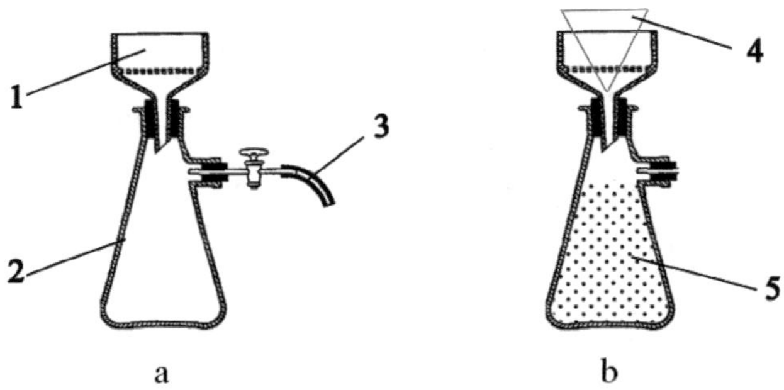

Figure 1. Installation assembly diagram.

After assembling the installation, the test sample is mixed, poured into a measuring cup, diluted with solvent and filtered through a paper filter. The laboratory technician ensures that the filtrate (5) does not exceed the critical level and does not get into the pump for air exhaust. At the end of the test, the paper filter with the mechanical impurities remaining on it is dried and weighed.

The mass fraction of mechanical impurities (*X*) in percent is calculated by the formula (1).

$$X = \frac{(m_1 - m_2)}{m_3} \cdot 100 \quad (1)$$

where m_1– the mass of the glass for weighing with a paper filter and mechanical impurities, g.;

m_2– the mass of the weighing cup with a clean prepared paper filter, g.;

m_3– the mass of the sample, g.

The result of the analysis is the arithmetic average of the two parallel measurements results.

The amount of time spent on the analysis depends on:

- Quality of bringing the filters to constant weight (re-drying the filters in a drying cabinet takes 30 minutes, and after that the dried filter is kept for another 30 minutes in a desiccator).
- Pump suction capacity (when there is a large amount of mechanical impurities, the filter becomes clogged and complicates the filtration process).

Journal of Physics: Conference Series **1384** (2019) 012059 doi:10.1088/1742-6596/1384/1/012059

The method complexity for determining the content of mechanical impurities by the existing method is 2.20 man-hours. At one installation it is possible to conduct only one test at a time. During the test, the laboratory technician must be directly located near the installation, which eliminates its multitasking.

3. Results and Discussion

Technique automation consists of incorporating the following elements into the existing funnel-bulb-pump chain: silicone hose, tripod, valve, strain gauge sensor, power regulator, wire, data acquisition board, personal computer (PC). The principle of collecting the installation is shown in Figure 2.

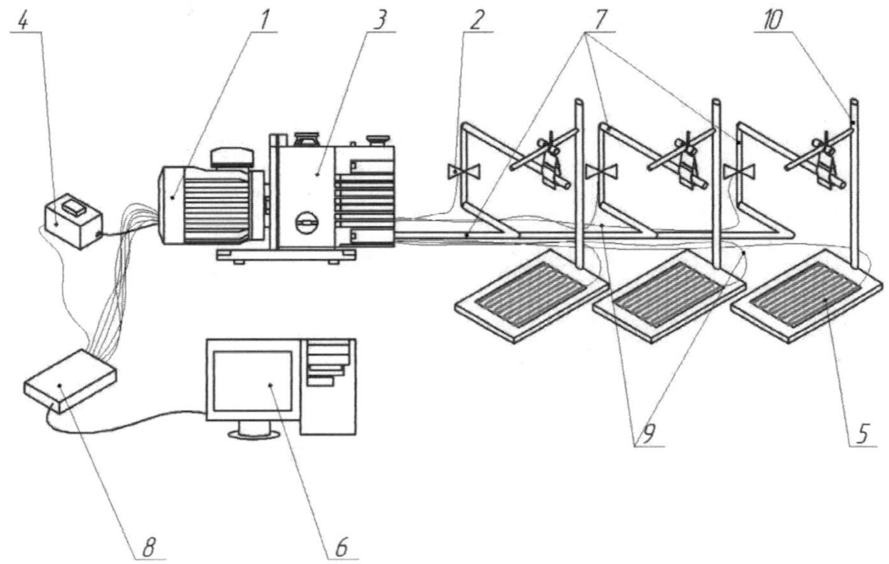

Figure 2. Automated installation for determination of mechanical impurities: 1 – engine, 2 – valve, 3 – vacuum pump, 4 – engine power regulator, 5 – bulb mass detection sensor, 6 – computer, 7 – suction line, 8 – data acquisition board, 9 – wire, 10 – tripod.

Principle of operation: a flask with a funnel joins the hose, a filter is inserted into it, and the whole system is installed on the base of the tripod. The mass sensor, located on the base of the tripod, is calibrated to the volume of the used flask. When the flask is installed on the sensor, the data acquisition board receives a signal, processes it and transfers it to the PC; the response signal comes in two directions: the valve opens the corresponding suction line, the power regulator transmits the optimum power to the pump motor. With a critical volume of filtrate in the flask, the mass sensor transmits the corresponding information to the PC, and the corresponding line stops participating in the work until the laboratory technician changes the flask upon receiving the message of its necessity.

The algorithm of operation of this installation is presented in Figure 3.

An option of solving the task of the determination of mechanical impurities is based on the ability to filter three samples at once, using one pump and an extensive suction line, and in the event of one or two samples being excluded from the process while the pump is running (or conducting the experiment initially with 1 or 2 samples), with the possibility of automatic shutdown of an unused suction line, and, therefore, with the ability to control the pump power in accordance with the number of participating suction lines in the filter process.

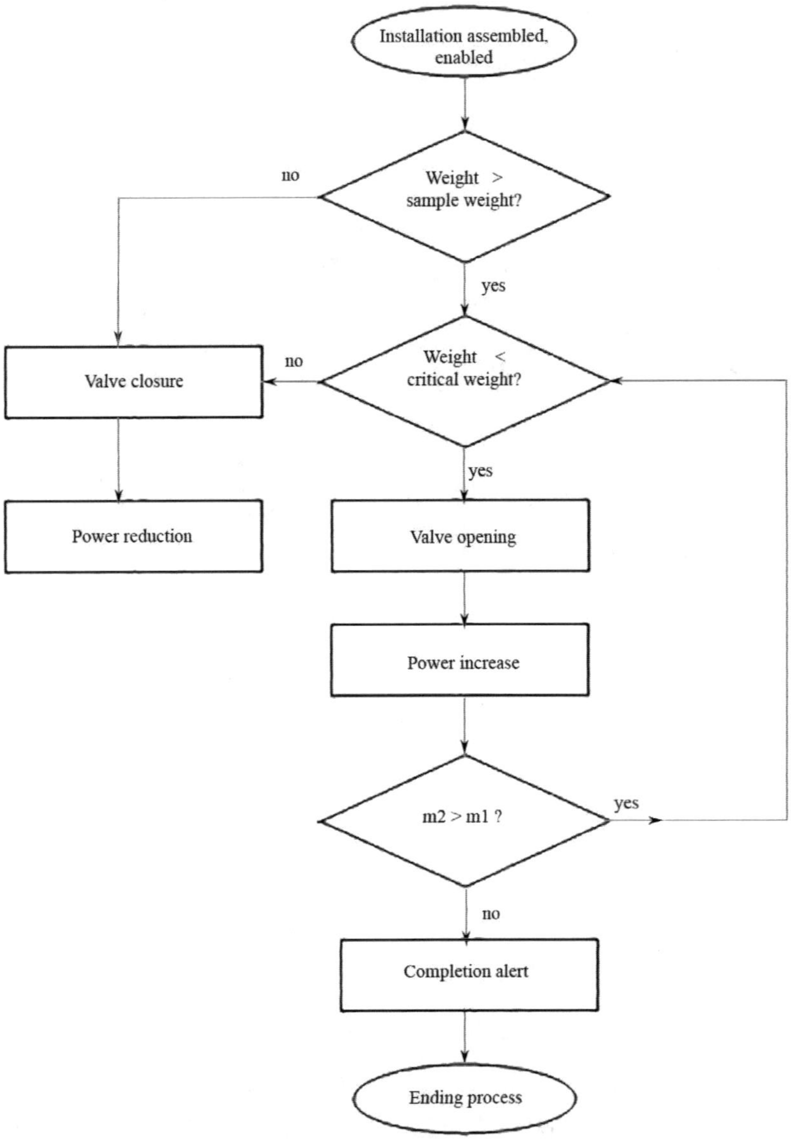

Figure 3. Installation working algorithm.

Before starting the test, the density of the tested oil is measured, this information is needed to calibrate the mass sensor and determine the critical mass. The critical mass is determined by calculation, as the product of volume and density.

4. Conclusion

In the process, an automated system for determining the mass fraction of mechanical impurities in petroleum was developed. The developed system has the following main functions:

- Automated control of the filtering process.
- Pump power control.

- Detection of inactive suction lines and their automatic shutdown.
- Laboratory study process visualization.

As a result of this work, the algorithm for automated laboratory research software for determining the mass of mechanical impurities in oil was developed. The development improves the quality of testing and minimizes the human factor.

References

[1] Hepbasli A 2008 A study on estimating the energetic and exergetic prices of various residential energy sources *Energy and Buildings* **40(3)** 308-315

[2] Speight J G 2012 The chemistry and technology of coal, third edition *The Chemistry and Technology of Coal, Third Edition* **1** 1-808

[3] Smidovich E V 1980 *Oil and gas processing technology. Part 2. Cracking of crude oil and processing of hydrocarbon gases* (Moscow, Chemistry)

[4] Bowen A R and Nash A W 1933 Petroleum and petroleum gases as chemical raw materials *World Petroleum Congress Proceedings* **1933-July** 853-861

[5] Sokolov A N 2011 Efficiency of energy resources and change of technological facilities *Oil and gas business* **5** 416-427

[6] Lord D, Luketa A, Wocken C, Schlasner S, Aulich T, Allen R and Rudeen D 2015 Literature survey of crude oil properties relevant to handling and fire safety in transport *Crude Oil Properties and the Hazards of Transport: Background, Data and Literature Summary* **01** 1-95

[7] Pinkovskii Y I 1974 Technique and economics of preparing crude oil for processing *Chemistry and Technology of Fuels and Oils* **10(5-6)** 463-467

[8] Chuhareva N B and Novikov A A 2009 *The study of hydrocarbon systems in determining the quality characteristics in the system of main pipelines* (Tomsk, Tomsk Polytechnic University)

[9] Tarasov M Yu 2007 Research of processes of mechanical impurities separation from high-viscosity production of wells *Neftyanoe khozyaystvo - Oil Industry* **11** 122-124

[10] Bukhtoyarov V, Bashmur K, Nashivanov I., Petrovsky E and Tynchenko V 2018 Magnetic impact dampening of vibrations in technological equipment for oil and gas production *International Multidisciplinary Scientific GeoConference Surveying Geology and Mining Ecology Management, SGEM* **18(1.4)** 573-580

[11] Pachezhertsev A A, Erofeev A A, Mitrushkin D A, Tsitsorin A I, Kaushansky D A, Demianovskiy V B and Dmitrievsky A N 2018 Determination of porosity and permeability of a porous medium as a result of polymer structuring of disintegrated quartz sand using the IPNG-PLAST 2 technology *Neftyanoe Khozyaystvo - Oil Industry* **8** 52-54

[12] Wilk A 2002 The modelling of movement of mechanical impurities during flow in pump and 1st separation in a pump separator of impurities *Mechanics and Mechanical Engineering* **6(3)** 295-302

[13] Zhang G A, Zeng L, Huang H L and Guo X P 2013 A study of flow accelerated corrosion at elbow of carbon steel pipeline by array electrode and computational fluid dynamics simulation *Corrosion Science* **77** 334-341

[14] Verbitsky V S, Drozdov A N, Dengaev A and Rabinovich A I 2007 New technology of the electrocentrifugal pump unit protection from harmful effect of mechanical impurities *Neftyanoe khozyaystvo - Oil Industry* **12** 78-81

[15] KalashnikovaY V 2009 The control of parameters of oil quality *Interexpo Geo-Siberia* 18-26

[16] Bukhtoyarov V, Tynchenko V, Petrovskiy E, Bukhtoyarova N and Zhukov V 2018 Investigation of methods for modeling petroleum refining facilities to improve the reliability of predictive decision models *Journal of Applied Engineering Science* **16(2)** 246-253

[17] Tynchenko V S, Kukartsev V V, Tynchenko V V, Chzhan E A and Korpacheva L N 2018 Automation of monitoring and management of conveyor shop oil-pumping station of coal industry enterprise *IOP Conference Series: Earth and Environmental Science* **194(2)** 022044
[18] Ivanovsky V N, Sabirov A A, Degovtsov A B, Beijing C S, Patrushev S G, Kachin E V and Popov S V 2015 Experience in the operation of wells with side holes of small diameter pumping units with the cable rod in LUKOIL-Perm LLC *Territory neftegaz* **3** 78-87

Similarity criteria for modeling the process of pipes pulse pressing into the tube sheet

V S Tynchenko[1,2], E A Petrovsky[1], V V Bukhtoyarov[1,2], K A Bashmur[1], V V Tynchenko[1,2] and V V Kukartsev[1,2]

[1] Siberian Federal University, 79, Svobodny pr., Krasnoyarsk, 660041, Russian Federation
[2] Reshetnev Siberian State University of Science and Technology, 31, Krasnoyarsky Rabochy Av., Krasnoyarsk, 660037, Russian Federation

E-mail: vadimond@mail.ru

Abstract. In the presented article, the phenomena that characterize the essence of the process of fastening pipes to the tube sheet by the method of electropulse pressing are described. The article presents the results relating to methods and means of measurement, testing and operation of products in engineering, including theoretical studies of methods for fastening heat-exchange tubes into a tube sheet. The disadvantages of traditional methods of fastening pipes to the tube sheet are revealed. The analysis of the prospects of pipes electropulse pressing into the tube sheet is conducted. Using the dimensional analysis method, similarity criteria were obtained for the process of electropulse press-fitting of pipes into a tube sheet, which allow extending the results obtained on a reduced-scale model to products of natural dimensions.

1. Introduction

Heat exchangers, as devices for transferring heat from one medium to another, are widespread in many industries and are used as elements of various heat power plants. In the chemical, petrochemical and oil refining industries, the single unit capacity increases and will increase by about 4 times every ten years [1].

Nowadays, there are a large number of heat exchangers types that differ in design, type of coolant, performance, heat transfer surface, working conditions, repair regulations, etc.

With this variety of heat exchangers designs, the types of pipe connections with tube sheets can be classified according to the technology of their implementation into press, combined and welded [2, 3].

The operational characteristics of the heat exchange equipment largely depend on the quality of fastening of the heat exchange pipes in the tube sheet. Surveys in the oil refining and petrochemical industries showed that the losses associated with the downtime of technological lines, as a result of the poor quality of pipe fasteners in tube sheets, are 10-15 times higher than the cost of the heat exchangers themselves [4 - 6]. It is also worth noting that the operations associated with the fastening of heat-exchange pipes into tube sheets are one of the most labor-intensive and make up about 40% of the laboriousness of the apparatus assembly work [7, 8].

Caustic and poisonous coolants are often used in heat exchangers; therefore, depressurization of pipes can lead to a serious accident, both from the side of the environment and industrial safety of the enterprise.

Therefore, the problem of the quality and reliability of pipe fastening in the tube sheets of heat exchangers becomes especially urgent and the pipe connections created in them are technological only in conditions of using new technological methods developing in parallel.

The most common is the use of traditional mechanical methods of fastening pipes, usually with a rolling tool. But despite its prevalence, this method is not deprived of many significant drawbacks. There are difficulties when expanding pipes in tube sheets with a thickness of more than 100 mm, in double intermediate tube sheets, as well as pipes with an inner diameter of less than 10 mm and pipes made of steels with increased strength properties [9, 10].

Due to the low level of mechanization and automation of the process, low labor productivity is inherent - the fastening time of one end of the pipe varies from 1.5 to 12 minutes [11].

It is not possible to fasten pipes in tube sheets with preliminary welding (according to GOST R 55601-2013) due to the occurring axial deformations of the pipes, which lead to the destruction of the weld. Due to local deformation and hardening on the inner surface of the pipe and tube sheet, the corrosion resistance of the joints decreases [12]. Everything else, the tool wears out quickly.

Therefore, further in the article, as a method of fastening pipes with tube sheets, we will consider the electric pulse method as the most technologically advanced and promising.

The purpose of this work is to obtain criteria for the similarity of the process of electropulse press fitting of pipes into a tube sheet, which allow reducing the results obtained on a reduced-scale model to nature. Often, heat exchangers have large overall dimensions and, accordingly, have a large number of pressed pipes into the tube sheet, which primarily affects the modeling and processing of the information under study, therefore, when researching and studying this process, the application of similarity criteria will reduce information processing time and optimize the process building models.

2. Prospects of pipes pulsed pressing methods into the tube sheet

One of the promising methods of fastening pipes in tube sheets is electric pulse pressing. The use of an electric pulse makes it possible to automatically control energy parameters, control the process of energy release, create an optimized process and mechanized high-performance equipment [13].

By changing the pressure pulse in the pipe by controlling the energy parameters of the discharge, it is possible to control the quality of the connection, taking into account the properties of its materials and sizes. With an increase in the discharge voltage, the pressure amplitude and the radius of propagation of the plastic strain zone in the compound increase.

By eliminating the possible sharp increase in the rate of pipe deformation, its hardening during deformation and rebound from the grating, it is possible to increase the tightness of the joints. Reducing the gap between the pipe and the hole of the tube sheet due to the use of high precision pipes and reducing the size of the hole increased the tightness by more than 1.5 times. It is possible to increase the efficiency of pressing in by increasing the load resistance, the relative length of the pressing in, and a higher discharge voltage.

The statistical estimate of the mean time between failure of joints obtained by electropulse pressing exceeds the corresponding index of joints obtained by mechanical expansion, sometimes by more than 25 times. This is also confirmed by the experience of operating heat exchangers at the corresponding facilities [14].

It is also worth noting that the industrial sites and workshops for the pulsed processing of high-strength metals based on explosive technology created in the former USSR made it possible to obtain an economic effect of about 1 billion rubles at defense enterprises of the Soviet country [14].

The first research results and their analysis confirmed the prospects of work on pulse pressing of pipes, predetermined the relevance of further developments in this direction [15].

3. Parameters characterizing the process of electropulse press fitting of pipes into the tube sheet

When modeling the process of electropulse press-fitting of pipes into a tube sheet, it is necessary to describe the following phenomena: generating an electric-discharge disturbance in the circuit of an electric pulse installation and transferring the corresponding charge to a working member (cartridge) of press-fitting tubes into a tube sheet; the interaction of the perturbed working body of the press fitting with the pressed pipe and the plastic deformation of the latter.

The generation and transmission of a certain electrical discharge disturbance to the working body of the press-fit is characterized by the possible operating modes of the corresponding installation. In turn, the operation modes of the installation are determined by parametric values, the main of which are: inductance of the excitation coil, installation frequency, charging voltage of the capacitor bank and current in the working circuit.

In turn, the interaction of a perturbed working body with an extruded pipe and its plastic deformation depends on the mode of the electropulse installation, as well as the following values, such as the length of the insert, the mass of the working body (cartridge), the specific work of deformation, the elastic modulus of the pipe.

Taken together, the phenomena described above allow characterizing the process of electropulse pressing of pipes into the tube sheet in sufficient volume to obtain the corresponding similarity criteria, so that in the subsequent simulation of the process of electropulse pressing, you have the necessary amount of information for analyzing the results.

4. Similarity criteria of electropulse press fitting of pipes into the tube sheet

Based on the previous section, we identify the necessary parameters characterizing the process of electropulse pressing of pipes into the tube sheet for further transformations in order to obtain the corresponding similarity criteria by the dimensional analysis method.

The parameters characterizing the process: M_{EBC} – weight of electric blasting cartridge [kg]; A_{SW} – specific work required for deformation of a pipe's unit length [J]; ν – frequency of the electric pulse installation [Hz]; L_C – coil inductance [GN]; U – voltage applied to the discharge circuit [V]; l – fitting length [m]; I – loop current [A]; E – pipe modulus [Pa].

Thus, the total number of participating quantities: m=8.

Let the existing relations between the parameters of the process of electropulse press fitting and the parameters of the elements of the system in which the process of electropulse press fitting takes place, can be represented as follows:

$$F(P_1, P_2, P_3, P_4, P_5, P_6, P_7, P_8) = 0, \tag{1}$$

$$F(M_{EBC}, A_{SW}, \nu, L_C, U, l, I, E) = 0. \tag{2}$$

We choose four (k = 4) independent variables that are applicable to the LMTI measurement system (length, mass, time, current):

$$P_1 = M_{EBC}\ (kg), \tag{3}$$

$$P_2 = A_{SW}\ \left(kg \cdot m^2/_{s^2}\right), \tag{4}$$

$$P_3 = \nu\ (s^{-1}), \tag{5}$$

$$P_4 = L_C\ \left(kg \cdot m^2/_{s^2 \cdot A^2}\right). \tag{6}$$

Next, we determine the dimension of each basic basis parameter:

$$X = L^{\alpha} M^{\beta} T^{\gamma} I^{\epsilon}. \tag{7}$$

$P_1 = [M_{EBC}] = [L]^0 [M]^1 [T]^0 [I]^0;$

$P_2 = [A_{SW}] = [L]^2[M]^1[T]^{-2}[I]^0;$
$P_3 = [v] = [L]^0[M]^0[T]^{-1}[I]^0;$
$P_4 = [L_C] = [L]^2[M]^1[T]^{-2}[I]^{-2}.$

We define the dimensions for the remaining related parameters.

$P_5 = [U] = [L]^2[M]^1[T]^{-3}[I]^{-1};$
$P_6 = [l] = [L]^1[M]^0[T]^0[I]^0;$
$P_7 = [I] = [L]^0[M]^0[T]^0[I]^1;$
$P_8 = [E] = [L]^{-1}[M]^1[T]^{-1}[I]^0.$

We compose a matrix of the basic basis parameters and find the determinant of this matrix.

$$D_{1-4} = \begin{matrix} P_1 \\ P_2 \\ P_3 \\ P_4 \end{matrix} \begin{vmatrix} 0 & 1 & 0 & 0 \\ 2 & 1 & -2 & 0 \\ 0 & 0 & -1 & 0 \\ 2 & 1 & -2 & -2 \end{vmatrix} = -4 \neq 0.$$

As $D_{1-4} = -4 \neq 0$, then the parameters are independent of each other.
We compose expressions for the remaining $n = m - k$ similarity criteria:

$$P_5 = \frac{[U]}{[\mathrm{M}_{EBC}]^{\alpha s}[A_{SW}]^{\beta s}[v]^{\gamma s}[L_C]^{\varepsilon s}};$$

$$P_6 = \frac{[l]}{[\mathrm{M}_{EBC}]^{\alpha k}[A_{SW}]^{\beta k}[v]^{\gamma k}[L_C]^{\varepsilon k}};$$

$$P_7 = \frac{[I]}{[\mathrm{M}_{EBC}]^{\alpha c}[A_{SW}]^{\beta c}[v]^{\gamma c}[L_C]^{\varepsilon c}};$$

$$P_8 = \frac{[E]}{[\mathrm{M}_{EBC}]^{\alpha t}[A_{SW}]^{\beta t}[v]^{\gamma t}[L_C]^{\varepsilon t}}.$$

We make the corresponding matrix calculations and find similarity criteria:

$$D_{5\alpha} = \begin{matrix} P_5 \\ P_2 \\ P_3 \\ P_4 \end{matrix} \begin{vmatrix} 2 & 1 & -3 & -1 \\ 2 & 1 & -2 & 0 \\ 0 & 0 & -1 & 0 \\ 2 & 1 & -2 & -2 \end{vmatrix} = 0, \quad D_{5\beta} = \begin{matrix} P_1 \\ P_5 \\ P_3 \\ P_4 \end{matrix} \begin{vmatrix} 0 & 1 & 0 & 0 \\ 2 & 1 & -3 & -1 \\ 0 & 0 & -1 & 0 \\ 2 & 1 & -2 & -2 \end{vmatrix} = -2,$$

$$D_{5\gamma} = \begin{matrix} P_1 \\ P_2 \\ P_5 \\ P_4 \end{matrix} \begin{vmatrix} 0 & 1 & 0 & 0 \\ 2 & 1 & -2 & 0 \\ 2 & 1 & -3 & -1 \\ 2 & 1 & -2 & -2 \end{vmatrix} = -4, \quad D_{5\varepsilon} = \begin{matrix} P_1 \\ P_2 \\ P_3 \\ P_5 \end{matrix} \begin{vmatrix} 0 & 1 & 0 & 0 \\ 2 & 1 & -2 & 0 \\ 0 & 0 & -1 & 0 \\ 2 & 1 & -3 & -1 \end{vmatrix} = -2.$$

$$\alpha_s = \frac{D_{5\alpha}}{D_{1-4}} = \frac{0}{-4} = 0;$$

$$\beta_s = \frac{D_{5\beta}}{D_{1-4}} = \frac{-2}{-4} = 0.5;$$

$$\gamma_s = \frac{D_{5\gamma}}{D_{1-4}} = \frac{-4}{-4} = 1;$$

$$\varepsilon_s = \frac{D_{5\varepsilon}}{D_{1-4}} = \frac{-2}{-4} = 0.5\,.$$

$$P_5 = \frac{[P_5]}{[P_1]^{\alpha s}[P_2]^{\beta s}[P_3]^{\gamma s}[P_4]^{\varepsilon s}} = \frac{[U]}{[M_{EBC}]^0[A_{SW}]^{0.5}[v]^1[L_C]^{0.5}} = \frac{[L]^2[M]^1[T]^{-3}[I]^{-1}}{[L]^1[M]^{0.5}[T]^{-1}[T]^{-1}[L]^1[M]^{0.5}[T]^{-1}[I]^{-1}} = 1$$

The obtained criterion P_5 is dimensionless ($P_5 = 1$), verification passed.
Further similarly we find the following criteria: P_6, P_7, P_8.

$D_{1-4} = -4;$
$D_{6\alpha} = 2,$
$D_{6\beta} = -2,$
$D_{6\gamma} = 4,$
$D_{6\varepsilon} = 0.$
$\alpha_k = \frac{2}{-4} = -0.5;$
$\beta_k = \frac{-2}{-4} = 0.5;$
$\gamma_k = \frac{4}{-4} = -1;$
$\varepsilon_k = \frac{0}{-4} = 0\,.$

$$P_6 = \frac{[l]}{[M_{EBC}]^{-0.5}[A_{SW}]^{0.5}[v]^{-1}[L_C]^0} = \frac{[L]^1}{[M]^{-0.5}[L]^1[M]^{0.5}[T]^{-1}[T]^1} = 1$$

The obtained criterion P_6 is dimensionless ($P_6 = 1$), verification passed.

$D_{1-4} = -4;$
$D_{7\alpha} = 0,$
$D_{7\beta} = -2,$
$D_{7\gamma} = 0,$
$D_{7\varepsilon} = 2.$
$\alpha_c = \frac{0}{-4} = 0;$
$\beta_c = \frac{-2}{-4} = 0.5;$
$\gamma_c = \frac{0}{-4} = 0;$
$\varepsilon_c = \frac{2}{-4} = -0.5\,.$

$$P_7 = \frac{[I]}{[M_{EBC}]^0[A_{SW}]^{0.5}[v]^0[L_C]^{-0.5}} = \frac{[I]^1}{[L]^1[M]^{0.5}[T]^{-1}[L]^{-1}[M]^{-0.5}[T]^1[I]^1} = 1$$

The obtained criterion P_7 is dimensionless ($P_7 = 1$), verification passed.

$D_{1-4} = -4;$
$D_{8\alpha} = -6,$
$D_{8\beta} = 2,$
$D_{8\gamma} = -8,$
$D_{8\varepsilon} = 0.$
$\alpha_t = \frac{-6}{-4} = 1.5;$
$\beta_t = \frac{2}{-4} = -0.5;$

$$\gamma_t = \frac{-8}{-4} = 2;$$
$$\varepsilon_t = \frac{0}{-4} = 0\,.$$

$$P_8 = \frac{[E]}{[M_{EBC}]^{1.5}[A_{SW}]^{-0.5}[v]^2[L_C]^0} = \frac{[L]^{-1}[M]^1[T]^{-1}}{[M]^{1.5}[L]^{-1}[M]^{-0.5}[T]^1[T]^{-2}} = 1$$

The obtained criterion P_8 is dimensionless ($P_8 = 1$), verification passed.
Next, a general check was made.

$$F(P_5, P_6, P_7, P_8) = 0$$

All criteria passed the test.

5. Conclusion

Obtaining one-piece pipe connections in the tube sheets of heat exchangers by creating a guaranteed interference fit on the surface of their contact due to the interconnected stress state consists of the following steps:

- Pipe deformation by the amount of radial clearance between the pipe and the tube sheet.
- Joint deformation of pipes and tube sheets to the maximum value, due to the dynamics of the process of pressing in.
- Process of elastic unloading of the joint from the maximum deformed equilibrium stress state, leading to the formation of an interference fit and contact pressure on the contact surface of the joint, which determines the residual stresses and operational characteristics of the resulting compound.

In the presented article, the above-mentioned phenomena were described that characterize the essence of the process of fastening pipes to the tube sheet by the method of electropulse pressing. Criteria were obtained for the similarity of the process of electropulse press fitting of pipes into a tube sheet, which allow reducing the results obtained on a scaled-down model to nature. This will facilitate the modeling process for further experiments on objects in natural dimensions.

References

[1] Mazurovsky B Ya 1980 *Electro-impulse press-fitting of pipes in tube sheets of heat exchangers* (Kiev, Sci. Dumka)
[2] Doroshenko P A 1972 *Production technology of ship steam generators and heat exchangers* (Lenongrad, Shipbuilding)
[3] Stepanov V G 1972 *Fundamentals of pipe expansion technology in ship heat exchangers* (Lenongrad, Shipbuilding)
[4] Bukhtoyarov V V, Petrovskyi E A, Tynchenko V S, Zhukov V G and Smirnov A I 2016 Workpiece surface technological quality assurance with levitation tool modules *Indian Journal of Science and Technology* **9(29)** 99441
[5] Berliner Yu I and Brief V M 1970 Analysis of the applicability of modern methods of fastening pipes to tube sheets *High-performance welding methods in chemical and petroleum engineering* **3** 3-12
[6] Bukhtoyarov V V, Tynchenko V S, Petrovskiy E A, Tynchenko V V and Zhukov V G 2018 Improvement of the methodology for determining reliability indicators of oil and gas equipment *International Review on Modelling and Simulations* **11(1)** 37-50
[7] Vygovsky V F 1966 Research on the performance of non-separable connections by an electric explosion of wires *Abstract of a candidate of technical sciences theses* 26

[8] Lukovkin A I 1968 *New rolling compounds obtained using explosion energy* (Leningrad, LDNTP)
[9] Mazurovsky B Ya, Ischenko Zh N and Convisher B Ya 1976 Status and prospects of using electrohydropulse expansion in industry Bull*etin of the Academy of Sciences of the Ukrainian SSR* **10** 58-63
[10] Fraas A P 1989 Heat Exchanger Design (New York, John Wiley & Sons)
[11] Mazurovsky B Ya and Sizev A N 1983 Electrohydraulic effect in sheet metal stamping (Kiev, Sci. Dumka)
[12] Incropera D W and Bergman L 2007 Fundamentals of Heat and Mass Transfer Sixth (New York, John Wiley & Sons)
[13] Sokolov V A 2001 *Fundamentals of the theory of similarity and dimensional analysis in oil and gas production* (Ukhta: USTU)
[14] Guhman A A 1967 *Application of the theory of similarity to the study of heat-mass transfer processes* (Moscow, Higher School)
[15] Gatapova N Ts, Koliukh A N, Orlova N V and Orlov A Yu 2014 *Fundamentals of the theory and technique of physical modeling and experiment* (Tambov: TSTU)

Journal of Physics: Conference Series **1384** (2019) 012061 doi:10.1088/1742-6596/1384/1/012061

Metrological support of an automated control system for a rocking machine

V S Tynchenko[1,2], V V Bukhtoyarov[1,2], V V Tynchenko[1,2], A V Milov[2], K A Bashmur[1] and V V Kukartsev[1,2]

[1] Siberian Federal University, 79, Svobodny Avenue, Krasnoyarsk, 660041, Russian Federation
[2] Reshetnev Siberian State University of Science and Technology, 31, Krasnoyarsky Rabochy Avenue, Krasnoyarsk, 660037, Russian Federation

E-mail: vadimond@mail.ru

Abstract. The article discusses the problem of control automation for wells equipped with sucker rod pumps. The main goals of such automation are presented, the achievement of which allows increasing well productivity, as well as improving working conditions of staff. As a result of the study, the analysis of the features of wells automation equipped with sucker rod pumps was carried out. The principles of automation of such an object were considered and a control system diagram was presented. The presented functional diagram of the SCADA system involves the use of a number of sensors to measure the current parameters of the well. Using the expert assessment method, a reasonable choice of specific measuring devices was made. Reviewed measuring devices are modern equipment of Russian and foreign production, which has a good service life and the necessary measurement accuracy.

1. Introduction

Many factors of the current economic situation, for example, the depletion of oil reservoir resources, the limited amount of funds for repairing wells, the high cost of electricity, as well as the high complexity of the work, necessitate the automation of oil production processes [1, 2]. Automation is one of the factors that can increase productivity and improve working conditions [3 - 5].

The automation of wells equipped with sucker rod pumps (SRP) consists of the control of such technological parameters as a dynamogram, wattmetergram, dynamic fluid level, oscillation frequency, current consumption, pressure at the wellhead [6, 7]. The control functions must provide remote switching on and off of the drive motor, periodic operation, emergency shutdown of the installation, smooth regulation of the motor rotation speed using a frequency converter [8, 9].

Many foreign and Russian enterprises are engaged in the development of SRP automation equipment: Lufkin Automation (USA), NPP Grant (Ufa), NPF Intek (Ufa), LLC Mikon (N. Chelny), TNPVO Siam (Tomsk) and others. Nevertheless, not a single company can boast of absolute success in this direction, and this is due to the rapid development of measuring, converting and microprocessor technologies, which provides developers the ability to solve more and more complex problems.

In this paper, the main task is to analyze the metrological support of an automated control system for a rocking machine (RM).

2. Principles of automation of a well equipped with a rocking machine

Rocking machine refers to the equipment that is involved in the operation of oil wells using sucker rod pumps. Moreover, such equipment is one of the surface drives of sucker rod pumps. Oil specialists give it such a definition as "individual balanced mechanical drives of a sucker rod pump" or simply "rocking chair". By the principle of action, the rocking machine is often compared with a hand-held bicycle pump, which converts the reciprocating motion into air flow. From a rocking machine, an oil pump converts such movements into a stream of liquid hydrocarbons that enter the surface through tubing.

Let us consider the principle and operation sequence of the rocking machine. The electric motor rotates the mechanisms of the rocking machine, the balancer moves on the principle of a swing as a result of the suspension of the wellhead rod receives reciprocating motion. The received energy is transmitted to steel rods twisted together with the special couplings. Through these steel rods, energy is transferred to the sucker rod pump, which begins to pump oil and feed it up.

Figure 1 shows a diagram of the SCADA-system of the well's functioning, equipped with SRP (ensuring the well's work in the regime of periodic supply of products to the surface).

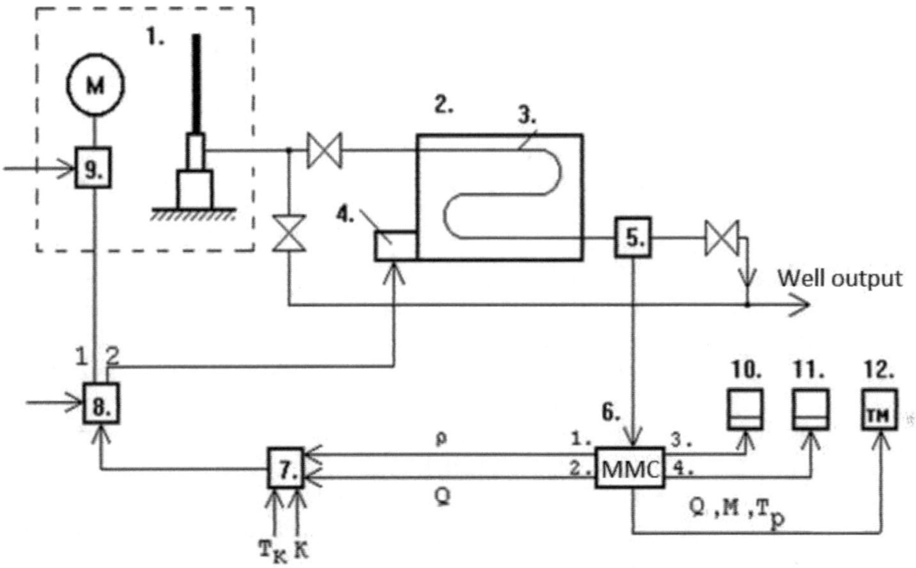

Figure 1. Control system of the well operating mode, where: 1 - well; 2 - oil-filled furnace (installed with a high content of paraffin in oil); 3 - coil; 4 - electric heater; 5 - measuring modular converter (MMC); 6 – MMC electronic device; 7 - analyzer of the well products supply nature; 8 - electric starting equipment; 9 - control station; 10, 11 - self-cleaning devices; 12 – telemechanic block (TM).

Functional diagram of the SCADA-system for the SRP-equipped well is shown in Figure 2.

3. Selection of metrological support for the automation system

To ensure measurements in the automated control system of the technological process of the SRP-equipped well's exploitation, it is necessary to make a reasonable choice of metrological support of the following types: force sensor, strain sensor, position sensor, viscosity sensor, flow meter, pressure sensor, temperature sensor, gas analyzer. The dynamometer sensor is the main element used for SRP automation, because this sensor allows us getting data to analyze the object operation and its condition.

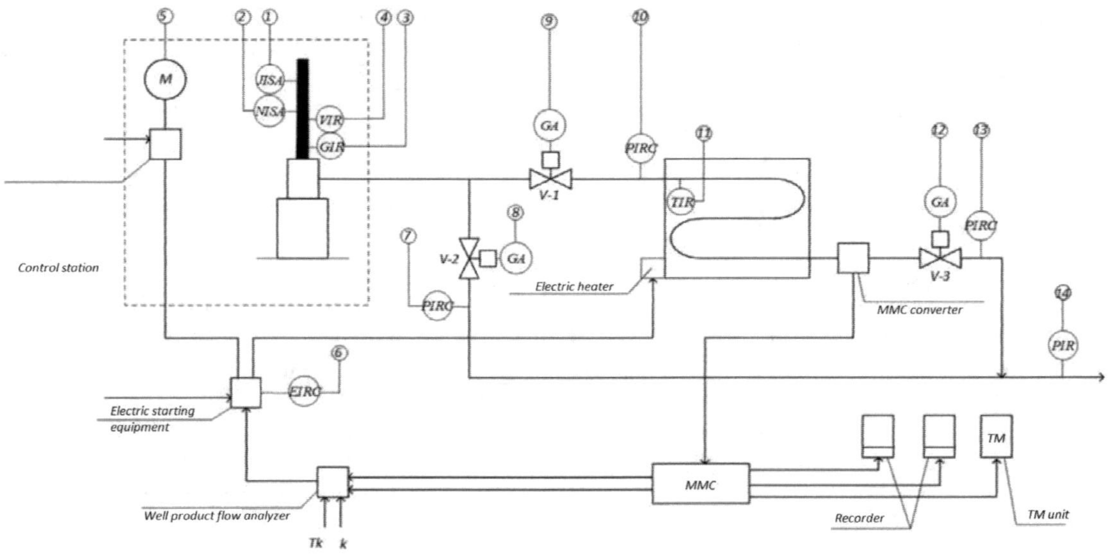

Figure 2. Functional diagram of the well's SCADA system, where: 1 - force sensor (JISA) is located on the rod above the upper traverse; 2 - balance deformation sensor (NISA); 3 - position sensor (GIR); 4 - viscosity sensor (VIR); 5 - electric motor; 6 – watt-metering sensor (EIRC); 7, 10, 13 - pressure sensor (PIRC); 8, 9, 12 - position sensors of electric valves (GA); 11 - temperature sensor (TIR); 14 - pressure sensor (PIR).

Force sensors on SRP are installed in the following positions: directly on the stock itself (overhead sensors); between traverses; between the upper traverse and the locks; on the SRP balancer.

Sensors that are located above the upper crosshead, perceive the load on the rod directly, and, therefore, provide maximum sensitivity to determine the force. As a result of expert analysis, a Loadtrol sensor from Lufkin (USA) was selected as a force sensor.

The strain gauges that are installed on the balancer are designed to measure the deformation of the balancer. However, the measurement accuracy of such sensors is not enough.

In order to increase the accuracy of strain measurements, it is possible to use the EPSI AX strain gauge (Figure 3), which is designed to measure the longitudinal strain of various structures under static or dynamic load.

Figure 3. Deformation sensor EPSI AX.

As a position sensor mounted on the reducer of the pumping unit, due to the expert analysis, Honeywell sensor 103SR12-A1 (HONEY) was selected (Figure 4).

This sensor operates on the Hall Effect. The position sensor is mounted with a bracket to the reducer of the SRP installation and is triggered when two magnets pass by it. The position sensor is mounted so that the magnets pass the Hall sensor at the moments when the rod of the SRP installation is in the extreme lower and / or upper positions.

Journal of Physics: Conference Series **1384** (2019) 012061 doi:10.1088/1742-6596/1384/1/012061

Figure 4. Position sensor 103SR12-A1.

Aveni Sense DEVIL sensor (Figure 5) was chosen as a viscosity sensor, which is designed for oils and pure oil and has strength and wide range of viscosity measurements.

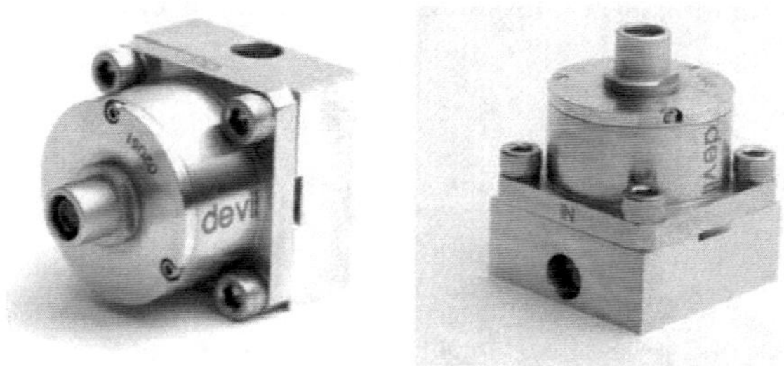

Figure 5. Sensor Aveni Sense DEVIL.

Such a sensor was chosen due to the following advantages:

- Built-in performance.
- Possibility of continuous measurement of density, dynamic and kinematic viscosity.
- High performance with small sizes.
- MesoScale and Selfbalancit technologies. The MesoScale design is used to reduce the internal volume of the sensor to less than 1 cc/cm, providing a clear and easy to clean path for fluid flow. "Selfbalancit" technology allows us taking measurements in accordance with ASTM requirements.
- Reliable solution for explosive atmospheres. Robust construction and ease of use are the key elements to ensure stable metrological performance. DEVIL is made from corrosion resistant materials and is certified intrinsically safe.

During the operation of the rocking machine, it is necessary to monitor the flow of pumped oil. To measure the flow rate, a Metran-350 flowmeter based on the Annubar was selected (Figure 6).

Figure 6. Flowmeter Метран-350.

This flowmeter was selected due to the following advantages:

- Integral design eliminates the need for impulse lines and additional devices, thereby reducing the number of potential leakage sites.
- Low irretrievable pressure loss in the pipeline reduces energy costs.
- 3051SMV multi-parameter transmitters as part of flowmeters provide the necessary calculation of the instantaneous mass flow rate of steam, liquid, gas or gas volumetric flow rate reduced to standard conditions.
- Installing a flowmeter is economical and less time consuming compared to installing a measuring complex based on a standard diaphragm.

As a result of the expert analysis for measuring pressure, the Elemer AIR-20/M2-N pressure sensor was selected (Figure 7), designed to continuously convert absolute, differential and overpressure, as well as overpressure, hydrostatic pressure, into a unified output current signal 4 ... 20 mA.

Figure 7. Pressure sensor Elemer AIR-20/M2-N.

In order to measure the temperature, a TPK 135 high-temperature thermocouple is used (Figure 8), designed to continuously measure the temperature of liquid, vapor and gaseous media.

Figure 8. Thermocouple TPK 135.

Optimus IR is used as a gas analyzer (Figure 9) - an infrared optical sensor for detecting the presence of explosive gases and is intended for continuous monitoring of explosive concentrations of methane, propane, butane, isobutane, pentane, cyclopentane, hexane, propylene, methanol, ethanol, as well as oil vapor and petroleum products in the area of the working zone in the range from 0 to 100%.

Figure 9. Gas analyzer Optimus IR.

The Optimus IR gas analyzer was selected due to a number of its advantages, such as:

- Ability to work in the temperature range from -60 °C to + 75 °C and relative humidity up to 98%.
- Robust design that allows the sensor operating in extreme climatic conditions, in aggressive atmosphere, in explosive areas of rooms and outdoor installations.
- Presence of a standard output current signal 4-20 mA, digital signal HART and RS-485 Modbus RTU, relay "dry contacts", LED light indication and display.

All sensors selected above have a unified 4-20 mA current signal.

In order to transmit signals from pressure sensors, flow meters, ammeters and a monitoring system to the shield of instrumentation and equipment, three wires must be used, and two wires for signaling devices. To connect the equipment, a control cable was selected with copper conductors with vinyl insulation in a polyethylene sheath, designed for fixed connection to electrical devices and switchgears with a rated alternating voltage of up to 660 V, frequency of up to 100 Hz or constant voltage of up to 1000 V at ambient temperature from -50 ° C to +50 ° C.

4. Conclusion

As a result of the study, an analysis of the automation features of the wells equipped with sucker rod pumps was carried out. The principles of automation of such an object were considered and a control system diagram was presented.

The presented functional diagram of the SCADA system involves the use of a number of sensors to measure the current parameters of the well. Using the expert assessment method, a reasonable choice was made of specific measuring devices, which are modern equipment of Russian and foreign production. The selected measuring devices have a good service life and the necessary measurement accuracy.

References

[1] Alboudwarej H, Muhammad M, Barrere C, Archer R, Hannan R and Jones T 2005 Viscosity of formation water: Measurement, prediction, and reservoir implications *Proceedings - SPE Annual Technical Conference and Exhibition* 2179-2186

[2] Schupak L V and Zhadan V V 2018 Modern problems of cost planning at the enterprise *Economy info* **15** 55-60

[3] Bukhtoyarov V V, Tynchenko V S, Petrovskiy E A, Tynchenko V V and Zhukov V G 2018 Improvement of the methodology for determining reliability indicators of oil and gas equipment *International Review on Modelling and Simulations* **11(1)** 37-50

[4] Sverdlik G V 2006 The automation of technological process is a source of steady economic advantage *Tselliuloza, Bumaga, Karton/Pulp, Paper, Board* **8** 82-83

[5] Klyuev A S, Glazov B V, Dubrovsky A Kh and Klyuev A A 1990 *Designing automation systems for technological processes: a reference guide* (Moscow, Energoatomizdat)

[6] Commissarchik V F 2001 *Automatic regulation of technological processes: a tutorial* (Tver, TSU)

[7] Komyagin A F 1983 *Automation of production processes and process control systems for gas and oil pipelines* (Moscow, Nedra)

[8] Popovich N G, Kovalchuk A V and Krasovsky E P 1986 *Automation of production processes and installations* (Kyiv, Vischashk publishing house)

[9] Tynchenko V S, Kukartsev V V, Tynchenko V V, Chzhan E A and Korpacheva L N 2018 Automation of monitoring and management of conveyor shop oil-pumping station of coal industry enterprise *IOP Conference Series: Earth and Environmental Science* **194(2)** 022044

Expert analysis of the automation equipment for the integrated oil treatment unit

V S Tynchenko[1,2], V V Bukhtoyarov[1,2], V V Tynchenko[1,2], E A Petrovsky[1], K A Bashmur[1] and V V Kukartsev[1,2]

[1] Siberian Federal University, 79, Svobodny pr., Krasnoyarsk, 660041, Russian Federation
[2] Reshetnev Siberian State University of Science and Technology, 31, Krasnoyarsky Rabochy Av., Krasnoyarsk, 660037, Russian Federation

E-mail: vadimond@mail.ru

Abstract. The article discusses the problem of control automation for integrated oil treatment unit. The main goals of such automation are presented, the achievement of which allows increasing oil treatment unit productivity, as well as improving working conditions of staff. As a result of the study, an analysis of the features of oil treatment unit automation was carried out. The principles of automation of such an object were considered and an oil treatment unit principal schema was presented. The presented features of the SCADA system involve the use of a number of sensors to measure the current parameters of the oil treatment unit. Using the expert study method, a reasonable choice of specific measuring and control devices was made. Reviewed measuring and control devices are modern equipment of Russian and foreign production, which have a good service life and the necessary measurement accuracy.

1. Introduction

The processes of dehydration, desalination and oil stabilization are carried out at the complex oil treatment facilities (COTF) [1]. The basic scheme of COTF with rectification is shown in Figure 1.

Consider the work of complex oil treatment facilities. Cold crude oil from the tanks pump 1 through the heat exchanger 2 is fed into the sedimentation tank 3. Here, most of the mineralized water settles to the bottom of the apparatus and is diverted for further preparation for injection into the reservoir (III). Next, fresh water (V) is added into the stream to reduce the salt concentration in the remaining mineralized water. In the electric dehydrator 4, the final separation of water from oil is carried out and the dehydrated oil through the heat exchanger 5 enters the stabilization column 6. By pumping oil from the bottom of the column through the furnace 10 by pump 11, its temperature is brought to 240 °C. In this case, light oil fractions evaporate, rise to the top of the column and then enter the condenser-cooler 7. Here the propane-butane and pentane fractions are mainly condensed, forming the so-called broad fraction, and non-condensed components are discharged for use as fuel. A wide fraction is pumped out by pump 9 for fractionation, and partially used for irrigation in column 6. Stable oil from the bottom of the column is pumped by pump 12 to commodity tanks. On this way, hot stable oil gives up part of its heat to crude oil in heat exchangers 1 and 5.

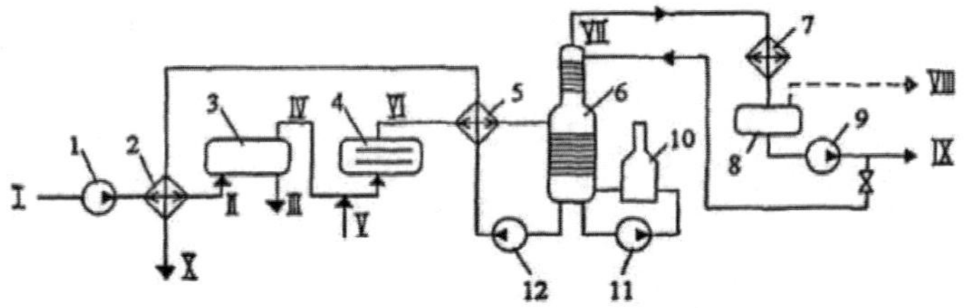

Figure 1. Schematic diagram of the integrated oil treatment installation, where: 1, 9,11,12 - pumps; 2,5 - heat exchangers; 3 - sedimentation tank; 4 - electric dehydrator; 6 - stabilization column; 7 - condenser-cooler; 8 - irrigation capacity; 10 - oven; I - cold crude oil; II - warmed up crude oil; III - drainage water; IV - partially dehydrated oil; V - fresh water; VI - dehydrated and desalted oil; VII - pairs of light hydrocarbons; VIII - non-condensing vapors; IX - broad fraction (condensed vapors); X - stable oil.

Thus, in COTF dehydration, desalination and oil stabilization are carried out. Moreover, for dehydration, heating, sedimentation and electrical exposure are used simultaneously.

The object of the study in this work is a part of the integrated oil treatment unit, consisting of a tank, a pumping station unit and an oil heater. The oil supplied for transportation to the highway must comply with the established parameters and be in a certain temperature range.

The automated COTF system, consisting of measuring devices, controllers and actuators, should provide [2, 3]:

- Automated monitoring and control in real time of the technological process of preliminary preparation and oil and water pumping [4].
- Safety of the technological process of oil and water preliminary preparation [5].
- Automatic and remote correction of the technological process to a safe state in case of emergency situations (fire, failure of technological equipment, etc.) [6].
- Monitoring the level of the product, the temperature of the heater, and their location within the specified regulatory limits and the transfer of COTF to a safe state when the level goes beyond the range [7].
- Control of the technological parameters of oil pumps and oil products [8, 9].
- Control of oil pumps and oil products [10].

In addition, associated gas is supplied to the oil heater for heating. According to current legislation, a strict accounting of associated gas consumption is required.

Thus, the urgent task is to choose effective means to control the parameters of the considered technological process in order to improve the quality of the complex oil treatment facilities automated control [11].

2. Choice of controller equipment

When choosing controller equipment, several widely used programmable logic controllers (PLCs) were considered: Emerson Delta V, Siemens SIMATIC S7-400, EMICON DSC-2000. Table 1 shows a comparison of the controller equipment characteristics.

As a result of the expert analysis according to the criteria presented in Table 1, a Siemens SIMATIC S7-400 PLC was selected, equipped with two communication processors (one local, and the second for communication with the upper level).

This controller provides the ability to solve automation problems in the optimal way. The advantages include high platform reliability, the ability to communicate via multiple protocols and interfaces with other systems, and ample expandability.

The appearance of the PLC is shown in Figure 2.

Table 1. Comparison of the controller equipment characteristics.

Characteristics	**PLC model**		
	Emerson Delta V	**EMICON DCS-2000**	**Siemens SIMATIC S7-400**
Programming languages	LD; FBD; ST; SCL; IL.	CONT-Designer; CoDeSys.	LD; FBD; ST; SCL; CFC; GRAPH; HiGraph.
Number of input / output channels	63488 discrete and not more than 3968 analog	1335 discrete and 199 analog	65536 discrete and not more than 4096 analog
Memory and logic data, Mb	48	8	0.5 for code, 3 for data
Communication options	FOUNDATION fieldbus; HART; WirelessHART; Profibus DP; DeviceNet.	RS-485; Modbus RTU; Ethernet	MPI; PROFIBUS; Industrial Ethernet; PROFInet; AS-I; BAC-net; Modbus TCP; Modbus RTU.
Life time, years	8	10	12
Price, rubles	250000	50000	180000
Architecture	modular	modular	modular
Reservation	full support	hardware redundancy	hot standby option

Figure 2. PLC Siemens SIMATIC S7-400.

3. Choice of measuring instruments

During operation of COTF, it is necessary to monitor the oil level in the tanks, the pressure on the suction manifold of the pump, the temperature in the heater, before and after heating the oil, the pressure of gas and nitrogen supplied to the heater. To control the indicated parameters, it is possible to use the following types of sensors: flowmeter, pressure sensor, temperature sensor, level gauge.

When choosing a flow meter, the following options were considered that were widely used in the oil and gas industry: Prowirl 72W vortex flow meter, DRG.M and Annubar Metran-350. A comparison of the selected flowmeter samples is shown in Table 2.

As a result of expert analysis, a Metran-350 flowmeter based on the Annubar averaging pressure tube (APT) was chosen to measure gas and nitrogen flow rates, since it has higher measurement accuracy, high degree of protection, and low price.

Table 2. Comparison of the flowmeters characteristics.

Characteristics	**Flowmeter model**		
	Prowirl 72W	**DRG.M**	**Metran-350 Annubar**
Medium	Liquid, steam, gas	Liquid, gas	Liquid, steam, gas
Measurement principle	Vortex	Vortex	Differential pressure
Output signal, mA	4 ... 20, pulse frequency status signal	4 … 20	4 … 20, HART
Intertesting interval, years	4	3	4
The limits of the standard error in the volumetric flow measurement	±1%	±1.5 %	±0.75 %
Degree of protection	IP67	IP57	IP68
Explosion protection	ATEX; FM; CSA; TIIS	IExdIICT6X	1ExdIICT6(T5) X
Price, rubles	65000	95000	29000

Pressure sensors are used to monitor the pressure on the intake manifold located before the pumps, because when the pressure drops below a predetermined level, gas from the oil will start to be released in the pipeline, which can cause the destruction of the pump units. It is also necessary to measure the pressure of the incoming gas and nitrogen into the heater. And finally, it is necessary to know the pressure of the heated oil at the outlet of the heater.

Three options for widespread gauge pressure sensors were considered: Metran-150CG, KVARTS-2 and Sapphire-22M. The comparison is shown in Table 3.

Table 3. Comparison of the pressure sensors characteristics.

Characteristics	**Pressure sensor model**		
	Metran-150CG	**KVARTS-2**	**Sapphire-22M**
Medium	Liquid, steam, gas	Liquid, steam, gas	Liquid, steam, gas
Output signal, mA	4 … 20, HART	4 … 20	4 … 20
Measuring range, Pa	25 … 68,000,000	0 … 100,000,000	80 … 2,400,000
Adjustment of ranges of measurements	25:1	-	-
Price, rubles	30000	23000	25000
Allowable error	0.075%	0.1%	0.25%
Explosion protection	1ExdIICT6X	ExiaIICT5X	Ex
Protection degree	IP66	IP54	IP65
Life time, years	12	6	12

As a result of expert analysis, pressure sensors Metran-150CG were selected for pressure monitoring. These sensors have a built-in protection against transients that occur in the communication line, which can be caused by lightning discharges or a heavy load caused, for example, by the operation of equipment for conducting electric welding. Metran sensors are resistant to electromagnetic interference, have high spark and explosion safety, high protection against environmental influences, and have a long service life.

When choosing temperature sensors, three options were considered: Rosemount-3144P, Metran-274 and WIKA TR10-C. A comparison of temperature sensors is given in Table 4.

Table 4. Comparison of the temperature sensors characteristics.

Characteristics	**Temperature sensor model**		
	Rosemount-3144P	**Метран-274**	**WIKA TR10-C**
Measured temperature ranges, °C	-200 … +850	-50 … +180	-30 … +150
Medium	Neutral and aggressive environments	Neutral and aggressive environments	Neutral and aggressive environments
Margin of error	0.10%	0.25%	0.1%
Output signal, mA	4 … 20, HART, Fieldbus	4 … 20, HART	4 … 20, HART
Protection degree	IP68	IP65	IP67
Life time, years	10	10	5
Explosion protection	ExdIICT5	ExdIICT6	ExdIICT6
Price, rubles	31500	2300	10000
Power consumption, W	Less than 4.5	Less than 0.5	-

As a result of expert analysis, a temperature sensor Metran-274 was selected to measure the temperature, since it is best suited for solving the task. The advantages of the sensor are: service life, high explosion protection, low price, low energy consumption and the absence of the need for additional standard converters.

In order to avoid emergency situations, the process of filling and emptying the oil tank must be controlled, for which level gauges are used. The following most popular level gauges on the market were considered: Rosemount 5600, Sapphire DU22, PMP-128. A comparison of the level gauges characteristics is given in Table 5.

For an oil tank, sensors with increased accuracy and a HART protocol are required, therefore the Sapphire DU22 level gauge does not meet the requirements. The output signal of the PMP-128 level meter does not meet the requirements of the system. Thus, the Rosemount 5600 is the best suited for level measurement, which also has higher explosion protection.

The Rosemount 5600 Series level gauges are smart meters for non-contact level measurement in tanks of various types and sizes. With high sensitivity, the 5600 Series gauges provide accurate and reliable measurements, even under very difficult process conditions. Level gauges can be used to measure the level of products, including those with low dielectric constant, can operate in a wide range of temperature, pressure and provide high measurement flexibility due to the large selection of various antennas and materials. These level gauges are very simple to manage and maintain during operation, which together reduces the costs of further maintenance.

Table 5. Comparison of the level gauges characteristics.

Characteristics	**Level gauge model**		
	Rosemount 5600	**Sapphire DU22**	**PMP-128**
Operating principle	Radar	Buoy	Reed switch
Measuring range, m	0 … 50	0.25 … 10	0.75 … 25
Medium	petroleum products, acids, solvents	petroleum products	petroleum products
Medium temperature, °C	- 40 … + 400	- 40 … + 80	- 50 … + 60
Mean time before failure, h	100000	100000	100000
Life time, years	12	12	15
Degree of protection	IP67	IP54	IP68
Price, rubles	90000	20000	20000
Basic error, mm	± 5	± 12	± 5
Output signal, mA	4 … 20, HART, Modbus	4 … 20	-
Explosion safety	2ExdeiaibIICT6	0ExiaIICT5	ExaidIIBT3

4. Conclusion

As a result of the study, an analysis of the automation features of the integrated oil treatment unit was carried out. The principles of automation of such an object were considered and an oil treatment unit principal schema was presented.

The presented features of the SCADA system involve the use of a number of sensors to measure the current parameters of the oil treatment unit. Using the expert study method, a reasonable choice of specific measuring and control devices was made. Reviewed measuring and control devices are modern equipment of Russian and foreign production, which have a good service life and the necessary measurement accuracy.

References

[1] Hadfield D 1994 A review of advanced technologies for the complex treatment of oily (produced) water from offshore oil and gas facilities *Environmental Protection Bulletin* **33** 13-24

[2] Zebzeev A 2010 Designing of complex information and control system of the deposit "Urubcheno-Tohomskoe" *Society of Petroleum Engineers - SPE Russian Oil and Gas Technical Conference and Exhibition 2010, RO and G 10* **2** 1203-1212

[3] Frenier W W and Ziauddin M 2010 A multifaceted approach for controlling complex deposits in oil and gas production *Proceedings - SPE Annual Technical Conference and Exhibition* **1** 285-308

[4] Tynchenko V S, Kukartsev V V, Tynchenko V V, Chzhan E A and Korpacheva L N 2018 Automation of monitoring and management of conveyor shop oil-pumping station of coal industry enterprise *IOP Conference Series: Earth and Environmental Science* **194(2)** 022044

[5] Bukhtoyarov V V, Tynchenko V S, Petrovskiy E A, Tynchenko V V and Zhukov V G 2018 Improvement of the methodology for determining reliability indicators of oil and gas equipment *International Review on Modelling and Simulations* **11(1)** 37-50

[6] Popovich N G, Kovalchuk A V and Krasovsky E P 1986 *Automation of production processes and installations* (Kyiv, Vischashk publishing house)
[7] Klyuev A S, Glazov B V, Dubrovsky A Kh and Klyuev A A 1990 *Designing automation systems for technological processes: a reference guide* (Moscow, Energoatomizdat)
[8] Sverdlik G V 2006 The automation of technological process is a source of steady economic advantage *Tselliuloza, Bumaga, Karton/Pulp, Paper, Board* **8** 82-83
[9] Commissarchik V F 2001 *Automatic regulation of technological processes: a tutorial* (Tver, TSU)
[10] Komyagin A F 1983 *Automation of production processes and process control systems for gas and oil pipelines* (Moscow, Nedra)
[11] Anon 1986 Colloquium on choice of equipment in the implementation of control systems *IEE Colloquium (Digest)* **132** var paging

Journal of Physics: Conference Series **1384** (2019) 012063 doi:10.1088/1742-6596/1384/1/012063

Improvement of the construction of vibration-centrifugal unit

V I Uralskiy, A V Uralskiy, E V Sinitsa, S V Karatsupa and L S Uralskaja

Belgorod State Technological University named after V.G. Shukhov, 46, Kostyukov str., Belgorod, 308012, Russia

E-mail: Alexx_1984.10@mail.ru

Abstract. The article presents the scientific and technical studies to create an effective system for balancing the lever mechanism of the centrifugal grinding unit with the possibility of automatic adjustment during the technological grinding process for changing the size of the grinding load in the working chambers.

1. Introduction

Grinding unit [1,2] is equipped with counterweights located at the ends of the eccentric shaft which is the input link of the crank-slider mechanism. Counterweights have the ability to manually move and install in the desired position which is determined depending on the mass of the moving parts of the unit including the grinding load in the working chambers. This principle of balancing allows static balancing of the lever mechanism which is ineffective since during operation of the unit the moving parts move with sufficiently high speeds, varying in size and direction. There are significant accelerations and inertial loads, the consequence of which is the vibration of the structure which negatively affects the strength characteristics of the unit (fatigue, additional stresses, etc.).

In addition, it may be necessary to change the parameters of the operating modes of the unit, for example, changing the load factors of the working chambers in the process of grinding materials with different physical and mechanical properties. Consequently, the magnitude of the mass and the position of the center of mass of the grinding unit. All this leads to the appearance of additional vibration which entails a reduction in the service life of the components and parts of the grinding unit and adversely affects its operation in industrial conditions.

2. Calculation of the dynamic characteristics of the grinding unit

Earlier studies of the dynamics of the mechanical system of the unit [3, 4] allow quantifying the dynamic characteristics of the structure.

To assess the impact on the power characteristics of CGU of the movement of the grinding load inside the grinding chambers, the operation of the unit was considered without taking into account the impact of the grinding load on the links of the lever mechanism. The design scheme of the lever mechanism is presented on Figure 1.

To solve the problem, we used the method of determining the reduced moments which consists in the equality of the capacities developed by the reduced moment (applied to the link of the reduction - link 1) and replaced by forces and moments applied to the links of the mechanism, i.e.

$$P_{\Pi}(\varphi)=\sum_{1}^{n} P_i \,. \tag{1}$$

We apply this method to the lever mechanism of the CGU for the steady motion mode (angular velocity of rotation of the link 1 ω = const).

Represent the power required to overcome the forces of resistance as an expression

$$P_{\Pi}(\varphi) = M_C(\varphi)\omega, \tag{2}$$

where M_C is the reduced moment of resistance forces, N·m; ω is the angular velocity of the link (eccentric shaft), rad/s.

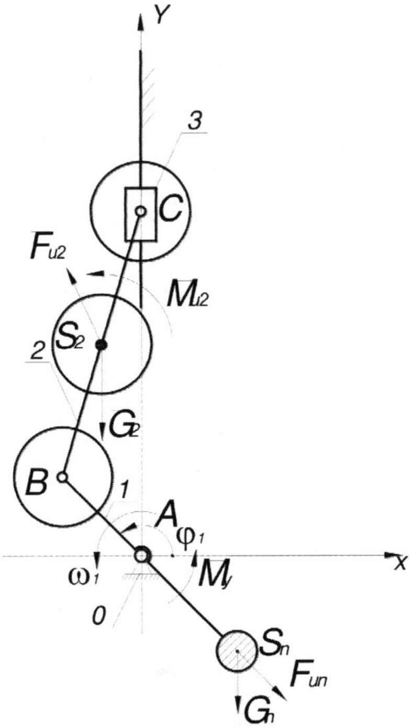

Figure 1. Design scheme of the lever mechanism without regard to the action of the grinding load

The magnitude of the given moment of M_C can be represented as follows

$$M_C(\varphi) = \frac{\sum_1^n P_i}{\omega}, \tag{3}$$

where $\sum_1^n P_i$ can be found from expression

$$\sum_1^n P_i = \sum_1^n F_i v_i \cos\alpha_i + \sum_1^n M_i \omega_i, \tag{4}$$

where F_i is force applied to the link i, H; M_i is the moment applied to the link i, N • m; v_i is the velocity of the point of application of force Fi, m / s; ω_i is the angular velocity of the link i, rad / s; α_i is the angle between the force and velocity vectors, degrees.

As a result, the expression for determining the reduced moment of force resistance M_C for various positions of the mechanism depending on the angle of rotation of the eccentric shaft was obtained. The origin of the angle is the positive direction of the X-axis (Figure 1). The mass of the sliders (link 3) is not taken into account since it is small compared to the masses of the eccentric shaft and the movable frame with grinding chambers fixed on it (link 2).

$$M_{Ci} = (G_1 \cdot V_{S1i} \cdot \cos\alpha_{1i} + G_2 \cdot V_{S2i} \cdot \cos\alpha_{2i} + F_{И2i} \cdot V_{S2i} \cdot \cos\alpha_{3i} + M_{E2i} \cdot \omega_{2i} + \\ +2 \cdot G_n \cdot V_{Sп} \cdot \cos\alpha_{4i} + 2 \cdot F_{En} \cdot V_{Sn} \cdot \cos\alpha_5 \;) / \omega, \quad (5)$$

where M_{Ci} is the reduced moment of resistance forces for the ith position of the mechanism, Nm; G_1 is gravity of the eccentric shaft, N; G_1 is gravity of the movable frame with grinding chambers fixed on it, N; $\boldsymbol{F_{E2i}}$ is the inertia force of the grinding unit for the ith position of the mechanism, N; $\boldsymbol{M_{E2i}}$ is moment of inertia forces of the grinding unit for the ith position of the mechanism, Nm; $G_{\text{п}}$ is gravity of the counterweight, N; $F_{\text{п}}$ is the inertia force of the counterweight, N; V_{S1i} is velocity of the center of mass of the eccentric shaft, m / s; $V_{S\text{п}}$ is speed of the center of mass of the grinding unit for the ith position of the mechanism, m / s; $V_{S\text{п}}$ is the speed of the center of mass of the counterweight, m / s; ω is the angular velocity of the eccentric shaft, rad / s; α_{2i} is the angle between the vectors $\overline{G}_2$ and $\overline{V}_{S2i}$, degrees; α_{3i} is the angle between the vectors $\overline{F}_{E2i}$ and $\overline{V}_{S2i}$, degrees; α_{4i} is the angle between the vectors $\overline{G}_n$ and $\overline{V}_{Sn}$, deg; α_5 is the angle between the vectors $\overline{F}_{En}$ and $\overline{V}_{Sn}$, degrees.

For the experimental model CGU with an eccentricity value of e = 0.02 m, the maximum possible value of the moment M_C = 19.55 Nm was determined.

According to the results of the study of the movement of grinding bodies in the CGU chambers a design diagram which is shown in Figure 2 was made.

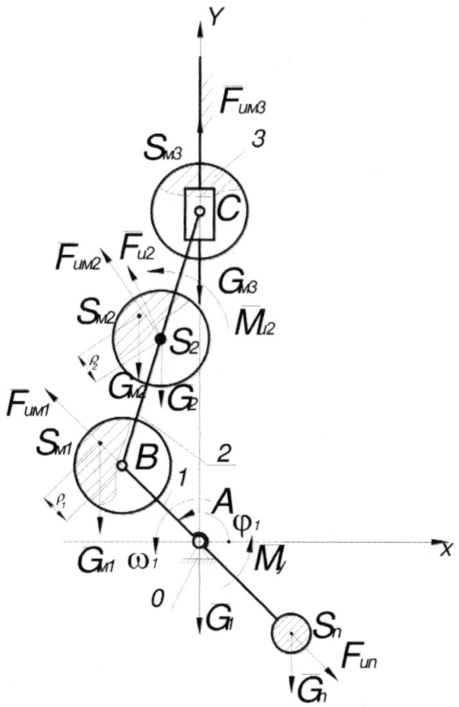

Figire 2. Design diagram of the lever mechanism, taking into account the action of the grinding load

For this case, the expression (3) has the form:

$$M_{Ci} = (G_1 \cdot V_{S1i} \cdot \cos\alpha_{1i} + G_2 \cdot V_{S2i} \cdot \cos\alpha_{2i} + F_{И2i} \cdot V_{S2i} \cdot \cos\alpha_{3i} + M_{И2i} \cdot \omega_{2i} + \\ + 2 \cdot G_n \cdot V_{Sn} \cdot \cos\alpha_{4i} + 2 \cdot F_{Иn} \cdot V_{Sn} \cdot \cos\alpha_5 \; + G_{M1} \cdot V_{SM1} \cdot \cos\alpha_{6i} + \\ + F_{ИM1} \cdot V_{SM1} \cdot \cos\alpha_{7i} + G_{M2} \cdot V_{SM2} \cdot \cos\alpha_{8i} + F_{ИM2} \cdot V_{SM2} \cdot \cos\alpha_{9i} + \\ + G_{M3} \cdot V_{SM3i} \cdot \cos\alpha_{10i} + F_{\Sigma И3i} \cdot V_{SM3i} \cdot \cos\alpha_{11i}) / \omega_1, \quad (6)$$

where G_{M1} is the gravity of the grinding load of the lower chamber, N; F_{IM1} is force of inertia of the grinding load of the lower chamber, N; G_{M2} is gravity of the grinding load of the middle chamber, N; F_{IM2} is the inertia force of the grinding load of the middle chamber, N; G_{M3} is gravity of grinding load of the upper chamber, N; F_{IM3} is the inertia force of the grinding load of the upper chamber, N; α_{6i} is the angle between the vectors $\overline{G}_{M1}$ and $\overline{V}_{SM1}$, degrees; α_{7i} is the angle between the vectors $\overline{F}_{IM1}$ and $\overline{V}_{SM1}$, degrees; α_{8i} is the angle between the vectors $\overline{G}_{M2}$ and $\overline{V}_{SM2}$, degrees; α_{9i} is the angle between the vectors $\overline{F}_{IM2}$ and $\overline{V}_{SM2}$, degrees; α_{10i} is the angle between the vectors $\overline{G}_{M3}$ and $\overline{V}_{SM3}$, degrees; α_{11i} is the angle between vectors $\overline{F}_{\Sigma E3}$ and $\overline{V}_{SM3}$, degrees.

The maximum possible value of the moment is MS = 28.90 Nm. The increase in the reduced moment of resistance forces by 47.8% indicates a significant effect of the grinding load on the dynamic properties of the mechanical system.

As a result, the need to create two balancing systems for units of this type becomes obvious. One system, the main one, balances directly the lever mechanism; the second, additional, balances the action of the grinding load in the grinding unit. Moreover, the additional system should be able to automatically control depending on changes in the mass and location of the center of mass of the grinding unit.

3. Design of the modular balancing device

Based on the task a modular balancing device which provides additional balancing of the mechanism and reduction of the vibration level during the operation of the unit [9, 10] was developed. The drive diagrams of the central control unit without an additional device and with a modular balancing device are shown on Figure 3.

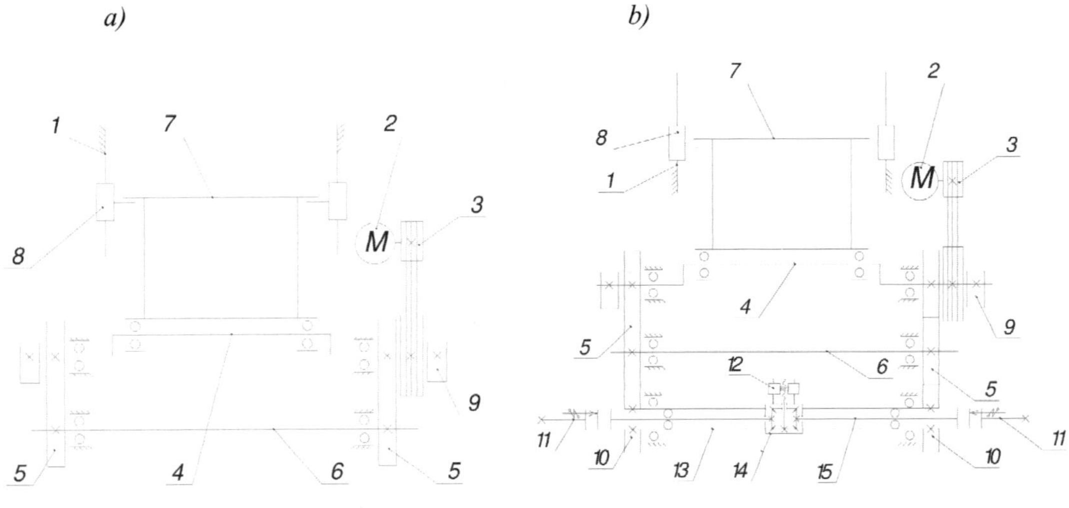

Figure 3. Driving diagram of CGU: a – without modular balancing device; b – with modular balancing device

The parts of the unit are mounted on the frame 1. The rotation from the electric motor 2 is transmitted through the V-belt transmission 3 to the eccentric shaft 4. To prevent elastic deformations of the eccentric shaft due to its considerable length, gears 5 and intermediate shaft 6 are used. The frame 7 is a connecting rod in a crank-slider mechanism; it is pivotally connected with the eccentric shaft 4 and the sliders 8. On the eccentric shaft 4 there are counterweights 9, balancing the lever mechanism. The modular balancing device which is a differential gear mechanism contains gears 10 for transmitting torque from the eccentric shaft 4 to an additional hollow shaft 13, which is connected to the carrier 14 of the differential mechanism. Inside the shaft 13 there are axle shafts 15 connected to the bevel gears of the half-axes of the differential mechanism and electromagnetic clutches 11. The counterweight 12 through the spindle and the screw-and-gear transmission is connected to the satellite of the differential mechanism and performs translational movement along the guides mounted on the carrier in one or the other side when changing the rotational speed of one of the semi-axes. Electromagnetic couplings are controlled by the control system, which includes a programmable controller, a counterweight position sensor and a power supply.

Modular balancing device works as follows. When the permissible level of vibration of the unit rack (set point) is set, the clutches are disconnected and the satellites are stationary relative to the cogwheels of the semi-axles 15. The semi-axles 15 and the intermediate shaft 13 rotate at the same angular velocity. When the vibration level changes, the control system includes one of the couplings, the angular velocity of the semi-axis decreases, and as a result, the relative rotation of the satellite of the differential mechanism and the associated lead screw are obtained. The counterweight 12 moves along the axis of the spindle until the vibration level reaches the set value. It should be noted that the mass of the counterweight 12 is substantially less than the mass of the main counterbalances 9, because calculated on the basis of the nominal mass of the grinding load in the working chambers.

The control circuit of the modular balancing device is shown on Figure 4.

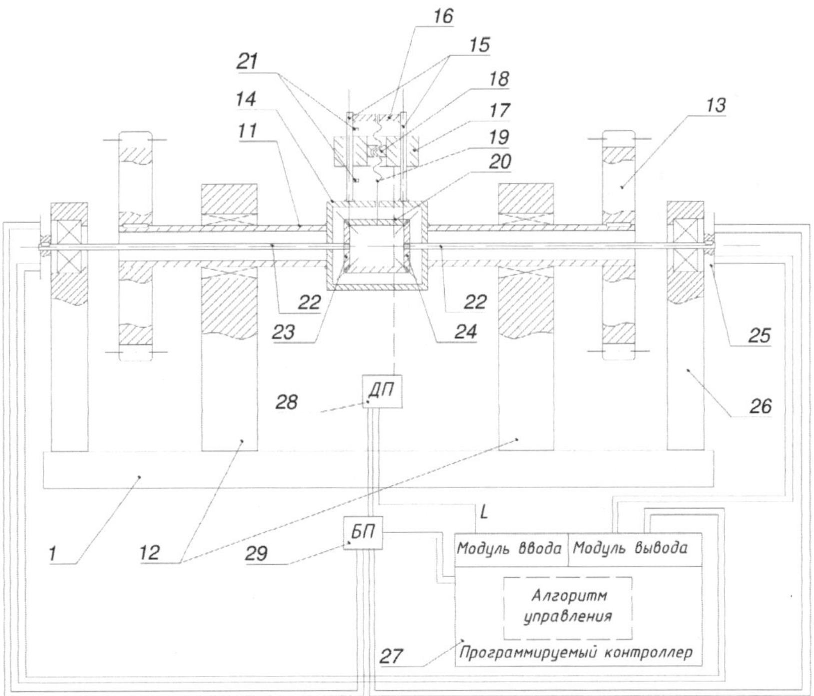

Figure 4. Modular balancing device: 1– frame; 11– shaft; 12, 26 – supports; 13 – gears; 14 – carrier; 15 – guides; 16 – transverse rod; 17 – counterweight; 18 – "screw-nut" transmission; 19 – spindle; 20 – satellite; 21 – stroke limiters; 22 – semi-axles; 23, 24 – gears; 25 – electromagnetic clutches; 27 – programmable controller; 28 – counterweight position sensor; 29 – power supply

4. Conclusion

Thus, the developed system of automatic vibration reduction, containing a modular balancing device, can maintain the specified vibration amplitude or change it periodically depending on the process conditions without stopping the unit.

5. Acknowledgments
The article was prepared within development program of the Flagship Regional University on the basis of Belgorod State Technological University named after V.G. Shukhov, using equipment of High Technology Center at BSTU named after V.G. Shukhov.

References
[1] Gridchin A M, Sevostyanov V S, Lesovik V S, Uralskii V I, Uralskii A V and Sinitsa E V 2005 *The grinding-mixing plant* Pat. 2277973 Russian Federation, B 02 C 17/08, applicant and patent owner of LLC "TC RECYCLE",no 2005118705/03, filed. 24.06.05; publ. 06.20.06, Bul. no 17, 8 p

[2] Gridchin A M, Sevostyanov V S, Lesovik V S, Uralskii V I, Uralskii A V and Sinitsa E V 2008 *The grinding-mixing plant* Pat. 2381837 Russian Federation, B02C17/08, applicant and patent holder of the BSTU V G Shukhova, LLC "TC RECYCLE", no 2008109444/03, filed. 11.03.08; publ. 20.02.10, Bul. no 5, 11 p

[3]. Sinitsa E V, Uralsky A V and Pletnev A V 2007 Influence of grinding loading on the dynamics of a centrifugal grinding / mixing unit *Sci. res., nanosystems and resource-saving technologies in the construction industry: coll. reports of the Int. Sci. and Pract. Conf.* (Belgorod: Publishing house BSTU V G Shukhova) pp 188–92

[4]. Sevostyanov V S, Uralsky V I, Sinitsa E V and Uralsky A V 2008 Questions of dynamic research of a centrifugal grinding-mixing plant *Vibration machines and technologies: Collection of sci. tr.* rare: Yatsun S F (editorial editors) (Kursk: Kursk gos. tehn. univ-t) pp 596–601

[5] Levitte M E 1986 *Balancing parts and assemblies* (Moscow: Engineering) 248 p

[6] Dorf R 2002 Modern control systems (Moscow: Laboratory of Basic Knowledge) 832 p

[7] Gusarov A A 2004 Balancing of rotors of machines Kn 1 (Moscow: Science) 267 p

[8] Gusaro A A 2002 Direct acting autobalancing devices (Moscow) 119 p

[9] Glagolev S N, Rubanov V G, Sevostyanov V S, Uralsky V I, Stativko A A, Stativko S A and Bushuev D A 2012 *Grinding and mixing unit with automatic balancing* Pat. 2494813 Russian Federation, B02C17/14, applicant and patent holder BSTU named after V G Shukhov, no 2012117656/13; declare 04/27/2012; publ. 10.10.13, Bul. no 28, 10 p

[10] Rubanov V G, Sevostyanov V S, Uralsky V I, Stativko A A, Bushuev D A and Stativko S A 2011 *Grinding and mixing unit with automatic balancing* Pat. 114875 Russian Federation, B02C17/00, applicant and patent holder BSTU named after V G Shukhov, no 2011144546/13; declare 11/02/2011; publ. 04/20/12, Bul. no 11, 10 p

Simple universal Kelvin equation valid in critical point vicinity, external-internal state correction, and their application in understanding of normal pentane capillary evaporation and condensation in mesoporous silica MCM-41

A A Valeev[1], E V Morozova[1,2]
[1] Kamyshin Technological Institute (branch) of Volgograd State Technical University, 6a, Lenina Street, Kamyshin, Volgograd Region, 403874, Russia
[2] Saint Petersburg Electrotechnical University "LETI", 5, Professora Popova Street, Saint Petersburg, 197376, Russia

E-mail: antonvaleev@gmail.com

Abstract. A new simple universal form of the Kelvin equation, which can be used near the gas-liquid phase transition critical point, and the correction of the pressure and density for gas phase fluid outside the porous medium are used for the normal pentane meniscus effective curvature radius calculation at the phase equilibrium in mesoporous silica MCM-41, on the basis of the capillary evaporation and condensation experimental data. Normal pentane is suggested as a promising fluid for characterizing the porosity that can be used even at room temperature.

1. Introduction

Porous media open the new opportunities for science and technology [1-3]. Considering mesoporous ones, we should take into account that theese are nanoscale materials. Many methods for studying adsorption-desorption and condensation-evaporation processes in mesoporous media have been developed: traditional techniques (see the review paper [4]), CARS spectroscopy [5,6], NMR study [7], and positron annihilation spectroscopy [8]. The works [9,10] will be used further on in this paper for having the data for n-pentane.

There is a bit of ambiguity in characterizing the situation at the moments of condensation and evaporation. Our treatment is based on the assumption that some of the processes happen in the state of the thermodynamic equilibrium and with hemispherical meniscus geometry. For them we estimate effective meniscus curvature radii on the basis of the experimentally known evaporation and condensation pressures [9,10].

The model that has already been tested with 4 substances (CO_2 (Ref. [3] in [12]), N_2 [12], Ar (Ref. [16] in [13]), and O_2 [13]) will be used for normal pentane in this work. MCM-41 is a controlled pore glass, which means the pores can be treated as almost the same size cylinders. The model consists of the 2 stages: the new Kelvin equation (Refs. [2,3] in [12]) and the state transformation of the gas phase from the center of the meniscus hemisphere to the bulk [12]. Joining the two model stages allows [12] one getting the gas-phase pressure outside the porous medium as a function of the meniscus effective curvature radius at the phase equilibrium inside the pores. Vice versa, one can find the radius as a function of the pressure in bulk. The latter is available in the works [9,10] for several temperatures.

The results produced in this way will be presented in the next section. The works [10,11] also have the results for the n-pentane adsorption in a sample with the pore radius of 1.45 nm. At the same time the authors of [11] conclude (and we agree with them) that the adsorption took place in a supercritical state where our model is not applicable.

2. Results

Normal pentane is a nonpolar fluid. Thus its Tolman’s length [14] should be smaller than the molecule average diameter in the states distant enough from the critical point, to which the scope of this work is limited.

PC-SAFT EOS used in the work [15] behaves in some sense correctly in the instability region and therefore is suitable for handling the ‘light phase’. In the present work, the treatment is limited to stability and metastability regions. Thus the choice of the EOS can be extended to high-accuracy empirical ones. The equation of state of [16] and the phase equilibrium equations of [17] will be chosen further on in this paper.

If we take some spherical meniscus effective curvature radius then using the integral equations (9), (13) in (Ref. [3] in [12]), the equations of state and the parachor equation [18] for pure fluids, the phase equilibriums can be calculated. Pressure outside the pores can be calculated using equations (6-9) of [12]. Fitting it to the measured one [9,10], we can find the radius.

The value 230.93 $\mathrm{dn}^{1/4}\mathrm{cm}^{11/4}/\mathrm{mol}$ of [15] will be used for the n-pentane parachor. For the Lennard-Jones potential, the authors will use the parameters [19] $\varepsilon_{FF} = 309.75$ K$\cdot k_B$ and $\sigma_{FF} = 0.6709$ nm.

Because it is not quite firmly established [15,20-28] yet, which (if either) of these processes (evaporation and condensation) takes place in the state of the thermodynamic equilibrium, the method used in this work is applied formally to the both of them, giving the possible meniscus radii at the moments of capillary evaporation and condensation (see figure 1). For the condensation branch of the $r = 2.285$ nm sample [9] open symbols are used. The same is for the temperatures, at which the hysteresis is not observed. Solid symbols show the results for the evaporation branch. Pentagons with horizontal lines show the results for the $r = 2.095$ nm sample [10]. For the single available [10] 297.95 K isotherm of the $r = 4.04$ nm sample we have the effective curvature radius of 3.51 nm, not shown in the figure 1. The model these results are obtained with, does not take the pore size into account. At the same time, as soon as we know the curvature radius and the average pore radius, we are able to calculate the λ-parameter [15], which is shown in figure 2 by pentagons. The shape of the symbols presented for n-pentane is the same as in figure 1. The λ-parameter of $r = 4.04$ nm sample is shown by a pentagon with a vertical line. Analogous results of [15] are shown by semisolid symbols. In figure 2 hexagons show the results for O_2 of [13] (closed symbols are used for the evaporation branch, otherwise are used open symbols, semisolid symbols show the results of [15] for comparison), rhombi show the similar way the results for Ar of (Ref. [16] in [13]) and [15], rounds show the results for N_2 of [12] and [15]. Triangles show the results of the works (Ref. [3] in [12]) and [15]. Open triangles show the results published in (Ref. [3] in [12]), where the capillary condensation pressures were taken indirectly from figure 3 of [29]. Closed triangles show the results (Ref. [16] in [13]) by the improved model (Ref. [4] in [12], [12]), where the corrected equations (7), (8) of [12] were used, and also the capillary condensation pressure data were taken from figure 4 of [29]. Semisolid triangles show the results of [15].

3. Discussion

One may notice that the results given by the present model for n-pentane are sinificantly different from these for the other 4 substances. In [15] this situation is not observed. Thus, we should look at it in detail. The present model requires [12] that the adsorbed layer is sufficiently thick, which is not valid now. This can be mended by adding an extra (accounting for the cylindrical pore wall attraction caused mainly by its oxygen atoms) term $\mu_{f,S}$ to μ_f [12]:

$$\mu_f = \mu_{f,S} + \frac{\mu_{f,sph} + \mu_{f,cyl}}{2}, \tag{1}$$

where

$$\mu_{f,S} = -\alpha \frac{3\pi}{2(R + \sigma_{SF}/2)^3}, \tag{2}$$

$$\alpha = (2\pi/3) N_A^2 (n_S \varepsilon_{SF} \sigma_{SF}^6 - n_L \varepsilon_{FF} \sigma_{FF}^6). \tag{3}$$

Here N_A is the Avogadro number, n_S, n_L are the silica bridging oxygen atoms and liquid molecules molar concentrations, respectively; ε_{SF} and σ_{SF} are the bridging-oxygen-fluid, ε_{FF} and σ_{FF} are the fluid-fluid Lennard-Jones potential parameters, respectively, $R + \sigma_{SF}/2$ is used instead of the pore radius R to account for the solid atom center shift from the pore surface. We don't take into account the attraction of non-bridging oxygen atoms of silica because their actual amount is not well known.

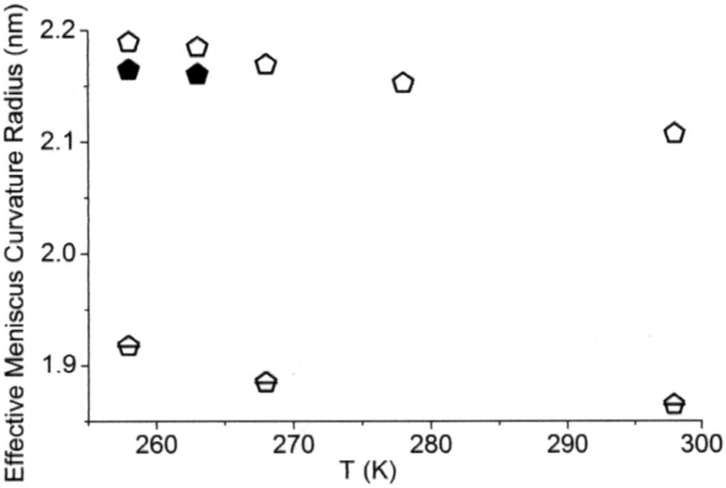

Figure 1. Spherical meniscus effective curvature radius at the moment of normal pentane capillary evaporation and condensation in MCM-41 as a function of the temperature.

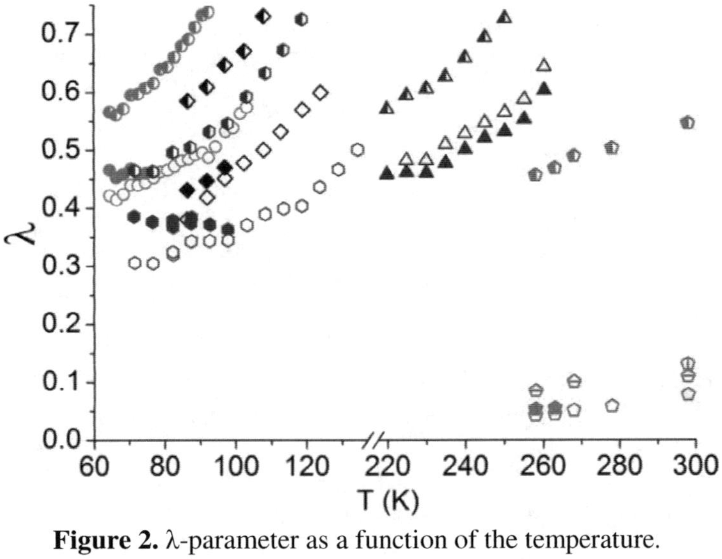

Figure 2. λ-parameter as a function of the temperature.

We may adopt the Lennard-Jones parameters of the bridging oxygen used in many molecular dynamics simulations [30,31,32]: $\varepsilon_{SS} = 230$ K$\cdot k_B$, $\sigma_{SS} = 0.27$ nm. Upon application of the Lorentz-Berthelot combining rules we obtain $\varepsilon_{SF} = 266.9129$ K$\cdot k_B$, $\sigma_{SF} = 0.47045$ nm.

If we look at two terms in the brackets in (3), we can see that the first one responsible for interaction of the gas phase fluid with the silica is $2.93\cdot10^{-72}$ J$\cdot$mol$\cdot$m^3 and the second one responsible for the interaction of the gas with the liqud phase fluid varies from $3.25\cdot10^{-72}$ J$\cdot$mol$\cdot$m^3 (for the lowest density achieved at 298 K) to $3.47\cdot10^{-72}$ J$\cdot$mol$\cdot$m^3 (for the highest density achieved at 258 K). The difference between the two terms in (3) is much less than their values. The interaction with the silica oxygen atoms appears to be weaker than the fluid-fluid one. At the same time, as far as we are not accounting for the stronger interaction with the non-bridging oxygen atoms concentrated at the pore surface, we cannot say with much certainty, which of the interactions is actually stronger in the $\mu_{f,S}$ term. We can only deduce that $\mu_{f,S}$ value should be relatively small. Therefore neglecting it in the present work results is justified. On the other hand, the weak pore walls attraction makes multiple layer adsorption at the phase equilibrium (with the hemispherical meniscus) questionable. Hence the curvature radius should be approximately equal to the pore radius, this fact making n-pentane an alternative to nitrogen for assessing the average pore radius even at room temperature.

4. Summary

On the basis of the available capillary evaporation and condensation data [9,10], the spherical meniscus effective curvature radius at the moment of the normal pentane possible thermodynamic equilibrium capillary evaporation/condensation in MCM-41 and the λ-parameter have been calculated for several temperatures. The λ values we have obtained for n-pentane are significantly closer to zero than those we had got for other fluids, which is not observed in [15]. The results of this work are justified by the fact that the attraction of the fluid molecule by the silica wall is of almost the same magnitude as the attraction by the adsorbed or liquid-phase fluid. Normal pentane is suggested as a prospective medium for characterizing the porosity.

References

[1] Kholodova S and Peregudin S 2015 *AIP Conference Proceedings* **1648** 450011
[2] Peregudin S and Kholodova S 2015 *International Conference on Mechanics - Seventh Polyakhov's Reading* 7106766
[3] Syrkov A G and Smirnov V M 1987 *Kinetics and Catalysis* **28** 965–969
[4] Barsotti E, Tan S P, Saraji S, Piri M and Chen J-H 2016 *Fuel* **184** 344–361
[5] Arakcheev V G, Bekin A N and Morozov V B 2017 *Laser Phys.* **27** 115701
[6] Arakcheev V G, Bekin A N and Morozov V B 2018 *Journal of Raman Spectroscopy* **49** 1945–52
[7] Omi H, Nagasaka B, Miyakubo K, Ueda T and Eguchi T 2004 *Phys. Chem. Chem. Phys.* **6** 1299–1303
[8] Alam M A, Clarke A P and Duffy J A 2000 *Langmuir* **16** 7551–3
[9] Russo P A, Ribeiro Carrott M M L and Carrott P J M 2012 *Micropor. Mesopor. Mat.* **151** 223–230
[10] Barsotti E, Saraji S, Tan S P and Piri M 2018 *Langmuir* **34** 1967−80
[11] Barsotti E, Tan S P, Piri M and Chen J-H, Phenomenological Study of Confined Criticality: Insights from the Capillary Condensation of Propane, n−Butane, and n−Pentane in Nanopores: to be published
[12] Valeev A A and Morozova E V 2018 *Solid State Phenomena* **284** 801–806
[13] Valeev A A and Morozova E V, Simple universal Kelvin equation valid in critical point vicinity, external-internal state correction, and their application in understanding of oxygen capillary evaporation and condensation in mesoporous silica MCM-41: to be published

[14] Tolman R C 1949 *J. Chem. Phys.* **17** 333–337
[15] Tan S P and Piri M 2015 *Fluid Phase Equilibr.* **393** 48–63
[16] Span R and Wagner W 2003 *Int. J. Thermophys.* **24** 41–109
[17] Kratzke H, Műller S, Bohn M and Kohlen R 1985 *J. Chem. Thermodynamics* **17** 283–294
[18] Poling B E, Prausnitz J M and O'Connell J P 2001 *The Properties of Gases and Liquids* (New York: McGraw-Hill)
[19] Cuadros F, Cachadiña I and Ahumada W 1996 *Molecular Engineering* **6** 319–325
[20] Morishige K and Ito M 2002 *J. Chem. Phys.* **117** 8036–41
[21] Evans R, Marini Bettolo Marconi U and Tarazona P 1986 *J. Chem. Phys.* **84** 2376–99
[22] Evans R, Marini Bettolo Marconi U and Tarazona P 1986 *J. Chem. Soc. Faraday Trans.* **82** 1763–87
[23] Ravikovitch P I, Domhnail S C Ó, Neimark A V, Schüth F and Unger K K 1995 *Langmuir* **11** 4765–72
[24] Neimark A V, Ravikovitch P I and Vishnyakov A 2000 *Phys. Rev. E* **62** R1493–96
[25] Vishnyakov A and Neimark A V 2001 *J. Phys. Chem. B* **105** 7009–20
[26] Heffelfinger G S, van Swol F and Gubbins K E 1988 *J. Chem. Phys.* **89** 5202–05
[27] Marini Bettolo Marconi U and van Swol F 1989 *Phys. Rev. A* **39** 4109–16
[28] Papandopulu A, van Swol F and Marini Bettolo Marconi U 1992 *J. Chem. Phys.* **97** 6942–52
[29] Morishige K and Nakamura Y 2004 *Langmuir* **20** 4503–6
[30] Brodka A and Zerda T W 1991 *J. Chem. Phys.* **95** 3710–18
[31] Yoshioka T, Tsuri T and Asaeda M 2004 *Molecular Physics* **102** 191–202
[32] Yang X and Yue X 2007 *Colloids and Surfaces A: Physicochem. Eng. Aspects* **301** 166–173

Journal of Physics: Conference Series **1384** (2019) 012065 doi:10.1088/1742-6596/1384/1/012065

Implementation of fuzzy logic in the smelting process of control algorithms of copper-nickel sulfide materials

N V Vasilyeva, P V Ivanov

Saint-Petersburg Mining University, Vasilyevsky Island 21st line, 2, Saint-Petersburg, 199106, Russia

E-mail: Vasileva_nv@pers.spmi.ru

Abstract. The analysis of control quality of technological process of melting of copper-Nickel sulfide materials is carried out. The structure of the automatic melting control system is proposed, information and control channels are described. The algorithm of process control using fuzzy logic, which allows stabilizing the copper content in the matte within the specified limits.

1. Introduction

It is quite difficult to imagine the functioning of any technological facility without an automated control system today. In a market competitive environment, enterprises are forced not only to ensure the safety and sustainability of technological processes, but also to constantly improve their economic efficiency. Certainly, the most obvious method of improving the efficiency of technological processes is the improvement of technological schemes, hardware design of a technology and modes of technological processes. However, under this approach, only a part of the savings reserves can be benefited. The most significant effect can be obtained by improving automated process control systems by integrating the intellectual component of the mathematical apparatus that operates on the basis of fuzzy logic algorithms, neural networks, etc. into the control system structure [1-3]. To obtain a flowchart of the process control algorithm, it is necessary to carry out a full analysis of a technological process, determine the place of the mathematical apparatus in the structure of the process control system and set limitations imposed on the basic parameters of the process.

2. Controlled object analysis in the system

The solution of the problem in question is observed on the example of processing the operational control data set of the actual process of smelting copper-nickel sulfide raw materials in the Vanukov furnace (Vanukov process).

The Vanukov process in copper metallurgy is used-on both in terms technology and hardware [4, 14] allowing processing raw materials with high productivity and has wide possibilities in terms of control of the technological mode of smelting.

The processed furnace charge (ore, concentrate) is fed into the furnace from the top using the charging device without any preliminary preparation (fine grinding, deep drying, etc.). Once on the surface of the bath, the furnace charge is moved deeper into the smelt, vigorously mixed with it and is smelted under the influence of high temperatures. There is an oxygen-air mixture (OAM) used as an oxidizer in the furnace, depending on the composition of the raw material. The blast (OAM) is fed into the smelt through special tuyeres located on both sides of the bath in the side walls of the furnace. Liquid smelting products are divided into matte (sulphide smelt) and slag (oxide smelt), which, as they accumulate, are removed from the unit from the end sides of the furnace.

The main parameters for controlling the smelting process are the following: the total consumption of the furnace charge, the total consumption of the blast (oxygen-air mixture – OAM), the consumption of technical oxygen and the oxygen content in the OAM. These variables virtually completely control the Vanukov process and determine the copper content in the matte.

Smelting control is currently carried out by technical personnel based on their own experience and a subjective analysis of the indications of instrumentation, data obtained from visual observations, results of chemical analyzes received with delay and other information the personnel have on the state of individual components of the technological process, as well as on the basis of preliminary calculations of material and heat balances [5,6].

The purpose of the smelting process control of copper-nickel sulphide raw materials is to obtain a matte of a given composition, since the quality of the matte (copper content in the matte) received for further conversion affects the technical and economic indicators of the further stages of copper production: blast consumption, purge duration, the amount fluxes used and the resulting converter slag, thermal mode of the process, etc.

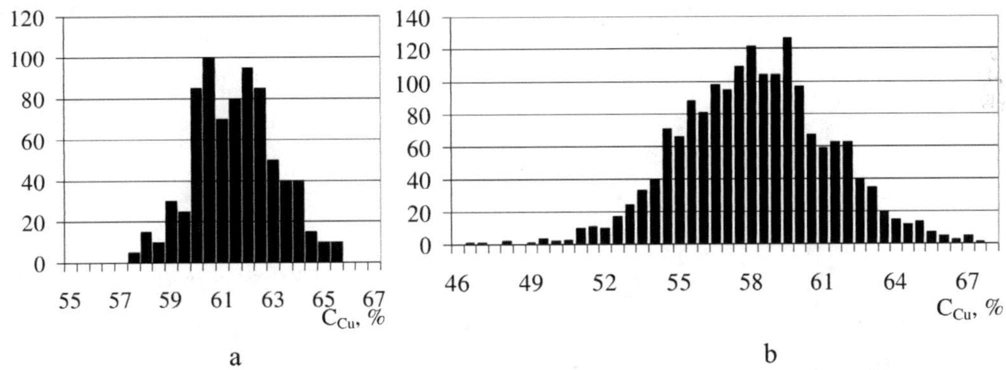

Figure 1. Histogram of the distribution of the copper content in the matte:
a) dates of 2016 year, b) dates of 2018 year

The composition of the matte in terms of copper content therein is regulated by changing the ratio of the oxygen in the blast and the amount of the furnace charge with a constant flow rate of natural gas.

The analysis of the operational control data of the Vanukov process for 2016 and 2018 showed (Figure 1) that the copper content in matte (C_{Cu}) varies from 50 to 68%. Therefore, there is a need to stabilize the copper content in the matte.

The state of the furnace is constantly changing, so when controlling the process, it is necessary to avoid imbalances inside the furnace between the charged materials and the pressure of the exhaust gases. Matte quality improvement (stabilization of the copper content in the matte within the specified limits) is possible by fairly tight coupling of the mass fluxes charged and blast modes by introducing an automated control system that includes a quantification model of the copper content in the matte in the control loop allowing reducing the influence of “human factor" to a minimum. In this regard, there was an automatic control system developed using fuzzy logic algorithms. The implementation of fuzzy logic algorithms involves the use of a computer control similar to that performed by a qualified operator by presenting his work methods with the use of control rules as a model [7-9]. The control rules relate the assessment of the controlled object state with the sequence of operations using the statements “IF ... THEN”, divide the space of input variables into areas and indicate the sequence of operations in each local area [12, 13].

3. Block diagram of the control system

To improve the quality of the smelting process control [10,11], a block diagram has been developed for an automated system for controlling the copper content in matte, shown in Fig. 2, which includes the Vanukov furnace 1, equipped with charge funnels 2. There are tanks 3 installed in the Vanukov furnace charge system to feed bulk materials (copper concentrate, ore, flux, coal and recycled materials). Under the tanks 3 there are belt feeders 4 installed to dose charge materials loaded into the furnace. Charge materials are fed from the feeders 4 to the assembly conveyors 5, with the help of which they are fed through the charge funnels 2 into the reaction zone of the Vanukov furnace.

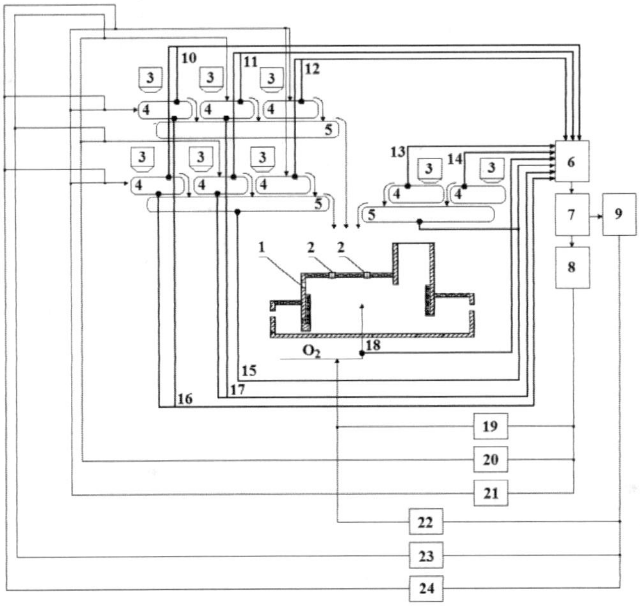

Figure 2. Block diagram of the control system.

Measuring channels 10-18 are designed to receive information about the instantaneous values of the relevant parameters (flow rate and loading rate of charge materials and blast flow rate) and have direct access to the information collection and preprocessing block 6. The information collection and preprocessing block 6 is connected to the switching unit 7, either activating the automatic process control mode using the control unit 8 according to the algorithm or deactivating it, and transfers all the information to the automated workplace (AWP) 9 of the operator. The control unit 8 according to the algorithm is connected with devices 19 for generating a control action on the flow rate of technical oxygen, 20 and 21 for generating a control action on the flow rate and loading speed of charge materials. Using the AWP 9 the operator also has the ability to influence the flow rate and loading rate of charge materials and technical oxygen through direct manual control of control output devices through the direct task setups 22 for technical oxygen consumption, 23 and 24 flow rate and load rate of charge materials.

Information on the values of the measured current smelting parameters is transferred to the information collection and preprocessing block **6** through the information transfer channels **10-18** in order to calculate the main parameter (total consumption of charge materials (G_s), t/h) and assigning the process to one of the established areas ($\mathbf{G_{-1}}$, $\mathbf{G_0}$, $\mathbf{G_{+1}}$) (Figure. 3).

In addition, the ratio of the flow rate of technical oxygen to blast per ton of charge materials (m^3/t of charge) is additionally determined, and in case of the change gradient over 10% of the prescribed value, the total consumption of charge materials and technical oxygen is adjusted depending on the fact in which area the main parameter is, until reaching the $\mathbf{G_0}$ area.

$\mathbf{G_{-1}}$ area (Figure 3) is characterized by insufficient consumption of technical oxygen per ton of charge materials for the oxidation of sulfides to proceed. In this regard, oxidation reactions do not proceed in full and the matte is obtained with a low copper content. In this case, the control in the automatic mode is stopped, the manual control mode is activated, and the process is carried out with an increase in the consumption of charge materials and the consumption of technical oxygen blast per ton of charge materials (m^3/t of charge) until reaching the $\mathbf{G_0}$ area.

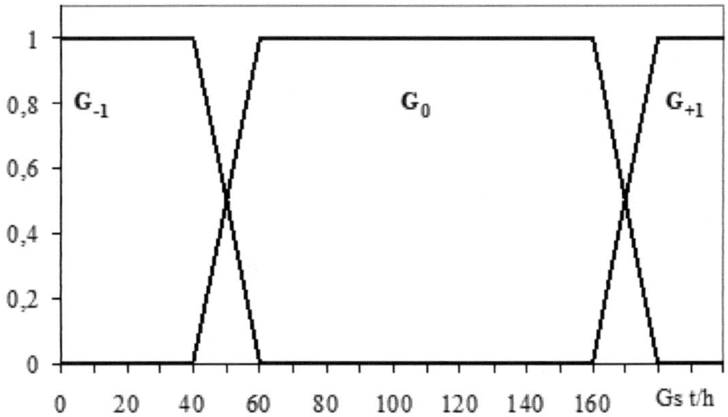

Figure 3. Layout of the main parameter areas – the total consumption of charge materials in the furnace (t/h)

The $\mathbf{G_0}$ area is characterized by the highest stability of all process parameters: hydrodynamic, energy, and physicochemical ones. In this case, the smelt is maintained in the initial state, the process proceeds without sharp spikes and local extrema which indicates the smoothness of the flow of all physicochemical reactions. Thus, the $\mathbf{G_0}$ area (Fig. 3) is effective for obtaining matte with a high copper content and stable composition and it should be sought during the smelting process. In this case, the control is carried out automatically according to the algorithm (Figure 4).

The $\mathbf{G_{+1}}$ area (Figure 3) is characterized by over-oxidation of sulphides due to the high consumption of technical oxygen from the blast per ton of charge materials, as well as an increase in the height of the smelting bath due to the high furnace productivity which can lead to the release of the smelt in the up-take and melt tuyeres. In this case, the control in the automatic mode is stopped, the manual control mode is activated, and the process is carried out with a decrease in the consumption of charge materials and the consumption of technical oxygen blast per ton of charge materials (m^3/t) until reaching the $\mathbf{G_0}$ area.

4. Control algorithm for the determination of copper content in matte

All the information obtained regarding the values of the measured current smelting parameters is displayed on the automated workplace (AWP) 9 of the operator. The operator, guided by the relevant technological regulations and one's own experience in relation to the current production and technological situation, selects one of two modes of Vanukov process control: "automatic" or "manual", selecting the appropriate control mode in the switching unit 7.

When the operator selects the "automatic" control mode, data from information transfer channels 10-18 (and others not considered in this paper) is delivered to the model to quantify the copper content in matte, developed using fuzzy logic methods. Depending on what value of copper content in the matte will be calculated by the model, either the process is carried out without changing the control settings or the recalculation and change of the control settings is made.

When the operator selects the "manual" control mode, data from information transfer channels 10-18 (and others not considered in this paper) is sent to the automated workplace (AWP) 9 of the operator.

The obtained data on the current state of the Vanukov process serves as a basis for the operator to select control settings.

If the tolerance of the system exceeds 10%, the model should be retrained.

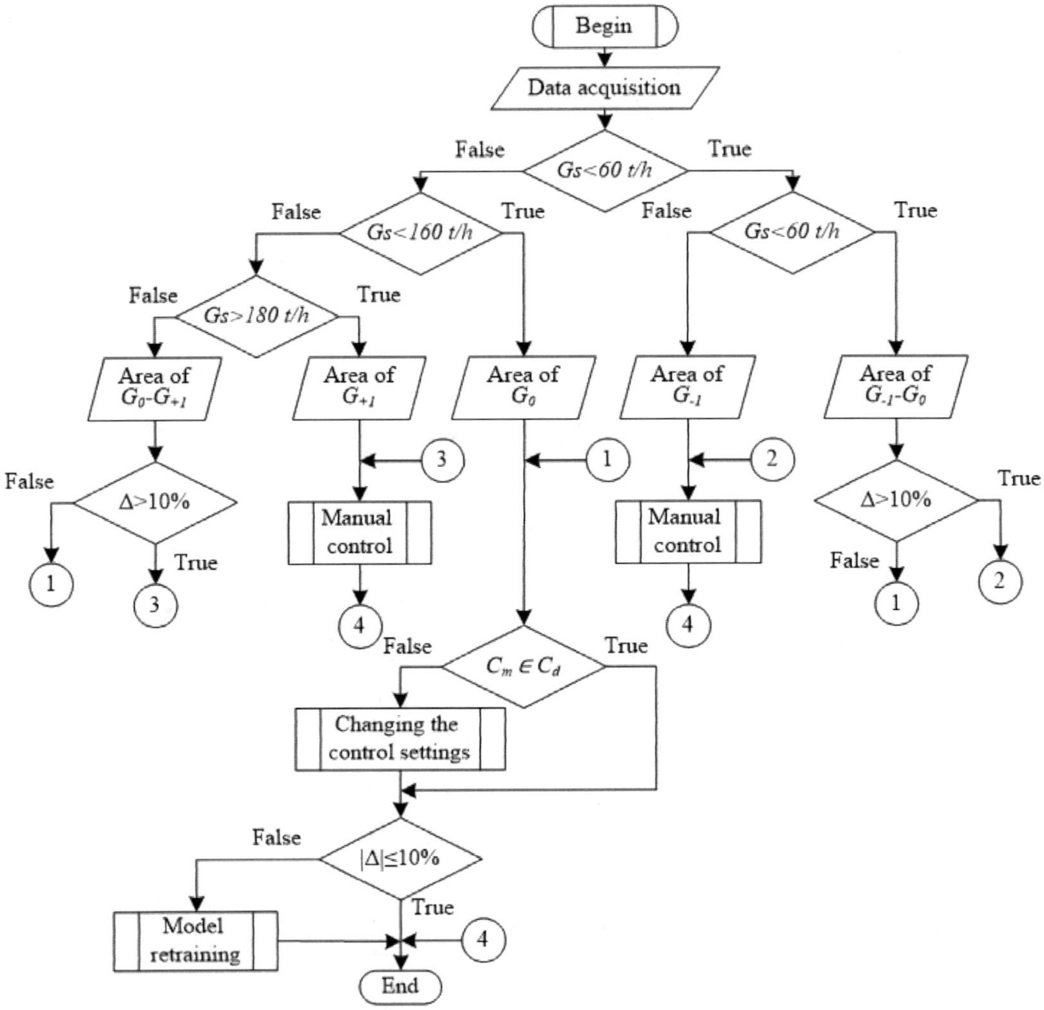

Figure 4. Block diagram of the algorithm for controlling the process of autogenous melting of copper-nickel sulfide raw materials in the Vanukov furnace

5. Testing the performance of the proposed algorithm for controlling the copper content in matte

In order to evaluate the efficiency of the developed algorithm for controlling the process of smelting copper-nickel sulfide raw materials in a Vanukov furnace, a numerical experiment was performed on the model. The experiment was carried out in three stages.

At the first stage, on the basis of the operational control data, the copper content in the matte was calculated using the quantification model for the copper content in the matte and C_m compared with the set (desired) one $C_d = \lfloor 59; 63 \rfloor \%$.

At the second stage, based on the data of operational control and the magnitude of the mismatch calculated on the basis of the model and the specified copper content in the matte, C_d the settings for the magnitude of control actions — consumption of charge materials (t/h) and consumption of technical oxygen per ton of charge materials — were calculated (m^3/t matte).

At the third stage, based on the operational control data, taking into account the change in the consumption of charge materials and the consumption of technical oxygen per ton of charge materials for the same point in time, the copper content in matte was again calculated.

Figure 5 presents the results of numerical simulation.

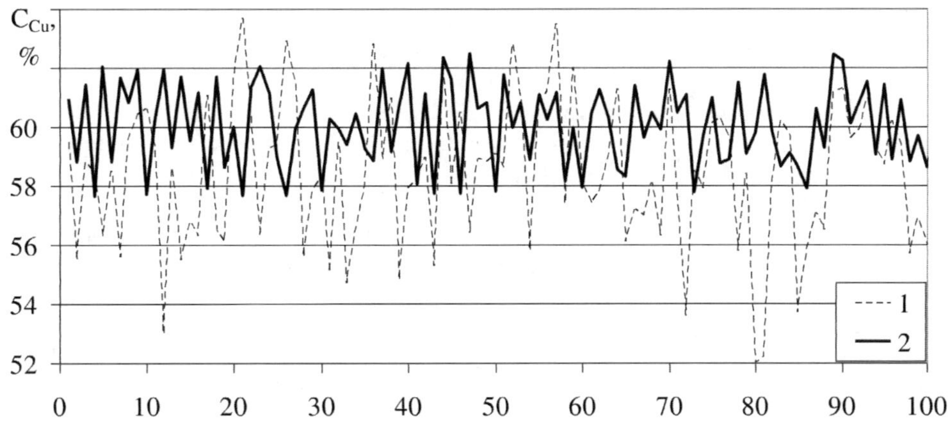

Figure 5. Results of numerical simulation of control of copper content in the matte. 1 – actual value, 2 – simulation result

As it follows from Figure 5, the change in the consumption settings of charge materials and the consumption of technical oxygen per ton of charge in accordance with the developed algorithm led to the stabilization of the copper content in the matte within the specified limits. Thus, the experiment carried out makes it possible to recommend a control system for the determination of the copper content in matte to be used in industrial conditions.

References

[1] Deryabin M Yu, Pisarev A I 2009 Automatic system for control of the process of sulphur extraction from waste gases in autogenous generators, *Russian J. of Non-Ferrous Metals.* **5** 86-89

[2] Gukovskiy Y L, Sychev Y A and Pelenev D N 2017 The automatic correction of selective action of relay protection system against single phase earth faults in electrical networks of mining enterprises *International J. of Applied Engineering Research* **12(5)** 833-838

[3] Il'yaschenko D P, Kievskaya E I, Ling V V, Gaibova T V, Kiyamova A G and Mironova A A 2018 The optimization of the competency model of bachelors of engineering specialties, *International J. of Mechanical Engineering and Technology* **9(9)** 1085-1093

[4] Leonidovich Z Y, Urievich V B 2015 The development and use of diagnostic systems and estimation of residual life in industrial electrical equipment *International J. of Applied Engineering Research*, **10(20)** 41150-41155

[5] Kniss S V, Naboichenko S S and Zhukov V P 2013 Simulation of the Kinetics of the Autoclave Dissolution of Copper and Nickel Sulfides in the Presence of Oxygen *Izvestiya VUZ. Tsvetnaya Metallurgiya* **5** 25–27

[6] Lutskiy D Litvinova T, Ignatovich A and Fialkovskiy I 2018 Complex processing of phosphogypsum - A way of recycling dumps with reception of commodity production of wide application *Journal of Ecological Engineering* **19 (2)** 221-225.

[7] Nanda P K, Sahu P K, Swain S K and Jagannath B K. 2000 Genetic Algorithm based adaptive line enhancement schemes for extraction of power frequency signal *Proceedings of International Conference on Evolutionary computing for computer, communication, control and power, ECCAP-2000 (India)* 91–98

[8] Semenova I N, Kirpichenkov I A 2009 Development of the control system for temperature

conditions of melting process in the Vanyukov furnace *Russian J. of Non-Ferrous Metals* **5** 59-62

[9] Ugolnikov A V, Makarov N V 2019 Application of automation systems for monitoring and energy efficiency accounting indicators of mining enterprises compressor facility operation *J. of Mining Institute* **236** 245-248

[10] Vasilyeva N V and Ivanov P V 2019 Development of a control subsystem to stabilize burden materials charging into a furnace *Journal of Physics: Conference Series* **1210** 012158

[11] Vasilyeva N V, Koteleva N V and Ivanov P V 2018 Quality analysis of technological process control *International J. for Quality Research* **12(1)** 111 – 128

[12] Nedosekin A O, Antonova Ya V 2016 Risk identification and analysis in a mining company using fuzzy logical methods *XIX IEEE International Conference on Soft Computing and Measurements (SCM)* 16160043

[13] Pimonov, A., Raevskaya, E., Sarapulova, T. 2017 Expert evaluation of innovation projects of mining enterprises on the basis of methods of system analysis and fuzzy logics *E3S Web of Conferences* **15** 01021

[14] Konovalov G V, Bazhin V Yu, Nikolaev A K 2014 Improving the blast supply in the conversion of copper-nickel mattes *Metallurgist* **57** 954-957

Non-contact methods and means of measuring the oil film thickness on the water surface

I E Ushakov, A A Vinogradova

Saint-Petersburg Mining University, 2, 21st Line, Saint-Petersburg, 199106, Russia

E-mail: vinogradova_aa@pers.spmi.ru

Abstract. The article discusses options for solving the problem of ocean pollution by oil products. Currently, it is important to measure the thickness of an oil film to obtain information on the volume of spills and operations to eliminate them. The features of non-contact methods and the main characteristics and conditions of the use of means for measuring the thickness of an oil film are discussed. The accuracy of measurements in real conditions (in the presence of surface waves) was identified.

1. Introduction

More than 4 billion tons of crude oil is produced annually. Currently, the number of drilling platforms is more than 6500. More than 3 thousand tankers are used for the transportation of petroleum products. Every year, 2-10 million tons of oil are discharged into the World Ocean (oil losses during transportation and pumping, offshore production, destruction of platforms). The pollution rates are increasing due to the growth in the volume of sea transportation of oil products.

The assessment of damages caused by oil films should be comprehensive, since oil spills have a harmful effect. Upon contact with water, petroleum products spread and form a film up to 1.0-0.1 mm in thickness. Light fractions are eroded, and aromatic hydrocarbons are partially dissolved in water. This entails the spread of oil over a large area, while thinning of the oil film may occur up to the monomolecular layer. Heavy fractions undergo various processes: photochemical oxidation due to the action of ultraviolet radiation, natural self-oxidation, biochemical oxidation by marine microorganisms. Due to continuous mixing and the above processes, an aqueous emulsion is formed on the water surface. They form clots that hold on the surface or move in the water thickness.

Oil products have a negative impact on the ocean flora and fauna, while they change the gas exchange between the ocean and the atmosphere, affect the thermal regime of water basins, upset the oxygen balance of the planet.

To solve the problem of ocean pollution, it is necessary to measure the degree of pollution and remove oil products from the surface and the aquatic environment. The most common methods for cleaning water areas are as follows: burning, collecting with special devices or using special chemical cleaning agents. Almost all existing methods can be used only under the condition that oil pollution exists on the water surface in the form of a film.

Currently, many countries are developing remote sensing oil spill detection methods and means [1, 2]. Oil products form films whose thickness can reach the monomolecular layer. Prompt information about the amount of oil spilled, the source of pollution and the most affected areas will help eliminate pollution.

The environmental monitoring of marine areas should provide the following information:
1. The presence of the film
2. The area and shape of the oil spot
3. Film thickness
4. Type of oil product
5. Age of pollution
6. Direction and speed of the spot

The urgent task is to design tools for measuring the thickness of an oil film on the water surface to assess the volume of spills and plan operations for their liquidation.

The most promising is the use of non-contact methods and measuring instruments. The article discusses the implementation of methods of measuring the thickness of an oil film, and assesses the accuracy of measurements in real conditions.

2. Results and Discussion

Means of measuring the thickness of an oil film, their main characteristics and application conditions are presented in Table 1.

In radiometric measuring instruments, the polarization method has been implemented - the thickness of the oil film is determined by the ratio of reflection coefficients on the vertical and horizontal polarizations.

The PRINT-3 three-frequency radiometric device [3] which implements the polarization method is a registered remote sensing instrument for measuring the thickness of an oil film on the water surface. Its work is based on passive radiometric remote sensing which allows you to measure the intrinsic radiation of the earth surface characterized by the brightness temperature of the radiation. The use of three frequencies eliminates the ambiguity in determining the thickness of a film caused by the interference frequency of the received signals reflected from the media: air-film - film - water. The use of the receiver response makes it possible to identify the source of surface inhomogeneity and calculate the film thickness without using additional information about weather conditions.

The KTS RMK-N device [4] is a portable radiometric meter for measuring the thickness of an oil layer on the water surface. The methodology for determining the thickness of the oil film is based on the dependence of the reflection coefficient of radio waves on mutually orthogonal polarizations (horizontal and vertical) on the oil film thickness.

The PRINT-3 and the KTS RMK-N were tested in order to approve the type of measuring instruments and were included in the State Register of Measuring Instruments.

Oil and oil products fluoresce when excited with light in the ultraviolet range, and the fluorescence spectra of different types of oil products have various intensity degrees and shapes. The multifunctional fluorescent lidar makes it possible to identify oil products by their composition, as well as evaluate the thickness of the oil film [5].

The four-wave and two-wave laser sensor measures the thickness of the oil film in the range (0.1 ... 140) μm [6, 7].

In a laser ultrasonic device, the film thickness is determined by the propagation time of the acoustic wave between the upper and lower boundaries of the oil slick [8]. To implement this method, three lasers are used, one of which is attached to the interferometer to measure the thickness of the oil film.

The multi-frequency radar complex Mars determines the thickness of the oil film by radar contrasts of oil spots obtained using radio waves of various lengths [9]. Radar tools allow you to work in various weather conditions and explore significant water bodies.

Table 1. Main characteristics of oil film thickness measuring tools

Name of the tool	Wave range	Measurement range	Measurement accuracy	Note
PRINT-3 Three-Frequency Radiometric Sensor	(10.7…11.7); (11.7…12.7); (33…35) HHz	(0.2…0.6) mm (0.6…4.0) mm (4.0…6.0) mm	±0.1 mm ±0.2 mm ±0.4 mm	1. The type of a measuring tool was approved. 2. Portable tool 3. Measurements are performed manually. 4. Excitement - up to 2 points.
Portable radiometric tool for measuring oil layer thickness on the water surface KTS RMK-N	(11.7…12.7); (33…35) HHz	(0.2…0.6) mm (0.6…4.0) mm (4.0…6.0) mm (6.0…12.0) mm	±0.1 mm ±0.2 mm ±0.4 mm ±0.6 mm	1. The type of a measuring tool was approved. 2. Portable tool
Multifunctional Fluorescent Lidar	Возбуждение светом УФ диапазона			1. Distance - up to 100 m (when placed at a height of 30 m).
Laser	1.53; 1.43; 1.406; 1.33 µm	(0.1...4.0) µm	20 %	1. Project.
	2.86 и 5.76 µm	(4...140) µm	20 %	
Laser ultrasound device				1. Three lasers are used, including powerful CO2 laser. 2. Research is conducted in the laboratory. 3. A prototype is planned to be installed on the plane.
Mars	0.8; 3; 23; 180 cm	от 200 µm		1. Complex procedure for calibrating radars and choosing the optimal shooting mode. 2. Complex information processing procedure.

3. Conclusion

The analysis of the above data allows for the following conclusions:
- to control pollution, the initial stages of the existence of an oil film are of interest;
- laser tools allow you to measure the thickness of the oil film at a distance of several hundred meters;

Journal of Physics: Conference Series **1384** (2019) 012066 doi:10.1088/1742-6596/1384/1/012066

- the most promising is the development of radiometric tools for measuring the thickness of an oil film on the water surface;
- the method for measuring the thickness of an oil film on the water surface using multi-frequency radars requires further research and development.

References

[1] Bondura V G 2012 *Aerospace monitoring of oil and gas facilities* (Moscow: Scientific World) 558 p
[2] Ushakov I E 2016 *Notes of the Mining Institute* **129** 421-427
[3] Portable three-frequency devices Print-3. Retrieved from: http://www.td-str.ru/file.aspx?id=30526
[4] Technology of multi-level environmental monitoring for ensuring information security of offshore oil and gas production. http://neftegas.ru/scince/view/181
[5] Lobkovsky L, Zatsepin A, Kovachev S 2007 *Technologies of the fuel and energy complex* 2 72-78
[6] Kozintsev V I, Belov M L, Gorodnichev V A 2006 *Vestnik of MSTU n.a. N.E. Bauman* **2** 56-64
[7] Belov M L, Gorodnichev V A, Kozintsev V I, Smirnova O A, Fedotov Yu V, Khrustaleva A M 2007 Patent No. 2298169 of the Russian Federation. Two-spectrum remote method for detecting oil pollution on the surface of the water. 04/27/07.
[8] Fingas M, Brown C 2014 *Mar Pollut Bull* **83(1)** 9-23
[9] Boev A G, Matveev A I 2005 *Radiophysics and radio astronomy* **10** 178-188
[10] Adamo M, De Carolis G, De Pasquale V, Pasquariello G 2005 *Proceedings of the SPIE* **5980** 153-164
[11] Brown C E, Fingas M F, Marois R 2004 *Environment Canada Arctic and Marine Oil Spill Program Technical Seminar (AMOP) Proceedings* **27 (1)** 317-334
[12] Kormac D 1989 *Combating marine pollution, oil and chemicals* (Moscow: Transport) 365 p

Research of the parameters of boundary friction of tribosystems in the introduction of surface-modified Al-samples

M O Silivanov, A A Vinogradova

Saint Petersburg Mining University, 2, 21 line, Saint-Petersburg, 199106, Russia
E-mail: Silivanov_MO@pers.spmi.ru

Abstract. In this work, the influence of the structure of the adsorption layer of the modified aluminum dispersed powder on the thickness of the boundary layer of the tribosystem with lubrication in the form of oil I-20 with the addition of surface-modified samples was studied. The thickness of the lubricating layer (according to Petrov formula), the value characterizing the antifriction properties under boundary friction conditions, was calculated. The data on the effect of adsorption of various quaternary ammonium compounds on the boundary layer are important for understanding the interaction of these additives with oil components and their ability to reduce friction in the system.

1. Introduction

The main causes of failure of transmissions of equipment and their control mechanisms are the wear of workers, the appearance of unacceptable gaps, etc. In order to ensure their smooth operation, it is necessary to create a stable oil film on the friction surface, which is achieved by the use of various lubricants. One of the simplest and most effective ways to regulate the properties of lubricant compositions is the introduction of additives that can improve the antifriction characteristics.

2. Materials and methods

The Mining University has developed a method of layering different-sized molecules of cationic surfactants, such as quaternary ammonium compounds (QAC), on the metal surface, which allows regulating the physico-chemical and tribochemical characteristics of dispersed metal powders [1, 2]. It is based on the application of the principles of molecular layering, quaternary ammonium compounds such as alkamone (A) and triamone (T) having different-sized alkyl radicals in the nitrogen atom, C_{16}-C_{18} and C_1-C_2, respectively. The selection of preparations was made on the basis of the stabilizing effect of low-molecular-weight QAC in the adsorbed state found earlier in the works [1, 2]. The force and coefficient of friction in the sliding bearing were measured on a tribological stand based on the DM-29M friction machine using a software package in the LabView environment, as well as using mechanical sensors. The experiment was carried out in isothermal conditions in a tribological pair St45 (GOST 1050-88) and Bronze (GOST 18175-78) containing oil I-20 with additives. The load on the bearing varied from 50 to 500 kgf.

3. Study of tribological characteristics of lubricant compositions

Along with the coefficient and friction force, one of the necessary characteristics of the friction process and the lubrication state is the thickness of the lubricant layer (C) at the point of greatest

Journal of Physics: Conference Series **1384** (2019) 012067 doi:10.1088/1742-6596/1384/1/012067

convergence of the surfaces [3]. The thickness of the oil layer between the rubbing surfaces affects the durability and performance of the mechanism, as well as the possibility of jamming mechanism as a result of too strong convergence rubbing surfaces. In this work the influence of the structure of the adsorption layer samples on the basis of aluminium (PAP-2, particle size less than 1 μm, smallest particles about 45 nm)) on the thickness of the lubricating layer of the tribosystem with the lubrication in the form of industrial oil (I-20) with the addition of a modified sample (0.5 masses. %) was studied. The thickness of the lubricating layer was calculated by the Petrov equation:

$$F = \eta \frac{VA}{c}$$

, where F - the friction force, η - the dynamic viscosity coefficient of the oil, V - the speed of rotation of the shaft ($V = \omega R$), A - the area of the support surface.

Previously obtained fundamental tribological characteristics, such as force and coefficient of friction (table. 1) showed that the best antifriction effect shows tribosystem with the addition of aluminum powder, processed from a mixture of alcamone and triamone (mass ratio 1:1) [4, 6].

Table 1. Effect of the type of dispersed Al – additive on the equation of the relationship $F_{fr}=\Phi(N)$, the change in F_{fr} (ΔF_{fr}) relative to the starting oil and the friction coefficient (f)

№	Al – additive or lubrication	Equation $F_{fr} = \Phi(N)$	$\Delta F_{тр}$ (medium), %	$\Delta F_{тр}$ (N = 5 kN), %	f (N = 3.5 kN)
1	Al/(A+T)	y = 0.037x + 12,47	-11.41	-15.92	0.0075
2	Al/T/A	y = 0.048x + 10,81	-7.75	-3.69	0.0079
3	Al/A	y = 0.050x + 12,05	-1.05	-1.4	0.0089
4	И-20 (without additives)	y = 0.050x + 12,29	0	0	0.0089
5	Al/T	y = 0.050x + 11,86	-1.52	0.13	0.0087
6	Al (ПАП-2)	y = 0.065x + 11,74	12.61	20	0.0101

In this paper, the thickness of the lubricating layer was calculated for these systems, depending on the cationic surfactant treatment program. It was observed that surface-modified metal powders increase the thickness of the lubricant layer, ceteris paribus. The values of *c* are in the range of 0.04 – 0.06 μm. These data are consistent with the opinion of Rosenberg that the total thickness of the entire possible boundary films, including the oxide, usually less than 0.1 μm [7]. In addition, the fact of boundary friction is confirmed by the fact that with a further increase in the load there is a hook of rubbing surfaces and the transition to dry friction. The values of the friction coefficients f presented in [4 - 6] are within the permissible limits for boundary friction.

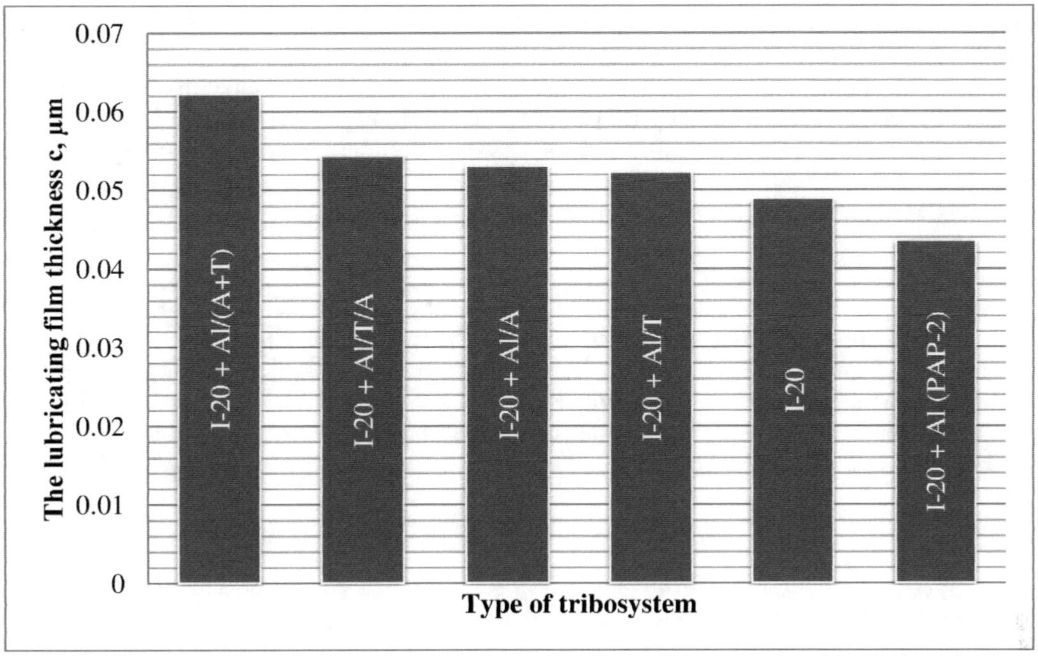

Figure 1. Dependence of the thickness of the lubricating layer of the composition of the investigated tribological system.

The fact that the introduction of surfactant into the lubricant leads to the increase in the thickness of the boundary layer and helps to reduce wear is a well known fact. For example, it is confirmed, for example, by the results of the work [8]. The analysis of figure 1 leads to a similar conclusion that the introduction of surface-modified aluminum-based powders into industrial oil I-20 increases the value of *c*. Interestingly, the addition of the original powder PAP-2, not only does not increase, but even reduces the thickness of the lubricant layer. The use of modifiers of alkamone and triamone, as well as their successive application to the sample, provides only a small thickening of the boundary film, relatively pure oil. When the sample Al/(A+T) Is introduced into the tribosystem, the greatest increase in the antifriction effect is observed, as well as the thickening of the lubricating layer *c*. This is most likely due to the conclusions obtained in the work[9] that the lubricating layer of the QAC on the sample of The Al/(A+T) type is most strongly retained on the surface of the solid particle, ensuring the stability of the tribosystem under high loads. Organic hydrophobic radicals in the nitrogen atom in the deposited cationic surfactants, having a small force field [10], provide thickening of the lubricating layer and the easiest sliding of rubbing surfaces. Tribological properties of lubricants with solid additives can be most accurately estimated in the study of boundary friction. Since when approaching the dry friction mode, the antifriction properties of the system are determined by the characteristics of the hardest additive. Selection of the necessary additive affects the behavior of the system under extreme conditions. These additives were also tested using the ARP-11 device in the work [4, 6], some of them are presented in table 2.

Table 2. Value of the integral friction index D for tribosystems with surface-modified aluminum powders

Additive (powder)	Al/T	Al/A	Al/(A+T)	Al	Al/T/A
D	780	910	300	1690	1000

Control of the nature of interaction in a pair of drill-plate was determined by fixing the integral index D, proportional to the magnitude of the acoustic emission arising in contact with friction. The change in the magnitude of the acoustic emission is proportional to the change in the friction force in the pair (coefficient of friction) and by its value it is possible to estimate the nature of friction. The data of table 2 and [4, 6] confirm that the greatest decrease in the tribological characteristics occurs in the system containing the additive of aluminum powder with adsorbed alkamon and tremona from the mixture, and the smallest system with aluminum PAP-2.

4. Conclusion

The process of increasing the thickness of the boundary layer of the lubricant composition in transmissions and other mechanisms is an important task, as it reduces the wear of friction parts, as well as pushes the boundary of dry friction, thereby improving the stability of the equipment. Despite this, the study of boundary friction is also important for the study of the effect of solid additives on the extreme properties of lubricants. In the work, modified samples were identified that can increase the thickness of the lubricant layer, thereby improving the anti-friction characteristics of the system.

References

[1] Syrkov A G 2013 *Russian Journal of General Chemistry* **83** **N**8 1392-1394
[2] Syrkov A G, Fadeev D V, Taraban V V, Silivanov M O 2014 *Condensed matter and interphase boundaries* **16** N2 215-219
[3] Raiko M V 1970 Gear lubrication 196
[4] Taraban V V, Syrkov A G, Silivanov M O, Nazarova E A 2015 *Chemical physics and mesoscopy* **17** N4 557-564
[5] Nazarenko M Y,, Kondrasheva N K,, Saltykova S N 2017 *Tsvetnye Metally* **7** 29-33
[6] Syrkov A G 2015 *Russ. J. Gen. Chem.* **85** N6 1538-1539.
[7] Rozenberg Yu A 1972 The effect of lubricating oils on the reliability and durability of machines 524
[8] Beliy V A 1989 The role of the structure of the surface layers in the external friction of polymeric materials 115
[9] Syrkov A G, Sychev M M, Silivanov M O, Rozhkova N N 2018 *Glass Physics and Chemistry* **44** N5 474-479
[10] Abramson A A, Shchukin E D 1984 *Surface phenomena and surfactants* (Leningrad)

Journal of Physics: Conference Series **1384** (2019) 012068 doi:10.1088/1742-6596/1384/1/012068

Fundamentals of ISO 9001:2015

S M Leontyuk, A A Vinogradova, M O Silivanov

Saint-Petersburg Mining University, 2, 21stLine, St Petersburg, 199106, Russia

E-mail: vinogradova_aa@pers.spmi.ru

Abstract. The choice of an effective quality management system is a non-trivial task which requires a special approach. We consider the main changes in the new edition of the international standard for quality management systems ISO 9001: 2015, its impact on the workflow and benefits.

1. Introduction

A characteristic feature of the modern market is production of quality goods and services in order to ensure the competitiveness of enterprises [3]. The implementation of a quality management system is a strategic decision for a company capable of improving the results of its activities and providing a solid basis for initiatives focused on sustainable development [4]. Although all companies certified for compliance with the ISO requirements should have changed them by September 2018 taking into account ISO 9001: 2015 requirements, not all the companies have done this due to internal problems and other reasons. There are companies that felt the need to implement a QMS. Therefore, the topic remains crucial due to the need for improving management systems in order to confirm compliance and ensure proper quality aimed at internal audit as a tool to find areas for improvement [2]. Improving the quality of products is one of the priority areas for the development of domestic industries [7-]. The most common model is a quality management system (QMS) whose requirements and recommendations are described in international standards ISO 9000 [8].

Consumer requirements for product quality are increasing due to globalization and development of international trade relations. To confirm the quality of products, international quality certificates are required. Until the 1980s, there were various international and national quality standards for goods and services.

2. Results and discussion

International ISO 9000 standards are the most common. Today, the number of certified quality management systems (QMS) exceeds 1.5 million, which tells us about the attractiveness of standards and their positive impact on the performance of companies.

QMS allows you to create a solid foundation for a productive and efficient company. Changes in the QMS caused by the changing edition of ISO 9001 allows you to view the general concept and trend. This consists in a gradual decrease in the degree of rigor in the requirements. The unification of standards is not correct, because each company has its own system and it is not necessary for the QMS to repeat the structure and section numbers.

This tendency is also observed in the transition from an elemental approach to a process approach to the QMS development; the company can set standards at its own discretion in the absence of a requirement to reproduce the structure of the ISO 9001 standard in the QMS structure; it can use "its" own terms and definitions (Table 1). Documentation of the QMS is necessary in order to indicate the procedure and rules for the implementation of existing processes, as well as record the results achieved [5].

Table 1. The main stages of ISO 9000 development

Year	Stages
1987	Development of the first edition of the ISO 9000 standards
1994	Clarification and updating of standards. The "elemental" principle of QMS GOST R ISO 9000-9004 development
2000	Radical revision of standards. Process and systematic approach to the QMS development. 2001 implementation of GOST R ISO 9000-9004
2005	Updating of ISO 9000 standards. 2008 - updating of GOST R ISO 9001-2008 standards.
2011	Updating of ISO 9001. Implementation of GOST ISO 9001- 2011 - Interstate standard
2015	Implementation of GOST R ISO 9001-2015. Significant changes in the structure and content of the standard

It is necessary to understand that such a "free" approach to QMS must be preceded by all previous stages of development, including formalized processes, procedures, personnel training. Only in this case, the company can take advantage of freedom and have an opportunity to build its QMS according to ISO 9001: 2015.

A fundamental change in the QMS models includes the transition from a deterministic approach to company quality management to a probabilistic risk-based approach. In the context of the new version of ISO 9001, risk analysis and management activities become mandatory for organizations [6]. The organization should plan measures to address risks and opportunities, as well as measures that are adequate to possible consequences of non-compliance of products and services [9]. Risks can be reduced [10]. An assessment of risks must be based on the assessment of risk consequences and probability classes. Tables 2, 3, 4 are presented below.

An important feature is the "exit" outside the company through the implementation of section 4 of the standard (ISO 9001: 2015) which tells us that senior management should formulate a mission, a strategy, a policy, goals and objectives taking into account the external and internal environment. An analysis of the internal and external environment should be considered as a process, including identification of factors that affect intents, purposes and sustainability. It takes into account legal, technological, competitive, market, cultural, social and economic factors. When developing a mission, a strategy and goals, company values, culture and knowledge are taken into account (Fig. 1).

Table 2. Identification of a risk consequence class

Class of consequences	Ranking	Criteria
1, 2	Can be neglected	Random factors whose influence is leveled over time without special events
3, 4	insignificant	affecting but easily eliminated
5, 6	observable	Changes that require active intervention and resources but not causing loss of stability
7, 8	significant	Significant (more than 30%) drop in turnover and profits, mass layoffs of employees, temporary suspension of business, the need to reprofile activities, partial loss of investments
9, 10	crucial	Bankruptcy or closure of an enterprise

This approach represents the company's QMS as a tool ensuring quality of its activities and managing the competitiveness which allows the company to ensure its stability and outstripping development. The QMS is a tool for solving internal problems and implementing strategic tasks. This development course enables the top management to create and formulate unique development strategies using various models, methods and means. A comprehensive analysis of all factors (counter-partners, competitors, consumers, supervisory organizations, suppliers, economic, social, political, demographic situation) affecting the activities of the company should be a condition for successful solution of strategic problems, etc. "... An effective strategy cannot be developed if managers do not understand what changes are taking place in the competitive environment and how these changes will affect future competitive advantages" [1].

Table 3. Identification of a risk probability class

Class of probability	Ranking	Criteria
A	Frequent	Very likely to happen
B	Possible	Likely to happen
C	Probable	Might happen
D	Unlikely	Unlikely to happen
E	Impossible	Very unlikely to happen

Table 4. Risk class identification

Class of consequences / Class of probability	1, 2	3, 4	5, 6	7, 8	9, 10
A	II	II	III	III	III
B	I	II	II	III	III
C	I	I	II	II	III
D	I	I	I	II	II
E	I	I	I	I	II

I – Allowable risks not requiring reduction. Such risks must be maintained at an acceptable level.
II – Moderate medium-term risks.
III – Significant risks requiring urgent management measures aimed at reducing them

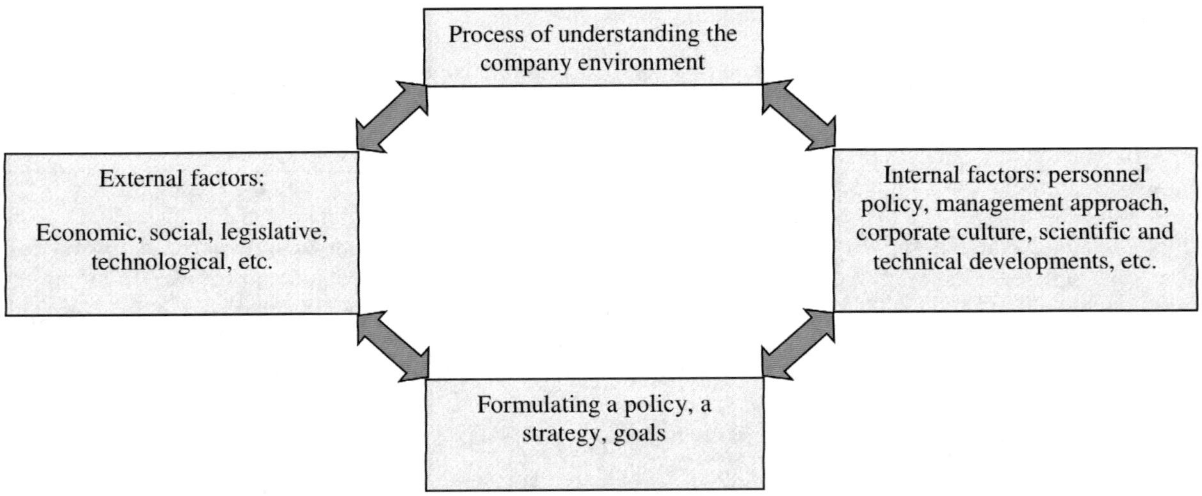

Figure 1. The components of the process of understanding the role of the corporate environment in implementing company goals

The company's strategy and mechanisms of its development occupy a special niche, realizing the connection between expectations and their actual results. The task of aligning the strategy and a possibility of its implementation is carried out by a management mechanism that combines a set of methods and tools for the development, adoption and implementation of management decisions based on the comparison of the vision and indicators of achievement of established goals.

3. Conclusion.

A creative approach to the QMS using widespread and new methods will allow the company to build its unique QMS, which will become a tool for effective work.

References

[1] Yashin N S, Chemyshova T V 2018 Features of Risk Management in the Oil and Gas Industry within the Framework of the Quality Management System
[2] Baurina S B 2016 Requirements for quality management systems: innovations of Russian standardization
[3] Frolova I 2016 Internal audit of the quality management system
[4] Popova L F 2017 Formation of a quality management system based on a process approach in the management of industrial enterprises
[5] Makhov V E, Sytko I I 2018 Shape and Relief Evaluation Using the Light Field Camera **194 N2** 1 - 6
[6] Smirnova E E, Lutskiy D S 2019 Improving the efficiency of purification in the technological cycles of limestone **14 N12** 2306-2309
[7] Kremcheeva D, Kremcheev E 2018 Use of a quality management system at the iron and steel enterprise **41 N1** 151-155
[8] Sytko I I, Kremcheeva D A 2017 Instrumentation for measuring the parameters and characteristics of four-poles **8 N108** 44-854
[9] Zhukovskiy Y L, Vasiliev B U, Koteleva N I 2017 Quality estimation of continuing profes-sional education of technical specialists 704 – 707
[10] Zamyatin E O, Rastvorova J V, Shklyarskiy A Y 2018 Developing of electric power quality indicators evaluation and monitoring intellectual system 761 – 762
[11] Smirnova N A 2015 The quality and safety of Russian food products is the basis of import substitution
[12] Churilov A S 2017 "Documentation of a quality management system in accordance with ISO 9001: 2015"
[13] Dmitriev A Ya, Mitroshkina T A 2016 Risk analysis of the processes of the company quality management system
[14] Kruglov V I 2015 Formation and development of quality systems of educational institutions
[15] Glukhova L V 2017 Organization and management of enterprises, industries, and complexes
[16] Seymour Tills 1985 Strategy and Learning. Strategies That Work: Karl Stern, George Stoke Jr. Strategies that work. (BCG M Approach: Mann, Ivanov, and Ferber)

Noise immunity of signal demodulation procedure in binary coded communication channels with a frequency-position multiplexing

Yu K Vyboldin and S V Borisov

Saint-Petersburg Mining University, 2, 21 line, Vasilievsky island, St. Petersburg, 199106, Russia

E-mail: y.vyboldin@mail.ru, borisov_sv@pers.spmi.ru

Abstract. The article presents the algorithms for demodulating binary channels of data transmission lines with frequency multiplexing used in remote control and warning systems. The noise immunity of the algorithms were investigated. It is shown that significant simplification and significant noise immunity increasing can be obtained by applying the majority decision rule.

1. Introduction

This paper discusses one of the ways to build a noise-immunity communication lines. These lines are designed to transmit information about the status of binary sensors of any devices. It is nesessary, for example, in fire and burglar alarm systems, discrete control systems, and in selective warning systems. Such systems may also include warning systems using the transmission of infra-low frequency signals through rock massives and preserving their operability in emergency situations. Basically, these systems use signals with discrete frequency modulation and amplitude modulation with a single sideband [1–7]. A comparative effectiveness estimate of using these signal generation methods in underground communication systems is given in [4], which shows the advantages of using discrete frequency modulation (FM) in underground warning systems. Discrete FM signals using at infralow frequencies and its coding for transmission through rock massives allows solving the problem of preserving warning system good condition after an accident. However, if a significant level of interference in the propagation medium is present, among which are both additive and multiplicative interference, it is necessary to apply signal processing techniques to improve the message transmission accuracy. In addition, narrow bandwidth of the channel force us to use two a bit differ one from another operating frequencies and the data transfer rate when generating an alert signal is rather small. This paper presents an investigation dedicated to research a noise-resistant warning system that uses multi-position discrete frequency shift keying method for transmitting information. Algorithms for demodulating signals have been synthesized with minimum of error probability in each binary channel and the algorithm was obtained that ensures the minimum error probability of the summary binary message of all binary channels.

2. Results and discussions

Total number of frequency positions required M for a data transmission system with the number of channels is defined as $M = 2^N$.

The signal at the transmission of position number r has the form

$$Z'(t) = U_c\, cos\omega_r + U_s sin\omega_r + n(t),$$

$$0 \le t \le T\,,\quad r = 1,2,\dots M\,.$$

Here $U_c = \cos\varphi$; $U_s = U\sin\varphi$ - quadrature components of the amplitude position at the input of the demodulator, U and φ are, respectively, the amplitude of the initial phase of this position; $n(t)$ is an additive interference approximated by white noise with a zero mean value and a finite variance;

$$\omega_r = \omega_1 + \frac{2\pi d(r-1)}{T}\ , d = 1,2,\dots \tag{1}$$

ω_r – frequency of position with number r; ω_1 – is the minimum signal frequency adopted for the first position (r = 1); d – integer; T – the duration of the symbol. The choice of frequency positions according to the relation (1) ensures their orthogonality.

Unlike the multichannel lines that uses the method of phase-positional or amplitude-phase-positional interleaving, for frequency-positional interleaving the error probability is the same for all frequency channels [8]. There are two approaches for the optimal decision schemes synthesis. The first approach is to minimize the error probability in each binary channel. The second approach involves minimizing the error probability in the summary of all message transmitted over all channels [9-13].

First we will consider a decision-making algorithm that minimizes the error probability in each binary channel. Assuming a priori known intensity of white additive noise, it can be represented as

$$\eta_{m_1} = \sum_{r\in<r>_{m_1}} \exp((2D)^{-1}V_r^2) - \sum_{r\in<r>_{m_0}} \exp((2D)^{-1}V_r^2) \underset{<\to 0}{\overset{>\to 1}{}} 0 \tag{2}$$

Here r and m are respectively the signal frequency position number and the binary channel number; $r = 1,2,\dots,M$; $m = 1,2,\dots,N$; $D = v^2/_T$; $r \in< r >_{m1}$ and $r \in< r >_{m0}$ - subsets of positions that carry information about 1 and 0 in the binary channel with the number m, the arrow indicates the decision to accept one or zero with a corresponding inequality sign, μ^2 one-sided noise power spectral density; $V_r = \sqrt{X_r^2 + Y_r^2}$ – sample of voltage at the end of the reception of the next element in the r-th chain of optimal incoherent processing, where

$$X_r = \frac{2}{T}\int_0^T Z'(\tau)\cos\omega_r\,\tau d\tau;\ Y_r = \frac{2}{T}\int_0^T Z'(\tau)\sin\omega_r\,\tau d\tau.$$

The implementation the algorithm (2) requires to know the magnitude of D. Overcoming a priori uncertainty is possible by using an asymptotic rule, which may be obtained from (2) by series expanding of the exponential functions and then limit the number of terms. In this case

$$\exp((^1/_{2D})V_r^2) \cong 1 + (^1/_{2D})V_r^2. \tag{3}$$

Using (3), we obtain the second demodulation algorithm, which is asymptotically optimal for low signal-to-noise ratio.

$$\eta_{m_1} = \sum_{r\in<r>_{m_1}} V_r^2 - \sum_{r\in<r>_0} V_r^2 \underset{<\to 0}{\overset{>\to 1}{}} 0. \tag{4}$$

The third algorithm is obtained from (2) by replacing the unknown value D with its maximum likelihood estimate $\widehat{D}$, which can be obtained from the prehistory or at the next element reception interval. For the calculation $\widehat{D}$, it is possible to use the fact that when processing orthogonal FM signals, the signal position is formed only at the output of one receiving path, while in the overs $(M-1)$th paths at these moments the signal value are zero and the sample values are determined by only noise intensity. These $(M-1)$th samples can be used to estimate the spectral density estimate of the noise power. The estimate can be found as

$$\widehat{D} = \frac{1}{2(M-1)} \sum_{r=1}^{M} V_r^2. \quad (5)$$

When finding the estimate (5), it is assumed that information contained in signal position and transmitted in sample may be identified with the known maximum likelihood algorithm [8, 9]

$$V_q > V_r, r \neq q, r, q = 1,2, \ldots \mathrm{M},$$

Then

$$\eta_{m1} = \sum_{r\epsilon<r>_{m1}} exp\left[\left(2\widehat{D}\right)^{-1} V_r^2\right] - \sum_{r\epsilon<r>_{m1}} exp\left[\left(2\widehat{D}\right)^{-1} V_r^2\right]_{<\to 0}^{>\to 1} 0. \quad (6)$$

The squaring operation in (6) can be omitted. It is saved here due to the fact that it is necessary in the formation of (5) and (6).

$$V_q > V_r \quad r \neq q \quad r, q = 1,2, \ldots, \mathrm{M} \quad (7)$$

Then, using the code converter, values "1" or "0" in all N binary channels are formed. The forming rule in mth binary channel is:

$$\begin{matrix} "1"\ if\ r \in <r>_{m1} \\ "0"\ if\ r \in <r>_{m0} \end{matrix}. \quad (8)$$

Then, the authors estimate the noise immunity of the reception according to the above algorithms in the absence and in the presence of Rayleigh fading. For a four-channel data transmission system with number of channel $N = 4$ and number of positions $M = 16$. Let assume that the duration of the signal element is $T = 0{,}1\ c$. Then the minimum frequency division between adjacent frequency positions is 10 Hz, and the frequency band occupied by the signal will be slightly more than 1600 Hz. Let estimate the ratio of the average signal energy to the spectral density of the average noise power as

$$H^2 = \frac{P_c T}{\nu^2},$$

and calculate the error probabilities by the rules (2), (4), (6), (7), which are denoted respectively by P_1, P_2, P_3, P_4, P_5. In mathematical modeling of algorithms, the results of [14, 15] were used. The calculation results are summarized in table 1 and 2.

The error probability when using the algorithm with minimum error in the summary message is not significantly different from the one in the optimum reception. Thus because this demodulation algorithm is more simple compared to the one presented in (6), it can be recommended for implementation.

The considered receiving methods with FM multi-position signals need hard clock synchronization, which significantly complicates the demodulator's circuit. These difficulties may be reduced by using the feature of the considered data transfer system. The things is that it does not require binary data continuous transmission for each binary channel, but only detection is necessary, and for quite a long time (of a second). Thus it is possible the asynchronous operation of the demodulator with a local clock generator which frequency is equal

$$F_{clock} = \frac{\omega_{r+1} - \omega_r}{2\pi} = \frac{d}{T} = \Delta F_{min}.$$

Table 1. Channel without Rayleigh fading

P	H			
	1	2	3	3,5
P_1	0,40	0,22	0,024	0,007
P_2	0,40	0,19	0,025	0,007
P_3	0,38	0,27	0,051	0,023
P_4	0,41	0,18	0,025	0,061
P_5	0,37	0,086	0,002	0,0001

Table 2. Channel with Rayleigh fading

P	H			
	1	2	3	3,5
P_1	0,30	0,060	0,017	0,06
P_2	0,30	0,060	0,017	0,06
P_3	0,033	0,093	0,020	0.09
P_4	0,41	0,063	0,017	0,010
P_5	0,37	0,010	0,0009	0,0003

Here d is an integer. The clock frequency matches the minimum frequency spacing of adjacent positions ΔF_{min}. In terms of bandwidth savings, it is recommended $d = 1$ when

$$F_{clock} = \frac{1}{T} = \Delta F_{min} = B, \text{Hz}.$$

Here B is manipulation speed. The clock generator phase may be arbitrary regarding changes in the information parameter in any binary channel. Due to the fact that the time interval after state change is much longer T, one can apply the majority principle of detecting the information parameter when the L parameters of elements are taken in series and decision is made by the majority of "votes". This means that when transmit, for example, "0" or "1" in any channel, a decision about presence is made if in L elements the number of decisions equal "1" is greater than $int(L/2)$, where $int(X)$ denotes the integer part of the number X. This method of reception improves the noise immunity compared to usual elementwise reception. Indeed, if the error probability in element-by-element reception is equal toP_4, then the error probability in the majority method with independent distortions in the elements can be obtained in the form

$$P_5 = 1 - \sum_{i=0}^{INT(L/2)} C_L^i P_4^i (1 - P_4)^{L-1}.$$

The calculation results for $P_5$with this ratio for the minimum number L-1 are presented in Table 1. With P_4 of the order $(1 \dots 5)10^{-2}$, the gain in error probability with the majority solution for $L = 3$ will be more than 10 times. In our case, when $T = 0.1\ s$, the detection time of the state changing will be no more than 0.3 s.

3. Conclusions

The noise immunity of binary demodulators of communication lines with frequency-positional division has been investigated by the method of mathematical modeling. It is shown that a significant simplification of demodulators with a significant increase in noise immunity can be obtained by applying the majority principle of decision making. In the presence of fluctuation and impulse noise, a scheme operating according to rule (7), (8) and using a limiter [7] may be recommended for use.

Acknowledgments

The reported study was funded by RFBR, according to the research project No.16-07-00540.

References

[1] Conti R S and Yewen R G 1997 Evaluation of a Signaling and Warning System for Underground Mines, U.S. department of health and human services *Report of Investigations* 9641
[2] *United States Patent Application Publication* Pub. no US 2006/0148514 Al. 06.07.2006
[3] *United States Patent Application Publication* Pub. no US 2009/0140852 A1. 04.06.2009
[4] Stolarczyk L G 1978 *Acomparison between SSB and FM modulation processes in a near field medium frequency wilieless communication system* (Inc. RATON: New Mexico)
[5] Fedosov D 2015 FERRA – A New Portable Underground Communication Transceiver Cave *Radio & Electronics Group J.* **89**
[6] Murphy J N and Parkinson H E 1978 Underground mine communications *Proc. IEEE* vol 66 p 26–50
[7] Kumar A, Chaulya S K, Kumar S and Badyopahyay L K 2003 Trapped miners detection? Location and communication system *Minetech* **24** 1–13
[8] Fink L M 1970 *Theory of the transfer of discrete messages* (Moscow: Sov. Radio)
[9] Vyboldin Yu K, Krivosheikin A V and Nurmukhamedov L Kh 2015 *Methods of processing discrete signals in information transfer systems* (St. Petersburg: Institute of Cinema and Television)
[10] Proakis J G and Salehi M 2008 *Digital Communications* (NewYork: McGraw-Hill).
[11] Feher K 2000 *Wireless Digital Communications: Modulation and Spread Spectrum Applications* (Moscow: Radio and Communication)
[12] Wilson S G 1996 *Digital Modulation and Coding* (London: Prentice Hall)
[13] Sklar B 2001 *Digital Communications: Fundamentals and Applications* (London: Prentice Hall)
[14] Vyboldin Yu K and Borisov S V 2018 Simulation of interference and signals in subsurface radar *IOP Conf. Series: Earth and Environmental Sci.* **194** 062036
[15] Belitskiy A A, Rastvorova I I and Denisova O V 2018 Nonlinear and unbalanced load as a basic factor of a neutral conductor current *Proc of the 2018 IEEE Conf of Russian Young Researchers in Electrical and Electronic Engineering, ElConRus*
[16] Vyboldin Yu K and Borisov S V 2017 Interference effect of impulse noise on noise immunity of communication and control channels in underground structures *IOP Conf. Series: Earth and Environmental Sci.* **87** 082054

Journal of Physics: Conference Series **1384** (2019) 012070 doi:10.1088/1742-6596/1384/1/012070

Application of ANSYS/FLUENT software package for analysis of thermohydraulic processes in the low-temperature main pipeline for liquified hydrocarbon

G S Zakirova[1], **E I Krapivsky**[2]

[1] Saint-Petersburg Mining University, 2, 21 Line, Saint-Petersburg, 199106, Russia
[2] Ukhta State Technical University, 13, Pervomayskaya St., Ukhta, Republic of Komi, 169300, Russia

E-mail: gulnur-live@mail.ru

Abstract. The article discusses examination automation of thermohydraulic processes in the low-temperature main pipeline for liquified hydrocarbon. The computerized simulation of the nonisothermal flow of liquified hydrocarbon in the low-temperature main pipeline is performed by means of ANSYS/FLUENT software package. The authors have performed the analysis of the nature how the temperature and the pressure is distributed in the flow of liquified hydrocarbon mixture along the pipeline length. The computational model of the low-temperature pipeline for examination of hydraulic and thermal loss of the pumped mixture of liquified hydrocarbon is designed. The possibility of application of ANSYS/FLUENT computational fluid dynamics software package is proved for automation of thermohydraulic calculation of the low-temperature pipeline for liquified hydrocarbon mixture taking its thermophysical properties into account.

1. Introduction

The simulation experiment during scientific examinations carrying out is one of the key directions during studying of heat and mass exchange tasks. However for better understanding of physical effects, high cost or impossibility of the physical or natural experiment performance it is preferable to use different software applications and software packages for scientific and technical tasks solution [1-3].

At the present time a large number of researchers are busy with designing of the calculation methods and developing of the software applications and packages for scientific and technical tasks solution. Such software packages as ANSYS/FLUENT, Star-CD, Comsol Multiphysics, CFdesign, ADINA, FlOWVISION, etc. are known for solution of the tasks to calculate liquid and gas flow. In this article the application of computational fluid dynamics tasks is described relating to the pipeline transportation of oil and gas and their calculation by means of ANSYS/FLUENT software environment.

2. Computerized simulation

ANSYS/FLUENT is the part of the generalized program system of the ANSYS finite element analysis. In the software package the simulation of hydrocarbon flow with the wide range of thermophysical properties changes by means of providing various parameters of simulation and application of multiple grid methods with improving convergence is performed [4].

In this project the virtual simulator of the low-temperature pipeline was built by means of ANSYS/FLUENT software environment. ANSYS/FLUENT software package is intergrated into ANSYS Workbench platform. Such intergartion allows performing data exchange with other software packages, for example ANSYS (Mechanical, Maxwell, etc.), as well as with the external CAD-systems. This environment provides high productivity and application simplicity. The structure of the project is consistent and includes the following stages: geometrical arrangement designing by means of ANSYS DesignModeler, grid model designing by means of ANSYS Meshing, preliminary processing by means of CFX-Pre, solver ANSYS CFX-Solver Manager and post processing by means of ANSYS CFD-Post [5].

For calculation by means of software package ANSYS/FLUENT the following RANS turbulence models are available: the system of k-ε models (standard; Renormalizated Group Model – RNG – on the basis of technology borrowed from the theory of renormalized groups), k-ω-model, SST (shear-stress transport model).

The standard k-ε turbulence model satisfies the tasks of pipeline transportation of multicomponent hydrocarbon which are oil and gas.

During transportation in the low-temperature pipelines the main problem is the nature of temperature and pressure distribution in the flow of transported product.

3. Temperature and pressure change in the flow of liquefied hydrocarbon mixture by means of ANSYS/FLUENT software package

In the published articles about the methods of hydrulic and thermal calculations of the pipelines for liquified gases analytical dependences were obtained for temperature and pressure distribution along the length of the pipeline during steady conditions of its flow, different allowances were adopted for some physical and chemical and thermodynamic properties of liquified hydrocarbon mixture [6].

In standard calculations thermophysical properties change along the length of the pipeline is not taken into account for oil. For the pipeline of liquified gases such accountance is necessary. We suggest taking into account dependences of the change of pumped product thermophysical properties (density, heat capacity, viscosity, heat conductivity, compressability, etc.) on main parameters of the system (pressure and temperature).

The characteristic property of the given below hydraulic and thermal calculations of the low-temperature main pipeline is:

1. Low temperature of the pumped mixture to exclude two-phase flows formation of the pumped medium.
2. Pipeline laying by underground method as well as if it is necessary by ground surface, underwater methods, etc.
3. Comparatively small allowable level of temperature rise (adopted more than enough for temperature nearly 10 °C) between the intermediate pumping station and the intermediate cooling station.
4. Turbulent flow (self-similar mode) of the pumped product.

For nature analysis of temperature and pressure distribution in the pipeline by means of ANSYS/FLUENT software package a virtual model of the pipe for the undergound low-temperature pipeline was designed having set all necessary reference geometrical data and boundary conditions. The pipeline model was consisted of four parts: the part of hydrocarbon mixture, the pipeline part, the insulation part and the ground part. As well as the hydraulic and thermal analysis of the pipeline model operation was carried out. This project includes the following stages: designing of the pipeline geometrical model, generation of the computational grid on the basis of the geometrical model, setting of the model boundary conditions and the model starting-up for calculation.

By means of ANSYS/FLUENT software package computerized simulation of the nonisothermal flow of liquified hydrocarbon in the low-temperature pipeline was performed using the software package, the nature analysis of temperature and pressure distribution in the flow of liquified hydrocarbon mixture during thermophysical parameters change (density, heat conductivity and heat capacity) was carried out. During transportation of liquified hydrocarbon mixture it is recommended to take the change of thermophysical parameters along the length of the pipeline into account.

For the purpose of the nature analysis of temperature and pressure distribution in the pipeline by means of ANSYS/FLUENT software package a virtual model of the pipe for the underground low-temperature pipeline was designed having set all necessary reference geometrical data and boundary conditions (the pipeline with the nominal diameter 720mm, with the length 100m, with foam polyurethane heat insulation). The pipeline model was consisted of four parts: the part of hydrocarbon mixture, the pipeline part, the insulation part and the ground part. As well as the hydraulic and thermal analysis of the pipeline model operation was carried out. This project includes the following stages: designing of the pipeline geometrical model, generation of the computational grid on the basis of the geometrical model, setting of the model boundary conditions and the model starting-up for calculation. The designing final result of the computational grid of the pipeline virtual model is shown in Figure 1.

For the purpose of time reducing it is reasonable to cut twice fully-symmetrical parts of the virtual model because all processes flowing in the cut part will be absolutely identical to those which flow in the designed part. Simulation of the quarter of the model volume greatly reduce the calculation time without any influence on the final result. In our case the symmetry planes (section) are situated in planes XZ and YZ.

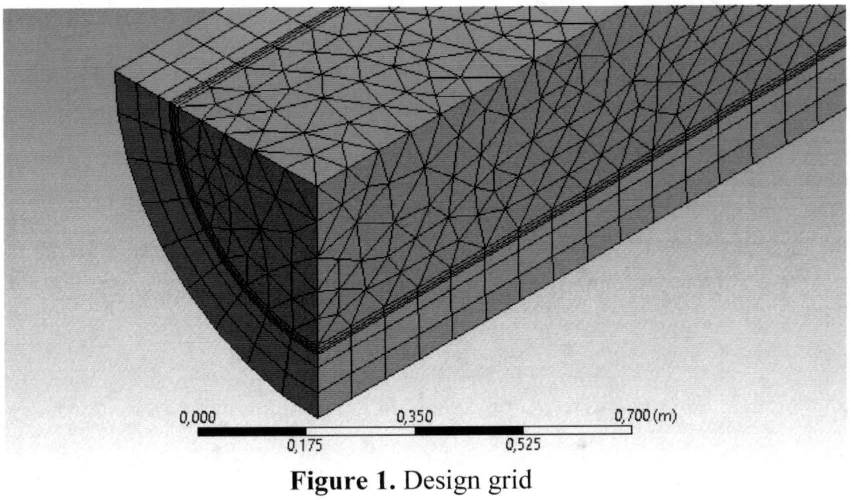

Figure 1. Design grid

The calculation results are listed below. In figure 2 the example of the temperature field for the cross section of the flow, the pipeline, thermal insulation and ground is shown.

To be noted that during transporation of liquified hydrocarbon on long distances the change of thermophysical parameters is great.

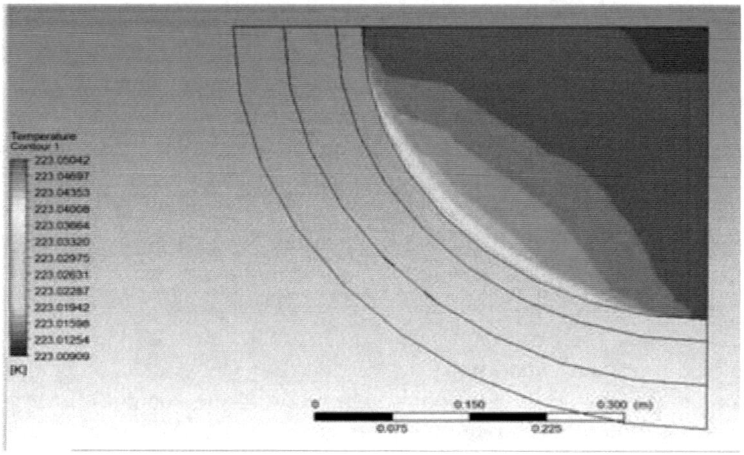

Figure 2. Temperature distribution in the cross section of the flow

Figures 3 and 4 shows the diagrams of temperature and pressure drop along the pipeline with the length of 100 meters.

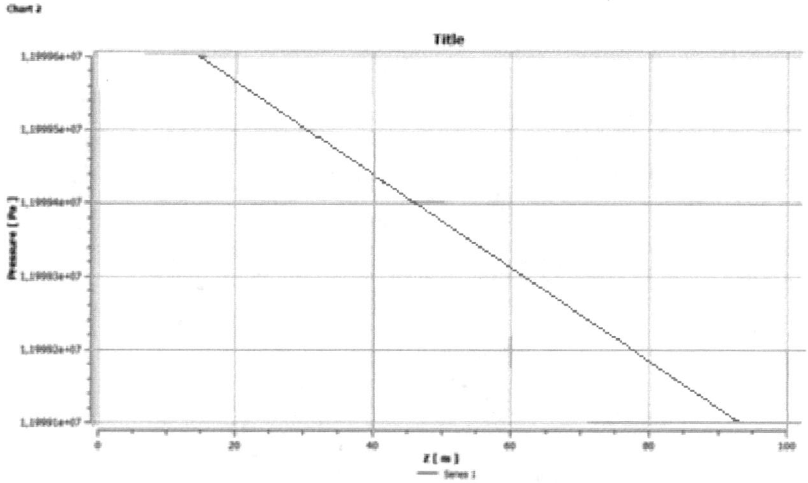

Figure 3. Hydraulic gradient line.

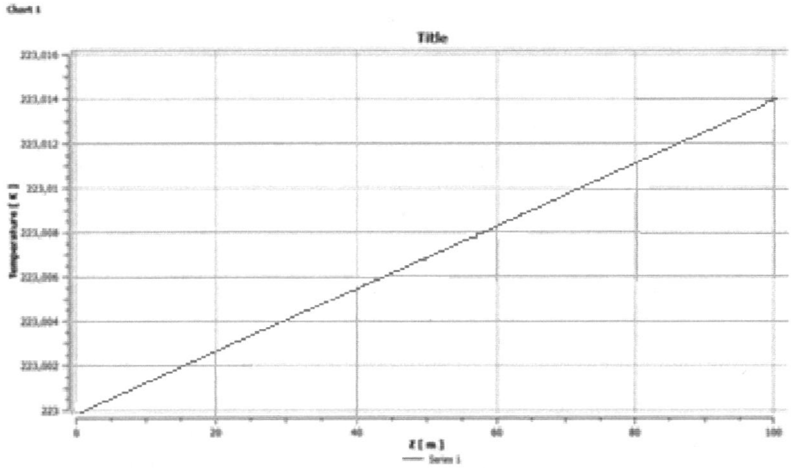

Figure 4. Diagram of temperature change of the pumped product along the pipeline length

The calculations for one part of the low-temperature pipeline between two transfer intermediate pumping stations (IPS) and intermediate cooling stations (ICS) were performed. The authors analyze the thermohydraulic calculation of the low-temperature pipeline with producing capacity 5 mln tons per year (the pipeline diameter D=0,72 m) by means of ANSYS/FLUENT software package. At that the following basic data were used: initial pressure $P_{\mathrm{i}} = 12$ MPa, allowable final pressure $P_{\mathrm{f}} = 10$ MPa, initial temperature of liquified gas mixture $\mathrm{T_i} = -50$ °C, allowable temperature of liquified gas mixture $\mathrm{T_н} = -40$ °C, the part length $L = 10$km , heat insulation – foam polyurethane, heat insulation thermal conductivity $\lambda_{\mathrm{ins}} = 0{,}03\ \frac{\mathrm{Вт}}{\mathrm{м{\cdot}К}}$. It is required to determine the change of the flow parameter (pressure, temperature) along the length of the pipeline.

The results of pressure and temperature distribution along the the length of the pipeline part of liquified gases between IPS and ICS are shown in Figures 5 and 6. On the basis of the performed calculation it is allowed using the ANSYS/FLUENT software package to calculate pressure and temperature differences (taking thermophysical parameters change into account). It is proved the necessity of application of ANSYS/FLUENT computational fluid dynamics software package for thermohydraulic calculation of the low-temperature pipelines of liquified hydrocarbon mixture if the calculation high accuracy is required.

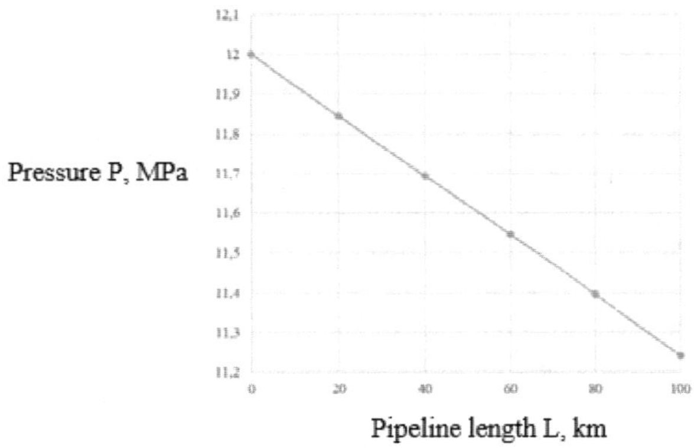

Figure 5. Pressure distribution along the length of pipeline part.

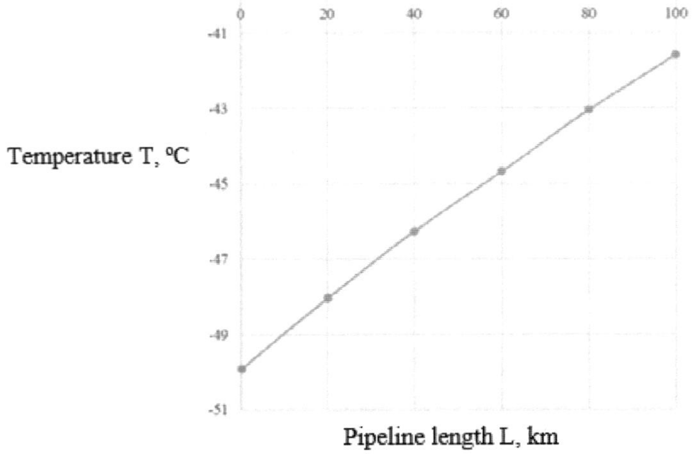

Figure 6. Temperature distribution along the length of pipeline part.

4. Conclusion

Consequently, for automation of thermohydraulic calculations the computerized simulation of the nonisothermal flow of liquified hydrocarbon in the low-temperature main pipeline is performed by means of ANSYS/FLUENT software package, as well asthe nature analysis of temperature and pressure distribution in the flow of liquified hydrocarbon mixture during thermophysical parameters change (density, heat conductivity and heat capacity) was carried out. During transportation of liquified hydrocarbon mixture it is recommended to take the change of thermophysical parameters along the length of the pipeline into account.

References

[1] Korobkov G E, Yanchushka A P and Zakiryanov M V Numerical modeling of a stress-strain state of a gas pipeline with cold bending offsets according to in-line inspection 2018 *Journal of Mining Institute* **234** 643–646

[2] Krotova S Y, Ilin A E and Chirgin A V 2019 *International Journal of Mechanical Engineering and Technology* **10** 1823-1832

[3] Protosenia A G, Karasev M A and Ochkurov V 2017 *Eastern-European Journal of Enterprise Technologies* **6** 11-18

[4] Krapivsky E I, Vishnyakov I A and Kozachok M V 2011 *Mining informational and analytical bulletin* **7** 334-338

[5] Krapivsky E I, Kabanov O V, Vishnyakov I A and Klimko V I 2013 *The computerized simulation of the nonisothermal flow of oil in the pipeline by means of ANSYS/FLUENT software package* (St. Petersburg: National Mineral Resources University Mining) p. 103

[6] Bleicher E M, Vladimirov A E, Ivantsov O M and Polskiy S M 1977 *Pipeline transportation of liquified natural gas* (Moscow: VNIIegazprom) p. 63

[7] Zakirova G S, Krapivsky E I and Minnegulova G S 2015 *Mining informational and analytical bulletin* **6** 332-338

[8] Taranov V A, Baranov V F and Aleksandrova T N 2013 *Obogashchenie Rud* **5** 3-7

Journal of Physics: Conference Series **1384** (2019) 012071 doi:10.1088/1742-6596/1384/1/012071

Nuclear magnetic resonance methods – the key to determining the structure of organic substances

A A Kuzhaeva[1], P V Zgonnik[1], N A Anisimova[2,3],

[1] Saint-Petersburg Mining University, 21-th Line of Vassilievsky Iceland, 2, Saint-Petersburg, 199106, Russian Federation
[2] Herzen State Pedagogical University of Russia, nab. r. Moiki 48, St. Petersburg, 191186, Russian Federation
[3]St. Petersburg State University of Industrial Technologies and Design, ul. Bol'shayaMorskaya 18, St. Petersburg, 191186, Russian Federation

E-mail: zgonnik@spmi.ru

Abstract. This article is intended to help an organic chemists and industrial engineers to choose a suitable method for the analysis of substances with complex structure and its control in industry. In the paper the current level of the NMR development and analytical abilities of the different NMR techniques is considered. It is noticed, that as a rule the results of the one-dimensional NMR are quite difficult to interpret in the case of a complex molecular structure. The main principles and different varieties of two-dimensional NMR techniques are described.

1. Introduction

At the present time various physicochemical methods of analysis are organic systems are widely developed [1-4]. The structural study of the organic compounds is a very special and complex task, and Nuclear Magnetic resonance spectroscopy is known to be an unique method that makes a substantial contribution to the solution of this problem.

NMR is a method based on the resonance absorption of electromagnetic waves by a substance in a constant magnetic field caused by nuclear magnetism. NMR was discovered in 1946 independently by two groups of researchers of American physicists headed by F. Bloch and E. Purcell. The NMR phenomenon is based on the fact that the nuclei of atoms of many elements (like electrons) have their own magnetic moment (spin). [5] From the description of the "traditional" NMR method and the experience of its practical use, certain limitations of the classical NMR experiment become obvious. The increase in the size of the molecule, the complication of its structure lead to a critical complication of the type of the observed NMR spectrum. As an example the ^{13}C and ^{31}P NMR spectra of bis-(2-chloroethyl)-9,10-dihydro-(11-bromine,11-nitroethano)-anthracene-12-ylphosphonate in $CDCl_3$ solution is included (Figure 1) [6].

Two-dimensional NMR is able to solve this problem. [7]

2. Overview of modern NMR techniques.

The history of two-dimensional NMR dates back to 1971, when George Jinner proposed and set up an experiment in which the spectral peak is determined not by one but by two frequencies. As a result of the first action, the excitation of one type of nuclei occurs, then, the second effect promotes the transfer of magnetization to neighboring nuclei, depending on whether the nuclei of one element or two different elements are exposed to the first and second effects, distinguish between homonuclear and heteronuclear two-dimensional NMR. In this case, the spins of the nuclei interact with each other through a chemical bond through a spin-spin interaction or through a dipole-dipole mechanism. The peak position is determined by the chemical shift or resonance frequencies of the interacting spins.[7]

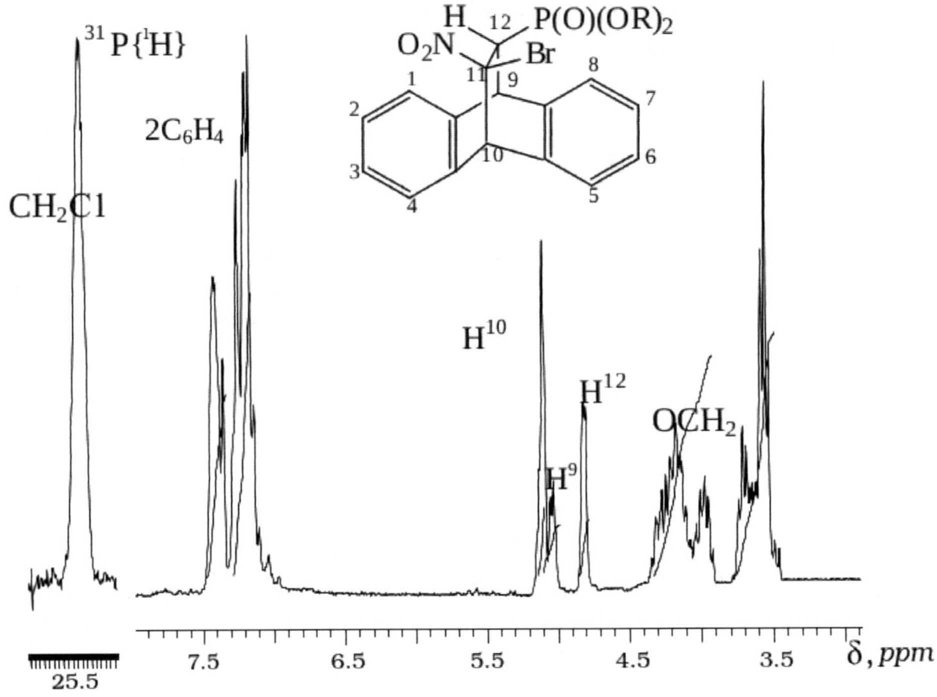

Figure 1. ^{13}C and ^{31}P NMR spectra of bis-(2-chloroethyl)-9,10-dihydro-(11-bromo-11-nitroethano)-anthracene-12-ylphosphonate in $CDCl_3$ solution.

The two-dimensional NMR experiment includes (Figure 2):

- the preparation period during which the spins come to a state of thermodynamic equilibrium; its duration is fixed; it ends with an radiofrequency impulse (p_1) that excites the first spin A;
- the period of evolution or the period of indirect measurement t_1, preceded by one or more exciting radiofrequency pulses (p_2), during which the total magnetization evolves;
- mixing period, during which several pulses are produced, which ensure the transfer of magnetization from one spin to another;
- free induction decay (FID), during which (duration of time t_2) registration takes place.

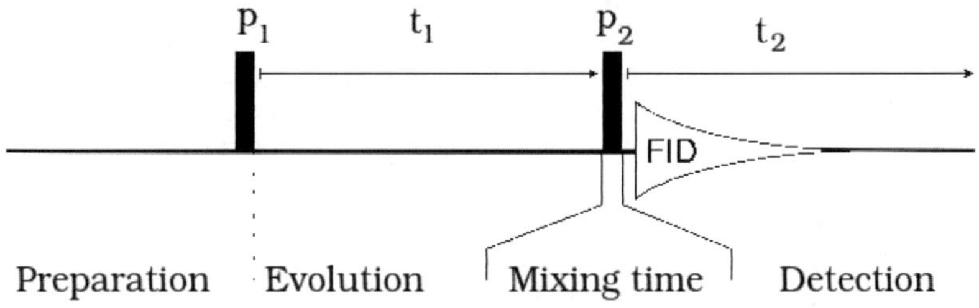

Figure 2. Graphic illustration of the 2D NMR experiment.

During the two-dimensional NMR experiment, several consecutive measurements of the same sample are made, while the time t_1 increases with each subsequent experiment according to the linear law. The initial value $t_1=0$, and with each subsequent experiment it increases by Δt. A feature of the period of evolution is that the excited state of the spin is not immediately detected by the receiving coils, so it is also called the period of "indirect measurement". During the mixing period, the magnetization of nuclei A is transferred to the nuclei B.

Two-dimensional NMR is represented by several methods, the main of which are:

- COSY (CorellationSpectroscopY) correlation ^1H – ^1H spectroscopy, which allows estimating magnetic interactions between protons through electrons of chemical bonds separating them.
- HSQC (Heteronuclear Single Quantum Correlation) – correlation ^1H – ^{13}C spectroscopy, which allows estimating the magnetic interactions of the proton and the carbon atom, with which it is directly connected, that is, for the C– H Bonds;
- HMQC (Heteronuclear Multiple Quantum Coherence) – correlation ^1H – ^{13}C (or ^{15}N) spectroscopy, which allows estimating the direct spin-spin interactions of the proton and the carbon atom, with which it is directly connected, that is, for the C– H Bonds. It is protons that are defined nuclei in this method;
- –COLOC (COrrelation of LOng range Coupling) — correlation ^1H – ^{13}C (or ^{15}N) spectroscopy in which interacting nuclei are separated by two or three bonds (rarely, by a large number of bonds). The detectable nucleus is the ^{13}C (or ^{15}N);
- HMBC (Heteronuclear Multiple Bond Correlation) —correlation ^1H – ^{13}C (or ^{15}N) spectroscopy in which interacting nuclei are separated by two or three bonds (rarely, by a large number of bonds). In this technique, the detected nucleus is protons;
- NOESY (Nuclear Overhauser Effect SpectroscopY) – spectroscopy of the nuclear Overhauser effect between ^1H nuclei located close in space (up to 3.5 E), but not necessarily connected by a chain of covalent bonds;
- ROESY (Rotating Overhauser Effect SpectroscopY) – variation of the nuclear Overhauser effect spectroscopy;

Nowadays, this area of analysis is intensively studied and new methods of two-dimensional NMR appear. [7]

Two-dimensional NMR NOESY methods can successfully solve quite complex problems, such as the study of the three-dimensional structure of proteins and proteins, complementing x-ray diffraction analysis [8,9,10]. The methods of two-dimensional spectroscopy are also successfully used in classical organic chemistry to determine and (or) confirm the structure of synthesized substances

At the Figure 3 we listed typical examples of heteronuclear experiments 4 -trichloromethyl-1-phenyl-1,2,3-triazole by the use of two methods.: ^1H – ^{13}C HMQC and. ^1H – ^{13}C HMBC [11] .

Traditionally, NMR spectra of solids have rather wide peaks, which greatly complicates their interpretation. [5] Homonuclear 2D NMR study of nanoparticles can successfully overcome this difficulty. The authors of the article [12] investigated diamonds in an amount not exceeding 10^{-21} g with quantum sensors in the form of vacancies occupied by nitrogen atoms by ^{1}H and ^{19}F NMR spectroscopy methods. Pulsed NMR sequences were used to measure homonuclear decoupling and spin diffusion.

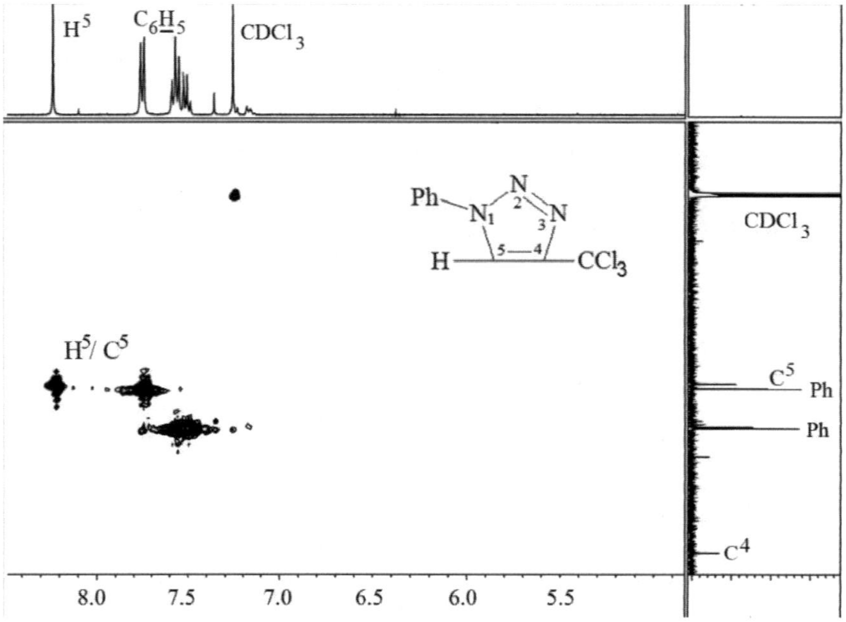

a

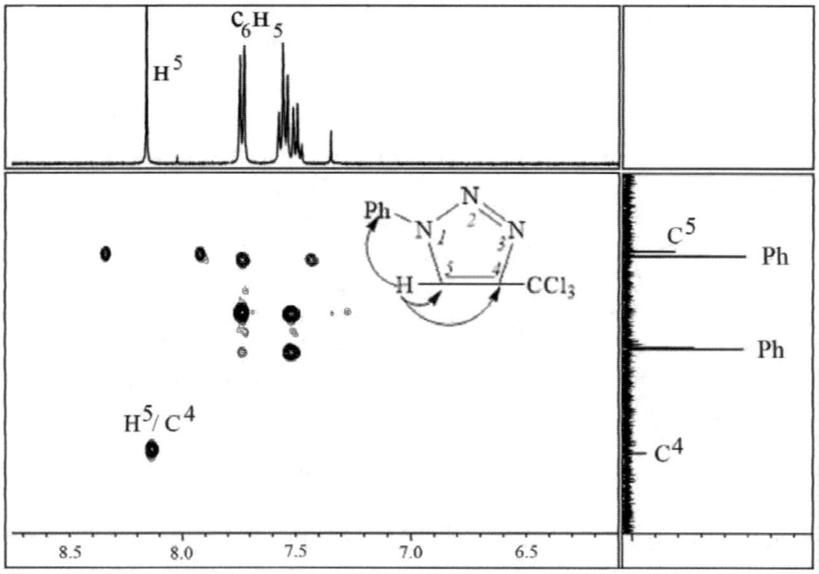

b

Figure 3. ^{1}H – ^{13}C HMQC (a), and HMBC (b) heteronuclear spectra of 4 -trichloromethybl-1-phenyl-1,2,3-triazole.

The best measured NMR line width of the liquid sample was ~1 ppm, mostly limited by molecular diffusion. In order to reduce the influence of diffusion, it was performed by solid-state high-resolution NMR with the use of homolateral interchange, which allowed 20 times narrowing the line width of the NMR [13]. It is noted however, that there is a resolution limit of a spectrum, determined by the ratio of signal to noise causes by errors in the Fourier transformation. The discrepancy in the spectral resolution of the present measurement of nanoscale NMR and the measured volume width of the NMR line can be overcome, for example, by immobilization of sample molecules in nanocapsules in the liquid state of NMR, such as liposomes, micelles or polymer shells [12]. This makes it possible to directly apply NMR in Microbiology. [12]

Since the 1980s, industrial control laboratories have used desktop NMR spectrometers with sub-ppm spectral resolution. They are much less demanding working conditions and the installation site.

3. Results and discussion

However, their significant drawback is the low sensitivity and significant overlap of peaks in the NMR spectra of ^1H due to the fact that the magnetic fields of these devices are relatively weak (1-2 T) [15]. The article [16] considers the difficulties of online control of production and laboratory processes using NMR spectroscopy. A brief theoretical consideration of the NMR method and the resulting main limitations for real-time analysis in production is presented. On the one hand, the existing NMR devices, dedicated to the needs of the industry, have low sensitivity, on the other – devices designed for research laboratories due to its technical parameters and high cost is not suitable for use in industry. At the same time, the significant progress of the last decade in the development of cheaper, small and light permanent magnets based on rare earth materials that use Halbach and one-sided configurations will be mentioned [16]. Currently, powerful cryogen-free superconducting magnets are available on the market, which opens up the prospect of developing cheaper and more compact devices. Based on this, it should be expected that in the near future NMR spectroscopy will be widely used for on-line monitoring of industrial raw materials and products, as well as the real-time chemical reaction progress. Flow-through NMR analyzers are already used in the oil industry for the analysis of crude oil and borehole fluid directly at the site [16].

Another key to solve the problem of the improvement of the quality of desktop NMR instruments for industrial use is the use of hyperpolarized NMR $^{13}C\{^1H\}$[14, 16]. Using the method of reversible exchange (SABRE) hyperpolarization based on hydrogen vapor increases the sensitivity of the 1D and 2D $^{13}C\{^1H\}$ NMR spectra of these devices. This calls for the recording of informative spectra for substances with low concentrations. It is reported [16, 17] that for 26 mmol of pyridine in deuterated methanol (CD_3OD), the increase in the ^1H NMR signal to 17,000 times was achieved within a few seconds, which corresponds to the polarization levels of ^1H P = 5,9%. The comparable levels of amplification can be achieved in both deuterated and proton solvents, but currently the enhanced (SABRE) signals of the analyte predominate due to the relatively weak thermally polarized solvent response. The use of SABRE also allows obtaining^{13}C NMR spectra in the natural presence of isotopes in a single scan, as in the case of hyperpolarized ^{13}C NMR spectra of 4-methylpyridine with a concentration of tens of millimoles. Now the corresponding signal amplification factors are increased up to 45 500 times (P = 4.0%) and are achieved in just 15 seconds [17].

Recently, the American company Thermo Fisher Scientific has launched a serial production of a new generation of compact NMR spectrometers [18]. It allows obtaining information about the structure of molecules, using less than 40 microliters of the studied solution. In addition, the new device costs $ 20,000, which is about a tenth of the cost of traditional NMR spectrometers. "We designed a device that really changes the "landscape" of NMR spectroscopy. Now scientists will be able to place the NMR system directly on his laboratory table," - said the head of the work on the creation of PicoSpin-45 John Price of a new generation of compact NMR spectrometers [19]. It allows obtaining information about the structure of molecules, using a very small volume of the studied solution – less than 40 microliters.

Adiabatic pulses [19, 20], ultrafast 2D NMR [21, 22], fast droplet size distribution (DSD) measurements [21, 22, 23] and reaction monitoring in medium-resolution NMR spectroscopy [24, 25] are among the potential methods that can expand NMR applications in the industrial sector.

Compact NMR equipment and the technique development, led to the creation of not only desktop but mobile NMR devices [26] allowing rapid analysis of water and milk for the presence of bacteria.

4. Conclusion.

Most organic chemists who began their work back in the 1980s remember bulky NMR spectrometers, which occupied anentire room. These instruments were difficult to maintain and not always reliable. Now the method has undergone radical improvements, based both on the revolutionary achievement of George Jinner, and latest advances in materials science, electronics and computer engineering. Today the NMR spectrometers have become more compact, and the era of desktop and portable instruments has come. These modern instruments allow solving difficult tasks associated with determining the complex structure of the big molecules, industrial, environmental and food analysis of different solids, solutions and disperse systems. Current progress in NMR technique makes it possible to solve a wider range of analytical tasks, previously inaccessible or very time-consuming.

References

[1] Boytsova A A, Kondrasheva N K, Ancheyta J.2018 Energy and Fuels 32(2), 1132-1138.

[2] Manukhov V V, Fedortsov A B, Ivanov A S. 2015 Semiconductors 49 (9) 1119–1124

[3] Tomaev V V, Levin K L, Stoyaniva T V, Syrkov A G 2019 Glass Physics and Chemistry 45(4), 291–297.

[4] Zaporozhets E P, Shostak N A. 2019 Journal of Mining Institute.. 238. 423-429

[5] Farrar T S, Becker E D 1971 Pulse and Fourier Transform NMR: Introduction to Theory and Methods (New York:Academic Press)

[6] Anisimova N A, Kuzhaeva A A, Berkova G A, Deiko L I, Berestovitskaya V M 2005 *Russian Journal of General Chemistry.***75(5)**689–693.

[7] Bax A *Two-dimentional nuclear magnetic resonance in liquids*. 1984 (Dellft: Delft University Press),1984

[8] Wüthrich K (2001). *Nuclear Magnetic Resonance (NMR) Spectroscopy of Proteins*. (New York: Wiley–Interscience)

[9] Mariasina S S, Efimov S V, Petrova O A, Rodina E V, Malyavko A N, Zvereva M I, Klochkov V V, Dontsova O A, Polshakov V I 2018 *Biomolecular NMR Assignments***12(1)** 57–62

[10] Blokhin D S, Kul'kova M A, Klochkov V V2012*Uchenye Zapiski Kazanskogo Universiteta. Seriya Estestvennye Nauki*, 2011, 154(4)80–91 (In Russian)

[11] Anisimova N A, Slobodchikova E K, Kuzhaeva A A, Stukan' E V, Bagryanskaya I Y, Berestovitskaya V M 2016 *Russian Journal of Organic Chemistry.* **52(10)** 1379–1384

[12] Aslam N, Pfender M, Neumann P, Reuter R, Zappe A, de Oliveira F F, Denisenko A, Sumiya H, Onoda S, JIsoya J, Wrachtrup J 2017 *//Science*. **357 (6346)**67–71

[13] Timko B P, Whitehead K, Gao K W, Kohane D S , Farokhzad O, Anderson D, Langer R *2011 Advances in drug delivery* **41** 1–20

[14] Richardson P M, Parrott A J, Semenova O, Nordon A, Duckett S B, and Halse M E 2018 *Analyst.***143(14)**: 3442–3450.

[15] Colnago L A, Souza A, Andrade F D, de VasconcellosAzeredo R B 2014 *Chemical Engineering & Technology* **37(2)** 191–203

[16] Kashaev R S, Kubango B E 20016 *Flow NMR analyzers for the control of borehole fluid, crude oil and oil disperse systems.* Kazan state energy University. Tambov-Kazan: LLC "Consulting company Ucom" (in Russian)

[17] Robinson A D, Peter M. Richardson P M, Halse M E 2019 *Appl. Sci.**9***(6)** 1173;

[18] https://www.thermofisher.com/order/catalog/product/912A0813

[19] Idiyatullin D, Suddarth S, Corum C A, Adriany G, Garwood M, 2012 *J. Magn. Reson.* **220 (1)** 26–31.

[20] Garwood M, DelaBarre L, 2001 *J. Magn. Reson.* **153 (2)**, 155–177.

[21] Donovan K J, Kupce E, Frydman L, 2013 *Angew. Chem., Int. Ed.* 2013, **52 (15)** 4152–4155.

Queiroz L H K Jr,. Queiroz D P K, Dhooghe L, Ferreira A G, Giraudeau P, 2012 *Analyst* **137 (10)** 2357–2361.

[22] Hollingsworth K G, Sederman A J, Buckley C, Gladden L F, M. L. Johns M L 2004, *J. Colloid Interface Sci.* **274 (1)** 244–250.

[23] Van Duynhoven J P M, Maillet B, Schell J, Tronquet M, Goudappel J-G V, Trezza E, Bulbarello A, Van Dusschoten D 2007 *Eur. J. Lipid Sci. Technol.* **109 (11)** 1095–1103.

[24] Mariette F 2009 /*Curr. Opin. Colloid Interface Sci.***14 (3)**, 203–211.

[25] Dalitz F, Cudaj M, Maiwald M, Guthausen G 2012 *Prog. Nucl. Magn. Reson. Spectrosc***60(1)** 52–70.

[26] Luo Y, Evangelyn C, Alocilja E C 2017 *J Biol Eng.* **11(14).** *11: 14. https://doi.org/10.1186/s13036-017-0053-8.*

Journal of Physics: Conference Series **1384** (2019) 012072 doi:10.1088/1742-6596/1384/1/012072

Influence of the content of caluminate on their physicochemical properties

I T Zhadovskiy[1], I V Berlinskiy[1], V V Vasiliev[1], A V Stogov[1]

[1] Saint-Petersburg Mining University, 2, 21st line, St. Petersburg, 19916, Russia

E-mail: sergaur@mail.ru

Abstract. The research of the properties of the complex aluminate solutions obtained through Bayer process or sintering process, will enable to determine the optimal parameters for the process, to increase the efficiency of the production and will be instrumental in high quality alumina production. The key stages of the alumina production through Bayer process which define its efficiency are digestion and decomposition processes. The main properties of the obtained products and amount of losses of the commercial components are established on these stages.

1. Introduction

Physicochemical methods of analysis are widely adopted for analysis of hydrometallurgical solutions, which contain different speciations of light, rare and heavy non-ferrous metals. However, in case of aluminate solutions, classic methods of analysis are very limited in its application or cannot be used. These restrictions occur due to the complexity of the system and its unique behaviour under different conditions. To develop modern methods of analysis and quality control it is necessary to continue to research the nature, properties and composition of alkaline solutions of these metals.

It is critical to review physicochemical properties of aqueous sodium aluminate solutions such as surface tension, density, metastable zone of aluminate solutions.

It is well-known that properties of any aqueous system are determined by its surface tension and viscosity, these properties are commonly used in different fields such as thermal physics, molecular physics, biophysics, etc. The values of the surface tension and viscosity coefficients are decreasing with the increase of the temperature, i.e. the change in temperature is followed by the same change in these parameters, therefore, it is necessary to determine the relationship between these properties.

The behaviour of the viscosity of the sodium aluminate solution in defined temperature zone and set Na_2O concentrations in the Na_2O – Al_2O_3 – H_2O system was considered by N. M. Baron and R. P. Matveev [1-3]. These researches were performed in the clear solutions with no additional impact of the impurities.

Experience of alumina production proved that real composition of an aluminate solution differed significantly from perfect Na_2O – Al_2O_3 – H_2O system. Industrial aluminate solutions contain different organic and mineral impurities. The main source of these impurities is the raw material – bauxite.

Theoretical analysis is presented in the paper written by I. I. Makarov and E. I. Hrinin. As a result of this analysis, the relationship between surface tension coefficient of the solution and its viscosity is established [4-5]. The authors used solutions, which contain ethanol, to validate their research, therefore, it is critical to validate these calculations for the aluminate solutions.

In the works [6-7], the properties and methods of using products obtained from Bayer solutions in Na_2O - Al_2O_3 - H_2O systems as sorption materials for composition correction were investigated.

In the works [8–12], the author considers the possibility of identifying compounds in technological processes. Bayer also takes into account the properties of systemic formation and various substances: phenolates, polyalcohols and additional carboxylates.

2. Materials and methods

Thermostat was used to maintain the constant value of the temperature. Surface tension was measured using semi-automatic du Nouy tensiometer (ring) Lauda TD1C. Density was measured using pycnometer in the volume of 50 ml. Dynamic viscosity was measured using viscometer VPZH-4, automatic viscometer Herzog HVM 472 and action camera Sony HDR as-20 capable of slow motion shot to accurately determine the time, moreover, professional non-linear editing program Adobe Premiere Pro was used.

Standard strong test solution of caustic was used in this research. Strong caustic solutions of the invariant point were prepared on the first stage of the experiment. The content of NaOH expressed as $Na_2O_к$ was around 300 g/l and $\alpha_к$ = 1,45 – 1,60. This solution is very close to the equilibrium composition solutions, stable and can be stored for a long period of time. The initial solutions were diluted with the calculated amount of water to prepare diluted solutions with the required composition; where applicable, alumina to caustic ratio can be changed by changing the amount of caustic. Sodium hydroxide (NaOH) AR and industrial aluminum hydroxide for alumina G-00 production were used as the original reagents.

The required volume of the initial strong solution, which contains equivalent amounts of $Na_2O_к$ and Al_2O_3, was calculated to prepare the diluted solutions. The measured volume of the solution was diluted using distilled water with constant stirring and adjusted to the required value. The sample of the prepared solution is collected to analyze the content of $Na_2O_к$ and Al_2O_3 in the diluted solution. This analysis was carried out in compliance with industry standard procedure for the analysis of aluminate solutions.

3. Experiment

The research was carried out using test solutions of sodium aluminate. Four main impurities of the alumina system were taken into consideration: sodium chloride, carbonate, sulphate and oxalate. The main physical properties of the aluminate solutions were reviewed in this research: dynamic viscosity, density and surface tension. The experiments to determine the impact of the impurities were carried out under the following conditions: the temperature was set on 60 ^{0}C and 80 ^{0}C, the impurities were added to the aluminate solutions (with the content of Al_2O_3 of 120 g/l and α_k = 1,6) in the amount corresponding to its content in industrial aluminate solutions. The analysis of the solutions was done using thermostatic temperature control. Results of the experiments are on Fig. 1-12.

4. Conclusion

1. The dependencies of main physicochemical parameters, such as viscosity, density, surface tension and ionic strength, on the content of the solution were determined. Some of these dependencies are non-linear functions; therefore, these parameters have a complex impact on the main parameters of the alumina production. Carbonate, sulphate and oxalate ions increase viscosity and density of the solution, which influence the stability of the aluminate solution.
2. The duration of the induction period of the Na_2O – Al_2O_3 – SiO_2 – H_2O system in the preferential thermodynamic state depends on the relationship between the surface tension coefficient and the crystal-forming coefficient of the medium.

Journal of Physics: Conference Series **1384** (2019) 012072 doi:10.1088/1742-6596/1384/1/012072

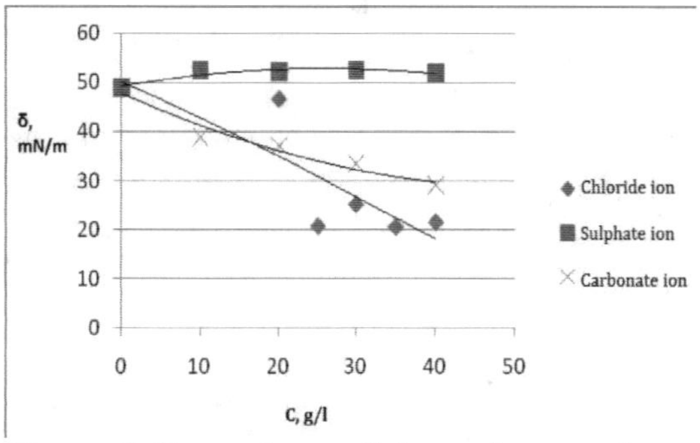

Figure 1. Dependence of the surface tension on the concentration of sodium chloride, carbonate and sulphate in the aluminate solution under 60 ^{0}C.

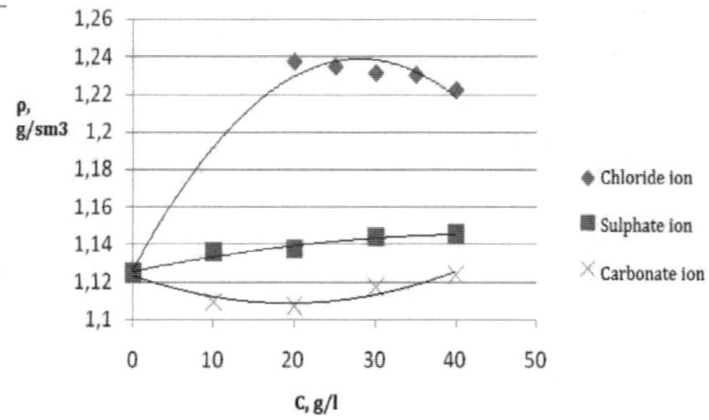

Figure 2. Dependence of the density on the concentration of sodium chloride, carbonate and sulphate in the aluminate solution under 60 ^{0}C.

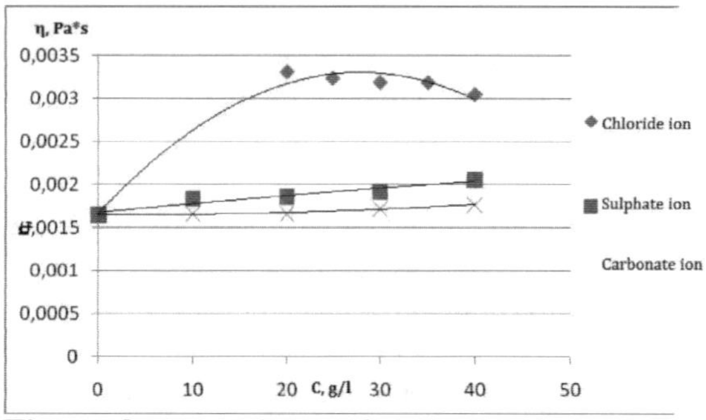

Figure 3. Dependence of the dynamic viscosity on the concentration of sodium chloride, carbonate and sulphate in the aluminate solution under 60 ^{0}C.

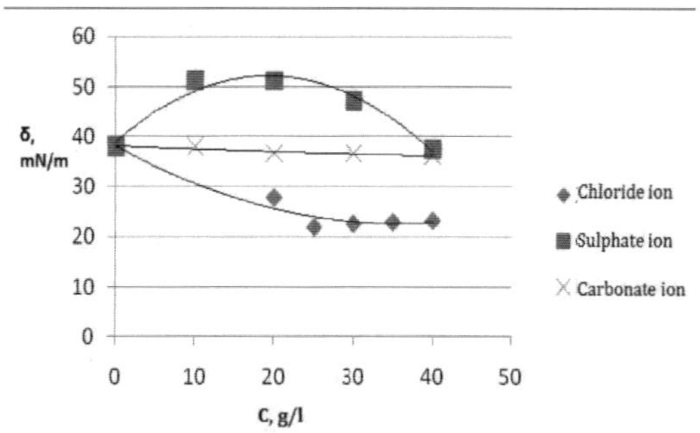

Figure 4. Dependence of the surface tension on the concentration of sodium chloride, carbonate and sulphate in the aluminate solution under 80 ^{0}C.

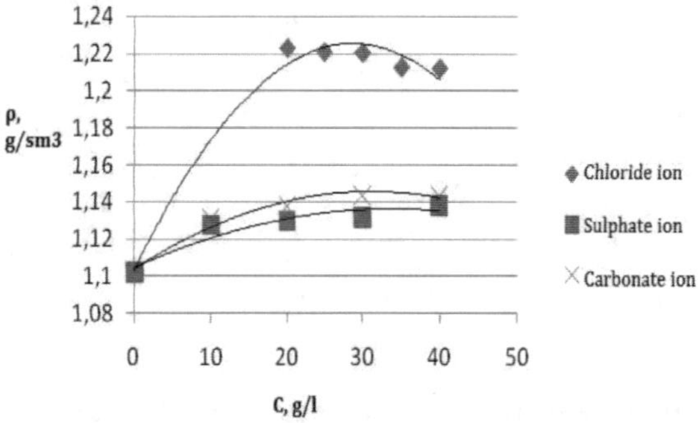

Figure 5. Dependence of the density on the concentration of sodium chloride, carbonate and sulphate in the aluminate solution under 80 ^{0}C.

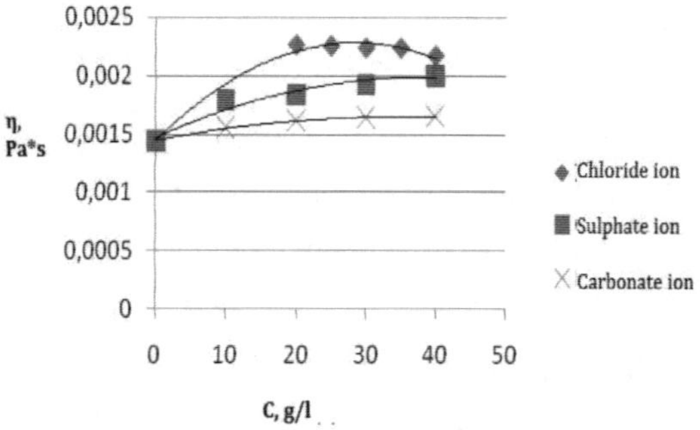

Figure 6. Dependence of the dynamic viscosity on the concentration of sodium chloride, carbonate and sulphate in the aluminate solution under 80 ^{0}C.

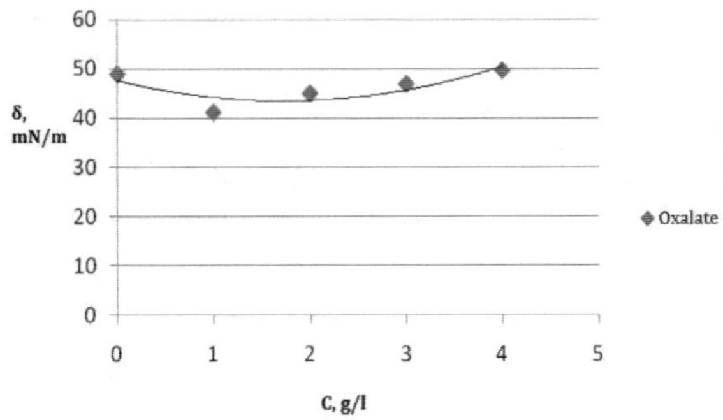

Figure 7. Dependence of the surface tension on the concentration of sodium oxalate in the aluminate solution under 60 ^{0}C.

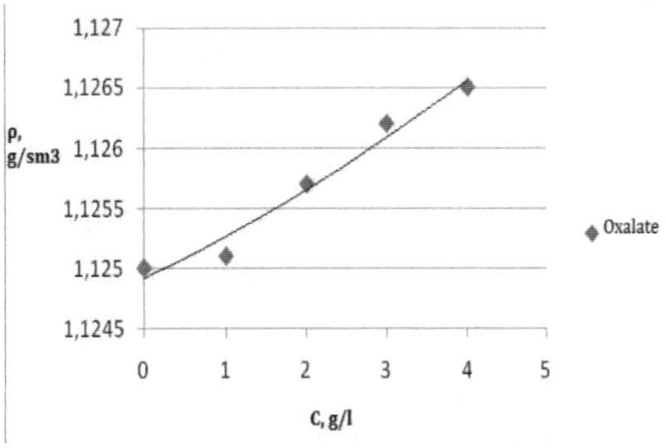

Figure 8. Dependence of the density on the concentration of sodium oxalate in the aluminate solution under 60 ^{0}C.

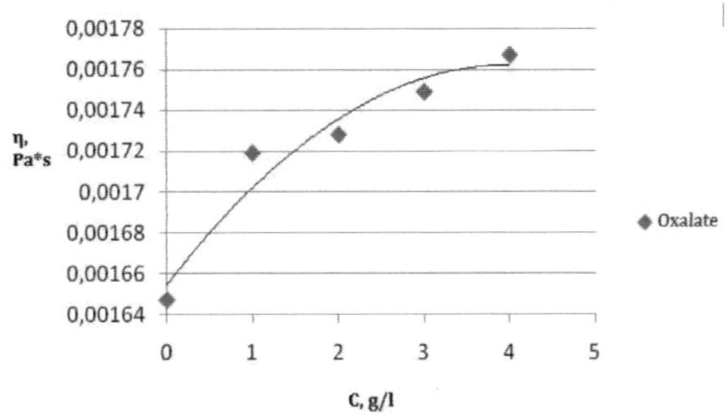

Figure 9. Dependence of the dynamic viscosity on the concentration of sodium oxalate in the aluminate solution under 60 ^{0}C.

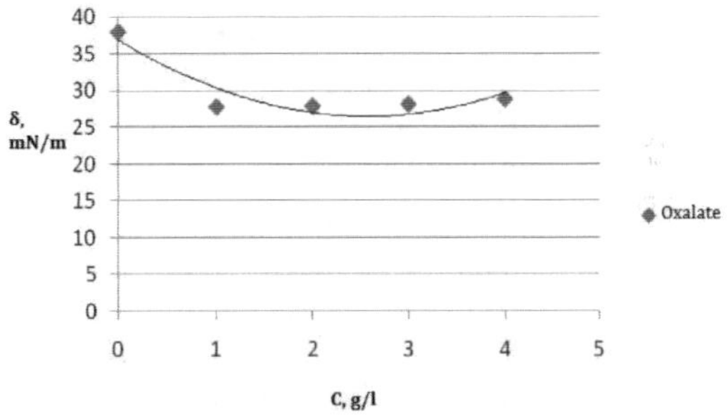

Figure 10. Dependence of the surface tension on the concentration of sodium oxalate in the aluminate solution under 80 ^{0}C.

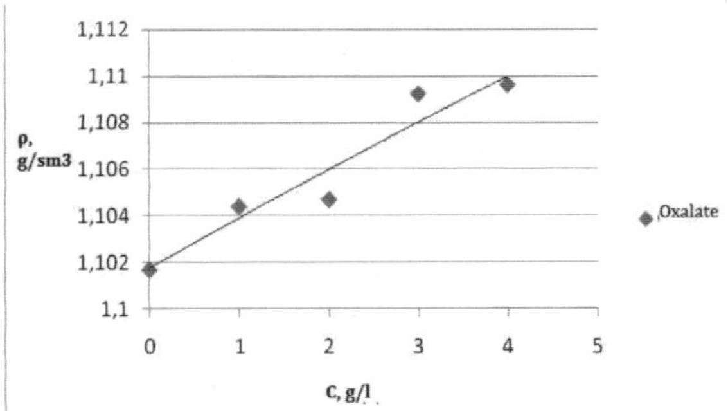

Figure 11. Dependence of the density on the concentration of sodium oxalate in the aluminate solution under 80 ^{0}C.

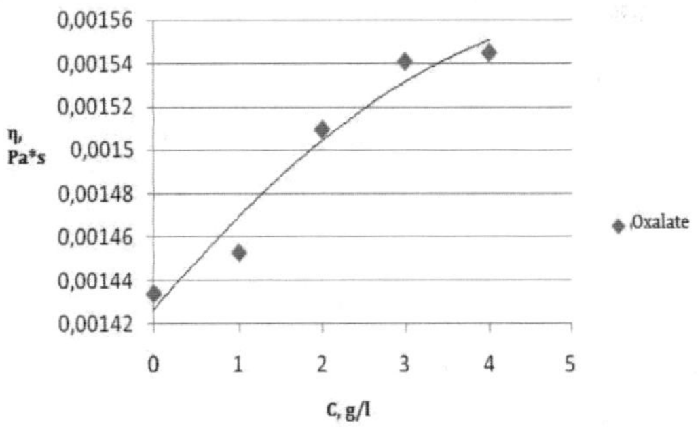

Figure 12. Dependence of the dynamic viscosity on the concentration of sodium oxalate in the aluminate solution under 80 ^{0}C.

5. Acknowledgements
The work was carried out with the financial support of the Russian Science Foundation under the Agreement No. 18-19-00577 dated April 26, 2018 on providing a grant for fundamental research and exploratory research.

References

[1] Baron N M and Matveev R P 1967 *Viscosity of sodium aluminate solutions at temperatures from 25 to 90 ^0C* (Leningrad: Lensovet Institute of Technology)

[2] Power G, Joanne S.C. Loh, Niemelä K 2011 Organic compounds in the processing of lateritic bauxites to alumina Addendum to Part 1: Origins and chemistry of organics in the Bayer process *Hydrometallurgy* **108** 149–151

[3] Sizyakov V. M., Tikhonova E. V., Cherkasova M. V.Efficiency of oxide compounds of magnesium in purification of alumina industry solutions from organic impurities (2) 2013 Non-ferrous Metals 23-26

[4] Markov I I and Hrinina E I 2004 *Bulletin of the N. Caucasus St. Tech. Uni.* **1**(8) 80-82

[5] Sizyakov V.M.,Tikhonova E.V. Physical and chemical transformations of calcium in aluminate solutions from alumina production. Russian Journal of Applied Chemistry Volume 85, Issue 11, November 2012, Pages 1658-1661

[6] Vasiliev V V, Zhadovskiy I T 2008 Application of calcium hydrocarbon-aluminates for waste water treatment in hydro-metallurgical production In *Proc. of the International Multidisciplinary Scientific GeoConference Surveying Geology and Mining Ecology Management* 1151-1158

[7] Brickin V N, Sizyakov V M, Litvinova T E, Fedorov A T 2018 Modern physico-chemical description of equilibria in the Na_2O – Al_2O_3 – H_2O system and its analogues *Journal of Mining Institute* **237** 307-316

[8] Power G, Joanne S.C. Loh et al 2011 A review of the determination of organic compounds in Bayer process liquors *Analytica Chimica Acta* **689** 8–21

[9] Shin H, Lee S 2004 Study on the effect of humate and its removal on the precipitation of aluminium trihydroxide from the Bayer process *Minerals Engineering* **17** 387–391.

[10] Machold T, Laird D W et al 2011 Decomposition of Bayer process organics: Phenolates, polyalcohols, and additional carboxylates *Hydrometallurgy* **107** 68–73

[11] Kuznetsov S I, Dereviankin V A 1964 *Physical chemistry of alumina production through Bayer process* (Moscow: Metallurgizdat)

[12] Petrov G.V., Boduen, A.Y., Fokina, S.B., Sidorova, M.S. Sorption recovery of rhenium from chromium-containing sulfate solutions by cybber low-basic anionites. Russian Journal of Non-Ferrous Metals Volume 56, Issue 5, 1 September 2015, Pages 500-504

Development of a three-level information model for processing diagnostic information for electric transport systems

B V Malozyomov[1], N S Druzhinina[2], I M Daudov[3]

[1] Novosibirsk State Technical University, 20, Karla Marksa ave., Novosibirsk, 630073, Russia
[2] Irkutsk National Research Technical University, 83, Lermontov str., Irkutsk, Russia, 664074
[3] Chechen State University, 32 Sheripova Street, Grozny, 634050, Russia

E-mail: mbv5@mail.ru

Abstract. The article proposes a way to improve the existing system for forecasting the technical condition of vehicles at electric transport enterprises. The authors propose a three-level architecture of the software and hardware of the information system and its implementation in Intranet and Internet technologies. The transition to a three-level system is primarily due to the desire to reduce the information system costs, which consists of the cost of equipment, software and operating expenditures. In addition, the work defines the structure of effective forecasting of the developed system operation.

1. Introduction

Currently, various information systems are actively developing at electric transport enterprises [1]. The systems used at these enterprises differ in the volume of existing software, tasks performed, processing power, branching, etc.

For the system forecasting their technical condition, the authors proposed a three-level architecture of the software and hardware of the information system. This system is implemented in Intranet and Internet technologies [2, 3]. The paper presents the developed architecture and hierarchy of network computing. Network computing is done using software of Oracle and Microsoft companies.

The basic principle underlying the management of a three-level information system is "devide et impera", i.e. "divide and conquer". This control principle is similar to the control principle of any complex system. According to it, a complex software system at the top level should consist of a small number of relatively independent components with clearly defined interfaces. Next it is decomposed into elements. The components selected in the first stage are subjected to it. Next, the same decomposition procedure is executed up to a specified level of detail. As a result, the system is represented by a hierarchy in which several levels of abstraction are distinguished.

Today, in software engineering, there are two main approaches to the development of software systems. The main difference between them is in the decomposition criteria [4]. The first approach is called structural or functional-modular. It is based on the principle of algorithmic decomposition, when the functional elements of the system are distinguished and a strict order of actions is established. The second, object-oriented approach, uses object decomposition. Moreover, the behavior of the system is described in terms of the interaction of objects [5].

Journal of Physics: Conference Series **1384** (2019) 012073 doi:10.1088/1742-6596/1384/1/012073

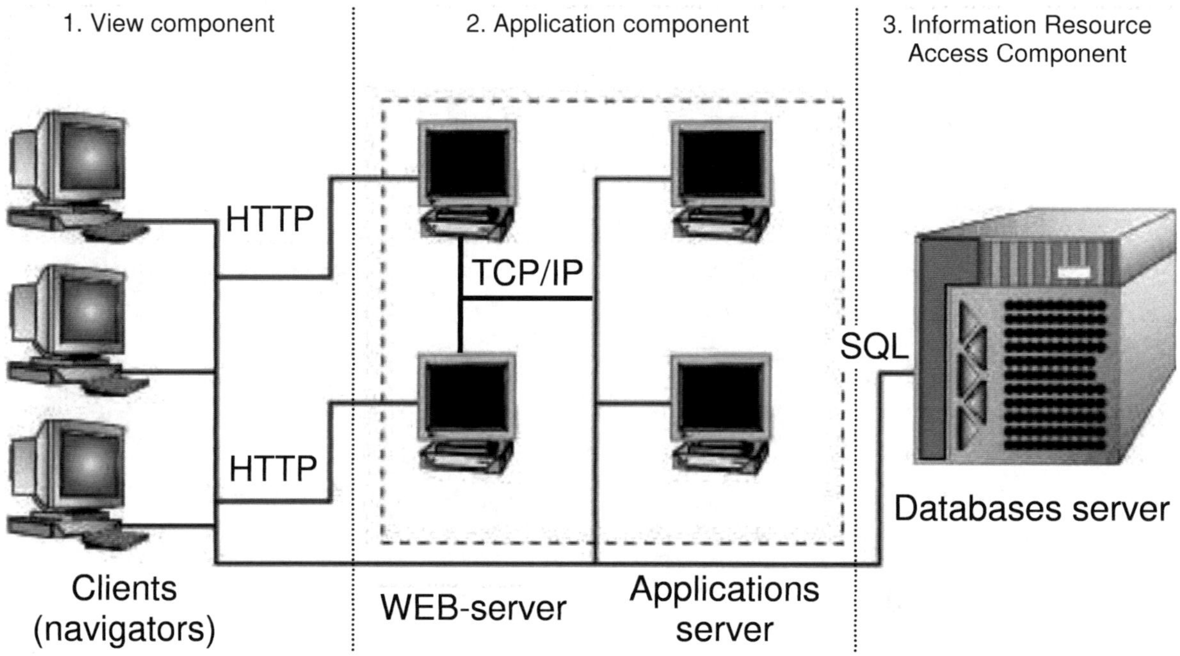

Figure 1. Structural and functional diagram of the three-level architecture of the system for forecasting the technical condition in Intranet technology

The need to move from a two-level system, which has already become classical, client-server architecture to a three-level one is primarily due to the desire to reduce the cost of ownership of an information system, which consists of the cost of equipment, software and operating expenditures. To date, according to analysts at Gartner Group, the average US cost of ownership per workplace is $8,000 per year [6].

2. Theory

Research and analysis of existing work in this area allowed developing a three-level information system using the principles of object decomposition:

- first level is presentation level (data input and display functions);
- second level is application level (universal services, as well as functions related to the decision on the operability of the object as a whole);
- third level is access to information resources (fundamental functions of storage and management of information and computing resources).

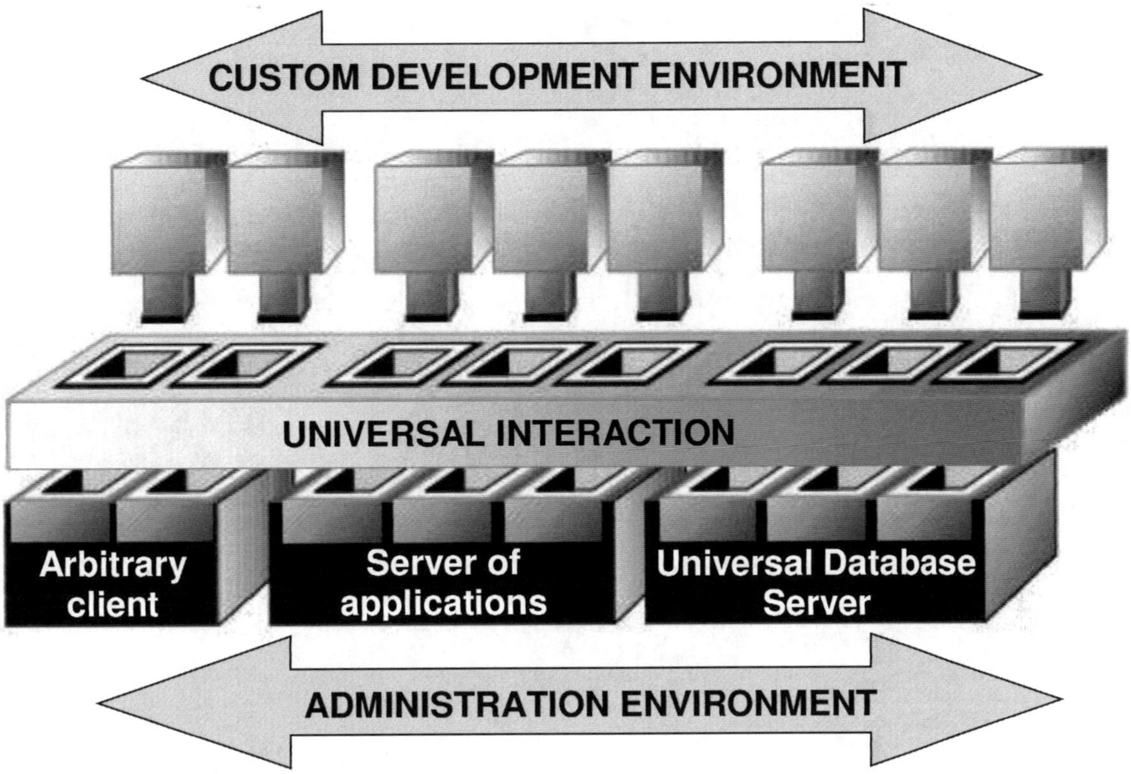

Figure 2. Open network computing architecture in the information system using Oracle Corporation software

Transaction and Communication Manager provides communication between levels. Intranet technology leaves its mark on this classic design. At the presentation level is a universal client, a WEB navigator. In this case, the WEB navigator is supplemented by application functions. The functions of the information concentrator are assigned to the WEB server [7, 8]. This combines the functions of an information hub with the functions of a transaction and communication manager. The resulting circuit is shown in Fig. 1.

3. Results and discussion

Based on the proposed architecture and software of Oracle Corporation, three key elements of the information system were proposed:

- Oracle 8 DBMS object-relational server;
- Oracle Application Server 6.0, as well as a number of specialized application servers (Oracle Payment Server, Internet Commerce Server, Video Server, etc.);
- a set of JDBC drivers [9], specially optimized for access from Java to the Oracle 8 DBMS, as well as SQLJ - SQL embedded in the Java language.

Using Oracle Corporation software gives the three-level architecture another feature. This architecture was originally conceived as extensible open architecture (Fig. 2). The main unit of expansion of such an architecture is a cartridge. This can be a client, application, or data cartridge. The system provides means for creating cartridges. Interfaces for connecting them with other components of the information system are also provided [10].

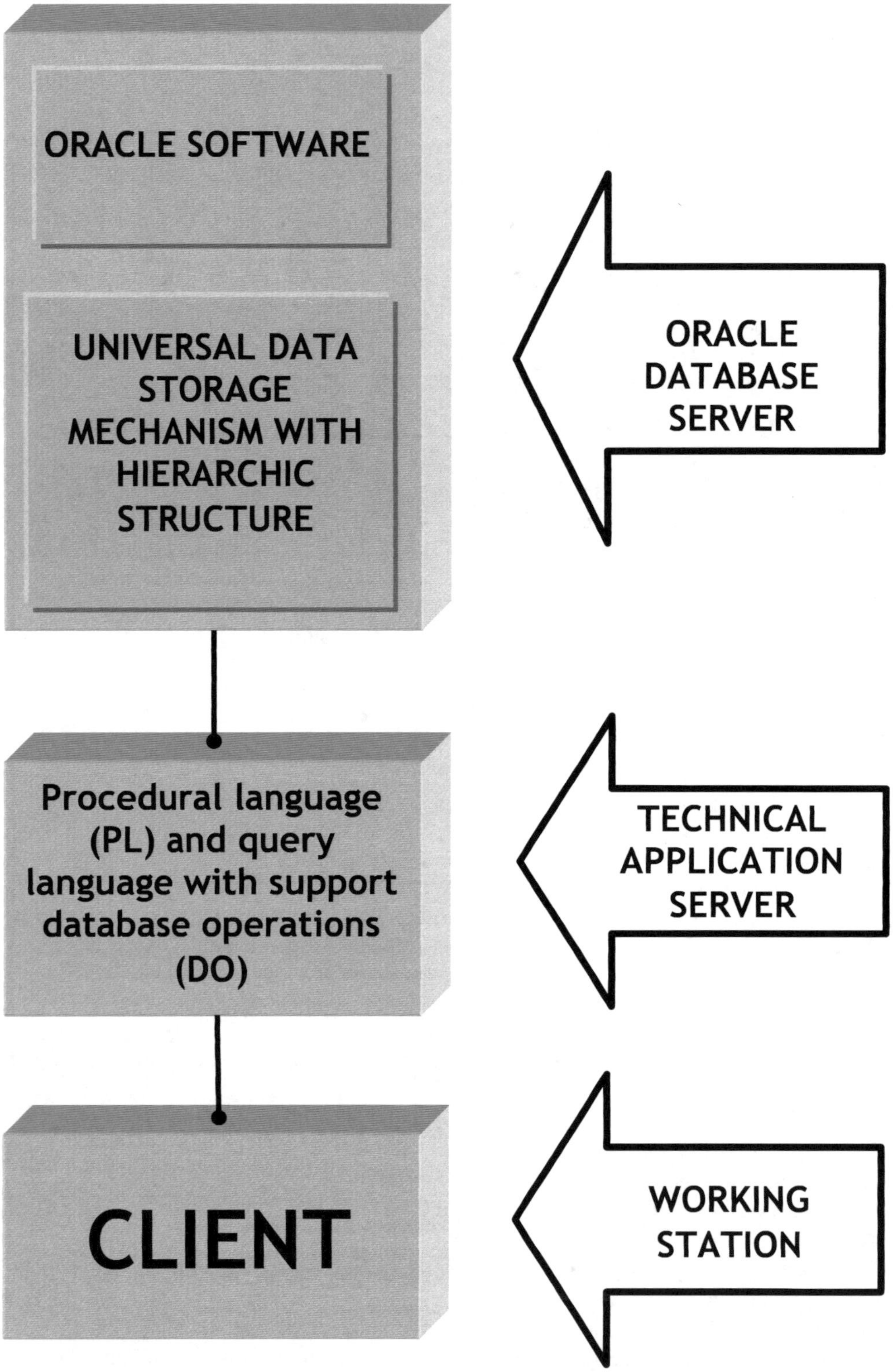

Figure 3. Information system diagram based on Oracle DBMS

The scheme of the developed information system is shown in Fig. 3. The concepts at the enterprises of electric transport are: employees of divisions, depots, diagnostic and repair divisions of the enterprise, warehouses.

All these concepts are in a certain hierarchical relationship. They have directional (upper and lower) connections. As a result, units are part of the enterprise. Employees are part of the unit, etc.

The properties of the concept of "enterprise" can be: the name of the company, the actual and legal address of the company. For each concept (category), an instance (object) can be created that has (or does not have) upper and lower connections with other objects and has (or does not have) properties (attributes of the object).

An element that ensures the distribution of information flows in the system is a document. A document in a system is a screen form designed by a developer. This form has the necessary functionality to select the necessary information from the database and create new objects in the database with relationships and attributes.

Within the framework of the developed system, a database structure was developed and a functional was created. The developed functionality includes all the necessary development tools: a designer of screen forms (documents), a tool for describing a category graph (accounting scheme of the application) and category properties, a procedural language (PL) designed to describe the functionality of a document, and a database query language (LI) [11, 12].

The combination of such an ideology with the corresponding functionality and structure of the system database provides a powerful tool for creating applications. The main area of work of such a tool will be the storage and processing of various kinds of information. In such a system, the developer is completely spared from creating the database structure and its design. Also, there is no need to tune its performance. In the case of the used Oracle database management system, this is a very time-consuming and complex process. When creating a specific enterprise information system, the corresponding Oracle database structure is formed. In this case, when changing the logic of the system, it is necessary to change the structure of the database, which entails significant costs.

The implementation of the work of such systems can significantly reduce costs and expenses for organizations or when using equipment at various enterprises of heavy industry [13-19].

The data storage scheme developed and presented in this article allows implementing any technical process in the information system, perform its calculation and forecasting regardless of its level of complexity. The same applies to the ideology associated with the information system of constructing mechanisms for working with data. The developed data storage and processing scheme does not impose any restrictions on changing their structure. This approach to developers puts forward the requirements for knowledge and experience of analytical work in the subject area in the first place. And only then come the requirements for technical programming.

4. Conclusions

A system for predicting the technical condition was developed. Within its framework, a three-level architecture of software and hardware for the information environment was proposed. This three-level system is implemented in Intranet and Internet technologies. Architecture and hierarchy of network computing were elaborated. To create this hierarchy, Oracle and Microsoft software were used. The need to move to a three-level system from a two-level one architecture of the "client-server", which has already become classic, is due to the desire to reduce the cost of the information system. Total costs are the sum of the cost of software, hardware, and operating expenditures. The structure of effective forecasting of the developed system was determined.

References

[1] Kuznetsov S M 2005 *Traction network protection of SC current.* (Novosibirsk: NSTU) p. 352

[2] Kuznetsov S M 2011 Setting of electronic security with simulation model corrected. *Transport Science, Technology, Management. VINITY* **12** 30-34

[3] Kuznetsov S M, Demidenko I S, Yaroslavtsev M V and Krivova A O 2009 Mathematical model

study of traction network dynamics of direct-current railway with train starting. / *Scientific transport problems of Siberia and Far East. NSAWT* **2** 324-327

[4] Malozyomov B V, Vorfolomeyev G N and Schurov N I 2005 Reliability and diagnosing electrotechnical systems. *In the collection: Proceedings - 9th Russian-Korean International Symposium on Science and Technology, KORUS-2005* 347-350

[5] Ivanov G Ja and Malozyomov B V 2005 Reliable power saving electric drive of wide application. *In the collection: Proceedings - 9th Russian-Korean International Symposium on Science and Technology, KORUS-2005* 330-332

[6] Filyushov Yu P, Zonov P V, Malozemov B V and Wilberger M E 2011 Energy efficient control of an alternating current machine. *The Polzunovsky Herald.* 2011 **2** 45-51

[7] Shchurov N I and Wilberger M E 2011 Spectral analysis of rectifier unit current for unbalance and non-sinusoidal supply voltage *Transport: science, technology, management* **12** 41-43

[8] Mischenko T M 2011 The mathematic stimulation of transient process in a.c. - system "electric-traction network - LOCOMOTIVE". *Transport: science, technology, management* **12** 105-109

[9] Anurov V I 2008 Modeling of transient processes in case of short circuit in the traction network and the presence of electric locomotives on the feeder zone. *Electro. Electrical engineering, electric power industry, electrotechnical industry* **2** 16-18

[10] Bykadorov A L, Zarutskaya T A and Muratova-Milekhina A S 2015 Increase of efficiency of short-circuits fault location in traction networks of alternating current on the basis of information technologies. *Bulletin of transport of the Volga region* **6(54)** 15-19

[11] Zvonarev I E, Folomkin A I, Muratbakeev E H 2017 Computer simulation of machining *Key Engineering Materials* **743** 248-251

[12] Solovev S V, Bardanov A I 2018 Efficiency estimation method of three-wired AC to DC line transfer, *Journal of Physics: Conference Series* **1015**

[13] Shishkin E V Shishkin P V 2017 Process duty effect on the vibration gyratory-cone crusher dynamics. Journal of Industrial Pollution Control **33(1)** 909-913

[14] Kibirev V, Dedushenko I, Avksentyev S, Aleksandrov V I 2017 Energy efficiency of application of pipes with polyurethane coating. *Obogashchenie Rud.* 54-59. 10.17580/or.2017.02.10.

[15] Maksarov V V, Makhov V E 2017 Method for studying shape of cutting tool by light field recorder. *AER-Advances in Engineering Research* **1(133)** 452 – 457

[16] Chupin S, Bolobov, V 2018 Influence of thermomechanical treatment modes on wear resistance of mining equipment material *Materials Science Forum* **945** 695-699

[17] Maksarov V V, Olt J, Krasnyy V A 2018 Pre-surface preparation features when applying wear resistant composite sprayed coatings. *Epitoanyag-journal of silicate based and composite materials* **2(70)** 34 – 37

[18] Nguyen L K, Gabov V V, Zadkov D A 2019 Improving efficiency of cleanup and coal flow formation on conveyor by shearer loader with accessorial blade. *Eurasian Mining* **1** 37-39 DOI: 10.17580/em.2019.01.09

[19] Nasonov M Y, Lykov Y V 2017 Setting up excavators with growing cracks in their metal structures for repairs. *IOP Conference Series: Earth and Environmental Science* **87** 022015

Presentation of distortion power in an alternating current network and analysis of factors of its occurrence

S V Myatezh[1], M V Novolodskiy[1], E T Kitova[1] V Y Konyukhov[2], I M Daudov[3]

[1] Novosibirsk State Technical University, 20, Karla Marksa ave., Novosibirsk, 630073, Russia
[2] Irkutsk National Research Technical University, 83, Lermontov str., Irkutsk, Russia, 664074
[3] Chechen State University, 32 Sheripova Street, Grozny, 634050, Russia

E-mail: max_nov94@mail.ru

Abstract. The article examines the conditions for the appearance of such components of inactive power in electric alternating current networks, which are known as reactive power and distortion power. Through simulation modeling, the validity of the expressions that were obtained analytically and determine the components of inactive power was verified. The validation of expressions based on theoretical concepts. The results are useful in optimizing modes in AC networks and creating effective technical means of compensating latent capacities.

1. Introduction

Optimization of modes in alternating current (AC) networks leads to the need to compensate for all power components that are not able to perform useful work, i.e., which are not active power by definition.

Compensation issues have been studied to the greatest extent for idealized conditions when electric networks have symmetry and sinusoidal phase voltages and balanced loads, while the loads themselves have linear current-voltage characteristics. Under such conditions, the appearance of a component of inactive power, which is known as reactive power (Q), explains the excess of apparent power (S) overactive power (P).

In real conditions, AC electric networks do not have a sinusoidal phase voltage, and most modern loads have sharply non-linear current-voltage characteristics. In these cases, non-sinusoidal currents begin to flow along with circuit elements with non-sinusoidal voltages. This process causes an excess of power S to overpower P, which can no longer be explained by the appearance of reactive power Q. In the scientific literature, this circumstance due to the appearance of another component of inactive power. These components are known as the distortion power, and T [1] or D (Distortions) [2] indicate them. Under certain conditions in an AC network, power D can form as the only component of inactive power.

Optimization of modes in AC networks leads to extremely important when working with various equipment [17, 18], stabilization of compensation of power components during operation of heavily loaded equipment is especially important [19, 20].

2. Statement of the problem

The task is to analyze the most characteristic static modes, which lead to the separate appearance of the power Q and D in electrical AC networks. The results are useful in studying the features of the operating modes of electric networks using simulation tools and creating further effective technical means for compensating latent capacities.

3. Theory

In the theory of electrical engineering, the causes of the appearance of power Q and its negative effects on the operation modes of electric networks with sinusoidal currents and voltages have a fairly good description. Currents and voltages in a significant part of electrical installations, thanks to especially taken measures, can indeed be considered sinusoidal. However, in some cases, currents and voltages are more or less different from sinusoidal. The reason for the appearance of non-sinusoidal voltages and currents can be both generators and energy receivers. Therefore, in practice, such technical means are successfully used, the compensating power of which allows reducing the value Q_k to the minimum value Q to $Q - Q_K \to 0$, and thereby bring the value of S closer to the numerical value P of the consumer.

In this case, one of the obvious criteria that form the objective function has the following form of notation

$$S = \sqrt{P^2 + (Q - Q_K)^2} \to \min . \qquad (1)$$

Power compensation is a more difficult task if non-sinusoidal voltages or currents operate in the electrical network. In this case, for calculations, they are traditionally represented by the Fourier series [3]

$$\begin{aligned} u &= U_0 + U_{1m}\sin(wt + \alpha_1) + U_{2m}\sin(2wt + \alpha_2) + \ldots + U_{Nm}\sin(Nwt + \alpha_N), \\ i &= I_0 + I_{1m}\sin(wt + \alpha_1 + \varphi_1) + I_{2m}\sin(2wt + \alpha_2 + \varphi_2) + \ldots + I_{Nm}\sin(Nwt + \alpha_N + \varphi_N) \end{aligned} \qquad (2)$$

We consider the constant component of series (2) as zero harmonic. Then the total power is calculated by the products of harmonics of voltages and currents with identical numbers

$$S = U \cdot I = \sqrt{\sum_{i=0}^{N} U_i^2 \cdot \sum_{i=0}^{N} I_i^2}, \qquad (3)$$

$U_i = U_{im}/\sqrt{2}$, $I_i = I_{im}/\sqrt{2}$ – the effective values of voltage and current harmonics, i – the harmonic sequence number.

The active component of power is calculated as the average value of the product $u \cdot i$ of all harmonics (2) for the period T

$$P = \frac{1}{T}\int_0^T (u \cdot i)dt = U_0 I_0 + U_1 I_1 \cos\varphi_1 + U_2 I_2 \cos\varphi_2 + \ldots + U_N I_N \cos\varphi_N . \qquad (4)$$

The above equation proves that all harmonics can do useful work in a non-sinusoidal AC network. It also shows that the active power is characterized by the sum of the active powers of the individual harmonics

$$P = P_0 + P_1 + P_2 + \ldots + P_N = \sum_{i=0}^{N} P_i . \qquad (5)$$

By analogy with power P, in the theory of electrical engineering, there is a perception [4] that power Q in AC networks is defined as the sum of the reactive powers of individual harmonics

$$\begin{aligned} Q &= U_1 I_1 \sin\varphi_1 + U_2 I_2 \sin\varphi_2 + \ldots + U_N I_N \sin\varphi_N , \\ Q &= Q_1 + Q_2 + \ldots + Q_N = \sum_{i=1}^{N} Q_i \end{aligned} \qquad (6)$$

Journal of Physics: Conference Series **1384** (2019) 012074 doi:10.1088/1742-6596/1384/1/012074

The component of the inactive power D, which is found [5] in networks with non-sinusoidal currents and voltages, is usually determined through the unbalance of the previously found powers P and Q in the inequality $S > \sqrt{P^2 + Q^2}$, using the property of their orthogonality

$$D = \sqrt{S^2 - P^2 - Q^2}\,. \tag{7}$$

It should be noted that in the system of state standards for terms and definitions of the basic concepts of electrical engineering (GOST 19880-74), units of physical quantities (GOST 8.417-81), letter designations of the main quantities in electrical engineering (GOST 1494-77), the term "distortion power", universally used in conversion technology, generally absent. There is no generally accepted designation for distortion power. In the theory of non-sinusoidal currents, it is denoted by "T." In this work, for the distortion power, the designation is made with the letter D (from the English distortion) and measurement in volt-amperes [15].

Moreover, among individual scientific schools there is no clear understanding of the physical meaning of the total reactive power, determined by expression (6) for electric networks in which non-sinusoidal voltages and currents operate. This circumstance is reflected in the adopted IEEE 1459-2010 standard [6], which introduced the concept of inactive power, however, the definition of power Q is limited only by the first harmonic.

Based on this, it is advisable to conduct a series of experiments. These experiments devote to modeling the most common static modes of electrical circuits. The results of these experiments determine the most consistent theory and answer the question of the task.

4. The choice of modeling environment and the peculiarity of creating measuring subsystems.

Modes that lead to the separate appearance of Q and D powers in electrical AC networks are the object of analysis. It is advisable to analyze the modes according to the results of model experiments for a series of individual cases, each of which differs in the harmonic composition of the supply voltages, currents, and current-voltage characteristics of the loads. To solve this problem, we used the simulation environment Simulink of the MATLAB package [12] and the experience described in [16].

The functional unit of measurements Power, available in the Simulink environment, confirms the current paradigm of power representation [6]. Because of this, it is not convenient as a tool for measuring the values of P and Q. Its algorithm based on the calculation of powers relative to the fundamental harmonic (fundamental frequency), and this requires preliminary settings before each experiment.

Integral methods characterize instantaneous power as $p = u \cdot i$, the integral of which determines the transmitted electric energy W over a time interval. This interval limited by the integration limits t_1 and t_2 and does not require preliminary consideration of the harmonic composition. For model experiments, it is advisable to calculate the power P for the load in the AC network as follows:

$$P = \frac{W_{t_2 - t_1}}{t_2 - t_1} = \frac{1}{t_2 - t_1}\int_{t_1}^{t_2}(u \cdot i)dt\,. \tag{8}$$

Using the resulting expression in the Simulink environment, calculating the power P implements a simple universal model as a measuring subsystem.

For the convenience of constructing the corresponding measuring subsystem, the power S is calculated similarly in the following model, taking into account expression (3):

$$S = \sqrt{\frac{1}{t_2 - t_1}\int_{t_1}^{t_2} u^2(t)dt} \cdot \sqrt{\frac{1}{t_2 - t_1}\int_{t_1}^{t_2} i^2(t)dt}\,. \tag{9}$$

To determine the value of S, the resulting subsystem (9) corresponds to the measuring subsystem.

This subsystem also requires preliminary settings in the simulation model.

5. Modeling the conditions of inactive power Q and D

Two previously known harmonics models the non-sinusoidally of the voltage in the AC network to simulate the analysis of the results obtained. These are such harmonics as the primary (50Hz) and highest. An example of modeling such harmonics is the tenth (500Hz) is modeled by changing the ratios of their amplitudes.

The ratio of the values of the elements R and L connected in parallel determines the load nature in the model of AC network. Giving the non-linear character of the current-voltage characteristic is carried out by including a semiconductor diode in the load circuit of the load. Then a simple half-wave rectifier circuit is implemented.

The Fourier measuring units determine the monitored values in the model:

u_{50}^{max}, u_{500}^{max} – amplitudes of voltage harmonics with a frequency of 50 Hz and 500 Hz;

α_{50}^{u}, α_{500}^{u} – the initial phase angle angles of voltage harmonics with a frequency of 50 Hz and 500 Hz;

i_{50}^{max}, i_{500}^{max} – amplitudes of current harmonics with a frequency of 50 Hz and 500 Hz;

α_{50}^{i}, α_{500}^{i} – the initial angle of the phase shift of the harmonics of the current frequency of 50 Hz and 500 Hz;

$\varphi_{50} = \alpha_{50}^{u} - \alpha_{50}^{i}$, $\varphi_{500} = \alpha_{500}^{u} - \alpha_{500}^{i}$ – phase angle between harmonics of voltages and currents.

The results of a series of model experiments are in Table 1.

Table 1. Parameters of electric circuits, modes, and power

№	R (Ω)	L (H)	u_{50}^{max}	u_{500}^{max}	i_{50}^{max}	i_{500}^{max}	φ_{50}	φ_{500}	P (W)	Q_1 (VA)	Q_{10} (VA)	D (VA)	S (VA)
1	100	-	100	0	1.000	0	0	-	50.0	0	0	0	50.0
2	100	-	100	100	1.000	1.000	0	0	100.0	0	0	0	100.0
3	-	0.1	100	0	3.183	0	-90.0	-	0	159.2	0	0	159.2
4	-	0.1	100	100	3.183	0.318	-90.0	-90.0	0	159.2	15.92	143.2	226.2
5	100	0.1	100	0	3.336	0	-72.5	-	50.0	159.2	0	0	166.8
6	100	0.1	100	100	3.336	1.049	-72.5	-17.6	100.0	159.2	15.92	143.2	247.3
7	100 (+VD)	-	100	0	0.500	-	0.4	-	25.2	0.2	0	25.0	35.7

An analysis of the results indicates the achievement of a balance of orthogonal capacities for all modes. The analysis confirms the validity of expression (7). The total reactive power of the load is determined by the sum of the reactive powers of the individual harmonics, as follows from expression (6). The power of distortion occurs when the shape of the load current curve is not identical to the shape of the supply voltage curve. Accordingly, combinations of the existing harmonics of voltage (h) and current (k), the product $u \cdot i$, due to the general properties of orthogonality over a period, give zero average power values (N):

$$N = \frac{1}{T}\int_0^T \left(U_{hm}\sin(h \cdot wt + \alpha_h) \cdot I_{km}\sin(k \cdot wt + \alpha_k)\right)dt = 0, \tag{10}$$

And therefore, the value of N does not apply to power P, and is inactive, as it is not able to perform useful work. By the definition $h \neq k$ of expression (6), this power does not form power Q. Under such conditions, the inactive power N should be considered as the power of distortions D.

6. Analysis of the results

At the end of the work, the most characteristic modes presented in Table 1 are analyzed.

Mode № 1

The supply network is strictly sinusoidal $U_{10} = 0$; the nature of the load is purely active. In this case $I_{10} = 0$, $\cos\varphi_1 = 1$, $\sin\varphi_1 = 0$. Therefore, it obtained that $Q_i = U_i I_i \sin(\varphi_i) = 0$ and $D = 0$.

Mode № 2

The supply network is non-sinusoidal $U_{10} \neq 0$; the nature of the load is purely active. In this case, we obtain $\cos\varphi_1 = 1$, $\cos\varphi_{10} = 1$, $\sin\varphi_1 = 0$, $\sin\varphi_{10} = 0$. It turns out that $D = \sqrt{(U_1 I_{10} - U_{10} I_1)^2} = U_1 I_{10} - U_{10} I_1$. However, in an AC network with a purely active load, the identity between the harmonics of voltages and harmonics of currents is satisfied

$$\frac{U_1}{U_{10}} = \frac{I_1}{I_{10}}. \tag{11}$$

Therefore, for this mode $U_1 I_{10} - U_{10} I_1 = 0$.

Thus, in an AC network, even with a non-sinusoidal voltage, but with an active load, the power D.

Mode № 6

The supply network is non-sinusoidal; the nature of the load is active-inductive. In this case $0 < \cos\varphi_1 < 1$, $0 < \cos\varphi_{10} < 1$, $1 > \sin\varphi_1 > 0$, $1 > \sin\varphi_{10} > 0$. For such a regime, having made a series of transformations of expression (11), next formula is relevant

$$D = \sqrt{\frac{U_1 I_{10}}{U_{10} I_1} + \frac{U_{10} I_1}{U_1 I_{10}} - 2(\cos\varphi_1 \cos\varphi_{10} + \sin\varphi_1 \sin\varphi_{10})}\text{, or}$$

$$D = \sqrt{\frac{U_1 I_{10}}{U_{10} I_1} + \frac{U_{10} I_1}{U_1 I_{10}} - 2\cos(\varphi_1 - \varphi_{10})}. \tag{12}$$

From this equation, it follows that for such a regime, where $D = 0$, if two conditions are satisfied: identity (12) and equality $\varphi_1 = \varphi_{10}$. However, from the electrical engineering theory, the resistance of reactive elements is determined by the frequency of the alternating current that flows through them.

So, for an inductive element $x_L = j2\pi f L$. Therefore, the harmonic value of the current flowing through the inductive element in the electrical circuit is

$$I_i = \frac{U_i}{x_L} = \frac{U_i}{j2\pi f L}.$$

Accordingly,

$$\frac{I_1}{I_{10}} = \frac{U_1}{j2\pi f L} \cdot \frac{j2\pi 10 f L}{U_{10}} = \frac{10\, U_1}{U_{10}}.$$

However, the result obtained contradicts the condition (11).

For a similar reason, condition (11) is violated for a circuit with a capacitive element, the resistance x_C of which depends on the frequency $x_C = \dfrac{1}{j2\pi f C}$.

Thus, if a load with reactance includes in the AC network with a non-sinusoidal voltage, then, in this case, both power Q and power D.

Mode № 7

The supply network is sinusoidal; the nature of the load is active but non-linear. In this case $U_1 \neq 0$, $U_i = 0$, where $i > 1$, and the alternating current network, only the load current turn out to be non-sinusoidal, decomposing into a series of harmonics that not useful work by expression (10).

If, as a first approximation, we neglect switching processes, then the main harmonic of the current not have a phase shift relative to the first and only harmonic of the supply voltage, i.e. $\cos\varphi_1 = 1$, and $\sin\varphi_1 = 0$. Under these conditions, the value $Q = 0$, and the value D from expressions (3), (4), (5) and (7) expressed as

$$D = \sqrt{S^2 - P^2} = \sqrt{\sum_{i=0}^{\infty} U_1^2 I_i^2 - U_1^2 I_1^2} = U_1 \sqrt{\sum_{\substack{i=0 \\ i\neq 1}}^{\infty} I_i^2} \tag{13}$$

However, if the current-voltage load characteristics create a primary harmonic of the current for which $\sin\varphi_1 \neq 0$, then a power Q is generated in the network with a sinusoidal voltage. For example, this is typical for the operation mode of controlled rectifiers, which form both power D and power Q, even without specific reactive elements in their circuits.Thus, the mode in which power $D=0$ according to the results of the analysis of expressions (12) and (13) achieves under the following conditions:

- equality of phase shifts of all harmonics of currents and voltages

$$\varphi_i = \varphi_j \tag{14}$$

- identity of the relations of the practical values for all harmonics of voltages and currents

$$\frac{U_i}{U_j} = \frac{I_i}{I_j} \tag{15}$$

If in the model we increase the number of higher harmonics of currents and voltages, then, by analogy with the derivation of expression (11), a general expression obtained for directly finding the quantity D

$$D = \sqrt{\sum_{i=0}^{\infty} \sum_{\substack{j=1 \\ i\neq j}}^{\infty} U_i I_j \left(U_i I_j - 2U_j I_i \cos\varphi_i \cdot \cos\varphi_j\right) + U_j I_i \left(U_j I_i - 2U_i I_j \sin\varphi_i \cdot \sin\varphi_j\right)} \cdot \tag{16}$$

In the scientific literature, variants of searching for similar expressions in a general form for finding the value of D also present. So, in [13] it is proposed to determine the value of D by the formula

$$D^2 = \sum_{\substack{k,l \\ k\neq l}}^{\infty} U_k^2 I_l^2 - \sum_{\substack{k,l \\ k\neq l}}^{\infty} U_k I_k U_l I_l \cos(\varphi_k - \varphi_l),$$

and the recording format closest to expression (17) is found in [14].

$$D = \sqrt{\sum_{i=1}^{N-1} \sum_{j=1}^{N} U_i I_j \left(U_i I_j - U_j I_i \cos(\varphi_i - \varphi_j)\right) + U_j I_i \left(U_j I_i - U_i I_j \cos(\varphi_i - \varphi_j)\right)} \cdot$$

Thus, the components of power N can be analyzed and calculated and, at least for simple alternating current circuits in static mode, can be unambiguously divided into components of power Q according to different harmonics and power D. Closest to the stated results obtained by analytical and confirmed Simulation tools include representations of power Q and power D described in [7].

7. Conclusion

The paper analyzes the main modes leading to the separate appearance of the Q and D capacities in electrical AC networks. This analysis allows us to conclude that the compensation of power Q and power D are most effective if the technical device provides such a correction of the load current, the shape of which as a result is close in shape to the voltage curve of the supply network, which helps to fulfill the condition $P/S \to 1{,}0$.

The technique of direct derivation of analytical expressions of power D in AC networks is developed. The use of simulation tools made it possible to confirm two simple essential criteria for achieving distortion power compensation, namely, the equality of the phase shifts of all harmonics of the current and load voltage and the equality of the ratios of the valid values of the harmonics of the voltage and load currents.

The obtained patterns and conclusions can be used to model the optimal operating modes of various equipment used in various industries. This opportunity is especially important for the mining industry and heavy-duty pumps [21-24].

References

[1] Zeveke G V, Ionkin P A, Netushil A V 1963 Fundamentals of circuit theory. (M-L .: Gosenergoizdat)

[2] Akagi H, Kim H 1999 *IEEE International Conference on Power Electronics and Drive Systems, PEDS'99* (Hong Kong)

[3] Ango A 1967 Mathematics for electrical and radio engineers. (M .: Science)

[4] Sharon D 1996 Power factor definitions and power transfer quality in nonsinusoidal situations. *IEEE Transactions on Instrumentation and Measurement* **45** 728-733

[5] Krogeris A F 1993 AC power (Riga, Phys.-Energetic Institute of Latvia. AN)

[6] 2010 *IEEE Standard Definition for the measurement of Electric Power Quantities under sinusoidal, nousinusoidal, balanced or unbalanced conditions* (IEEE std. 1459TM – 2010). (IEEE Power and Energy Society, New York)

[7] Budenau C 1927 *Puissance reactives et fictives. Inst. Romain de l'Energie.* (Buharest, Rumania)

[8] Czarnecki L S 1988 Orthogonal decomposition of the currents in a 3-phase nonlinear asymmetrical circuit with a nonsinusoidal voltage source. *IEEE Transactions on Instrumentation and Measurement* **37** 30-34

[9] Fryze S 1931 Active and Apparent power in non-sinusoidal systems. *Przeglad Elektrot* **7** 193-203.

[10] Mayevsky O A 1978 Energy performance of valve converters / OA Mayevsky. (M .: Energy)

[11] Akagi H, Nabae A 1993 *The p-q Theory in Three-Phase System under Non-Sinusoidal Conditions.* (ETEP)

[12] German-Galkin S G 2013 Virtual laboratories of semiconductor systems in the environment of Matlab - Simulink. (St. Petersburg: Doe)

[13] Bezikovich A A, Shapiro E Z 1980 Measurement of electrical power in the sound frequency range. (Leningrad: Energy)

[14] Shklyarsky Ya E, Bragin A A, Dobush V S 2012 Impact of harmonic components of current and voltage on distortion power. Oil and gas business **4** 26-32

[15] Zeveke G V, Ionkin P A, Netushil A V, Strakhov S V 1963 Fundamentals of circuit theory. (M-Leningrad: Gosenergoizdat)

[16] Korolev N A, Solovev S V 2017 AC motor diagnostics system based on complex parametric analysis. *Materials Science and Engineering* **177**

[17] Ershov D Y, Zlotnikov E G, Koboyankwe L E 2017 Analysis of causes and mathematical description of process of manufacturing errors and local defects in mechanical system nodes, *IOP Conference Series: Earth and Environmental Science* **87(8)** 082015

[18] Avksentiev S Y, Avksentieva E Y 2018 Determining the Parameters of the Hydraulic Transport of Tailings for Processing Iron Ore. *IOP Conference Series: Earth and Environmental Science* **194 (3)** 032003

[19] Serzhan S L, Trufanova I S, Malevannyi D V 2019 Substantiation of the draghead application as a mining unit in conditions of solid minerals deep-sea mining. *IOP Conference Series: Earth and Environmental Science* **272** 1 - 6

[20] Maksarov V V, Keksin A I 2018 Technology of magnetic-abrasive finishing of geometrically-complex products. *IOP Conference Series: Materials Science and Engineering* **4(327)** 42 – 45

[21] Nasonov M Y, Lykov Y V 2017 Setting up excavators with growing cracks in their metal structures for repairs *IOP Conference Series: Earth and Environmental Science* **87** 022015

[22] Lykov Y V, Gorelikov V G, Gantulga B 2017 Analytical research and classification of mechanism of diamond drilling-bits contact with rocks during well sinking. *IOP Conference Series: Earth and Environmental Science* **87** 022012

[23] Krasnyy V A 2018 Application of seals made of directed reinforced polymeric composite materials to improve wear resistance of friction units of oil well pumps. *IOP Conf. Series: Earth and Environmental Science* **194** 042008 doi:10.1088/1755-1315/194/4/042008

[24] Aleksandrov V I, Kibirev V 2018 The Kachkanarsky MCC iron ore processing tailings slurry hydraulic transport parameters determination. *Obogashchenie Rud.* 56-63 10.17580/or.2018.01.10

AUTHOR INDEX

AUTHOR INDEX